# Nano-Oncology Applications for Sustainable Synthesis and Precision Therapeutics

Ahmed El-Mallul
*University of Radom, Poland*

Dilawar Hassan
https://orcid.org/0000-0001-9422-8580
*University of South Africa, South Africa*

Rashid Abbas Khan
https://orcid.org/0000-0002-5438-6274
*University of South Africa, South Africa*

Amjid Khan
*University of South Africa, South Africa*

IGI Global
Scientific Publishing
Publishing Tomorrow's Research Today

| | |
|---|---|
| Vice President of Editorial | Melissa Wagner |
| Director of Acquisitions | Mikaela Felty |
| Director of Book Development | Jocelynn Hessler |
| Production Manager | Mike Brehm |
| Cover Design | Jose Rosado |

Published in the United States of America by
IGI Global Scientific Publishing
701 East Chocolate Avenue
Hershey, PA, 17033, USA
Tel: 717-533-8845 | Fax: 717-533-7115
Website: https://www.igi-global.com E-mail: cust@igi-global.com





Library of Congress Cataloging-in-Publication Data

LCCN: 2026028353 (CIP Data Pending)
ISBN13: 979-8-3373-9848-8
Isbn13Softcover: 979-8-3373-9849-5
EISBN13: 979-8-3373-9850-1
British Cataloguing in Publication Data
A Cataloguing in Publication record for this book is available from the British Library.

# Table of Contents

# Detailed Table of Contents


*Ayesha Sani, School of Science, College of Science, Engineering, and Technology, University of South Africa, South Africa*
*Hotaf Hassan Makki, Biology Department, Faculty of Science, University of Tabuk, Saudi Arabia*
*Andrew Tapiwa Kugedera, Morgesnter Teachers College, Zimbabwe*

The clinical effectiveness of conventional chemotherapy is often hindered by off-target toxicity, a narrow therapeutic index, and the development of multidrug-resistant cancer cells. While legacy synthetic nanomedicines do manage to overcome macro-scale biological barriers by targeted drug delivery strategies, they are still hindered in their clinical translation by a chemical bottleneck. The reduction process used in conventional chemical manufacturing methods involves toxic reducing agents and detergents, leading to high residual in vivo toxicity, rapid clearance via denatured protein coronas, and an unsustainable global environmental footprint. The chapter explores the clinical and ecological rationale of Green Nano-Oncology to address these fundamental limitations. This eco-friendly method uses biogenic fabrication pathways, such as plant extracts, and microbial secretomes to produce highly stable and biocompatible therapeutic nanocarriers. Green shift offers an intrinsically sustainable, biologically collaborative solution for next-generation cancer therapeutics.

This chapter reviews the paradigm shift in nano-oncology from conventional, heavy-metal-based nanoparticles to sustainable carbon-based nanostructures (CNMs). Although currently, orthodox cancer therapies are limited because of systemic toxicities and lack of selectivity, carbon dots, carbon nanotubes, and graphene quantum dots can serve as a biocompatible substitute to orthodox cancer treatments, as they were green-synthesized using a range of different carbon-based materials. In addition, they can be targeted to tumor receptors through an active direct approach using their versatile surface chemistry, and stimuli-responsive therapeutic release, in particular, pH, enzyme-triggered, and near-infrared light-triggered. This summary paper underscores the various uses of sustainable CNMs in high-resolution optical imaging, photo-induced cancer therapies (PTT/PDT), advanced gene or immunotherapy delivery, and ultra-sensitive biomolecular diagnostics to precision theranostics.



This chapter provides a detailed analysis of the green synthesis of metallic nanoparticles (NPs) i.e. silver (Ag), gold (Au) and selenium (Se) using medicinal plant extracts and microbial systems in the framework of precision oncology. The green biosynthetic method utilizes bioactive phytochemicals including flavonoids, terpenoids and polyphenols, as natural reductants and stabilizers/capping agents, and, thus, enables the environmentally friendly synthesis of NPs. The characterization protocols, include UV- Vis spectroscopy, energy-dispersive X-ray spectroscopy,

Nano-gene therapy is an emerging approach for precision cancer treatment that involves integration of Advanced nanotechnologies with genome editing. Compared to the other existing platforms, the CRISPR/Cas9 system has proven to be a revolution, which allows selective disruption of oncogenes, correction of pathogenic mutations, bypass of drug resistance, and development of immune cells with stronger anticancer effects. The clinical application of CRISPR/Cas9 has been hindered, however, due to delivery inefficiency, lack of stability of the nucleases, off-target editing, and immune-related toxicity. Nanoparticle-based delivery systems can effectively address these issues, including protecting the gene-editing cargo, enhancing the accumulation at tumor sites, achieving intracellular delivery and controlling release in the tumor microenvironment.

Therapeutic resistance remains a major challenge in effective cancer management, leading to treatment failure and disease progression. This study highlights the integration of artificial intelligence (AI), multi-omics technologies, and Nano-CRISPR platforms as a novel strategy to overcome resistance mechanisms. AI-driven analytics enable the identification of key molecular drivers, predictive biomarkers, and resistance pathways from complex biological data. Concurrently, Nano-CRISPR systems facilitate precise gene editing and targeted delivery to tumour cells, enhancing therapeutic efficacy. The convergence of these technologies supports personalized and adaptive treatment approaches. Despite challenges such as off-target effects and ethical concerns, this integrated framework offers significant potential for advancing precision oncology.

Traditional preclinical tumor models have limited predictive capabilities, hindering effective translation in nano-oncology. To address this, integration of advanced in vitro systems with patient-derived xenograft (PDX) models provides a robust pathway. This chapter presents an overview of two-dimensional and three-dimensional in vitro models, microfluidic tumor-on-chip systems, and in vivo approaches, with emphasis on sustainability-driven strategies. These platforms assess efficacy, biodistribution, pharmacokinetics, and safety of green-synthesized and biodegradable nanomaterials. PDX models are highlighted for their ability to preserve tumor heterogeneity and patient-specific molecular features, enabling more precise therapeutic evaluation. Key challenges, including scalability, ethical considerations, and regulatory requirements, are discussed alongside emerging trends such as humanized PDX models and artificial intelligence-based translational approaches. Overall, this chapter highlights a continuum bridging preclinical research and clinical application in nano-oncology.



This chapter per the authors explores the emerging field of cryo-nanomedicine as an advanced strategy in precision oncology for improving cancer treatment outcomes. The chapter discusses how the integration of therapeutic hypothermia and nanotechnology enables targeted drug delivery, enhanced tumor ablation, and reduced systemic toxicity. By applying the Q10 principle, controlled cooling suppresses metabolic activity and oxidative stress within the tumor microenvironment. The chapter further examines the use of advanced nanomaterials, including phase-change nanoparticles and thermally conductive systems, for temperature-responsive therapeutic action. In addition, Cryo-Electron Microscopy (Cryo-EM) is presented as an essential tool for validating the structural stability and functionality of nanoformulations. Overall, the chapter highlights cryo-nanomedicine as a sustainable and precise approach for next-generation cancer interventions.



Hyperthermia in nano-oncology integrates controlled heat therapy with nanotechnology for precise cancer treatment. This chapter reviews nano-enabled hyperthermia systems, including magnetic, photothermal, radiofrequency, microwave, and ultrasound approaches, along with metallic, magnetic, carbon-based, semiconductor, and polymeric nanoplatforms. It highlights green synthesis strategies that improve biocompatibility and sustainability. The synergistic use of hyperthermia with chemotherapy, radiotherapy, immunotherapy, and gene therapy is discussed, as are theranostic applications enabling simultaneous imaging and treatment. Challenges related to toxicity, biodistribution, and clinical translation remain, while AI-driven designs and smart nanoplatforms offer promising directions for future precision oncology

# Preface

## INTRODUCTION

The field of oncology stands at a pivotal moment, driven by the urgent need to develop cancer therapies that are not only more effective and precise but also safer, more sustainable, and better suited for clinical translation. While remarkable advances in nanotechnology have transformed the landscape of cancer diagnosis and treatment, many conventional nanomedicine platforms continue to rely on chemically intensive synthesis methods that present challenges related to environmental impact, biocompatibility, reproducibility, scalability, and regulatory approval. These limitations have highlighted the necessity of rethinking how nanomaterials are designed and produced to support the next generation of precision cancer therapeutics.

*Nano-Oncology Applications for Sustainable Synthesis and Precision Therapeutics* was developed to address this critical need by presenting green nano-oncology as an emerging and transformative framework that integrates sustainable nanomaterial synthesis with precision oncology. As editors, our goal has been to assemble a collection of scholarly contributions that demonstrate how environmentally responsible approaches can be successfully combined with cutting-edge nanotechnology to improve therapeutic efficacy while minimizing toxicity and supporting long-term clinical applicability. The chapters in this volume collectively illustrate that sustainability and innovation are not competing priorities but complementary principles that can shape the future of cancer treatment.

This volume explores the scientific foundations and practical applications of sustainable nanotechnology across a broad spectrum of nano-oncology research. Contributors examine green synthesis strategies for metallic nanoparticles, the development of biopolymer-based nanocarriers, targeted and stimuli-responsive delivery systems, gene therapy vectors, nano-enabled thermal medicine, and other advanced therapeutic platforms designed to enhance treatment precision while reducing adverse effects. Together, these topics provide readers with a comprehensive understanding of how sustainable nanomaterials can be engineered to address the complex biological challenges associated with cancer diagnosis and therapy.

A distinguishing feature of this book is its emphasis on translation from laboratory discovery to clinical implementation. Beyond discussing material design and therapeutic performance, the chapters examine critical considerations such as nanotoxicology, safety assessment, good manufacturing practice (GMP) scale-up, quality control, and evolving regulatory frameworks. By integrating these translational dimensions, the volume offers a practical roadmap for advancing sustainable nano-oncology innovations from experimental research toward real-world clinical applications. This perspective reflects the growing recognition that successful cancer nanomedicine requires not only scientific excellence but also regulatory readiness, manufacturing feasibility, and patient safety.

The interdisciplinary nature of green nano-oncology is another defining strength of this collection. Progress in this rapidly evolving field depends upon collaboration among researchers in nanotechnology, materials science, biomedical engineering, pharmaceutical sciences, molecular biology, toxicology, clinical oncology, and regulatory science. The contributors to this volume represent these diverse areas of expertise, offering complementary perspectives that illustrate how multidisciplinary research can accelerate innovation while addressing the technical, biological, environmental, and clinical challenges facing modern cancer therapeutics.

As editors, we have carefully organized the chapters to provide readers with a logical progression from the fundamental principles of sustainable nanomaterial synthesis to advanced therapeutic applications and translational considerations. This structure enables readers to appreciate both the scientific mechanisms underlying green nanotechnology and its practical implications for precision medicine. Whether examining nanoparticle fabrication, targeted drug delivery, immunotherapeutic strategies, or clinical translation, each chapter contributes to a cohesive understanding of sustainable nano-oncology as an integrated scientific discipline.

This reference volume is intended to serve a broad audience, including researchers, graduate students, postdoctoral fellows, clinicians, pharmaceutical scientists, biomedical engineers, and industry professionals engaged in cancer research and nanomedicine development. It is equally valuable for interdisciplinary academic programs and institutional libraries seeking comprehensive resources that bridge fundamental nanoscience with translational oncology. Readers will find both foundational knowledge and emerging research directions that support continued innovation in sustainable precision therapeutics.

We extend our sincere appreciation to all chapter authors whose expertise, dedication, and scholarly contributions have made this volume possible. Their work reflects the remarkable progress being achieved across the fields of nano-oncology and sustainable nanotechnology while identifying important opportunities for future research and clinical advancement. We also acknowledge the reviewers whose thoughtful evaluations strengthened the quality and scientific rigor of each contribution.

It is our hope that *Nano-Oncology Applications for Sustainable Synthesis and Precision Therapeutics* will serve as both a valuable reference and a catalyst for continued interdisciplinary collaboration. By advancing sustainable approaches to nanomaterial design alongside precision therapeutic strategies, this book aspires to contribute to the development of safer, more effective, and environmentally responsible cancer treatments that will ultimately improve patient outcomes and shape the future of oncology.

## ORGANIZATION OF THE BOOK

Chapter 1, "The Paradigm Shift in Oncology," establishes the foundation for the volume by examining the transition from conventional nanomedicine toward green nano-oncology. The chapter highlights the clinical limitations of traditional chemotherapy and chemically synthesized nanocarriers, including systemic toxicity, multidrug resistance, and environmental concerns. By introducing biologically derived nanomaterials produced through plant- and microbe-based synthesis pathways, the authors present a compelling case for sustainable, biocompatible, and environmentally responsible nanotherapeutics that redefine the future direction of precision oncology.

Chapter 2, "Sustainable Carbon-Based Nanostructures," explores the growing importance of carbon-based nanomaterials as safer and more sustainable alternatives to conventional metallic nanoparticles. The chapter examines carbon dots, carbon nanotubes, and graphene quantum dots, emphasizing their green synthesis, exceptional biocompatibility, and versatility in targeted drug delivery. The discussion extends to advanced applications in molecular imaging, photothermal and photodynamic therapies, gene and immunotherapy delivery, and precision theranostics, demonstrating the expanding role of sustainable carbon nanostructures in next-generation cancer care.

Chapter 3, "Green Synthesis of Metallic Nanoparticles (Ag, Au, Se)," provides a comprehensive examination of environmentally friendly methods for producing silver, gold, and selenium nanoparticles using medicinal plant extracts and microbial systems. The authors discuss the underlying biosynthetic mechanisms, characterization techniques, and biological activities that support their antimicrobial, antioxidant, and anticancer properties. By demonstrating selective tumor cytotoxicity alongside reduced toxicity toward healthy tissues, the chapter illustrates how green synthesis bridges fundamental nanomaterial development with clinically relevant precision oncology applications.

Chapter 4, "Green Biopolymer-Based Nanocarriers for Precision Cancer Therapeutics," focuses on biodegradable biopolymers as sustainable platforms for advanced drug delivery. The chapter reviews polysaccharide-, protein-, and microbe-derived nanocarriers, highlighting their environmentally friendly fabrication methods, targeting capabilities, and stimuli-responsive release mechanisms. By emphasizing enhanced therapeutic performance alongside improved safety and biodegradability, the authors present these natural materials as promising candidates for clinically translatable precision cancer therapies.

Chapter 5, "Polymeric Nanocarriers for Controlled Drug Release," examines the design and therapeutic potential of polymer-based nanocarriers capable of improving drug delivery precision and minimizing systemic toxicity. The chapter discusses mechanisms of controlled drug release, including diffusion, degradation, and stimuli-responsive activation, while also exploring diverse targeting strategies that enhance tumor specificity. Looking beyond chemotherapy, the authors consider applications in gene delivery, combination therapies, and theranostics, while addressing current translational challenges and emerging opportunities involving artificial intelligence and hybrid nanomaterial systems.

Chapter 6, "Stimuli-Responsive 'Smart' Nanomaterials in Oncology," presents an in-depth overview of intelligent nanoplatforms capable of responding to both internal tumor microenvironment signals and externally applied stimuli. The chapter explores smart delivery systems activated by pH, enzymes, redox conditions, reactive oxygen species, temperature, light, magnetic fields, and ultrasound to achieve highly controlled therapeutic release. The authors further examine multifunctional theranostic platforms, sustainable nanoparticle fabrication, and the growing role of artificial intelligence in designing personalized nanomedicine solutions for future oncology.

Chapter 7, "Surface Functionalization and Active Targeting in Nano-Oncology," highlights the critical role of nanoparticle surface engineering in enhancing therapeutic precision. The chapter reviews polymer modifications, ligand-mediated targeting, biomimetic coatings, and stimuli-responsive functionalization strategies that improve nanoparticle stability, circulation time, tumor accumulation, and cellular uptake while minimizing off-target toxicity. Emerging innovations, including AI-assisted nanoparticle optimization, CRISPR-based delivery systems, and sustainable nanotechnology approaches, further illustrate the evolving sophistication of targeted cancer therapeutics.

Chapter 8, "Nano-Gene Therapy: CRISPR/Cas9 and Beyond," explores the convergence of nanotechnology and genome editing as a transformative approach to precision oncology. The authors discuss how nanoparticle delivery systems overcome many of the barriers limiting CRISPR/Cas9 therapies, including poor stability, inefficient intracellular delivery, off-target effects, and immune responses. By improving gene-editing efficiency and tumor-specific targeting, nano-enabled gene therapy offers significant promise for correcting cancer-associated mutations and overcoming therapeutic resistance.

Chapter 9, "AI-Guided Omics and Nano-CRISPR Platforms for Overcoming Therapeutic Resistance in Cancer," examines how artificial intelligence, multi-omics technologies, and Nano-CRISPR platforms can be integrated to address one of oncology's greatest clinical challenges: therapeutic resistance. The chapter demonstrates how AI-driven data analysis identifies predictive biomarkers and resistance pathways while Nano-CRISPR technologies enable highly precise gene editing and targeted therapeutic delivery. Together, these complementary technologies advance personalized treatment strategies capable of adapting to the molecular complexity of individual tumors.

Chapter 10, "From In-Vitro Systems to Patient-Derived Xenografts," focuses on the translational models that support the clinical development of sustainable nano-oncology therapies. The chapter reviews advanced two-dimensional and three-dimensional cell culture systems, tumor-on-chip technologies, and patient-derived xenograft models that preserve tumor heterogeneity and patient-specific characteristics. By evaluating efficacy, biodistribution, pharmacokinetics, and safety within increasingly sophisticated preclinical platforms, the authors present an integrated framework that strengthens the transition from laboratory research to clinical application.

Chapter 11, "The Role of Hypothermia and Cryo-Nanomedicine in Precision Oncology," introduces cryo-nanomedicine as an innovative therapeutic strategy that combines controlled hypothermia with nanotechnology to improve cancer treatment outcomes. The chapter examines how temperature-responsive nanomaterials, phase-change nanoparticles, and thermally conductive systems enable targeted drug delivery and enhanced tumor ablation while reducing systemic toxicity. The discussion also highlights the important role of Cryo-Electron Microscopy in validating nanoparticle structure and functionality, reinforcing the potential of cryo-nanomedicine as a sustainable precision oncology platform.

Chapter 12, "Hyperthermia in Nano-Oncology," explores the complementary application of controlled heat therapy and nanotechnology for targeted cancer treatment. The chapter reviews multiple hyperthermia modalities, including magnetic, photothermal, radiofrequency, microwave, and ultrasound-based systems, alongside diverse nanoplatforms engineered to maximize therapeutic effectiveness. By integrating green synthesis strategies with combination therapies and theranostic capabilities, the authors demonstrate how nano-enabled hyperthermia continues to expand the possibilities for precise and minimally invasive cancer treatment.

Chapter 13, "Lab to Clinic Scale-Up and Global Regulatory Frameworks," concludes the volume by addressing the essential challenges associated with translating nano-oncology innovations into routine clinical practice. The chapter examines manufacturing scale-up, quality assurance, Good Manufacturing Practice requirements, and the regulatory expectations of major international agencies. By discussing reproducibility, sustainability, engineering considerations, and global regulatory harmonization, the authors provide a comprehensive roadmap for transforming promising laboratory discoveries into safe, scalable, and clinically approved nano-oncology therapies.

## CONCLUDING REMARKS

As editors, we believe that the chapters presented in this volume collectively demonstrate that the future of cancer therapeutics will not be defined by scientific innovation alone, but by the deliberate integration of sustainability, safety, and clinical translatability into every stage of nanomaterial design. While green synthesis strategies establish the foundation for environmentally responsible nanomedicine, meaningful clinical impact ultimately depends on the capacity of researchers, clinicians, regulatory bodies, and industry partners to collaboratively translate laboratory innovation into safe, scalable, and patient-centered therapies. The diverse contributions represented throughout this book illustrate that the challenges of clinical translation are complex and multidisciplinary, yet they also reveal innovative strategies and emerging platforms that can guide future progress in precision oncology.

The chapters in this volume reinforce the importance of advancing biocompatible nanomaterial synthesis, refining stimuli-responsive and actively targeted delivery systems, integrating artificial intelligence into nanoparticle design and resistance prediction, and developing coherent regulatory and manufacturing frameworks capable of supporting global clinical adoption. Collectively, these contributions encourage readers to view green nano-oncology not as a static research niche, but as a continuously evolving discipline shaped by iterative innovation, rigorous safety evaluation, and interdisciplinary collaboration. They also emphasize that achieving truly sustainable and precise cancer therapeutics requires ongoing dialogue among nanotechnologists, molecular biologists, clinicians, toxicologists, and regulatory scientists who share responsibility for advancing patient outcomes.

We hope that this volume serves not only as a valuable scholarly reference but also as a catalyst for continued research, translational innovation, and evidence-based clinical practice. By bringing together international perspectives spanning nanomaterial synthesis, gene-editing technologies, thermal medicine, and regulatory science, this book contributes to the growing body of knowledge dedicated to narrowing the gap between laboratory discovery and clinical reality. It is our sincere hope that the ideas presented here will inspire continued innovation, foster safer and more sustainable nanomedicine platforms, and support the development of precision cancer therapeutics that ultimately improve outcomes for patients worldwide.

**Ahmed El-Mallul**
*University of Radom, Poland*

**Dilawar Hassan**
*University of South Africa, South Africa*

**Rashid Abbas Khan**
*University of South Africa, South Africa*

**Amjid Khan**
*University of South Africa, South Africa*

# Chapter 1
# The Paradigm Shift in Oncology:
## Shift in Oncology

**Ayesha Sani**
http://orcid.org/0000-0001-9718-8131
*School of Science, College of Science, Engineering, and Technology, University of South Africa, South Africa*

**Hotaf Hassan Makki**
*Biology Department, Faculty of Science, University of Tabuk, Saudi Arabia*

**Andrew Tapiwa Kugedera**
http://orcid.org/0000-0002-1700-6922
*Morgesnter Teachers College, Zimbabwe*

## ABSTRACT

*The clinical effectiveness of conventional chemotherapy is often hindered by off-target toxicity, a narrow therapeutic index, and the development of multidrug-resistant cancer cells. While legacy synthetic nanomedicines do manage to overcome macro-scale biological barriers by targeted drug delivery strategies, they are still hindered in their clinical translation by a chemical bottleneck. The reduction process used in conventional chemical manufacturing methods involves toxic reducing agents and detergents, leading to high residual in vivo toxicity, rapid clearance via denatured protein coronas, and an unsustainable global environmental footprint. The chapter explores the clinical and ecological rationale of Green Nano-Oncology to address these fundamental limitations. This eco-friendly method uses biogenic fabrication pathways, such as plant extracts, and microbial secretomes to produce highly stable and biocompatible therapeutic nanocarriers. Green shift offers an intrinsically sustainable, biologically collaborative solution for next-generation cancer therapeutics.*

DOI: 10.4018/979-8-3373-9848-8.ch001

## 1. INTRODUCTION: THE MODERN ONCOLOGICAL DILEMMA

A grave paradox characterizes the twenty-first century of oncology: our molecular comprehension of malignant transformation has never been so deep, and yet cancer continues to be among the major sources of human death all over the world (Molla & Bitew, 2025). Cancer is not just increasing in numbers in its global burden, but it's also evolving. With the aging of populations and the convergence of both environmental and lifestyle risk factors and genetic risk factors, the rates of solid tumors and hematological malignancies are ever-increasing (Ki et al., 2025). This continuous epidemiological transition causes an appalling burden on the healthcare systems, economies, and, above all, human lives across the world.

Clinical oncology has long been associated with such traditional modalities as surgery and radiotherapy. Surgical intervention has transformed from very invasive operations to more precise and minimally invasive laparoscopic and robot-assisted operations (Rojas Burbano et al., 2025). Technologies such as fluorescence-guided surgery (FGS) can provide real-time visualization of the margins, and liquid biopsies are used to monitor the development of the disease by means of circulating tumor DNA (Nagaya et al., 2017). Nevertheless, surgery is a local treatment of a systemic disease. Its main failure is instigated by Minimal Residual Disease (MRD), microscopic malignant clusters remaining or floating throughout the system (Abdo et al., 2026). Moreover, the stress and acute immunosuppression that occur during surgery and the healing process inhibit natural killer (NK) cell cytotoxicity and unintentionally form a permissive physiological environment of metastatic colonization (Tai et al., 2013).

Similar localized maximization has been attained using radiotherapy, shifting to much more precise paradigms such as Intensity-Modulated Radiation Therapy (IMRT), Stereotactic Body Radiotherapy (SBRT), and ultra-high-dose FLASH radiotherapy . They are used with chemical radioprotectors such as Amifostine or radiosensitizers to control tissue damage (J. B. Yu et al., 2014). Nonetheless, radiotherapy's efficacy is continually restricted by the tumor microenvironment. Intratumoral hypoxia is the main cause of radio-resistance; since ionizing radiation needs the presence of molecular oxygen to repair DNA damage through free radicals, the oxygen-free cores of tumors are resistant (Wu et al., 2023). Genetic and metabolic changes to survive this oxidative stress also include the up-regulation of the KEAP1-NRF2 antioxidant pathway to counteract free radicals and up-regulation of GLUT1 to provide glycolytic DNA repair (Verza et al., 2025). Finally, the molecular strength of the cancer ecosystem systematically outmatched localized physical interventions.

The inability of the localized therapy to ensure long-term disease-free survival prompts the necessity of systemic intervention, which has traditionally been dominated by classical chemotherapy. Chemotherapeutic agents take advantage of the unregulated, rapid cell division in cancer by one or more of the following mechanisms during the cell cycle:

- Alkylating Agents: Bond to DNA bases by forming covalent bonds and cross-linking strands, making replication and transcription impossible (Andrés et al., 2024).
- Antimetabolites: Are rogue nucleotide analogs, which are incorporated into the produced DNA and RNA, causing replication stress (Kovalev et al., 2022).
- Antitumor Antibiotics: Bind to the DNA helix and produce local free radicals to separate the strands (Chifiriuc et al., 2022).
- Topoisomerase Inhibitors: Freeze critical nuclear enzymes, halting replication forks and producing lethal double-strand breaks (Delgado et al., 2018).
- Antimitotic Agents: Alter microtubule networks and the mitotic spindle, halting the cell in metaphase (Salinas et al., 2025).

Although this seems attractive in theory, systemic blindness sabotages clinical implementation. Standard chemotherapeutic molecules cannot distinguish between malignant cells and healthy and fast-growing non-malignant cells (Verma et al., 2023). This leads to severe, dose-limiting toxicities in multiple vital organ systems, characterized by bone marrow suppression, gastrointestinal mucosal erosion, cardiotoxicity, hepatotoxicity, and irreversible nephrotoxicity (Thölke, 2025). The therapeutic window is so small, and more often than not, clinicians have to lower dosages or even discontinue the cycles, giving the tumor time to adjust and recover.

This inherent weakness is exacerbated by the fact that multi-drug resistance (MDR) is bound to develop. The selection of clones with strong cellular defense networks in heterogeneous tumor populations subjected to sub-lethal concentrations of drugs occurs in response to the environment. It is driven by selection pressures (Lu et al., 2026). The most striking network is the upregulation of ATP-binding cassette (ABC) efflux transporters, the most common of which is P-glycoprotein (P-gp), which uses cellular energy to cause compounds to exit the cytoplasm, maintaining intracellular concentrations well below the therapeutic level (Tia et al., 2025). At the same time, tumors increase the rate of metabolic detoxification via cytochrome P450 enzymes (especially the CYP3A family) and change phenotypes, including the Epithelial-to-Mesenchymal Transition (EMT), which silences apoptotic signaling and provides cancer cells with physical motility and a very drug-resistant phenotype (Hossam Abdelmonem et al., 2024).

Developing nanocarriers reduced to one-billionth of a meter entraps cytotoxic cargos, preserve them until they reach the cancer site, and avoid the exposure of healthy tissues to the therapies. Target-specific ligands (monoclonal antibodies, peptides, or aptamers) that bind to certain receptors overexpressed on tumor membranes can be surface-modified onto these architectures, enabling receptor-mediated endocytosis (Wei et al., 2022). This focused delivery circumvents membrane-bound ABC efflux pumps and introduces high concentrations of therapeutic doses into the intracellular compartment (Wei et al., 2022). But with the development of nanomedicine for clinical practice, this technology has hit a safety and sustainability crisis due to nano-specific toxicities and industrial bottlenecks:

- Immune System Elimination: Foreign surface chemistries are identified by the mononuclear phagocyte system, resulting in rapid opsonization and elimination of nanoparticles, which may cause acute systemic immunotoxicity or immune (Hassan et al., 2023).
- Anatomical Barrier Disruption: Pushing nanocarriers across highly selective membranes such as the Blood-Brain Barrier presents extreme safety risks, whereby structural dysfunctions of endothelial tight junctions are broken, and the body becomes neurotoxic (Hassan et al., 2023).
- Intracellular Stresses: Chemical nanoparticles cause cytotoxicity by inducing enormous Reactive Oxygen Species (ROS) metabolism that compromises the electron transport chain in the mitochondrion and destabilizes cell membranes. This spreads to genotoxicity since nanoparticles penetrate the nuclear envelope or spill over toxic ions, which destroy DNA and lead to a double-strand break (Khanna et al., 2015).
- Organ Accumulation: Since nanoparticles are taken up through the macrophages, more than 30 percent of the given chemical nanomedicines are accumulated in filtering organs (liver, lungs, kidneys, spleen, eyes), substituting chemotherapeutic side effects with long-term, nano-specific organ toxicity (Hassan et al., 2023).

In addition to biological risks, the production poses a harsh industrial bottleneck. The classical top-down and bottom-up approaches to synthesis use inorganic solvents that are harsh, dangerous reducing agents, and consume a lot of energy, producing poisonous environmental waste . Moreover, these chemical processes are extremely challenging to scale to an industrial manufacturing process; small changes in temperature or pressure can change the morphology of nanoparticles, making it impossible to reproduce batch-to-batch in exact quantities and apply them to the clinical setting (Hassan et al., 2023). Thus, an effective atomic-based intervention cannot merely exchange chemical toxicity with nanotechnology-based toxicity.

Oncology is gravitating towards biogenic Green Nano-Oncology to sustainably break the oncological deadlock. Green chemistry produces nanoparticles by synthesizing under mild, aqueous conditions by substituting dangerous chemical agents with natural, biological systems, namely plant extracts, microbial secretomes, and agricultural bioproducts (Hassan et al., 2024). This provides a biomimetic coating of proteins, polyphenols, and carbohydrates, which is eco-friendly and highly biocompatible as a natural coating on nanostructures. This endogenous biological coating reduces immune response, eliminates ROS-caused genotoxicity, decreases the deposition of damaging organs, and offers a highly reproducible, scalable, non-toxic platform for future-generation cancer therapies (Omran & Baek, 2021).

Before the knowledge of molecular pathways, cancer was regarded as an anatomical deviation, which needed complete physical elimination, leading to the radical surgery era . Such a philosophy was led by people such as William Stewart Halsted and prescribed that survival directly depended on the quantity of tissue that was excised (Marquez, 2026). The radical mastectomy of Halsted was the typical example of such an aggressive practice.

Next to surgery was the accidental birth of radiation oncology. After X-rays and radioactivity were discovered, early practitioners found that surface tumors could be shrunk with ionizing radiation (Kardamakis et al., 2023). Non-selective delivery, however, reigned during this period. The early clinicians used low-energy orthovoltage machines, which could not be modulated to change the depth or intensity of the beam (Bell et al., 2022). Radiation was required to penetrate deep tumors, and therefore the healthy tissue was subjected to catastrophic collateral damage in the form of skin burns, soft-tissue necrosis, and secondary malignancies (Yang et al., 2020).

Although these breakthroughs revolutionized previously fatal diseases, the use of such discoveries in clinical practice is compromised by total systemic blindness. Since these drugs take advantage of the general division of cells instead of their inherent cancer traits, they are blindly targeting normal cells multiplying rapidly and thus creating normal cells (Supple, 2026). This leads to acute bone marrow suppression, sloughing of the gastrointestinal mucosa, cardiotoxicity, hepatotoxicity, and nephrotoxicity (Kaye et al., 2025). The result of this was that empirical oncology hit a frustrating plateau in which the therapeutic dose needed to destroy a tumor is in a very close range of the lethal dose that will kill the patient.

In the late twentieth century, recombinant DNA, genomic sequencing, and cellular biology brought about a conceptual revolution in oncology. Cancer was redefined as an ambiguously rapidly growing mass into a complicated genetic illness characterized by driver mutations, chromosomal translocations, and abnormal signaling pathways (Brown et al., 2023). This marked the beginning of the Targeted Era, whereby emphasis was laid on the random destruction of tissues, and instead, a rational design of drugs that specifically attack certain oncogenic proteins was emphasized.

The first successful proof-of-concept was the Chronic Myeloid Leukemia (CML) Imatinib (Gleevec) (Brown et al., 2023). CML is caused by the reciprocal translocation between chromosomes 9 and 22, which forms the BCR-ABL fusion gene that expresses a constitutively active tyrosine kinase (Amarante-Mendes et al., 2022). Imatinib is specifically targeted at the ATP-binding site of the enzyme, disrupting its energy supply and inducing the process of cell apoptosis with minimal systemic toxicity as compared to traditional chemotherapy (Iqbal & Iqbal, 2014). The desired landscape was to grow in two main directions:

- Small-Molecule Inhibitors (SMIs): Low-molecular-weight molecules that pass across lipid bilayers to hit intracellular receptor regions or signaling centres, e.g., Erlotinib in EGFR in lung cancer or Vemurafenib in mutated BRAF in melanoma (Hojjat-Farsangi, 2014).
- Monoclonal Antibodies (mAbs): Large, engineered extracellular proteins that bind tumor antigens to prevent ligands or neutralize growth factors or induce immune clearance through antibody-dependent cellular cytotoxicity, like Trastuzumab and Rituximab (Selewski et al., 2010).

Modern oncology research is now at the frontier of Precision Medicine and Macromolecular Design, which are limited by the Empirical and Targeted Eras. Cancer eradication requires a multi-dimensional strategy that combines comprehensive deep genomic, transcriptomic, and proteomic profiling of the disease with complex macromolecular therapeutic architectures at the nanoscale (Yan et al., 2025). Precision medicine doesn't accept the idea that cancer should be approached in a one-size-fits-all manner based on its organ of origin. Rather, it uses next-generation sequencing (NGS) and spatial transcriptomics to precisely characterize the mutational landscape of a tumor, mapping co-existing driver mutations, tumor mutational burden, and tumor immune-evasion mechanisms (Parvizpour et al., 2024). Oncology requires complex spatially controlled delivery operations to translate this diagnostic data into cures, which are carried out by using macromolecular design and nanomedicine:

- Antibody-Drug Conjugates (ADCs): Combining targeted and empirical paradigms as “guided missiles”. Target-specific monoclonal antibodies are used to internalize the drug by endocytosis, where it undergoes endosomal/lysosomal biogenesis to release a potent cytotoxic payload, which is kept away from healthy tissue (Wang et al., 2025).
- Engineered Polymeric and Liposomal Nanocarriers: Synthetic nanocarriers capable of carrying a wide range of payloads such as hydrophobic chemotherapeutics or fragile genetic payloads (siRNA, mRNA). They can be designed to have stimuli-responsive surface chemistries that do not affect normal blood but change their surface structure in response to the acidic, hy-

poxic, or enzyme-rich tumor microenvironment to release payloads (Arshad et al., 2024).

- Biogenic, Green Nanoparticles: Nanoparticles fabricated using plant extracts and microbial secretomes, which are to be nontoxic to the cell, non-genotoxic, and will not suffer from scale-up bottlenecks, as would synthetic carrier particles. These organic architectures are naturally surrounded by a biological corona, composed of protective proteins and polyphenols, that allows them to escape immune clearance and avoid accumulation in organs and cross physiological barriers safely (Khan et al., 2026).

The combination of precision medicine with the advanced design of macromolecules allows the engineering of delivery systems at exactly the nanometer scale of oncogenic mutations, cellular signaling, and drug-resistance pumps. From radical surgeries to intentional, biogenic nanoparticles, there has been a gradual shift towards greater precision, less toxicity and biological sustainability, confirming that to solve the cellular problem of cancer, one must intervene at the atomic level, and in a way which is inherently sustainable.

## 2. SYSTEMIC LIMITATIONS OF CONVENTIONAL CHEMOTHERAPY

The clinical administration of conventional cytotoxic chemotherapy remains a foundational pillar of modern oncological protocols. However, the long-term curative potential of these traditional interventions is severely restricted by fundamental biochemical and physiological bottlenecks. Classical chemotherapeutic agents, encompassing alkylating agents, antimetabolites, topoisomerase inhibitors, and antimitotics, were historically designed to disrupt the universal macromolecular processes of DNA replication, transcription, and mitotic spindle assembly (Anand et al., 2023). Because these mechanisms are inherent to all actively dividing cells, traditional systemic therapy is intrinsically devoid of tumor-specific localization. This section establishes the biochemical and pharmacological rationale driving the urgent need for a paradigm shift in cancer therapeutics, detailing the molecular roots of off-target toxicity, the pharmacokinetic barriers impeding efficient drug delivery, and the evolutionary pathways driving multi-drug resistance.

### 2.1 Off-Target Toxicity and the Narrow Therapeutic Index

Systemic toxicity of traditional small molecule chemotherapy is a major drawback (Hassan et al., 2023). The free drug molecules passively diffuse and cause significant collateral damage to healthy tissues with high mitotic indices (Arshad et al., 2024). This biochemical blindness results in serious, dose-limiting toxicities. The hematopoietic compartment of the bone marrow is particularly vulnerable (Gencheva et al., 2013). Rapidly dividing stem cells are destroyed, and this leads to dose-dependent myelosuppression, which in turn means a high risk of infection due to profound neutropenia, a high risk of bleeding due to thrombocytopenia, and severe anaemia (Crawford et al., 2024).

Highly differentiated organ systems are also permanently injured by traditional chemotherapeutics. Anthracycline-induced cardiomyopathy (e.g., from doxorubicin) is a prominent example. Doxorubicin's quinone group can undergo redox cycling in the mitochondria of the heart, leading to the production of large amounts of ROS, which is beyond the capacity of low antioxidant defence, ultimately causing cardiomyocyte apoptosis (Osataphan et al., 2020). Furthermore, doxorubicin also interacts with a topoisomerase-II-Beta (Top2β), which causes persistent DNA breaks and irreversible heart failure in quiescent heart tissue (Saroj et al., 2025). For instance, platinum-based alkylators (cisplatin) and taxanes (paclitaxel) have significant neurotoxicity (da Costa et al., 2020). Chemotherapy-Induced Peripheral Neuropathy (CIPN) happens when neurotoxins enter the dorsal root ganglia (DRG) (Ege et al., 2024). Cisplatin cross-linkages cause degradation of sensory neurons, and taxanes hyper-stabilize the microtubules of β-tubulin, which freezes axonal transport (Ege et al., 2024).

These off-target effects result in a therapeutic index (TI) that is dangerously close, defined as the ratio between the median effective dose ($ED_{50}$) and median toxic dose ($TD_{50}$) (Chen et al., 2024). Doctors and other health professionals often have to administer sub-optimal doses to prevent organ damage, which inadvertently leaves behind malignant sub-populations that can survive and cause the disease to recur.

## 2.2 Pharmacokinetic Hurdles: Delivery and Degradation

Clinical efficacy means that drug concentration in the tumor needs to be maintained at levels of therapeutic activity but not prematurely cleared from the system. Free drugs based on conventional methods are beset by very large biophysical challenges.

The first challenge is their poor aqueous solubility. The highly hydrophobic agents such as paclitaxel need toxic synthetic excipients such as Cremophor EL that cause severe hypersensitivity and systemic histamine release (Chen et al., 2024). In addition, Cremophor EL itself creates a micellar structure, which allows the drug to be physically trapped within the micelle which changes the pharmacokinetics of the drug and prevents it from diffusing into the interstitial space of the tumor (Serras et al., 2024). Small molecules (<1,000 Daltons) are very vulnerable to rapid clearance by the kidneys, which is below the 30- 50 K glomerular filtration limit, and this gives rise to a short plasma half-life ($t_{1/2}$), requiring high bolus doses with high plasma concentrations, leading to high systemic toxicity peaks (Wang et al., 2026).

At the same time, free drugs are subjected to premature enzymatic degradation in the circulation by the action of phase I and II enzymes (Zhao et al., 2021). For instance, gemcitabine is rapidly deactivated by cytidine deaminase, and camptothecin undergoes a pH-dependent spontaneous hydrolysis at the physiological blood pH of 7.4, before reaching the tumor site (Almeida et al., 2021).

Lastly, there are barriers to be overcome in the solid tumor microenvironment. In the core of the tumor, Interstitial Fluid Pressure (IFP) is abnormally high because of chaotic angiogenesis and absence of lymphatic drainage. This outward convective force results in the creation of an impenetrable biophysical barrier that causes free drugs to be washed back into the systemic circulation and not penetrate the tumor sufficiently (Simonsen et al., 2012).

## 2.3 Multi-Drug Resistance (MDR): Evolution and Evasion:

The biggest hurdle for classical chemotherapy is the ability of malignant cells to resist many different types of drugs; evolved resistance to multiple drugs, called MDR, violates the therapeutic payload by rapidly neutralizing it.

Over-expression of ATP-binding cassette (ABC) transmembrane efflux pumps, in particular P-glycoprotein (P-gp) (encoded by the ABCB1 gene), is the major cause of MDR (Tian et al., 2023), as shown in **Figure 1**. P-gp is an ATP-dependent transporter that has a broad substrate specificity for hydrophobic drugs, such as anthracyclines and taxanes. The cell membrane is where the drug is distributed when it enters the cell, and P-gp physically pumps the drug back out into the extracellular space, preventing the cell from being exposed to a therapeutic concentration of the drug and making the cell resistant to multiple drugs at once (Sharom, 2014).

*Figure 1. Multi-scale representation of chemotherapy barriers: (1) off-target passive diffusion toxicity, (2) pharmacokinetic hurdles (rapid clearance, physical barriers), and (3) P-glycoprotein efflux-mediated tumor evasion*

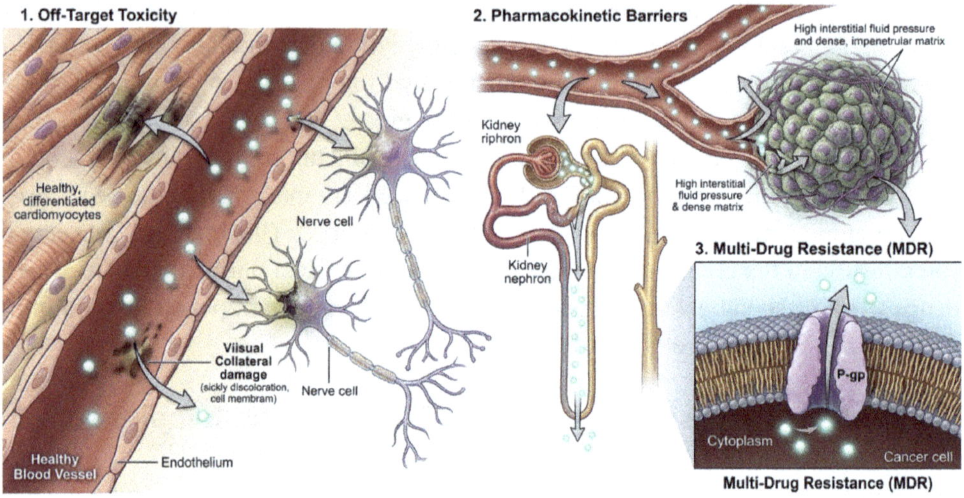


There is also increased metabolic detoxification in parallel with active efflux. Resistant tumors over-express Glutathione S-transferases (GSTs) that can conjugate glutathione (GSH) to chemotherapeutics, such as platinum compounds, thereby rendering the drugs inactive and marking them for extrusion (Ang et al., 2005). In addition, tumors exhibit phenotypic changes such as EMT, which rewire signaling networks to increase the amount of anti-apoptotic proteins (Bcl-2, Bcl-xL) and silence pro-apoptotic executioners (Ulrike Keitel et al., 2014).

## 3. NANOTECHNOLOGY – THE TRANSFORMATIVE SOLUTION

The current chemotherapy techniques are extremely ineffective, which necessitates a different approach to therapy. Interventions need to be designed at the same dimensional scale as the biological mechanisms that they modulate to overcome off-target toxicity and rapid clearance. These fundamental changes in the biodistribution and therapeutic index of cytotoxic agents can be achieved by tailoring macromolecular delivery vehicles to the 1–100 nm size range.

## 3.1 The Nano-Scale Advantage: Overcoming Macro-Scale Biological Barriers

A nanocarrier is a colloidal structure, typically between 1 – 100 nm in size, although sometimes up to 200 nm for complex payloads (Lu et al., 2021). These are at a scale with extremely large surface-area to volume ratios and very tunable surface chemistries.

The first benefit is to avoid pharmacokinetic limitations (Arshad et al., 2024). However, enclosing a payload in a 50 nm liposome or micelle will physically protect the payload from rapid clearance from the renal glomeruli, and thus dramatically improve the plasma half-life (Attri et al., 2024). Another advantage of encapsulation is that solubility problems are eliminated; hydrophobic drugs such as paclitaxel can be sequestered in a lipophilic core without any toxic excipients such as Cremophor EL (Ma & Mumper, 2013). It also shields sensitive payloads (siRNA, camptothecin) against enzymatic degradation.

But there are problems with immune surveillance at the nanoscale. The mononuclear phagocyte system (MPS) quickly opsonizes and removes circulating nanocarriers (Gustafson et al., 2015). As a way around this, surfaces are stealth engineered by PEGylation, which is the covalent grafting of polyethylene glycol (PEG) chains (Makharadze et al., 2025). This hydrophilic corona forms a dynamic steric barrier that protects the nanoparticle from macrophage recognition and allows for long circulation times in the system (Arshad et al., 2024).

Last but not least, nanoscale delivery neutralises MDR determinants such as P-gp efflux pumps (Thomas & Weber, 2019). Nanocarriers cannot passively diffuse into cells; they must be endocytosed (Thomas & Weber, 2019). They are contained in an endosomal vesicle where they avoid the membrane-bound efflux pumps, and are subsequently degraded in the acidic endolysosome, giving the particles direct access to the perinuclear space.

## 3.2 Passive vs. Active Targeting: Engineering Precision Homing

The goal of oncological nanomedicine is to selectively deliver therapeutic payloads to the malignant microenvironment by two targeting approaches (Bazak et al., 2014).

### 3.2.1 Dissection of Enhanced Permeability and Retention (EPR) Effect

Passive targeting is based upon tumor pathophysiology. Tumors respond to localized hypoxia by producing Vascular Endothelial Growth Factor (VEGF) that causes chaotic angiogenesis . The tumor neovasculature also exhibits architectural abnormalities: it lacks basement membranes and has larger fenestrations in the endothelial cells (100 to 800 nm) . Solid tumors, at the same time, do not have a functional lymphatic drainage . The Enhanced Permeability and Retention (EPR) effect allows nanoparticles to take advantage of these defects (Bazak et al., 2014). Nanocarriers with long circulation properties are able to extravasate into the tumour interstitium through large endothelial fenestrations and are selectively retained by the poor lymphatic clearance (Saw et al., 2021). Nanomedicines that are FDA-approved, such as Doxil and Abraxane, are based on the EPR effect and show much lower toxicity (Hassan et al., 2018).

But the clinical translation presents a problem: the EPR effect is very heterogeneous in humans, and the accumulation of nanoparticles is limited to the periphery of the tumour, where the IFP is high and opposes the diffusion of nanoparticles. Active targeting mechanisms have to be introduced to attain true intracellular precision.

### 3.2.2 Active Targeting: Surface Functionalization for Pinpoint Cellular Homing

Active targeting is the combination of the EPR effect with the surface functionalization of the nanocarrier using high-affinity ligands that bind selectively to the cancer cell's uniquely overexpressed receptor. This results in very selective internalization of the drug through receptor-mediated endocytosis (RME) (Nikhar et al., 2026). There are several classes of targeting ligands that have been developed:

- Whole antibodies (Trastuzumab, Rituximab) are very specific but are large (150kDa) and may be immunogenic and have limited penetration. Smaller fragments (scFv, nanobodies) have specificity but with a smaller footprint for denser functionalization and better stability (Asaadi et al., 2021).
- Aptamers: Short, single-stranded oligonucleotides (DNA or RNA) that have been designed to adopt a specific three-dimensional structure. These “chemical antibodies” are non-immunogenic, very stable, and simple to synthesize and target receptors such as EpCAM. Tumors have metabolic profiles that are aggressively up-regulated (small molecules/vitamins). Nanocarriers can be functionalized with the nutrients they require to act as a “Trojan horse” (such as folic acid targeting Folate Receptors or transferrin facilitating cellular iron uptake) (Narwade et al., 2023).
- Peptides: Short amino acid chains with high binding affinity. The RGD motif is selectively bound by the $\alpha v\beta 3$ integrins that are heavily expressed by tumour cells and sprouting neovasculature, offering a dual-targeting ability (Y.-P. Yu et al., 2014).

These ligands can be used to decorate surfaces, and in this way, the biodistribution profile is fundamentally altered. It actively drives the nanoparticles into the intracellular space of the cancer cell to maximize payload delivery and limit exposure to neighbouring stromal cells.

## 3.3 Theranostics: The Integration of Diagnostics and Therapeutics

The trend in oncology towards precision medicine has driven the emergence of the concept of 'theranostics', which is a nanoscale architecture that combines therapeutics and diagnostics in a single platform and provides real-time clinical feedback. Theranostic nanomedicine enables the simultaneous delivery of targeted cytotoxic payloads, real-time monitoring of biodistribution, quantification of nanomedicine accumulation, and assessment of early therapeutic responses to treatment (Yasir et al., 2024).

Nanocarriers (such as mesoporous silica, polymeric micelles) provide this integration due to their high surface-area-to-volume ratio and unique structural compartments (Arshad et al., 2024). A simple theranostic nanocarrier consists of a nanocarrier, therapeutic agent, and diagnostic imaging contrast agent. The contrast agent used will be incorporated and vary depending on the imaging modality chosen:

- Magnetic Resonance Imaging (MRI): Providing high spatial resolution with depth of penetration in the absence of ionizing radiation. A highly efficient T2 contrast agent that can be used by platforms to visually confirm tumor penetration with sub-millimeter resolution is gadolinium-based chelates or Superparamagnetic Iron Oxide Nanoparticles (SPIONs, such as Fe3O4) (Hassan et al., 2018).
- Positron Emission Tomography (PET) and Single-Photon Emission Computed Tomography (SPECT): Best for very high sensitivity and quantitative biodistribution. Conjugation of chelators to isotopes (such as $^{64}$Cu, $^{89}$Zr) allows PET to measure real-time pharmacokinetic profiles, such as the precise amount of injected dose that is delivered to the tumor versus other organs (such as the liver) (Kang et al., 2022; Krasnovskaya et al., 2023).

Although light does not penetrate deeply, optical and fluorescence imaging are important for the guidance of surgery, such as fluorescence-guided surgery (FGS), in the operating room. Tumors can be outlined with carriers filled with near-infrared dyes or quantum dots, which would enable exact surgical removal along with simultaneous chemotherapy.

## 4. THE SYNTHETIC "BOTTLENECK": ENVIRONMENTAL & BIOLOGICAL HAZARDS OF STANDARD NANOTECH

Nanomedicine is a paradigm with huge promise to overcome the macroscopic challenges of conventional chemotherapy, but a daunting "synthetic bottleneck" has hampered clinical translation (Pourmadadi et al., 2023). The inherent risks of traditional physical and chemical manufacturing processes are a major limiting factor for transition to global clinical deployment. Conventional synthetic routes based on caustic chemicals and onerous reaction conditions induce significant biological and ecological risks and obligations that are contradictory to the nanocarrier's therapeutic goal (Hassan et al., 2023).

### 4.1 The Cost of "Top-Down" and "Bottom-Up" Chemical Synthesis

Generally, fabrication of nanomaterials occurs in two ways: "top-down" and "bottom-up . The top-down approach is a physical process that reduces macroscopic materials into nanoscale materials. But irregular size distributions, defective surface topologies, and high energetic costs prevent their use in medical applications. Hence, the majority of therapeutic nanocarriers have been prepared using "bottom-up" chemistry, by building up atomic precursors into structured architectures (Hassan et al., 2018). The main drawback of bottom-up synthesis is the use of very toxic chemical reagents. Sodium borohydride, $NaBH_{4,}$ and hydrazine, $N_2H_{4,}$ are used to reduce dissolved forms of metal cations (e.g., $Au^{3+}$, $Ag^+$ to zero-valent states ($Au^0$ or $Ag^0$) (Abuzeid et al., 2023). These compounds are extremely toxic: $NaBH_4$ is corrosive, and $N_2H_4$ is a potent hepatotoxin and suspected carcinogen (Ahmed et al., 2023). The nanoclusters formed have an extraordinarily high surface free energy, and thus require artificial stabilisation by means of surfactants to avoid macroscopic agglomeration. The most common is cetyltrimethylammonium bromide (CTAB). CTAB is intrinsically cytotoxic and is morphologically effective, readily intercalating into mammalian cell membranes to cause catastrophic lysis and necrosis. Moreover, nanocarriers are often synthesized through energy-consuming evaporation or by using volatile organic solvents (VOCs), which makes conventional chemical synthesis very inefficient and unsafe (Park, 2014).

### 4.2 The Biological Backlash: Residual Toxicity and the Protein Corona

The need to use harmful chemicals results in a significant "biological backlash" when nanocarriers are introduced into a biological environment. Even with intensive purification, it is virtually impossible to eliminate toxic synthetic byproducts. Unreacted precursors, solvents, and surfactants are always present in trace quantities and are gradually released in the systemic circulation. With this in vivo release, severe off-target toxicities also occur that are similar to the side effects of conventional chemotherapy. Heavy metal ions or CTAB are leached and rapidly lead to oxidative stress in cells, which results in a huge influx of reactive oxygen species (ROS) (Cheng et al., 2025). This overloads the antioxidant mechanisms, leading to lipid peroxidation, protein denaturation, and genotoxic damage of DNA . In addition to chemical leaching, the artificial surface programs a direct interaction with the host immune system. Once in the blood plasma, the high surface free energy of the particle will induce the spontaneous coating of the particle by plasma proteins, known as a "protein corona". This adsorption, if performed through the use of aggressive surfactants, creates enormous conformational stress resulting in significant denaturation of key plasma proteins. This leads to exposure of "cryptic epitopes" that are recognized by the immune system as Damage-Associated Molecular Patterns (DAMPs) (Fadeel, 2022). The nanocarriers are then rapidly cleared by the mononuclear phagocyte system (MPS), completely nullifying the EPR effect and active targeting mechanisms (Mills et al., 2022). Further, bound opsonins may overactivate complement pathways leading to a possible life-threatening immune response called Complement Activation-Related Pseudoallergy (CARPA). The problem of scaling-up has been discussed in terms of the environmental footprint (Salmaso & Caliceti, 2013).

### 4.3 The Environmental Footprint: The Crisis of Scale-Up

Normal nanotechnology risks are also a serious threat to the global ecological balance. To take the milligram results from the research lab to the clinic, the volume of raw materials needs to be scaled up from the milligrams to metric tons, highlighting an unsustainable environmental footprint. The E-factor (mass of waste/mass of product (Rani et al., 2022)) often can be larger; more than one hundred kilograms of very poisonous waste are produced per kilogram of therapeutic nanocarrier. Scaling-up bottom-up synthesis involves using huge industrial bioreactors containing VOCs, heavy metal salts, and toxic reducing agents . Heavy metal nanoparticles are always released into the watersheds via industrial run-off and bioaccumulate in the watershed, thereby affecting agricultural microbiota and killing aquatic organisms (Hassan et al., 2023). In addition, VOC emissions are constantly vented to the atmosphere, and the high temperatures required for synthesis add to the industry's carbon footprint (Betke & Kickelbick, 2014). In this lies a great paradox: trying to cure cancer with manufacturing techniques that pollute the environment, as shown in **Figure 2**. In the end, the biological response and ecological impacts of the synthetic crutch necessitate a transition to inherently biocompatible and ecologically friendly approaches to oncology.

*Figure 2. Overarching hazards of conventional nanomanufacturing: (A) toxic chemical synthesis, (B) biological immune clearance via protein corona and ROS, and (C) unsustainable environmental footprint (E-factor >100:1)*

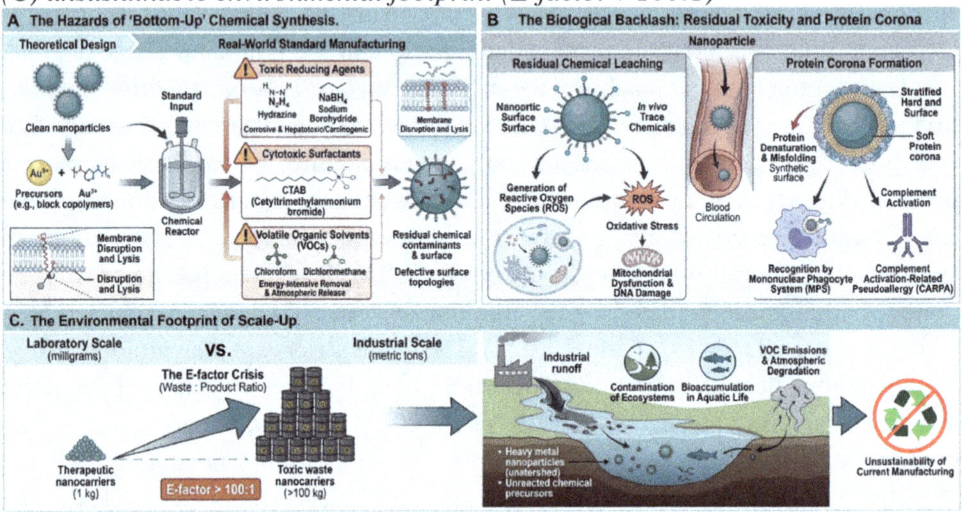


## 5. THE RISE OF "GREEN NANO-ONCOLOGY": A NEW PARADIGM

Although nanomedicine can address the macroscopic problems encountered by conventional chemotherapy, it does face a daunting "synthetic bottleneck" for clinical translation. Limited global deployment due to inherent risks of traditional physical and chemical manufacturing processes. Synthesis of nanomaterials is typically done through either "top-down" or "bottom-up" approaches. **Figure 3** contains details on all approaches and methods used to synthesized nano scaled materials.

*Figure 3. Foundational green nano-oncology paradigm: (A) green shift from toxic chemicals to biogenic one-pot synthesis with biocompatible corona, (B) bio-factories (phytochemical, microbial NADH reductase, chitosan pH-responsive), and (C) multi-functional advantages (immune evasion, antioxidant, P-gp inhibition, antimicrobial protection)*

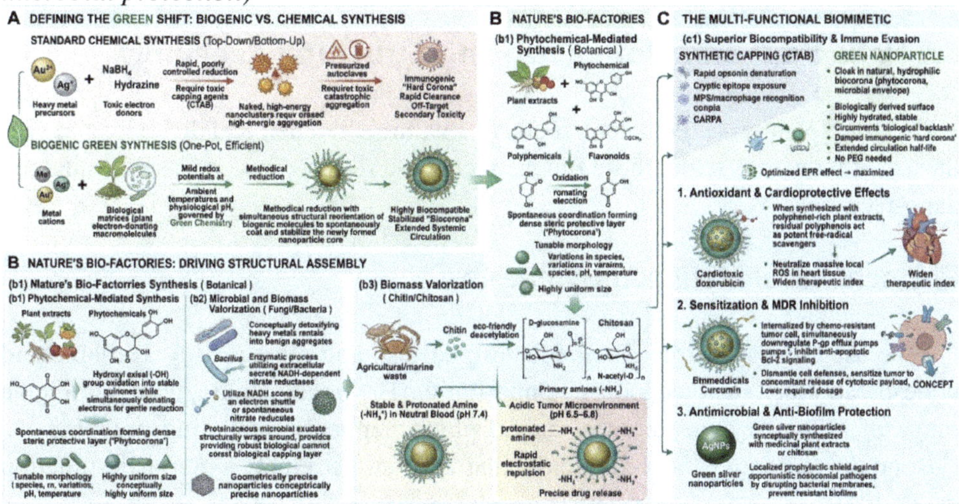


## 5.1 Defining the Green Shift: Transitioning to Biogenic Synthesis

The green shift in nanomedicine is the elimination of the hazardous bottom-up chemical reagents used in the reduction, assembly, and stabilization of nanoscale architectures, and the replacement of them by biological matrices which can perform the same functions. This transition entails a change from artificially induced thermodynamics to finely tuned, biology-controlled kinetics (Ayub et al., 2025).

The reduction of the heavy metal precursors (Au3+, Ag+, and $Pt^{2+}$) to zero-valent states ($Au^0$, $Ag^0$, and $Pt^0$) is based on highly reactive and toxic electron donors, such as $NaBH_4$ and hydrazine in standard synthesis (Dikshit et al., 2021). These are poorly controlled, highly exergonic reactions that quickly create high-energy nanoclusters, which are uncovered and need to be capped with toxic synthetic capping agents (CTAB), extreme heat, pressurized autoclaves, and complex purification.

In contrast, the biogenic synthesis is a single-step, highly efficient "one-pot" synthesis, which is based on Green Chemistry. Complex, electron-donating macromolecules are naturally occurring in biological systems (plant extracts and microbial cultures). They act at room temperature, physiological pH, and reduce metal cations step-by-step, using natural redox potentials (Syed et al., 2026). These macromolecules spontaneously orient themselves around the newly created high-energy surface, covering it and thereby eliminating the need for artificial surface-active agents. The nanocarrier is immediately surrounded by a highly biocompatible, biologically derived corona, thus changing its interactions with the body in a radical way (Syed et al., 2026).

## 5.2 Nature's Bio-Factories: Driving Structural Assembly

In order to increase the potential of using nature's biochemical diversity and synthesize at industrial scale, these biochemical systems are viewed as self-contained nano-factories.

### Phytochemical-Mediated Synthesis

The use of plant extracts is the fastest, easiest, and most readily available and widespread green nanotechnology opportunity. The botanical kingdom has a huge range of secondary metabolites, called phytochemicals, including polyphenols, flavonoids, terpenoids, and alkaloids, that have an extremely synergistic function in synthesis:

- **Simultaneous Reduction:** Flavonoids (quercetin) and polyphenols (EGCG) possess some hydroxyl (-OH) groups that are reactive and electron-rich and are arranged on aromatic rings. These are subjected to a very fast oxidation process during the synthesis and also act as free electron donors to the dissolved metal precursors, thus reducing metal ions to form the solid core of the nanoparticles (Sysak et al., 2023).
- **In Situ Capping and Stabilization:** The oxidation of phytochemicals alters the polarity of their structure and the oxidized biomolecules (quinones, carboxylates) have highly electrostatic affinity with the newly formed nanoparticle surface. They organize themselves around the core which has no charge, creating a protective "phytocorona". This physically inhibits agglomeration in the adjacent nanoparticles to provide a highly uniform monodisperse size distribution without toxic surfactants (Zuhrotun et al., 2023).

Researchers vary the species of plant extract, the pH and the temperature to carefully manipulate the morphology of the nanocarriers, spheres, nanorods and triangles, for optimized tumor penetration.The conversion of microbial and biomass resources to products of value.

### 5.2.1 Microbial and Biomass Valorization

Green Nano-Oncology actively exploits special metabolic pathway of microorganisms and agricultural biopolymers.

- **Fungal and Bacterial Bio-Factories:** Microorganisms (fungi, bacteria) are able to grow in heavy-metal rich environments because they have developed an intracellular detoxification mechanism that converts toxic metal ions into innocuous nanoscale aggregates. Microorganisms will release extracellular enzymes such as NADH-dependent nitrate reductases when they come into contact with metal precursors. These enzymes function as specific biological catalysts, that reduce metal ions into stable nanoparticles through the electron shuttle properties of NADH. The exudate then forms a thick, biological capping layer of structural proteins around the particle (Kumar et al., 2026).
- **Biomass Valorization (Chitin and Chitosan):** The green shift is based on renewable biopolymers from agricultural and marine waste, named biomass valorization (Chitin and Chitosan). Crustacean exoskeletons are used to produce chitin, which undergoes eco-friendly deacetylation to yield a linear biopolymer, chitosan. The primary amine groups ($-NH_2$) of the chitosan have a specific dissociation constant ($pK_a$ ~6.5). These amine groups protonate actively ($-NH^{3+}$) under slightly acidic conditions of solid tumor (pH 6.5-6.8). Chitosan nanocarriers exhibit this pH responsive behavior, which can help them to be stable in the neutral bloodstream (pH 7.4) and rapidly change their electrostatic repulsive force, swell, and release drugs exactly in the acidic tumor microenvironment (Zaiki et al., 2023).

## 5.3 The Multi-Functional Biomimetic Advantage

Adopting a biogenic fabrication paradigm alters the therapeutic profile of the nanocarrier that is generated by the process. The surface chemistry and the structural fluidity that is manifested by green nanomaterials is called as biomimetic advantage. This avoids the traditional nanotechnology backlash from biology and allows for intrinsic secondary therapeutic effects.

### Superior Biocompatibility and Lower Immunogenicity

The most important benefit of biogenic synthesis is to reduce the protein corona and to avoid immune clearance. In the presence of synthetic capping agents, the proteins in blood serum denature, revealing cryptic epitopes and causing massive immune recognition, macrophage engulfment and CARPA.

By contrast, a green nanoparticle enters the bloodstream as wrapped in a natural, hydrophilic, biological envelope (phytocorona or proteinaceous envelope). This is a very hydrated surface, and is the natural biological surface that does not violently unfold circulating opsonins. The mononuclear phagocyte system (MPS) identifies the biomimetic surface as non-threatening, not a pathogen. Biogenic nanocarriers have much longer systemic circulation half lives (without any synthetic polymers such as PEG) due to the naturally occurring "hard corona". This improved biocompatibility leads to greater delivery of the dose to the tumor vasculature and therefore to the most efficient use of the EPR effect (Sushnitha et al., 2020).

## Intrinsic Secondary Therapeutic Properties

With Green Nano-Oncology, the delivery vehicle is no longer clinically inert! The reduction and the capping of biomolecules do not destroy their pharmacological activity, thus generating a theranostic platform which works synergistically with the main chemotherapeutic payload..

- **Intrinsic Antioxidant and Cardioprotective Effects:** Synthetic phytocorona is highly redox-active and shows intrinsic antioxidant and cardioprotective properties if it is made with polyphenol-rich plant extracts. The biogenic vehicle actively reduces the severity of off-target side effects if it is loaded with a cardiotoxic drug such as doxorubicin. Residual polyphenols serve as a powerful free-radical scavenger and can intercept doxorubicin-induced ROS before they trigger irreversible injury to the cardiomyocytes, thus enhancing the therapeutic index of the entrapped drug (Iqbal et al., 2023).
- **Sensitization and MDR Inhibition:** Many phytochemicals that are used as ingredients in biogenic surfaces contain a natural MDR inhibition property (e.g., curcumin, resveratrol). They are internalized by chemo-resistant tumor cells and at the same time down-regulate the P-glycoprotein (P-gp) efflux pumps and inhibit the anti-apoptotic Bcl-2 signalling cascades. The biogenic vehicle can then actively break down the defense mechanisms, making the aggressive tumor more sensitive and drastically reducing the amount of tumor that can be completely eradicated (Trofin et al., 2025).
- **Antimicrobial and Anti-Biofilm Protection:** The chemotherapy-induced myelosuppression makes patients receiving systemic treatment for their underlying condition extremely vulnerable to microbial infections and the establishment of biofilms. Also, some biogenic nanomaterials have extraordinary broad spectrum antimicrobial properties (e.g., plant extract-based biogenic

silver nanoparticles or chitosan). They weaken bacterial cell membranes and prevent the formation of very resistant bacterial biofilms, offering a natural barrier against opportunistic nosocomial pathogens (Wozeak et al., 2026).

## CONCLUSION

Green Nano-Oncology is a breakthrough from the unscalable, toxic methods of the past and uses nature's fine biochemical machinery as its basis. This green shift results in smart biomimetic nanocarriers that can avoid immune destruction, protect healthy tissues, and systematically dismantle tumor resistance pathways to provide a comprehensive solution to the classical limitations of cancer therapy. A constant challenge in clinical oncology is to keep up with the changing evolution of malignant disease. Systemic blindness, low therapeutic window, devastating off-target side effects, and inevitable emergence of MDR are all problems with conventional chemotherapy. To overcome the biophysical barriers, protect the payloads, and actively target the tumour cells, nanomedicine was born. But there is a big "synthetic bottleneck" to the use of traditional chemical nanocarriers. The use of toxic reducing agents leads to huge ecological losses and to serious biological feedback. Green Nano-Oncology addresses both the macro-environmental footprint and the micro-cellular need for precision by synthesizing biomimetic nanocarriers that work synergistically with the human environment.

### The Vision Forward

In order to overcome the oncological deadlock, next-generation therapeutic agents should be inherently sustainable, extremely biocompatible, and able to deliver with high precision. The "whole picture" of the nanomedicine, from how it is made in an environmentally friendly way to how it is cleared from the body without being detected by immune system cells, should be considered. The following chapters will discuss particular biogenic platforms, pathways for phytochemical reduction, advanced stimuli-responsive biopolymers such as chitosan, and their clinical translation. By creating this sustainable model, the groundwork is laid for an oncology world in which the treatment is itself the treatment.

## REFERENCES


Abdo, T., Alhalabi, A., Yaghi, S., Aloran, M., Li, Y., Herrán, M., Tfayli, R., Samuel, T. A., & Nahleh, Z. (2026). Minimal residual disease in solid tumors: Clinical applications and future directions. *Cancer, 132*(3), e70286. 10.1002/cncr.70286. DOI: 10.1002/cncr.70286 PMID: 41604278

Abuzeid, H. M., Julien, C. M., Zhu, L., & Hashem, A. M. (2023). Green Synthesis of Nanoparticles and Their Energy Storage, Environmental, and Biomedical Applications. *Crystals, 13*(11), 1576. DOI: 10.3390/cryst13111576

Ahmed, M. J., Perveen, S., Hussain, S. G., Khan, A. A., Ejaz, S. M. W., & Rizvi, S. M. A. (2023). Design of a facile, green and efficient graphene oxide-based electrochemical sensor for analysis of acetaminophen drug. *Chemical Papers, 77*(4), 2275–2294. DOI: 10.1007/s11696-022-02628-9 PMID: 36589858

Almeida, A., Fernandes, E., Sarmento, B., & Lúcio, M. (2021). A biophysical insight of camptothecin biodistribution: Towards a molecular understanding of its pharmacokinetic issues. *Pharmaceutics, 13*(6), 869. DOI: 10.3390/pharmaceutics13060869 PMID: 34204692

Amarante-Mendes, G. P., Rana, A., Datoguia, T. S., Hamerschlak, N., & Brumatti, G. (2022). BCR-ABL1 Tyrosine Kinase Complex Signaling Transduction: Challenges to Overcome Resistance in Chronic Myeloid Leukemia. *Pharmaceutics, 14*(1), 215. DOI: 10.3390/pharmaceutics14010215 PMID: 35057108

Anand, U., Dey, A., Chandel, A. K. S., Sanyal, R., Mishra, A., Pandey, D. K., De Falco, V., Upadhyay, A., Kandimalla, R., Chaudhary, A., Dhanjal, J. K., Dewanjee, S., Vallamkondu, J., & Pérez de la Lastra, J. M. (2023). Cancer chemotherapy and beyond: Current status, drug candidates, associated risks and progress in targeted therapeutics. *Genes & Diseases, 10*(4), 1367–1401. 10.1016/j.gendis.2022.02.007. DOI: 10.1016/j.gendis.2022.02.007 PMID: 37397557

Andrés, C. M., Pérez de la Lastra, J. M., Munguira, E. B., Andrés Juan, C., & Pérez-Lebeña, E. (2024). Dual-Action Therapeutics: DNA Alkylation and Antimicrobial Peptides for Cancer Therapy. *Cancers (Basel), 16*(18), 3123. DOI: 10.3390/cancers16183123 PMID: 39335095

Ang, W. H., Khalaila, I., Allardyce, C. S., Juillerat-Jeanneret, L., & Dyson, P. J. (2005). Rational Design of Platinum(IV) Compounds to Overcome Glutathione-S-Transferase Mediated Drug Resistance. *Journal of the American Chemical Society, 127*(5), 1382–1383. DOI: 10.1021/ja0432618 PMID: 15686364

Arshad, R., Hassan, D., Sani, A., Mustafa, G., Rahdar, A., Fathi-karkan, S., Kharaba, Z., Medina, D. I., & Pandey, S. (2024). Nano-engineered solutions for ibuprofen therapy: Unveiling advanced co-delivery strategies and nanoparticle systems. *Journal of Drug Delivery Science and Technology*, *98*, 105815. DOI: 10.1016/j.jddst.2024.105815

Asaadi, Y., Jouneghani, F. F., Janani, S., & Rahbarizadeh, F. (2021). A comprehensive comparison between camelid nanobodies and single chain variable fragments. *Biomarker Research*, *9*(1), 87. DOI: 10.1186/s40364-021-00332-6 PMID: 34863296

Attri, N., Das, S., Banerjee, J., Shamsuddin, S. H., Dash, S. K., & Pramanik, A. (2024). Liposomes to cubosomes: The evolution of lipidic nanocarriers and their cutting-edge biomedical applications. *ACS Applied Bio Materials*, *7*(5), 2677–2694. DOI: 10.1021/acsabm.4c00153 PMID: 38613498

Ayub, A., Wani, A. K., Malik, S. M., Ayub, M., Singh, R., Chopra, C., & Malik, T. (2025). Green nanoscience for healthcare: Advancing biomedical innovation through eco-synthesized nanoparticle. *Biotechnology Reports (Amsterdam, Netherlands)*, *47*, e00913. 10.1016/j.btre.2025.e00913. DOI: 10.1016/j.btre.2025.e00913 PMID: 40895589

Bazak, R., Houri, M., El Achy, S., Hussein, W., & Refaat, T. (2014). Passive targeting of nanoparticles to cancer: A comprehensive review of the literature. *Molecular and Clinical Oncology*, *2*(6), 904–908. DOI: 10.3892/mco.2014.356 PMID: 25279172

Bell, B. I., Vercellino, J., Brodin, N. P., Velten, C., Nanduri, L. S. Y., Nagesh, P. K. B., Tanaka, K. E., Fang, Y., Wang, Y., Macedo, R., English, J., Schumacher, M. M., Duddempudi, P. K., Asp, P., Koba, W., Shajahan, S., Liu, L., Tomé, W. A., Yang, W.-L., & Guha, C. (2022). Orthovoltage X-Rays Exhibit Increased Efficacy Compared with γ-Rays in Preclinical Irradiation. *Cancer Research*, *82*(15), 2678–2691. DOI: 10.1158/0008-5472.CAN-22-0656 PMID: 35919990

Betke, A., & Kickelbick, G. (2014). Bottom-Up, Wet Chemical Technique for the Continuous Synthesis of Inorganic Nanoparticles. *Inorganics*, *2*(1), 1–15. DOI: 10.3390/inorganics2010001

Brown, J. S., Amend, S. R., Austin, R. H., Gatenby, R. A., Hammarlund, E. U., & Pienta, K. J. (2023). Updating the Definition of Cancer. *Molecular Cancer Research*, *21*(11), 1142–1147. DOI: 10.1158/1541-7786.MCR-23-0411 PMID: 37409952

Chen, Y.-S., Jin, E., & Day, P. J. (2024). Use of drug sensitisers to improve therapeutic index in cancer. *Pharmaceutics*, *16*(7), 928. DOI: 10.3390/pharmaceutics16070928 PMID: 39065625

Cheng, Y., Zhao, Y., Chen, C., & Zhang, F. (2025). Heavy Metals Toxicity: Mechanism, Health Effects, and Therapeutic Interventions. *MedComm*, *6*(9), e70241. 10.1002/mco2.70241. DOI: 10.1002/mco2.70241 PMID: 40843132

Chifiriuc, M. C., Filip, R., Constantin, M., Pircalabioru, G. G., Bleotu, C., Burlibasa, L., Ionica, E., Corcionivoschi, N., & Mihaescu, G. (2022). Common themes in antimicrobial and anticancer drug resistance [Review]. *Frontiers in Microbiology, Volume 13 - 2022*. DOI: 10.3389/fmicb.2022.960693

Crawford, J., Herndon, D., Gmitter, K., & Weiss, J. (2024). The impact of myelosuppression on quality of life of patients treated with chemotherapy. *Future Oncology*, *20*(21), 1515–1530. DOI: 10.2217/fon-2023-0513 PMID: 38587388

da Costa, R., Passos, G. F., Quintão, N. L. M., Fernandes, E. S., Maia, J. R. L. C. B., Campos, M. M., & Calixto, J. B. (2020). Taxane-induced neurotoxicity: Pathophysiology and therapeutic perspectives. *British Journal of Pharmacology*, *177*(14), 3127–3146. 10.1111/bph.15086. DOI: 10.1111/bph.15086 PMID: 32352155

Delgado, J. L., Hsieh, C.-M., Chan, N.-L., & Hiasa, H. (2018). Topoisomerases as anticancer targets. *The Biochemical Journal*, *475*(2), 373–398. DOI: 10.1042/BCJ20160583 PMID: 29363591

Dikshit, P. K., Kumar, J., Das, A. K., Sadhu, S., Sharma, S., Singh, S., Gupta, P. K., & Kim, B. S. (2021). Green Synthesis of Metallic Nanoparticles: Applications and Limitations. *Catalysts*, *11*(8), 902. DOI: 10.3390/catal11080902

Ege, E., Briggi, D., Vu, P., Cheng, J., Lin, F., & Xu, J. (2024). Targeting dorsal root ganglia for chemotherapy-induced peripheral neuropathy: From bench to bedside. *Therapeutic Advances in Neurological Disorders*, *17*, 17562864241252718. DOI: 10.1177/17562864241252718 PMID: 39318973

Fadeel, B. (2022). Understanding the immunological interactions of engineered nanomaterials: Role of the bio-corona. *Wiley Interdisciplinary Reviews. Nanomedicine and Nanobiotechnology*, *14*(6), e1798. 10.1002/wnan.1798. DOI: 10.1002/wnan.1798 PMID: 36416023

Gencheva, M., Hare, I., Kurian, S., Fortney, J., Piktel, D., Wysolmerski, R., & Gibson, L. F. (2013). Bone marrow osteoblast vulnerability to chemotherapy. *European Journal of Haematology*, *90*(6), 469–478. 10.1111/ejh.12109. DOI: 10.1111/ejh.12109 PMID: 23551534

Gustafson, H. H., Holt-Casper, D., Grainger, D. W., & Ghandehari, H. (2015). Nanoparticle uptake: The phagocyte problem. *Nano Today*, *10*(4), 487–510. 10.1016/j.nantod.2015.06.006. DOI: 10.1016/j.nantod.2015.06.006 PMID: 26640510

Hassan, D., Khalil, A. T., Saleem, J., Diallo, A., Khamlich, S., Shinwari, Z. K., & Maaza, M. (2018). Biosynthesis of pure hematite phase magnetic iron oxide nanoparticles using floral extracts of Callistemon viminalis (bottlebrush): their physical properties and novel biological applications. *Artificial Cells, Nanomedicine, and Biotechnology, 46*(sup1), 693–707. DOI: 10.1080/21691401.2018.1434534

Hassan, D., Sani, A., Chanihoon, G. Q., Antonio Pérez, A., Ehsan, M., & Torres Huerta, A. L. (2024). Environmentally Sustainable and Green Polymeric Method for Chitosan (CH) Film Synthesis Using Natural Acids and Impact of Zinc Ferrite Nanoparticles (NPs) on Water Solubility (WS) and Physical Properties. *Polymers, 16*(24), 3466. DOI: 10.3390/polym16243466 PMID: 39771318

Hassan, D., Sani, A., & Medina, D. I. (2023). Limitations of Nanocarriers Such as Cell and Tissue Toxicity, Genotoxicity, Scale-Up of Nanomaterials. In Khan, F. A. (Ed.), *Nano Drug Delivery for Cancer Therapy: Principles and Practices* (pp. 149–171). Springer Nature Singapore., DOI: 10.1007/978-981-99-6940-1_8

Hojjat-Farsangi, M. (2014). Small-Molecule Inhibitors of the Receptor Tyrosine Kinases: Promising Tools for Targeted Cancer Therapies. *International Journal of Molecular Sciences, 15*(8), 13768–13801. DOI: 10.3390/ijms150813768 PMID: 25110867

Hossam Abdelmonem, B., Abdelaal, N. M., Anwer, E. K. E., Rashwan, A. A., Hussein, M. A., Ahmed, Y. F., Khashana, R., Hanna, M. M., & Abdelnaser, A. (2024). Decoding the Role of CYP450 Enzymes in Metabolism and Disease: A Comprehensive Review. *Biomedicines, 12*(7), 1467. DOI: 10.3390/biomedicines12071467 PMID: 39062040

Iqbal, I., Wilairatana, P., Saqib, F., Nasir, B., Wahid, M., Latif, M. F., Iqbal, A., Naz, R., & Mubarak, M. S. (2023). Plant Polyphenols and Their Potential Benefits on Cardiovascular Health: A Review. *Molecules (Basel, Switzerland), 28*(17), 6403. DOI: 10.3390/molecules28176403 PMID: 37687232

Iqbal, N., & Iqbal, N. (2014). Imatinib: A Breakthrough of Targeted Therapy in Cancer. *Chemotherapy Research and Practice, 2014*(1), 357027. 10.1155/2014/357027. DOI: 10.1155/2014/357027 PMID: 24963404

Kang, C. S., Zhang, S., Wang, H., Liu, Y., Ren, S., Chen, Y., Li, J., Bandara, N., Rogachev, A. Y., Rogers, B. E., & Chong, H.-S. (2022). Novel Chelating Agents for Zirconium-89-Positron Emission Tomography (PET) Imaging: Synthesis, DFT Calculation, Radiolabeling, and In Vitro and In Vivo Complex Stability. *ACS Omega, 7*(42), 37229–37236. DOI: 10.1021/acsomega.2c03478 PMID: 36312419

Kardamakis, D., Baatout, S., Bourguignon, M., Foray, N., & Socol, Y. (2023). History of Radiation Biology. In Baatout, S. (Ed.), *Radiobiology Textbook* (pp. 1–24). Springer International Publishing., DOI: 10.1007/978-3-031-18810-7_1

Kaye, A. D., Shah, S. S., Johnson, C. D., De Witt, A. S., Thomassen, A. S., Daniel, C. P., Ahmadzadeh, S., Tirumala, S., Bembenick, K. N., Kaye, A. M., & Shekoohi, S. (2025). Tacrolimus- and Mycophenolate-Mediated Toxicity: Clinical Considerations and Options in Management of Post-Transplant Patients. *Current Issues in Molecular Biology*, *47*(1), 2. DOI: 10.3390/cimb47010002 PMID: 39852117

Keitel, U., Scheel, A., Thomale, J., Halpape, R., Kaulfuß, S., Scheel, C., & Dobbelstein, M. (2014). BclxL mediates therapeutic resistance of a mesenchymal breast cancer cell subpopulation. *Oncotarget*, *5*(23), 11778–11791. DOI: 10.18632/oncotarget.2634 PMID: 25473892

Khan, A., Khan, R. A., Anas, M., Qadri, T. A., Ashiq, B., Younis, T., Ullah, N., Khan, A. M., Nasreen, Z., & Shinwari, Z. K. (2026). Biogenic Nanoparticles and Green Nanocomposites as Sustainable Antimicrobial Strategies Against Bacteria, Fungi and Viruses. *Nano-Horizons: Journal of Nanosciences and Nanotechnologies, 5*, 31 pages. DOI: 10.25159/3005-2602/20247

Khanna, P., Ong, C., Bay, B. H., & Baeg, G. H. (2015). Nanotoxicity: An Interplay of Oxidative Stress, Inflammation and Cell Death. *Nanomaterials (Basel, Switzerland)*, *5*(3), 1163–1180. DOI: 10.3390/nano5031163 PMID: 28347058

Ki, M.-R., Kim, D. H., Abdelhamid, M. A. A., & Pack, S. P. (2025). Cancer and Aging Biomarkers: Classification, Early Detection Technologies and Emerging Research Trends. *Biosensors (Basel)*, *15*(11), 737. DOI: 10.3390/bios15110737 PMID: 41294748

Kovalev, I. S., Zyryanov, G. V., Santra, S., Majee, A., Varaksin, M. V., & Charushin, V. N. (2022). Folic Acid Antimetabolites (Antifolates): A Brief Review on Synthetic Strategies and Application Opportunities. *Molecules (Basel, Switzerland)*, *27*(19), 6229. DOI: 10.3390/molecules27196229 PMID: 36234766

Krasnovskaya, O. O., Abramchuck, D., Erofeev, A., Gorelkin, P., Kuznetsov, A., Shemukhin, A., & Beloglazkina, E. K. (2023). Recent Advances in 64Cu/67Cu-Based Radiopharmaceuticals. *International Journal of Molecular Sciences*, *24*(11), 9154. DOI: 10.3390/ijms24119154 PMID: 37298101

Kumar, S., Banerjee, S., Ghosh, A., Pramanik, C., Dey, I., Choudhury, N., Sarkar, S., Das, A., Ghosh, S., Samanta, S., Das, S., Chattopadhyay, N., Bag, M., Talukdar, P., & Das, S. (2026). Biological synergies for heavy metal bioremediation across microbial, algal, and phytoremediation perspectives. *Discover Green Chemistry, 1*(1), 3. DOI: 10.1007/s44509-026-00003-5

Lu, H., Zhang, S., Wang, J., & Chen, Q. (2021). A review on polymer and lipid-based nanocarriers and its application to nano-pharmaceutical and food-based systems. *Frontiers in Nutrition, 8*, 783831. DOI: 10.3389/fnut.2021.783831 PMID: 34926557

Lu, Y., Zhang, B., Wang, Z., He, Y., Ge, H., Ma, H., & Cui, P. (2026). From Host-Derived Pressures to the Environmental Anti-Antimicrobial Peptides Resistome: Mechanisms, Reservoirs and Implications for Therapeutic Peptide Design. *Marine Drugs, 24*(2), 76. DOI: 10.3390/md24020076 PMID: 41745479

Ma, P., & Mumper, R. J. (2013). Paclitaxel nano-delivery systems: A comprehensive review. *Journal of Nanomedicine & Nanotechnology, 4*(2), 1000164. DOI: 10.4172/2157-7439.1000164 PMID: 24163786

Makharadze, D., del Valle, L. J., Katsarava, R., & Puiggalí, J. (2025). The Art of PEGylation: From Simple Polymer to Sophisticated Drug Delivery System. *International Journal of Molecular Sciences, 26*(7), 3102. DOI: 10.3390/ijms26073102 PMID: 40243857

Marquez, J. M. (2026). William Halsted's Radical Mastectomy Patients: A Historical Analysis Through Patient Narratives. *Annals of Surgery Open : Perspectives of Surgical History, Education, and Clinical Approaches, 7*(1), e651. DOI: 10.1097/AS9.0000000000000651 PMID: 41890517

Mills, J. A., Liu, F., Jarrett, T. R., Fletcher, N. L., & Thurecht, K. J. (2022). Nanoparticle based medicines: approaches for evading and manipulating the mononuclear phagocyte system and potential for clinical translation []. *Biomaterials Science, 10*(12), 3029–3053. DOI: 10.1039/D2BM00181K

Molla, G., & Bitew, M. (2025). The Future of Cancer Diagnosis and Treatment: Unlocking the Power of Biomarkers and Personalized Molecular-Targeted Therapies. *Journal of Molecular Pathology, 6*(3), 20. DOI: 10.3390/jmp6030020

Nagaya, T., Nakamura, Y. A., Choyke, P. L., & Kobayashi, H. (2017). Fluorescence-Guided Surgery [Review]. *Frontiers in Oncology, Volume 7 - 2017*. DOI: 10.3389/fonc.2017.00314

Narwade, M., Shaikh, A., Gajbhiye, K. R., Kesharwani, P., & Gajbhiye, V. (2023). Advanced cancer targeting using aptamer functionalized nanocarriers for site-specific cargo delivery. *Biomaterials Research*, *27*(1), 42. DOI: 10.1186/s40824-023-00365-y PMID: 37149607

Nikhar, S., Jahan, A., Mahato, T., & Giri, A. S. (2026). Peptide and Antibody Functionalized Nanoparticles: Active Targeting Strategies for Next-Generation Contrast Agents and Drug Delivery Systems. *ACS Applied Nano Materials*, *9*(23), 10448–10477. Advance online publication. DOI: 10.1021/acsanm.5c05449

Omran, B., & Baek, K.-H. (2021). Nanoantioxidants: Pioneer Types, Advantages, Limitations, and Future Insights. *Molecules (Basel, Switzerland)*, *26*(22), 7031. DOI: 10.3390/molecules26227031 PMID: 34834124

Osataphan, N., Phrommintikul, A., Chattipakorn, S. C., & Chattipakorn, N. (2020). Effects of doxorubicin-induced cardiotoxicity on cardiac mitochondrial dynamics and mitochondrial function: Insights for future interventions. *Journal of Cellular and Molecular Medicine*, *24*(12), 6534–6557. DOI: 10.1111/jcmm.15305 PMID: 32336039

Park, Y. (2014). A New Paradigm Shift for the Green Synthesis of Antibacterial Silver Nanoparticles Utilizing Plant Extracts. *Toxicological Research*, *30*(3), 169–178. DOI: 10.5487/TR.2014.30.3.169 PMID: 25343010

Parvizpour, S., Beyrampour-Basmenj, H., Razmara, J., Farhadi, F., & Shamsir, M. S. (2024). Cancer treatment comes to age: From one-size-fits-all to next-generation sequencing (NGS) technologies. *BioImpacts*, *14*(4), 29957–29957. DOI: 10.34172/bi.2023.29957 PMID: 39104623

Pourmadadi, M., Aslani, A., Hassan, D., Sani, A., Rahdar, A., & Medina, I. D., Abdouss, M., & Romanholo Ferreira, L. F. (2023). Recent advancements in the targeted delivery of Gemcitabine: Harnessing nanomedicine for enhanced cancer therapy. *OpenNano, 13*, 100177. 10.1016/j.onano.2023.100177

Rani, M. Keshu, & Shanker, U. (2022). Chapter 3 - Green nanomaterials: An overview. In U. Shanker, C. M. Hussain, & M. Rani (Eds.), *Green Functionalized Nanomaterials for Environmental Applications* (pp. 43–80). Elsevier. 10.1016/B978-0-12-823137-1.00026-9

Rojas Burbano, J. C., Ruiz, N. I., Rojas Burbano, G. A., Guacho Inca, J. S., Arragan Lezama, C. A., & González, M. S. (2025). Robot-Assisted Surgery: Current Applications and Future Trends in General Surgery. *Cureus*, *17*(4), e82318. DOI: 10.7759/cureus.82318 PMID: 40376327

Salinas, Y., Chauhan, S. C., & Bandyopadhyay, D. (2025). Small-Molecule Mitotic Inhibitors as Anticancer Agents: Discovery, Classification, Mechanisms of Action, and Clinical Trials. *International Journal of Molecular Sciences, 26*(7), 3279. DOI: 10.3390/ijms26073279 PMID: 40244152

Salmaso, S., & Caliceti, P. (2013). Stealth Properties to Improve Therapeutic Efficacy of Drug Nanocarriers. *Journal of Drug Delivery, 2013*(1), 374252. 10.1155/2013/374252. DOI: 10.1155/2013/374252 PMID: 23533769

Saroj, N., Dholaniya, P. S., Alvi, S. B., Sridharan, D., Soni, N., Ashraf, S. A., Choudhry, A., Ashraf, Y. A., Mikula, S. K., Singla, D. K., & Khan, M. (2025). SiRNA-mediated knockdown of TOP2B protects hiPSC-derived cardiomyocytes from doxorubicin-induced toxicity. *Life Sciences, 371*, 123595. DOI: 10.1016/j.lfs.2025.123595 PMID: 40158615

Saw, W. S., Anasamy, T., Foo, Y. Y., Kwa, Y. C., Kue, C. S., Yeong, C. H., Kiew, L. V., Lee, H. B., & Chung, L. Y. (2021). Delivery of Nanoconstructs in Cancer Therapy: Challenges and Therapeutic Opportunities. *Advanced Therapeutics, 4*(3), 2000206. DOI: 10.1002/adtp.202000206

Selewski, D. T., Shah, G. V., Mody, R. J., Rajdev, P. A., & Mukherji, S. K. (2010). Rituximab (Rituxan). *AJNR. American Journal of Neuroradiology, 31*(7), 1178–1180. DOI: 10.3174/ajnr.A2142 PMID: 20448016

Serras, A., Faustino, C., & Pinheiro, L. (2024). Functionalized polymeric micelles for targeted cancer therapy: Steps from conceptualization to clinical trials. *Pharmaceutics, 16*(8), 1047. DOI: 10.3390/pharmaceutics16081047 PMID: 39204392

Sharom, F. J. (2014). Complex interplay between the P-glycoprotein multidrug efflux pump and the membrane: Its role in modulating protein function. *Frontiers in Oncology, 4*, 41. DOI: 10.3389/fonc.2014.00041 PMID: 24624364

Simonsen, T. G., Gaustad, J.-V., Leinaas, M. N., & Rofstad, E. K. (2012). High interstitial fluid pressure is associated with tumor-line specific vascular abnormalities in human melanoma xenografts. *PLoS One, 7*(6), e40006. DOI: 10.1371/journal.pone.0040006 PMID: 22768196

Supple, W. F. (2026). *Cancer Is a Parasite: Kill it with the Safe, Over-the-Counter Antiparasitic Fenbendazole*. Simon and Schuster.

Sushnitha, M., Evangelopoulos, M., Tasciotti, E., & Taraballi, F. (2020). Cell Membrane-Based Biomimetic Nanoparticles and the Immune System: Immunomodulatory Interactions to Therapeutic Applications [Review]. *Frontiers in Bioengineering and Biotechnology, Volume 8 - 2020*. DOI: 10.3389/fbioe.2020.00627

Syed, S. M., Kulkarni, S., Patil, M., & Satpute, K. (2026). A comprehensive review of green synthesis methods and applications of nanoparticles derived from plant extracts and microorganisms. *Discover Green Chemistry*, *1*(1), 6. DOI: 10.1007/s44509-026-00006-2

Sysak, S., Czarczynska-Goslinska, B., Szyk, P., Koczorowski, T., Mlynarczyk, D. T., Szczolko, W., Lesyk, R., & Goslinski, T. (2023). Metal Nanoparticle-Flavonoid Connections: Synthesis, Physicochemical and Biological Properties, as Well as Potential Applications in Medicine. *Nanomaterials (Basel, Switzerland)*, *13*(9), 1531. DOI: 10.3390/nano13091531 PMID: 37177076

Tai, L.-H., Zhang, J., & Auer, R. C. (2013). Preventing surgery-induced NK cell dysfunction and cancer metastases with influenza vaccination. *OncoImmunology*, *2*(11), e26618. DOI: 10.4161/onci.26618 PMID: 24404430

Thölke, D. J. C. (2025). *Diagnostic and Prognostic Role of Circulating Hsp70 Levels and Lymphocyte Subpopulation Frequencies as Tumor Biomarkers in Patients with Glioblastoma*. Technische Universität München.

Thomas, O. S., & Weber, W. (2019). Overcoming Physiological Barriers to Nanoparticle Delivery—Are We There Yet? [Review]. *Frontiers in Bioengineering and Biotechnology, Volume 7 - 2019*. DOI: 10.3389/fbioe.2019.00415

Tia, S. T., Luo, M., & Fan, W. (2025). Mapping the Role of P-gp in Multidrug Resistance: Insights from Recent Structural Studies. *International Journal of Molecular Sciences*, *26*(9), 4179. DOI: 10.3390/ijms26094179 PMID: 40362415

Tian, Y., Lei, Y., Wang, Y., Lai, J., Wang, J., & Xia, F. (2023). Mechanism of multidrug resistance to chemotherapy mediated by P-glycoprotein. *International Journal of Oncology*, *63*(5), 119. DOI: 10.3892/ijo.2023.5567 PMID: 37654171

Trofin, A.-M., Scripcariu, D. V., Filipiuc, S.-I., Neagu, A.-N., Filipiuc, L.-E., Tamba, B.-I., Palaghia, M. M., & Uritu, C. M. (2025). From Nature to Nanomedicine: Enhancing the Antitumor Efficacy of Rhein, Curcumin, and Resveratrol. *Medicina (Kaunas, Lithuania)*, *61*(6), 981. DOI: 10.3390/medicina61060981 PMID: 40572668

Verma, J., Warsame, C., Seenivasagam, R. K., Katiyar, N. K., Aleem, E., & Goel, S. (2023). Nanoparticle-mediated cancer cell therapy: Basic science to clinical applications. *Cancer Metastasis Reviews*, *42*(3), 601–627. DOI: 10.1007/s10555-023-10086-2 PMID: 36826760

Verza, F. A., Silva, G. C. D., & Nishimura, F. G. (2025). The impact of oxidative stress and the NRF2-KEAP1-ARE signaling pathway on anticancer drug resistance. *Oncology Research*, *33*(8), 1819–1834. Advance online publication. DOI: 10.32604/or.2025.065755 PMID: 40746900

Wang, R., Hu, B., Pan, Z., Mo, C., Zhao, X., Liu, G., Hou, P., Cui, Q., Xu, Z., Wang, W., Yu, Z., Zhao, L., He, M., Wang, Y., Fu, C., Wei, M., & Yu, L. (2025). Antibody–Drug Conjugates (ADCs): Current and future biopharmaceuticals. *Journal of Hematology & Oncology*, *18*(1), 51. DOI: 10.1186/s13045-025-01704-3 PMID: 40307936

Wang, Y., Chang, J., & Liu, J. (2026). Molecular engineering approaches to half-life extension of therapeutic biomolecules. *Frontiers in Pharmacology*, *17*, 1778569. DOI: 10.3389/fphar.2026.1778569 PMID: 42038289

Wei, Z., Zhou, Y., Wang, R., Wang, J., & Chen, Z. (2022). Aptamers as Smart Ligands for Targeted Drug Delivery in Cancer Therapy. *Pharmaceutics*, *14*(12), 2561. DOI: 10.3390/pharmaceutics14122561 PMID: 36559056

Wozeak, D. R., Pereira, I. L., Cardoso, T. L., Panagio, L. A., Nakazato, G., Acosta, I. B., Varela, A. S., & Hartwig, D. D. (2026). Biogenic Silver Nanoparticles: An Antibacterial and Antibiofilm Approach to Control Carbapenem-Resistant Klebsiella pneumoniae. *Current Microbiology*, *83*(4), 232. DOI: 10.1007/s00284-026-04747-y PMID: 41801463

Wu, Y., Song, Y., Wang, R., & Wang, T. (2023). Molecular mechanisms of tumor resistance to radiotherapy. *Molecular Cancer*, *22*(1), 96. DOI: 10.1186/s12943-023-01801-2 PMID: 37322433

Yan, L., Duan, B., Wang, T., Wang, T., Chi, H., & Gao, H. (2025). Omics-enhanced nanomedicine: Integrating multi-omics for precision cancer diagnosis and therapy. *Discover. Oncology*, *17*(1), 97. DOI: 10.1007/s12672-025-04088-5 PMID: 41379246

Yang, X., Ren, H., Guo, X., Hu, C., & Fu, J. (2020). Radiation-induced skin injury: Pathogenesis, treatment, and management. *Aging*, *12*(22), 23379–23393. DOI: 10.18632/aging.103932 PMID: 33202382

Yasir, M., Mishra, R., Tripathi, A. S., Maurya, R. K., shahi, A., Zaki, M. E. A., Al Hussain, S. A., & Masand, V. H. (2024). Theranostics: A multifaceted approach utilizing nano-biomaterials. *Discover Nano*, *19*(1), 35. DOI: 10.1186/s11671-024-03979-w PMID: 38407670

Yu, J. B., Cramer, L. D., Herrin, J., Soulos, P. R., Potosky, A. L., & Gross, C. P. (2014). Stereotactic Body Radiation Therapy Versus Intensity-Modulated Radiation Therapy for Prostate Cancer: Comparison of Toxicity. *Journal of Clinical Oncology : Official Journal of the American Society of Clinical Oncology*, *32*(12), 1195–1201. DOI: 10.1200/JCO.2013.53.8652 PMID: 24616315

Yu, Y.-P., Wang, Q., Liu, Y.-C., & Xie, Y. (2014). Molecular basis for the targeted binding of RGD-containing peptide to integrin αVβ3. *Biomaterials*, *35*(5), 1667–1675. 10.1016/j.biomaterials.2013.10.072. DOI: 10.1016/j.biomaterials.2013.10.072 PMID: 24268666

Zaiki, Y., Iskandar, A., & Wong, T. W. (2023). Functionalized chitosan for cancer nano drug delivery. *Biotechnology Advances*, *67*, 108200. 10.1016/j.biotechadv.2023.108200. DOI: 10.1016/j.biotechadv.2023.108200 PMID: 37331671

Zhao, M., Ma, J., Li, M., Zhang, Y., Jiang, B., Zhao, X., Huai, C., Shen, L., Zhang, N., He, L., & Qin, S. (2021). Cytochrome P450 enzymes and drug metabolism in humans. *International Journal of Molecular Sciences*, *22*(23), 12808. DOI: 10.3390/ijms222312808 PMID: 34884615

Zuhrotun, A., Oktaviani, D. J., & Hasanah, A. N. (2023). Biosynthesis of Gold and Silver Nanoparticles Using Phytochemical Compounds. *Molecules (Basel, Switzerland)*, *28*(7), 3240. DOI: 10.3390/molecules28073240 PMID: 37050004

# Chapter 2
# Sustainable Carbon-Based Nanostructures

**Abdul Momin Rizwan Ahmad**
*Department of Health Sciences, University of York, UK*

**Tauqeer Ahmed Qadri**
http://orcid.org/0009-0000-9617-4512
*University of South Africa, South Africa*

**Bushra Ashiq**
http://orcid.org/0009-0002-2975-3121
*College of Artificial Intelligence, Taiyuan University of Technology, China*

**Hotaf Hassan Makki**
*Biology Department, Faculty of Science, University of Tabuk, Saudi Arabia*

**Ishtiaq Hussain**
*Department of Botany, University of Baltistan, Skardu, Pakistan*

**Farhat Yasmeen**
http://orcid.org/0000-0003-1865-2296
*University of Wah, Pakistan*

## ABSTRACT

*This chapter reviews the paradigm shift in nano-oncology from conventional, heavy-metal-based nanoparticles to sustainable carbon-based nanostructures (CNMs). Although currently, orthodox cancer therapies are limited because of systemic toxicities and lack of selectivity, carbon dots, carbon nanotubes, and graphene quantum dots can serve as a biocompatible substitute to orthodox cancer treatments, as they were green-synthesized using a range of different carbon-based materials. In addition, they can be targeted to tumor receptors through an active direct approach using their versatile surface chemistry, and stimuli-responsive therapeutic release, in particular, pH, enzyme-triggered, and near-infrared light-triggered. This summary paper underscores the various uses of sustainable CNMs in high-resolution optical imaging, photo-induced cancer therapies (PTT/PDT), advanced gene or immunotherapy delivery, and ultra-sensitive biomolecular diagnostics to precision theranostics.*

DOI: 10.4018/979-8-3373-9848-8.ch002

# 1. INTRODUCTION TO CARBON-BASED NANOSTRUCTURES (CNMS) IN ONCOLOGY

Cancer has been among the most popular worldwide health crises that do need urgent paradigm shift, which not only requires an abandonment of the conventional oncological therapies, but also precision medicine. Whereas various clinical interventions, including the use of surgery or radiotherapy and systemic chemotherapy, have traditionally spurred decreases in the number of cancer-related mortality, clinical efficacy is necessarily restricted due to considerable downsides (Madani et al., 2011; Randviir et al., 2014). Traditional chemotherapeutic drugs are typically characterized by pathetic aqueous solubility, and an utter deficiency of selectivity, resulting in ineffective bioavailable distribution, and unselective damage to normal and cancerous tissues (Lukyanov & Torchilin, 2004). To avoid these adverse systemic toxicities and to enhance patient outcomes, oncology today has paid considerable attention to the development of biocompatible and targeted nanoplatforms, which can be controlled in terms of drug delivery (Peer et al., 2020). In this area, carbon-based nanomaterials (CNMs) or carbon-based nanostructures (CBNSs) have become significantly promising scaffolds through their unprecedented mechanical, electrical, thermal, optical, and chemical properties and considerably changed the way nanomedicine will progress compared to conventional bulk carbon allotropes such as graphite or diamond (Ahmed et al., 2025; Bagheri et al., 2022).

The singular beauty of the CNMs in nano-oncology is that they have a diverse structural variety and highly tuneable dimensions comprising in zero-dimensional (0D), one-dimensional (1D) and two-dimensional (2D) geometries that exist purely within the nanoscale of less than 100 nm diameter. This diverse family of architecture includes allotropic variants graphene (Randviir et al., 2014), carbon nanotubes (Madani et al., 2011), carbon nano-onions (Mohan et al., 2022), nanodiamonds (Ye et al., 2020), carbon nanohorns (Ma et al., 2014), fullerenes, and carbon/graphene quantum dots (Ahmed et al., 2025). With intrinsic multi-functional properties such as high drug-loading capacity, efficient cellular uptake routes, and intrinsic thermal ablation capabilities, these nanomaterials are very attractive in biosensing and tissue engineering, bioimaging, and multi-modal cancer therapies (Bagheri et al., 2022). More importantly, through the choice of particular environmentally friendly and innocuous production schemes, the physical size of such nanostructures can be fine-tuned to the size of the biological macromolecules (Bartkowski & Giordani, 2021). This is the optimum size that enables CNMs to passively take advantage of the leaky, compromised vasculature that encircles the solid tumors through the passive enhanced permeation and retention (EPR) effect that concentrates therapeutic payloads directly in the tumor microenvironment (Shinde et al., 2022).

Nonetheless, due to the infamously heterogeneous nature of the EPR effect among various patient groups and the ability to sometimes trigger drug resistance, it is impossible to passively accumulate the effect (Peer et al., 2020). The highly dynamic surface chemistry of CNMs can be utilized to develop more precise therapies by acting as a versatile canvas to surface engineering and active targeting. Surface oxygen-group functionalization (through simple oxidation) of the underlying nanostructures drastically enhances the aqueous solubility of the underlying nanostructures, permitting them to serve as highly soluble hosts in other, soluble-insoluble drug systems (e.g. the solubilization of tamoxifen using graphene oxide scaffolds) (Keklikcioglu Cakmak & Eroglu, 2023). These strong scaffolds make the conjugation of multiple-component systems, both covalent and non-covalent, possible. Researchers can design very versatile, single-particle theragnostic systems with many possibilities by linking them to a specific targeting agent (folic acid) (Ghosh et al., 2022), a detection imaging agent (BODIPY) (Su et al., 2020), or an elimative drug (cisplatin or doxorubicin) (Yang et al., 2022).

These nanocarriers vary in ability to be precisely localized in drug release, so the accuracy of localized drug release profile maximizes the therapeutic index of nanocarriers. Although covalent functionalization approaches (including amidation, alkylation, or fluorination) provide stable functionalizations, reducing an untimely leakage of drug at neutral pH (Bahuguna et al., 2018), covalent functionalization can sometimes cause damage to the intact π-system of the carbon core, uncritically eliminating the untailed electronic and physical properties of the material (Mohan et al., 2022). On the other hand, supramolecular non-covalent decoration that is facilitated by π- π stacking and hydrophobic effects is effective to maintain the integrity of the CNM core (Bartkowski and Giordani, 2021). This reversible binding architecture is more specifically adapted to developing smart, stimuli-responsive drug delivery systems that initiate the delivery of drugs in response to local external or internal molecular stimuli, e.g., redox potential (Zhang et al., 2013) or near-infrared (NIR) light irradiation (Gao et al., 2021). Since the microenvironment in tumors is characteristically acidic, carbon-based nanoplatforms that are responsive to pH are especially game-changing in cancer therapy (Arpicco et al., 2020), as they guarantee the safe delivery of cytotoxic cargoes when exposed to the rest of the body but unleash them locally at the lesionic site. Nevertheless, in spite of such deep benefits, to bring these sophisticated nanostructures to the clinic, a rigorous, continual study of their long-term biosafety, pharmacokinetics, and final physiological pathway in multifaceted and intricate biological systems is required (Bagheri et al., 2022). To accommodate these vital paradigms, this chapter will fully assess the transition into green and sustainable synthesis methods of carbon-based nanostructures wherein we will discuss their further surface engineering, stimuli-responsive properties as well as elucidate their multi-functional capabilities in precision nano-oncology.

# 2. GREEN AND SUSTAINABLE SYNTHESIS PARADIGMS

## 2.1. The Need for Sustainability: Moving Away from Conventional Methodologies

Depending on the conventional methods of production of carbon-based nanomaterials has posed great ecological and biological bottlenecks, preventing their clinical translation to nano-oncology. Traditional top-down processes, e.g., the Hummer process to extract graphene oxide, rely on the use of dangerous oxidizing species, volatile organic solvents, concentrative toxic acids (such as sulfuric and nitric acids) that produce waste streams harmful to the environment and produce toxic residues on the faces of the nanomaterials (Sajjadi et al., 2021). In a similar manner, bottom-up engineering through chemical vapor deposition (CVD) or laser ablation require high-energy supply, high-temperature, and special equipment (Bagheri et al., 2022). Such harsh processing not only undermines the ideals of green chemistry but also produce nanomaterials with inherent cytocompatibility hazards as a consequence of remaining chemical impurities or trace quantities of heavy metal catalysts (Ahmed et al., 2025). This urgent calls a shift to new sustainable, low-energy synthesis paradigms which exploit benign reagents and mild processing environments to produce highly bio-compatible nanostructures that can be used in precision oncology.

## 2.2. Precursors from Nature: Biomass Valorization

One approach to realizing this sustainability ideal is biomass valorization, being one of the most popular strategies that involves the use of natural precursors and agricultural wastes to form high-value carbon frameworks. The use of renewable sources of carbon, such as citrus peels, Azadirachta indica (neem) leaf extracts, coffee grounds, and pure natural carbohydrates (as glucose, starch, and citric acid), avoids the petroleum products as starting materials (Sajjadi et al., 2021; Venkataraman et al., 2024; Voss et al., 2024). These natural precursors already have high concentrations of organic molecules, proteins, and polyphenols rich profiles which have multifunctionality as the primary carbon skeleton and in situ capping or stabilization agents (A. Khan, M. Anas, T. A. Qadri, et al., 2025; A. Khan, M. Anas, N. Ullah, et al., 2025; Yu et al., 2020). As an example, naturally synthesizing synthesis pathways with Azadirachta indica extract achieves the same outcome of bioactive phytochemical passivation of the resulting carbon nanostructures, and this can have synergistic antioxidant properties or inherent anticancer properties prior to targeted drug loading (Ghosh et al., 2022; Sanjeev et al., 2024). The green transformation of rich biomass that is based on this basic sustenance of redefining lifecycle of oncological nanocarriers is essential to the convergence of clinical efficacy and environmental accountability.

## 2.3. Eco-Friendly Synthesis Strategies

These are environmentally-friendly synthesis methods that can be broadly classified as green top-down and bottom-up methods. Top-down approaches to green creatively apply green techniques to the mild and chemical-free exfoliation or cutting of bulk sustainable carbon sheets, whereas bottom-up green approaches have become immensely popular by virtue of being simple and scalable (Agrawal, 2022; A. Khan, B. Ashiq, et al., 2025; Khan, Khan, et al., 2026; Khan, Qadri, et al., 2026; A. Khan, T. Younis, et al., 2025; Ullah et al., 2026). Hydrothermal and solvothermal carbonization are high efficiency, closed-system methods in which biomass and water are placed under moderate temperatures and self-generated pressures, which cause dehydration, polymerization, and carbonization of the biomass into a water-soluble carbon quantum dot or fragment of fractured graphene, without producing toxin-forming gases (Esmaeili et al., 2022). Microwave-assisted green synthesis is further based on the principle of maximizing energy usage to use the nucleation of carbon cores in minutes, radically changing reaction times of hours down to a few seconds (Mohan et al., 2022). Controlled, low-energy thermal pyrolysis of natural precursors also generates pure carbonized frameworks, without the need to introduce synthetic surfactants or other dangerous surface-passivating agents, to provide clean topology of surface (Bartkowski and Giordani, 2021).

## 2.4. Atom Economy and Heteroatom Doping

A key strength of these bottom-up biomass routes is the easy incorporation of high atom economy via the homo-molecule of green heteroatoms doping. Unless post-functionalized, pristine carbon nanostructures typically have low quantum yields and the surface is inactive to be functionally functionalized at the desired active site (Ahmed et al., 2025). Those heteroatoms can be incorporated directly into the sp2 carbon lattice in a single-pot thermal or microwave process by the addition of natural co-precursors that are high in nitrogen, sulfur, or phosphorus (e.g., amino acids, egg whites, or plant waste) into the starting reaction mixture (Mohan et al., 2022). This phenolphthalein heteroatom doping changes the electronic structure and the localization of charges of the nanomaterial, and greatly increases the up-conversion photoluminescence, quantum yield, and production of reactive oxygen species (ROS) in reaction to light (Su et al., 2020). Finally, this one-pot execution maximizes atoms economy, minimizes multi-step purifications and provides structurally optimized, highly fluorescent, and multi-functional nanostructures that are specifically designed to perform advanced theragnostic and precise tumor targeting. As shown in a schematic overview in Figure 1, a traditional approach to CNMs making uses oxidizing agents that are hazardous, concentrated acids, and petroleum precursors, producing highly cytotoxic nanocarriers, whereas green strategies use biomass precursors through low-energy methods (hydrothermal, microwave, or pyrolysis) to generate inherently passivated, highly cytocompatible nanocarriers that are optimized to require low-energy processes in nano-oncology applications.

*Figure 1. Comparative synthesis pathways and biological profiles of traditional vs. green-synthesized Carbon Nanomaterials (CNMs)*

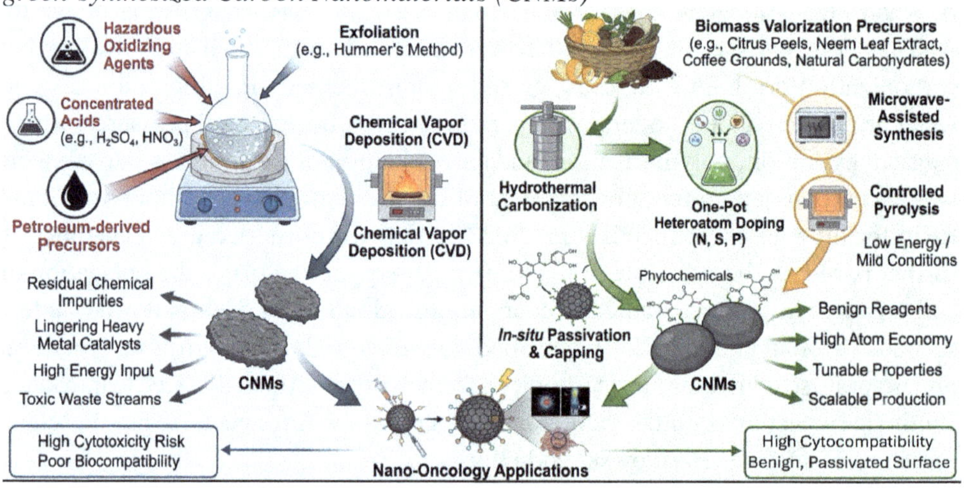


## 3. PHYSICOCHEMICAL PROPERTIES AND SURFACE ENGINEERING FOR TUMOR TARGETING

### 3.1. Optical and Thermal Phenomena

The unique structural design of sustainable carbon nanomaterials provides a complete range of outstanding optical and thermal events that do not exist in bulk carbon allotropes (Al-Jumaili et al., 2017; Bagheri et al., 2022). The fundamental building block of their structure and the ability to add or remove drugs is the π-π stacking. The vast system of delocalized, sp2-hybridized convectively carbon atom structures covering the surfaces of graphene, graphene oxide and carbon nanotubes provides a concentrative hydrophobic platform (Sajjadi et al., 2021). The system can be readily loaded with planar, aromatic chemotherapeutic molecules in high capacity and be non-covalent using supramolecular stacking interactions without disrupting the carrier structure (Bartkowski & Giordani, 2021). At the same time, quantum confinement effects can be activated by reducing these structures to the size of 10 nm seen in carbon quantum dots (CQDs) and graphene quantum dots (GQDs) (Ahmed et al., 2025; Anas et al., 2025; Falak et al., 2025). This size-dependent yet non-zero bandgap of excitons results in a size-dependent, non-zero photoluminescence (PL) with which researchers can easily tune the wavelengths of their emission by simply varying the diameter of the core of the nanoparticles during synthesis.

Besides typical down-conversion fluorescence, sustainable carbon structures are often exhibiting up-conversion photoluminescence (UCPL). The nanomaterials can absorb numerous low-energy, long-wavelength NIR photons through this anti-Stokes process and emit a single, higher-energy and shorter-wavelength photon in the visible spectrum (Y. Chen et al., 2023). This effect is very beneficial oncological bioimaging in deep-tissue theragnostics because NIR light can with ease penetrate dense biological tissues with minimal scattering and minimal background autofluorescence (Sajjadi et al., 2021).

Moreover, these nanostructures have high efficiency of photothermal conversion (Cui et al., 2023). The delocalized electrons respond to external light sources, typically within the first or second NIR biowindow, by vibrating quickly, emitting absorbed energy as localized heat, instead of light (Su et al., 2020). This highly efficient photothermal conversion could be used in targeted photothermal therapy (PTT), which can achieve local hyperthermic temperatures (≥42 °C) that can ablate cancer cells and disrupt the integrity of adjacent solid tumors (Yu et al., 2020).

## 3.2. Surface Functionalization: The Key to Precision

They do not get easily aggregated in physiological fluids because unfunctionalized carbon nanostructures are strongly attracted by hydrophobic and van der Waals interactions (Dubey et al., 2021). The desired step to cross this threshold, determined by either the choice of covalent versus non-covalent modification, is surface functionalization (Bartkowski & Giordani, 2021). The two routes enhance the hydrophilicity of the material greatly and enhance the blood circulation time of the material by incorporating hydration shells, which are resistant to non-specific opsonization by the RES (Keklikcioglu Cakmak and Eroglu, 2023). In order to promote the simple stability to active, receptor-mediated precision, the outer corona of such green nanocarriers is conjugated with select biogenic targeting ligands (Mohamed et al., 2025). This proactive target is a regular approach that makes use of the application of natural peptides, nucleic acid aptamers, and clinical-grade monoclonal antibodies (Zhu et al., 2026). The nanoconjugates are selective to receptors overexpressed in malignant cells, but not in healthy tissue, by targeting them through anchoring the biomolecules to the carbon scaffold, including the Epidermal Growth Factor Receptor (EGFR) and the Human Epidermal Growth Factor Receptor 2 (HER2) (Fatima et al., 2021; Spada & Gerber-Lemaire, 2025). This affinity through receptors triggers a quick endocytosis, accumulating the drug into the cell and causing the least side effects to the body due to off-target in the system (Peng et al., 2025).

### 3.3. Biocompatibility, Biodegradation, and Pharmacokinetics

Depending on the biological course of sustainable carbon nanostructures in the human body, the starting point is the local cellular uptake processes (Kumar et al., 2022). These particles enter through cellular membranes, based on their dimensional classification and surface engineering, using energy-dependent mechanisms, including clathrin- or caveolae-mediated endocytosis, or through direct lipid-bi-layer translocation (Varma et al., 2022). The green-synthesized CNMs are generally expected to invoke a much-weaker immune response than chemically-synthesised, with higher cytocompatibility due to the presence of a biogenic passivation layer on their surface made of natural phytochemicals or carbohydrates (Paveethra et al., 2024). After their therapeutic/diagnostic mission, the intrinsic degeneration of such nanostructures is also essential to prevent tissue accumulation and toxic retention in the long run. Sustainable carbon nanostructures cannot be truly biodegraded as in the case of metallic nanoparticles (Rozhin et al., 2022). Immune cells (particularly neutrophils and macrophages) play a dominant role in vivo breakdown pathways. These cells release oxidative enzymes, e.g., myeloperoxidase (MPO) and horseradish peroxidase (HRP) and intracellular reactive oxygen species (H2O2) to assault structural defects in the graphitic carbon lattice (Rizo-Téllez et al., 2022). Long nanotubes or large graphene sheets are gradually disintegrated by this enzyme digestion to produce benign and ultra-small carbon fragments highly carboxylated. These breakdown kinetics control the resultant pharmacokinetics: large aggregates are captured in the liver and the spleen and are cleared by enzymes gradually, but sub-10 nm sized fragments are small enough to be cleared by the renal excretion mechanism, thus enabling non-toxic, safe elimination of the aggregates in the body. In Figure 2, to facilitate rapid absorption, comprehensive schematic illustration shows that (1) Engineered carbon nanocarriers can be biologically modified in the tripartite compartment of the lifecycle (including receptor-mediated active endocytosis) to deliver active therapeutics to the malignant cells (EGFR/HER2 overexpressing) (2) with simultaneous (NIR) control delivery of the NIR-generated photoluminescence, or local heat in the form of thinking machine, of the cancer cells (42 o C), (3) through the degradation assistance of macrophages wherein myeloperoxidase (MPO) assists in breaking down the engineered carbon nanocarrier into sub-10 nm-size fragments, which can be safely eliminated by the kidney system.

*Figure 2. Mechanistic overview of green-synthesized carbon nanomaterials in oncological theragnostic*

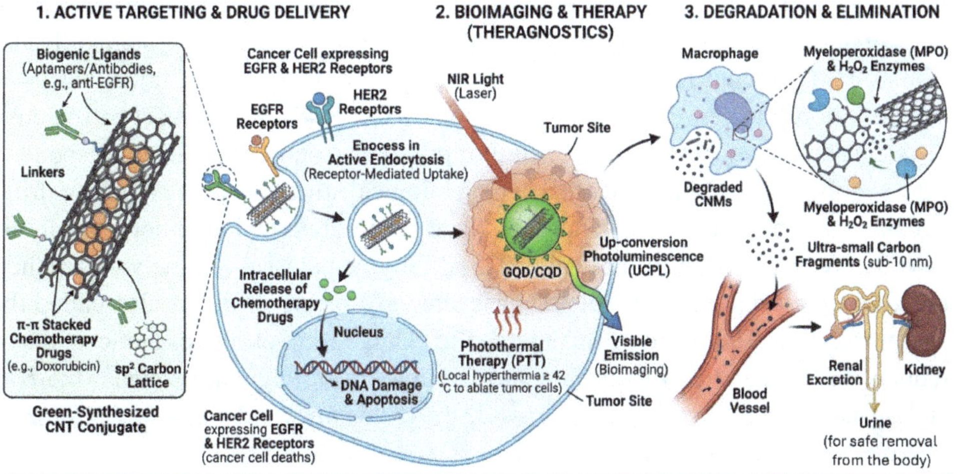


## 4. PRECISION THERAPEUTIC APPLICATIONS IN NANO-ONCOLOGY

### 4.1. Targeted and Stimuli-Responsive Drug Delivery Systems (DDS)

Smart nano-carriers need to develop sufficient capacity to carry large drug loads, circulate throughout the body without early leakage, and release therapeutic agents specifically to malignant tumors to enable the development of precision oncology (Gupta et al., 2026). Biomass-derived graphene, CQDs and green-functionalized CNTs provide a great platform on which these jobs can be performed. They are characterized by their unique surface morphology, which is comprised of large domains of sp2-hybridized carbon atoms being the core of their drug-loading performance. This geometric organization offers high-density, unhindered network of delocalized ions ions consisting of 6d of 5d ions, which are involved in non-covalent, supramolecular interfaces 6d, 5d of 6d, 5d of 6d (Sasidharan et al., 2024). More conventional synthetic carriers would have payload limitations as an internal volume or structurally unstable at high drug concentration (Ajorlou & Khosroushahi, 2017). Instead, graphitic carbon matrices have great drug-loading figures that are sometimes more than 200 per cent-300 per cent of the carrier weight. This facilitates the direct conjugation of hydrophobic chemotherapeutic agents, including doxorubicin, paclitaxel, methotrexate, and camptothecin to the carbon core without the use of complex, multi-step conjugation regimens (Sun et al., 2022). The drug-armed carbon nanostructures should be stable in structure once in the bloodstream so as to prevent non-specific toxicities of other healthy organs in the entire system (Alvandi et al., 2025). In order to realize selective therapeutic treatment, sustainable CNMs can be tailored to serve as a stimuli-responsive gate which can react to selective physical and chemical stimuli in the tumor microenvironment (TME) or respond to external physical stimuli. One of such physiological stimuli is pH-responsive drug-delivery, which is an efficient approach since of rapidly growing tumors, metabolic changes are observed (Z. Chen et al., 2023). Malignant masses routinely switch their metabolic pattern to an upregulated anaerobic glycolysis, which is also called the Warburg effect. The result of this metabolic change is an excess of lactic acid that reduces the extracellular environment of tumors to mildly acidic levels (6.5-6.8) and further to lower levels (4.5-5.5) inside the endosomes and lysosomes (Hamzah, 2026; Sezer et al., 2025).

As a drug-conjugated carbon nanocarrier is absorbed by these acidic areas through receptor-mediated endocytosis, a high concentration of the hydrogen ions causes a chemical change in the nanoconjugate. As an illustration, the primary amino groups on doxorubicin molecules become protonated and hence the hydrophilicity of the molecules is increased under acidic conditions, and results in the formation of positive charges. This establishes a strong electrostatic repulsion with the functionalized edges of the carbon nanostructures to destabilize the non-covalent bond of $\pi$- $\pi$ stacking assembly, leading to a quick and intense release of drugs within a target cancer cell (Gao et al., 2021). In addition to taking advantage of easily adjustable hydrogen ion concentration, sustainable carbonaceous matrices may be designed to take advantage of higher enzymatic levels in the tumor microenvironment. Malignant masses also excessively express certain proteolytic and hydrolytic enzymes including metalloproteinases (MMPs), cathepsin B and hyaluronidase, which essentially degrade the extracellular matrix in the area and cause tissue invasion. Using biogenic natural entities, scientists can encapsulate a green-produced carbon tightly with natural structural polymers such as chitosan, hyaluronic acid or even certain peptide sequences (Sezer et al., 2025). This outer protective layer is not destroyed when passing through healthy tissue (pH 7.4). But when they come into contact with the enzyme-enriched tumor matrix, they are cleaved by targeted enzyme-mediated cleavage of these biogenic coatings. This liberates the carbon core and discharges the therapeutic cargo in a targeted location, that is, the tumor. It is possible to combine this internal triggering approach with glutathione (GSH)-mediated redox-responsive mechanisms, which capitalize on the high levels of GSH in the cancerous cell versus in healthy blood plasma to specifically cleave disulfide cross-linkages that can further be controlled with respect to precision (Fu et al.,

2022). When the photons hit the target zone, the carbon nanostructures transform the photon into local thermal vibrations. This nanoscale temperature raise breaks the physical non-covalent bonds or heat-sensitive linkers attaching the chemotherapeutic drugs onto the graphitic carbon lattice, inducing an instant burst release of the chemotherapeutic payload on command and enabling clinicians to coordinate the delivery of the drugs with the activated laser (Zhang et al., 2021).

## 4.2. Photo-Induced Cancer Therapies

Due to the special electronics of carbon nanostructures that are sustainable, they can be used in photo-induced methods of cancer therapy, which are Photothermal Therapy (PTT) and Photodynamic Therapy (PDT) (Zhang et al., 2024). These modes provide the least invasive form of surgery or radiation, eliminating cancerous tissues through local physical treatments without the development of chemical resistance profiles, over time. Photothermal therapy takes advantage of high photothermal conversion efficiency of carbon allotments (single-walled carbon nanotube, graphene oxide nanoplatelets, multi-layered carbon nano-onions, etc.) (Sahu and Tiwari, 2023). When these materials are excited using NIR light the incident photons are resonant with the overall electronic vibrations or electronic transitions of the collective pi-electron cloud of the graphitic structure. The resonance induces fast non-radiative relaxation processes, transforming the electromagnetic energy absorbed into a high-energy, local thermal energy (Shamsian et al., 2025). This speedy heat production causes localized hyperthermia of tumorous tissue. The solid tumor and healthy cells exhibit vastly different tolerance to thermal stress because a disorganized, leaky and poorly clearable network of blood vessels renders them highly sensitive to the change of temperature (Ying et al., 2026). Intense light can elevate the local temperature to above 50C, inducing a rapid coagulative necrosis that is able to fully ablate the bulk of the primary tumor in a few minutes (Kok et al., 2020). Since these green-syntheses using carbon structures may be functionalized with targeted tumor-bound ligands, this region of high thermal destruction can be focused to the malignancy, sparing other healthy periphery to collateral heat damage (Paveethra et al., 2024). Parallel photodynamic Therapy (PDT) provides an alternative route to localized malignancies through the production of ROS in light activation, however. Conventional molecular photosensitizers are characterized by low water solubility, quick photobleaching and low quantum yields of singlet oxygen. Photosensitizers with structures based on carbon, especially those by carbon quantum dots (which are nitrogen- or sulfur-doped and are high-performance), and biomass-extracted (graphene) quantum dots, can be considered as stable, high-performance photosensitizers (Alvarez and Sevilla, 2024). Electrons in the valence bands in these quantum-confined carbon structures are excited to higher energy levels, when subjected to wavelengths of light. As an alternative to recombining to the ground state through conventional fluorescence, a large fraction of these excited electrons cross to triplet states by intersystem crossing.

## 4.3. Gene and Immunotherapy Delivery

In addition to the delivery of small-molecule chemotherapeutic agents, carbon nanostructures like graphene quantum dots and functionalized carbon nanotubes are sustainable delivery vectors due to their effectiveness in the delivery of gene therapies and advanced macromolecular immunotherapies (Tang et al., 2022). Gene therapy is also likely to be useful in the therapy of various cancers through silencing oncogenes or by reinstating functional tumor-suppressor pathways. Nevertheless, the naked genetic material, such as small interfering RNA (siRNA), microRNA (miRNA) and plasmid DNA (pDNA), have enormous biological obstacles. These are high degradation rate by circulating intravascular nucleases, recognition by the innate immune system and failure to penetrate the negatively charged lipophilic cellular membrane because of the large size and the presence of an anionic phosphate group. Green carbon nanostructures help overcome these challenges by providing a stable, protective carrier platform (Ahmed et al., 2025; Mostafavi et al., 2022). Exposure of the outer edges of biomass-derived GQDs or CNTs to biocompatible and positively charged molecules, e.g., short polyethylenimime chains or natural amino acids such as lysine can functionalize their outer edges and electricityally bind nucleic acid strands with negative charges on their surfaces. The product of this condensation is a stable polyplex which protects the delicate genetic cargo against enzyme degradation by serum nucleases when the polyplex is transported throughout the body system. Also, the long structure of carbon nanotubes or small size of GQDs allow cells to absorb the drug efficiently by directly translocation or by endocytosis through the cell membrane (Gárate-Vélez & Quintana, 2025). At that point, being within the endosomal vesicle, the carbon structure with the functionalized high buffering capacity causes a “proton sponge” effect. This mechanism leads to the influx of water into the vesicle up to the extent the vesicle ruptures, launching the genetic cargo into the cytoplasm. Such a mechanism enables siRNA molecules to become part of the RNA-induced silencing complex (RISC) to silence certain target oncogenes (like STAT3 or KRAS) or provides pDNA the ability to be transferred to the nucleus and activate functional p53 tumor-suppressor proteins. Along with the delivery of nucleic acids, sustainable carbon nanostructures are being used to enhance the delivery of cancer immunotherapies and their efficacy (Bagheri et al., 2022). These materials can be used as vehicles to deliver the more intricate macromolecular structures, including DNA origami arrays that are capable of assembling several immunostimulatory signals in specific spatiogeometries. Carbon cores made using biomasses may even be modified to deliver advanced T-cell engagers, tumor-associated antigens, and adjuvant molecules, to target antigen-presenting cells (APCs) in the lymph nodes. This local delivery induces dendritic cell maturation and leading to an improved antigen cross presentation which provokes a systemic cytotoxic T-lymphocyte response that targets the metastatic cancerous cells. Moreover, since these carbon cores are synthesized through the green synthesis of natural precursors, they do not see the systemic, non-specific, inflammatory reactions that are commonly triggered by metallic nanocarriers (El-Khawaga et al., 2026). Instead, they are neutral delivery vehicles that do not release their immunotherapeutic payloads in the immunosuppressive tumor microenvironment but only at the specific site on the vehicle known as the Integrin ligand. This assists in breaking down local immunity evasion systems, as well as changing the tumor related macrophages of the protumor M2 phenotype into the anticancer M1 phenotype, improving immunotherapy usage precision and safety. Figure 3 on the versatile application of functionalized CQDs in the cancer microenvironment such as lysine/polyethylenimine (PEI) coating of siRNAs to protect against serum nucleases, acidic pH-responsive (pH 4.55) release of π- π stacked aromatic chemotherapeutic cargoes, the production of hyperthermia on demand at 50 o C by photoresonance, and the immunotherapeutic re-education of tumor-associated macrophages (TAM) on a protumor M2 to an active, anticancer M1 phenotype.

*Figure 3. Multi-tiered intracellular precision and therapeutic modalities of green carbon CQD vectors*

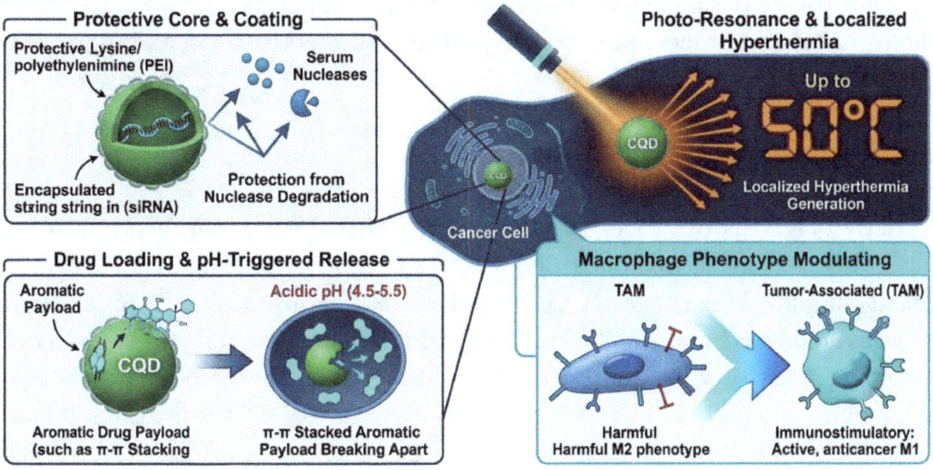

# 5. ADVANCED CANCER THERAGNOSTIC AND BIOSENSING

## 5.1. Imaging and Diagnostics

The theragnostic approach to cancer treatment is based on combining diagnostic imaging properties and targeted therapeutic action on one nanoparcel allowing to monitor nanocarrier biodistribution, localization of tumor tumor locally, and therapeutic response, in real-time (Sergeeva et al., 2025). Green synthesis to shape sustainable CDs and graphene quantum dots can be very appropriate to this application because of their unique optical and electronic characteristics. In comparison to other common synthetic semiconductor quantum dots (including cadmium selenide or lead sulfide, which are highly cytotoxic due to leaking of heavy metals), both biomass-derived carbon dots are naturally heavy-metal-free and biocompatible and also exhibit first-rate photostability against photobleaching. The photoluminescence properties of these sustainable carbon dots are intrinsic, which is controlled by the quantum confinement effects and surface state defects. Through tweaking green production conditions and using desirable natural precursors, scientists can/ can get the emission profiles into the visible spectrum and the near-infrared biowindows. The resulting tunable fluorescence provides high-resolution in vitro confocal microscopy of internal transport, cellular uptake, and sub-cellular localization of nanocarriers within malignant cell lines without the need to use a second chemical dye. In-vivo uses can be made of the multiphoton and anti-Stokes upconversion photoluminescence of these dots to provide deep-tissue optical imaging applications. It removes the shortages of tissue autofluorescence and light scattering to offer high-contrast photos of tumor borders and secondary clusters of metastases in real time (Wu et al., 2025).

## 5.2. Early Detection Biosensors

Early diagnosis of cancer is important in enhancing the survival rates of the patients. Nevertheless, traditional diagnostic procedures, including standard enzyme-linked immunosorbent assays (ELISA), polymerase chain reaction (PCR), and tissue biopsies could have constraints due to high cost of operation, long turnaround time, complicated use of sample preparation, and lack of sensitivity to detect low levels of biomarkers at early disease progression (Lee et al., 2026). The use of sustainable carbon nanostructures will facilitate the above challenges by offering electrochemical, colorimetric and fluorometric biosensing platforms based on highly sensitive, selective and cost-effective, point-of-care clinical diagnostics. Using carbon platforms based on biomass is highly advantageous in electrochemical biosensors due to their high electrical conductivity, rapid kinetics of electron-transfer, and rich surface chemistry (Martins et al., 2026). Graphene sheets, carbon nanohorns and multi-walled carbon nanotubes grown green can be coated over surfaces of working electrodes to exponentially increase the total electroactive surface area and speed up electron transfer between redox species and the electrode construct. Researchers can construct very sensitive biosensing interfaces by chemically attaching certain biorecognition factors, e.g. monoclonal antibodies, complementary single-stranded DNA strands, or synthetic peptide aptamers, onto the functionalized carbon interface. Upon binding to the immobilized probes, the localized biochemical interaction will change the electrical impedance, voltammetric response, or amperometric current at the carbon-modified interface when they are these target cancer biomarkers (including Carcinoembryonic Antigen, Prostate-Specific Antigen or specific microRNA) (Nadeem-Tariq et al., 2026). This detectable electrical change can be used to conveniently measure tumor markers in picomolar or femtomolar levels and measure at earliest in blood or saliva samples. Moreover, these sustainable carbon matrices are designable colorimetric and fluorometric biomarkers to monitor circulating tumor cells (CTCs) and circulating tumor DNA (ctDNA) shedding on masses of primary tumors into the systemic blood circulation (Mishra et al., 2024). A fluorophore labeled DNA probe can be loaded on the carbon surface with the use of long-range Förster Resonance Energy Transfer (FRET) (non-covalent) of intrinsic fluorescence quenching properties of graphene oxide sheets, inhibiting it and shutting off light emission. When the sample has certain cancer-specific ctDNA sequences the probe hybridizes with target and binds to the target strand using a rigid type of the double-helix formation.

*Figure 4. Advanced bioimaging diagnostics and early-stage biosensing configurations enabled by sustainable carbon nanomaterials*

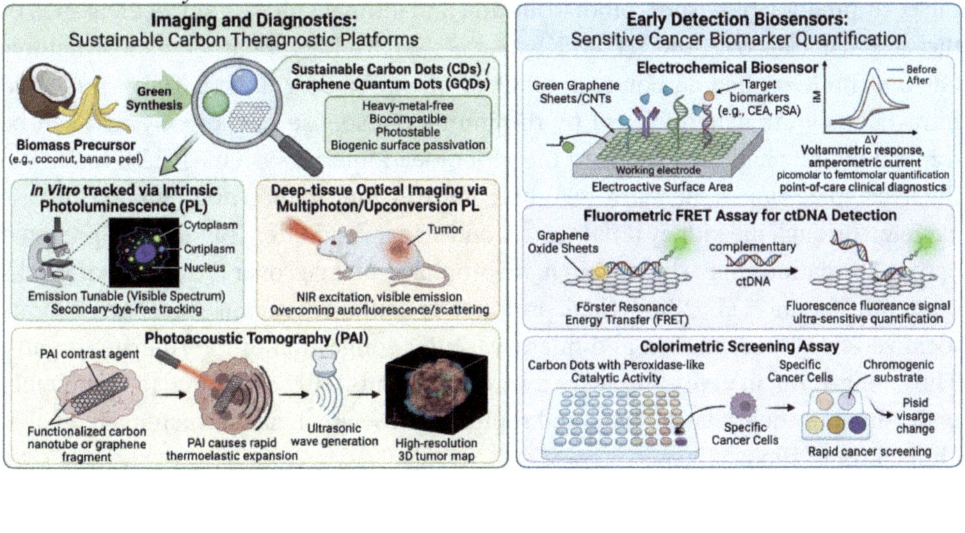

# 6. CHALLENGES, TOXICITY, AND CLINICAL TRANSLATION

## 6.1. The "Green" Toxicity Debate

Since sustainable and biomass-based carbon nanomaterials are emerging as the new, safer alternatives to the traditional, heavy-metal-based quantum dots and synthetic metallic nanoparticles, their long-term safety profile in biology has been investigated with a close eye (Hou et al., 2025; NV et al., 2026). One of the core principles of green nano-oncology is that there are natural precursors, namely plant extracts, fruit wastes, and pure carbohydrates that would produce ultimately non-toxic nanomaterials. Although it is admitted that green-synthesized CDs and graphene fragments tend to exhibit lower acute cytotoxicity in cell cultures than the structures prepared using harsh chemical techniques, this simplification of their complex interactions with their biology. The surface corona, structural sizes and internal purity of biomass-derived carbon nanomaterials are crucial to their biological behavior. The raw organic matrices are transformed to crystalline or amorphous cores of carbon using the high temperatures and pressures during green synthesis, especially through hydrothermal or microwave-assisted carbonization. Nevertheless, it is also in this way that a deep layer of leftover phytochemicals, organic molecules and proteins are stored on the surface of the particles (Chand et al., 2025). This biological passivation coating layer is essential in preliminary water solubility and cellular absorbing. It may also however complicate evaluations of toxicity in the long run. Provided these surface-bound biomolecules detach slowly into systemic circulation or inside the acidic lysosomes, they can cause unanticipated cellular responses, such as low-grade inflammatory responses or localized oxidative stress.

Moreover, to monitor the long term biosafety profile, it is necessary to monitor the accumulation of tissues, biodistribution and ultimate physiological clearance of the said materials over time (I. A. Khan et al., 2025). Larger carbon nanostructures (such as multi-walled carbon nanotubes or unexfoliated graphene flakes) injected intravenously are usually noted by the immune response of the body and can be trapped by the phagocytic cells of the reticuloendothelial system (RES), mainly in the liver, spleen and bone marrow. Whereas ultra-small carbon dots (sub-10 nm) are excreted through the kidney renal system and urinary excretion, bigger or aggregated carbon formations may stay entrap in macrophage vesicles over the course of weeks or months (Azmi et al., 2025). Such prolonged retention may cause focal granuloma, localized lipid peroxidation and micro-vascular congestion of the filtering organs. Thus, long-term, in vivo chronic toxicity experiments will be required to completely understand the degradation rates and routes of these green nanostructures to enable clinical translation.

## 6.2. Technical Bottlenecks

In spite of the encouraging laboratory findings, multiple technology bottlenecks such as the scale-up of and commercial feasibility of green-synthesized carbon nanomaterials to the clinic continue to exist. The most pronounced problem is that it is impossible to have batch-to-batch reproducibility when it comes to large-scale production (Dhumal et al., 2024). Green synthesis does not need the high quality, standardized synthetic reagents as compared to the conventional pathways of chemical synthesis, which uses natural precursors, biomass waste streams. Chemical composition of a natural precursor, like an extract of Azadirachta indica leaves or citrus peels, will be variable in nature. It is also based on the geographic origin, harvesting per season, soil profile in the location and the post-harvest storage. As such, carbon nanostructures synthesized in two different runs using biomasses obtained at varying times can have carbon nanostructures that are characterized by varying core diameters, surface functional definitions, and doping efficiencies. This heterogeneity of the material results in differences in the optical, mechanical, and electronic properties underpinning the material (Liu et al., 2025). Such variations are not tolerated in the field of precision medicine where nanocarriers need to sustain constant therapeutic doses and have predictable fluorescence properties in the form of diagnostics. Minute alterations in heteroatom doping content (nitrogen, sulfur, or phosphorus) may modify the photoluminescence quantum efficiency or are capable of modifying the strength of the photoacoustic signal with laser ignition. Moreover, the process of refining and purifying, green-synthesized mixtures is also a significant technical challenge. The output of the hydrothermal process of crude biomass is usually a complicated mixture with the desired quantum dots, and unreacted organic matter, burnt molecular fragments, and dissolved chemical byproducts. The conventional purification processes in the laboratory, including dialysis, ultrafiltration and size-exclusion chromatography, are time-consuming, costly and cannot be easily expanded to an industrial scale. The high throughput levels of extraction and purification essential to volume manufacturing of uniform and grade-certified carbon nanostructures have yet to be standardized, complicating the production of large amounts of the structure.

## 6.3. Regulatory and Clinical Outlook

The crucial step towards bringing sustainable carbon-based nanostructures to the next level in moving scientific laboratory designs into proven clinical countermeasures needs to overcome tricky regulatory environments and standardized testing protocols. The multi-functional theragnostic nanoplatforms are the subject of regulation as a combination product by regulatory agencies such as the U.S. Food and Drug Administration (FDA) and the European Medicines Agency (EMA). This classification indicates that the material has to be assessed independently in the diagnostic imaging component, and the active delivery of therapeutic drugs component of its action. In the case of carbon matrices based on biomass, this assessment is complicated by the fact that they have many functions. Since the carbon core is commonly used in two different ways simultaneously as an intrinsic photothermal therapy agent, a diagnostic fluorophore and as a delivery vehicle to carry chemotherapeutic payloads, regulators need explicit independent validation data to be documented associated with each individual functional application. Researchers should make a priority of developing a complete set of mass balance profiles and explicit pharmacokinetic models in order to resolve this translation gap (Coppola et al., 2019). This involves finding out the precise rate and path of elimination of each milligram of the carbonaceous material introduced, and confirming that its degradation products are completely non-toxic, non-immunogenic. Also, the industrial and academic communities need to work together to develop Good Manufacturing Practices (GMP) specific to biogenic nanomaterials (Souto et al., 2020). This implies the establishment of extreme tolerance thresholds to structural changes, the establishment of automated purification cascades and validated analytical methods to establish that batches are completely devoid of bacterial endotoxins or unaltered harmful organic molecules. A shift to the standardized and high-throughput models of animal animals that could closely recap human tumor vascular dynamics will aid in speeding up clinical translation of such eco-friendly nanostructures into human clinical trials (Mak et al., 2014).

## 7. CONCLUSION AND FUTURE PERSPECTIVES

Sustainable carbon based nanostructures are a significant improvement in precision nano-oncology, and offer an environmentally friendly alternative to heavy-metal and synthetic polymer carriers. These materials enable a high drug- loading capacity, surface-controlled drug release in response to the tumor microenvironment, and multi-modal photo-induced therapies (PTT/PDT) by giving bridging between the green synthesis paradigms and high level surface engineering. Intrinsic, stable photoluminescence of them allows tracking diagnostic and high-contrast bioimaging in real time without endangering the leaching of heavy-metals. Such dual-mode feature offers the opportunity to create unified theragnostic solutions that can detect, image and treat multifocal malignancies at a cellular scale and therefore they are capable of clinical translation.

In the future, inorganic carbon nanomaterials should be made sustainable through the implementation of Artificial Intelligence (AI) and machine learning algorithm into the process of discovery and optimization. Machine learning algorithms can be used to optimize large amounts of data sets of biomass chemical data and synthesis results to predict the specific parameters of hydrothermal process, microwave, or heteroatom doped settings to obtain desired quantum yield and size as well. This method of computation can greatly minimize the trial and error, which involves the use of experiments, to identify a solution, and assist in breaking the limits of reproducibility between batches, currently experienced by batch computation. In addition, combining AI-organized manufacturing with automated loop purifications, the area can proceed to high-throughput, personalized nanomedicine (Ahmed et al., 2026). Through this model, specialized carbon nanocarriers can be designed on demand which is a personalized biomolecular profile and tumor dynamics. With ongoing research to streamline biodegradation pathways and standardize green manufacturing of large quantities, sustainable carbon nanostructures will become a mainstay of personalized, zero-waste cancer therapies that is adequately aligned with clinical efficacy and ecological responsibility.

## REFERENCES


Agrawal, A. (2022). *Top-down synthesis of 2D nanomaterials. In 2D Nanomaterials*. CRC Press.

Ahmed, N., Abusalah, M. A. H. A., Absar, M., Noor, M. S., Bukhari, B., Anjum, S. A., Singh, K. K. B., & Yean, C. Y. (2025). Carbon-based Nanomaterials as Multifunctional Particles for Cancer Diagnosis and Treatment. *Nano LIFE, 15*(04), 2430005. DOI: 10.1142/S179398442430005X

Ajorlou, E., & Khosroushahi, A. Y. (2017). Trends on polymer-and lipid-based nanostructures for parenteral drug delivery to tumors. *Cancer Chemotherapy and Pharmacology, 79*(2), 251–265. DOI: 10.1007/s00280-016-3168-6 PMID: 27744564

Al-Jumaili, A., Alancherry, S., Bazaka, K., & Jacob, M. V. (2017). Review on the antimicrobial properties of carbon nanostructures. *Materials (Basel), 10*(9), 1066. DOI: 10.3390/ma10091066 PMID: 28892011

Alvandi, H., Shafie, A., Najafi, F., Sabzini, M., Mashayekhi, M., Omami, S. H., Eskandarisani, M. M., Dashti, S., Javanmard, A., Tajik, M., Bräse, S., & Farmani, A. R. (2025). Carbon-based nanostructure drug delivery systems and their biologic applications–a review. *Carbon Letters, 35*(4), 1529–1587. DOI: 10.1007/s42823-025-00927-1

Alvarez, N., & Sevilla, A. (2024). Current advances in photodynamic therapy (PDT) and the future potential of PDT-combinatorial cancer therapies. *International Journal of Molecular Sciences, 25*(2), 1023. DOI: 10.3390/ijms25021023 PMID: 38256096

Anas, M., Khan, A., Falak, A., Shinwari, Z. K., Mohamed, H. E. A., Hkiri, K., Maaza, M., Fahad, S., & Quraishi, U. M. (2025). Harnessing Biomedical Applications of Ascorbic Acid, Berberine, and Glutathione-Capped Quantum Dots Through Multifunctional Physico-chemical, Biochemical, and Anti-microbial Assays. *BioNanoScience, 15*(2), 259. DOI: 10.1007/s12668-025-01864-z

Arpicco, S., Bartkowski, M., Barge, A., Zonari, D., Serpe, L., Milla, P., Dosio, F., Stella, B., & Giordani, S. (2020). Effects of the molecular weight of hyaluronic acid in a carbon nanotube drug delivery conjugate. *Frontiers in Chemistry, 8*, 578008. DOI: 10.3389/fchem.2020.578008 PMID: 33381490

Azmi, F., Xu, X., Duong, H., Ye, P., Chen, T., Li, H., Chen, J., Ardekani, S. M., Dehghani, A., Zheng, G., Harris, D., Lu, H., Wang, Y., & Cao, Q. (2025). Renal clearable sucrose carbon dots for doxorubicin delivery to treat renal carcinoma. *Nanoscale Advances, 7*(9), 2751–2760. DOI: 10.1039/D4NA01082E PMID: 40160256

Bagheri, B., Surwase, S. S., Lee, S. S., Park, H., Rad, Z. F., Trevaskis, N. L., & Kim, Y.-C. (2022). Carbon-based nanostructures for cancer therapy and drug delivery applications. *Journal of Materials Chemistry. B, 10*(48), 9944–9967. DOI: 10.1039/D2TB01741E PMID: 36415922

Bahuguna, S., Kumar, M., Sharma, G., Kumar, R., Singh, B., & Raza, K. (2018). Fullerenol-based intracellular delivery of methotrexate: A water-soluble nanoconjugate for enhanced cytotoxicity and improved pharmacokinetics. *AAPS PharmSciTech, 19*(3), 1084–1092. DOI: 10.1208/s12249-017-0920-0 PMID: 29159749

Bartkowski, M., & Giordani, S. (2021). Carbon nano-onions as potential nanocarriers for drug delivery. *Dalton Transactions : An International Journal of Inorganic Chemistry, 50*(7), 2300–2309. DOI: 10.1039/D0DT04093B PMID: 33471000

Chand, S., Kumar, R., Thakur, N., Kumar, K., Umar, A., Almas, T., & Baskoutas, S. (2025). Sustainable synthesis and multifunctional applications of biowaste-derived carbon nanomaterials and metal oxide composites: A review. *Chemosphere, 385*, 144540. DOI: 10.1016/j.chemosphere.2025.144540 PMID: 40609513

Chen, Y., Wang, S., & Zhang, F. (2023). Near-infrared luminescence high-contrast in vivo biomedical imaging. *Nature Reviews Bioengineering, 1*(1), 60–78. DOI: 10.1038/s44222-022-00002-8

Chen, Z., Wang, X., Zhao, N., Chen, H., & Guo, G. (2023). Advancements in pH-responsive nanocarriers: Enhancing drug delivery for tumor therapy. *Expert Opinion on Drug Delivery, 20*(11), 1623–1642. DOI: 10.1080/17425247.2023.2292678 PMID: 38059646

Coppola, P., Andersson, A., & Cole, S. (2019). The importance of the human mass balance study in regulatory submissions. *CPT: Pharmacometrics & Systems Pharmacology, 8*(11), 792–804. DOI: 10.1002/psp4.12466 PMID: 31515957

Cui, X., Ruan, Q., Zhuo, X., Xia, X., Hu, J., Fu, R., Li, Y., Wang, J., & Xu, H. (2023). Photothermal nanomaterials: A powerful light-to-heat converter. *Chemical Reviews, 123*(11), 6891–6952. DOI: 10.1021/acs.chemrev.3c00159 PMID: 37133878

Dhumal, P., Chakraborty, S., Ibrahim, B., Kaur, M., & Valsami-Jones, E. (2024). Green-synthesised carbon nanodots: A SWOT analysis for their safe and sustainable innovation. *Journal of Cleaner Production, 480*, 144115. DOI: 10.1016/j.jclepro.2024.144115

Dubey, R., Dutta, D., Sarkar, A., & Chattopadhyay, P. (2021). Functionalized carbon nanotubes: Synthesis, properties and applications in water purification, drug delivery, and material and biomedical sciences. *Nanoscale Advances*, *3*(20), 5722–5744. DOI: 10.1039/D1NA00293G PMID: 36132675

El-Khawaga, A. M., Youssef, O. I., El-Dydamoni, O. A., Ragab, W. M., & Mohammed, R. A. E. (2026). Advances in green-synthesized magnetic nanoparticles for targeted cancer therapy: Mechanisms, applications, and future perspectives. *European Journal of Medical Research*, *31*(1), 347. DOI: 10.1186/s40001-025-03804-9 PMID: 41654877

Esmaeili, Y., Ghaheh, H. S., Ghasemi, F., Shariati, L., Rafienia, M., Bidram, E., & Zarrabi, A. (2022). Graphene oxide quantum dot-chitosan nanotheranostic platform as a pH-responsive carrier for improving curcumin uptake internalization: In vitro & in silico study. *Biomaterials Advances*, *139*, 213017. DOI: 10.1016/j.bioadv.2022.213017 PMID: 35882115

Falak, A., Anas, M., Khan, A., Hayat, A., Shaheen, Z., Saleem, M. H., Fahad, S., & Quraishi, U. M. (2025). Efficacy of ascorbic acid coated quantum dots in alleviating lead-induced oxidative damage and enhancing growth parameters in rice (Oryza sativa L.) for sustainable cultivation. *Journal of Trace Elements in Medicine and Biology : Organ of the Society for Minerals and Trace Elements (GMS)*, *88*, 127603. DOI: 10.1016/j.jtemb.2025.127603 PMID: 39847985

Fatima, I., Rahdar, A., Sargazi, S., Barani, M., Hassanisaadi, M., & Thakur, V. K. (2021). Quantum dots: Synthesis, antibody conjugation, and HER2-receptor targeting for breast cancer therapy. *Journal of Functional Biomaterials*, *12*(4), 75. DOI: 10.3390/jfb12040075 PMID: 34940554

Fu, S., Rempson, C. M., Puche, V., Zhao, B., & Zhang, F. (2022). Construction of disulfide containing redox-responsive polymeric nanomedicine. *Methods : A Companion to Methods in Enzymology*, *199*, 67–79. DOI: 10.1016/j.ymeth.2021.12.011 PMID: 34971759

Gao, Q., Gao, J., Ding, C., Li, S., Deng, L., & Kong, Y. (2021). Construction of a pH-and near-infrared irradiation-responsive nanoplatform for chemo-photothermal therapy. *International Journal of Pharmaceutics*, *593*, 120112. DOI: 10.1016/j.ijpharm.2020.120112 PMID: 33259903

Gárate-Vélez, L., & Quintana, M. (2025). Carbon-based nanomaterials: Interactions with cells, brain therapies, and neural sensing. *Journal of Materials Science. Materials in Engineering*, *20*(1), 23. DOI: 10.1186/s40712-025-00236-5

Ghosh, A., Ghosh, A. K., Chowdhury, M., & Das, P. K. (2022). Folic acid-functionalized carbon dot-enabled starvation therapy in synergism with paclitaxel against breast cancer. *ACS Applied Bio Materials*, *5*(5), 2389–2402. DOI: 10.1021/acsabm.2c00235 PMID: 35452214

Gupta, V., Kumar, D., Gupta, S., Tanwar, R., Jaiswal, N. K., Islam, M. M., Singh, S., Choudhary, N., Gowri, S., Webster, T. J., & Faiyazuddin, M. (2026). Nanotechnology-driven cancer therapies for precision oncology: Advances and clinical outlook. *International Journal of Nanomedicine*, *21*, 568254. DOI: 10.2147/IJN.S568254 PMID: 41869396

Hamzah, H. (2026). Targeted therapy and biomarker-guided applications of ecofriendly silver nanoparticles in precision oncology [Review]. *Journal of Pharmacy & Pharmaceutical Sciences, Volume 28 - 2025*. DOI: 10.3389/jpps.2025.15403

Hou, Y., Zhu, C., Shen, Z., Xu, Y., Zhou, S., & Zhou, X. (2025). Carbon Nanomaterials in Biomedicine: Opportunities and Toxicological Concerns. *International Journal of Nanomedicine*, *20*, 14707–14730. DOI: 10.2147/IJN.S552319 PMID: 41393084

Keklikcioglu Cakmak, N., & Eroglu, A. (2023). Doxorubicin and tamoxifen loaded graphene oxide nanoparticle functionalized with chitosan and folic acid for anti-cancer drug delivery. *Polymer Bulletin*, *80*(2), 2171–2185. DOI: 10.1007/s00289-022-04549-9

Khan, A., Anas, M., Qadri, T. A., Algopishi, U. B., Fayad, E., Binjawhar, D. N., Khan, R. A., Maaza, M., & Shinwari, Z. K. (2025). Eco-friendly synthesis of copper oxide-silver bimetallic nanoparticles using Withania coagulans: Characterization and biomedical potential through antioxidant, antibacterial, and catalytic activities. *Inorganic Chemistry Communications*, *181*, 115131. 10.1016/j.inoche.2025.115131. DOI: 10.1016/j.inoche.2025.115131

Khan, A., Anas, M., Ullah, N., Khattak, W. A., Ali, M., Maaza, M., & Shinwari, Z. K. (2025). Harnessing Green Nanotechnology: Exploring the Potential of Silver, Copper, and their Nanocomposites for Biomedical and Agricultural Applications. *Pakistan Journal of Botany*, *57*(6), 6. DOI: 10.30848/PJB2025-6(5)

Khan, A., Ashiq, B., Qadri, T. A., Khattak, W. A., Anas, M., Khan, R. A., Abbas, S., Shinwari, Z. K., & Maaza, M. (2025). Nanotechnology Solutions For Advancing Sustainable Development Goals (SDGs): Integrating Nanomaterials For Global Sustainability. *BioNanoScience*, *15*(4), 590. DOI: 10.1007/s12668-025-02214-9

Khan, A., Khan, R. A., Anas, M., Qadri, T. A., Ashiq, B., Younis, T., Ullah, N., Khan, A. M., Nasreen, Z., & Shinwari, Z. K. (2026). Biogenic Nanoparticles and Green Nanocomposites as Sustainable Antimicrobial Strategies Against Bacteria, Fungi and Viruses. *Nano-Horizons: Journal of Nanosciences and Nanotechnologies, 5*, 31 pages. DOI: 10.25159/3005-2602/20247

Khan, A., Qadri, T. A., Khan, R. A., Hassan, D., Ashiq, B., Sani, A., Shinwari, Z. K., & Maaza, M. (2026). Sustainable green synthesis of CeO 2–ZnO nanocomposites using Rhazya stricta: Enhanced synergistic solutions for biomedical challenges and environmental remediation. *RSC Advances*, *16*(26), 23370–23390. DOI: 10.1039/D6RA02267G PMID: 42100458

Khan, A., Younis, T., Anas, M., Ali, M., Shinwari, Z. K., Khalil, A. T., Munawar, K. S., Mohamed, H. E. A., Hkiri, K., Maaza, M., Seleiman, M. F., & Khan, N. (2025). Withania coagulans-mediated green synthesis of silver nanoparticles: Characterization and assessment of their phytochemical, antioxidant, toxicity, and antimicrobial activities. *BMC Plant Biology*, *25*(1), 574. DOI: 10.1186/s12870-025-06533-7 PMID: 40316892

Khan, I. A., Yu, T., Yang, M., Liu, J., & Chen, Z. (2025). A systematic review of toxicity, biodistribution, and biosafety in upconversion nanomaterials: Critical insights into toxicity mitigation strategies and future directions for safe applications. *BME frontiers, 6*, 0120.

Kok, H. P., Cressman, E. N., Ceelen, W., Brace, C. L., Ivkov, R., Grüll, H., Ter Haar, G., Wust, P., & Crezee, J. (2020). Heating technology for malignant tumors: A review. *International Journal of Hyperthermia*, *37*(1), 711–741. DOI: 10.1080/02656736.2020.1779357 PMID: 32579419

Kumar, N., Chamoli, P., Misra, M., Manoj, M., & Sharma, A. (2022). Advanced metal and carbon nanostructures for medical, drug delivery and bio-imaging applications. *Nanoscale*, *14*(11), 3987–4017. DOI: 10.1039/D1NR07643D PMID: 35244647

Lee, J., Kim, J., Cho, E., Jeong, J., Song, J., & Kang, M. (2026). Nanotechnology-Enhanced Approaches for Early Cancer Diagnosis. *Nano Select*, *7*(1), e70105. DOI: 10.1002/nano.70105

Liu, Q., Chen, H., Mi, R., Min, X., Fang, M., Wu, X., Huang, Z., & Liu, Y. (2025). Biomass-derived carbon dots: Preparation, properties, and applications. *Nanomaterials (Basel, Switzerland)*, *15*(16), 1279. DOI: 10.3390/nano15161279 PMID: 40863859

Lukyanov, A. N., & Torchilin, V. P. (2004). Micelles from lipid derivatives of water-soluble polymers as delivery systems for poorly soluble drugs. *Advanced Drug Delivery Reviews*, *56*(9), 1273–1289. DOI: 10.1016/j.addr.2003.12.004 PMID: 15109769

Ma, X., Shu, C., Guo, J., Pang, L., Su, L., Fu, D., & Zhong, W. (2014). Targeted cancer therapy based on single-wall carbon nanohorns with doxorubicin in vitro and in vivo. *Journal of Nanoparticle Research*, *16*(7), 2497. DOI: 10.1007/s11051-014-2497-9

Madani, S. Y., Naderi, N., Dissanayake, O., Tan, A., & Seifalian, A. M. (2011). A new era of cancer treatment: Carbon nanotubes as drug delivery tools. *International Journal of Nanomedicine*, •••, 2963–2979. PMID: 22162655

Mak, I. W., Evaniew, N., & Ghert, M. (2014). Lost in translation: Animal models and clinical trials in cancer treatment. *American Journal of Translational Research*, *6*(2), 114. PMID: 24489990

Martins, G., Luiz Gogola, J., Gevaerd, A., Kalinke, C., de Oliveira, P. R., Caetano, F. R., Janegitz, B. C., Bergamini, M. F., & Marcolino-Junior, L. H. (2026). Exploring the Potential of Zero-Dimensional Carbon Nanomaterials in Photoluminescent, Electrochemiluminescent and Electrochemical Sensors. *Advanced Materials Interfaces*, *13*(11), 01018. DOI: 10.1002/admi.202501018

Mishra, M., Ahmed, R., Das, D. K., Pramanik, D. D., Dash, S. K., & Pramanik, A. (2024). Recent advancements in the application of circulating tumor DNA as biomarkers for early detection of cancers. *ACS Biomaterials Science & Engineering*, *10*(8), 4740–4756. DOI: 10.1021/acsbiomaterials.4c00606 PMID: 38950521

Mohamed, R. G. A., Ali, S. M., Ahmed, I. S., Rawas-Qalaji, M., & Hussain, Z. (2025). Next-generation nanocarriers for colorectal cancer: Passive, active, and stimuli-responsive strategies for precision therapy. *Biomaterials Science*, *13*(20), 5626–5664. DOI: 10.1039/D5BM01176K PMID: 40928395

Mohan, H., Bincoletto, V., Arpicco, S., & Giordani, S. (2022). Supramolecular functionalisation of B/N co-doped carbon nano-onions for novel nanocarrier systems. *Materials (Basel)*, *15*(17), 5987. DOI: 10.3390/ma15175987 PMID: 36079368

Mostafavi, E., Iravani, S., Varma, R. S., Khatami, M., & Rahbarizadeh, F. (2022). Eco-friendly synthesis of carbon nanotubes and their cancer theranostic applications. *Materials Advances*, *3*(12), 4765–4782. DOI: 10.1039/D2MA00341D PMID: 35812837

N v, N. K., Mathew, A. A., Jafri, R. I., & Balachandran, M.NV. (2026). Biomass-derived Carbon Dots for Fluorescence-based Pollutant Sensing, Cellular Imaging, Antibacterial and Anticancer Therapy. *BioNanoScience*, *16*(4), 291. DOI: 10.1007/s12668-026-02527-3

Nadeem-Tariq, A., Rafanan, J. R., Kang, N., Zhang, S., Kanniyappan, H., & Merchant, A. (2026). Electrochemical Detection of Cancer Biomarkers: From Molecular Sensing to Clinical Translation. *Biosensors (Basel)*, *16*(1), 44. DOI: 10.3390/bios16010044 PMID: 41590296

Paveethra, S., Manisekaran, H., & Sasidharan, S. (2024). Medicinal plants derived green carbon dots: Synthesis, characterization and their potential applications in cancer therapy. *Asian Pacific Journal of Cancer Prevention : APJCP*, *25*(10), 3393–3411. DOI: 10.31557/APJCP.2024.25.10.3393 PMID: 39471005

Peer, D., Karp, J. M., Hong, S., Farokhzad, O. C., Margalit, R., & Langer, R. (2020). Nanocarriers as an emerging platform for cancer therapy. *Nano-enabled medical applications*, 61-91.

Peng, X., Zhang, M., Gao, Y., Zhao, C., Li, S., Liao, S., Zhao, X., Huang, R., Wang, J., & Zeng, L. (2025). Targeted endocytosis: Strategies in drug delivery systems for enhancing antitumor efficacy. *Colloids and Surfaces. B, Biointerfaces*, ***•••***, 115276. PMID: 41242027

Randviir, E. P., Brownson, D. A. C., & Banks, C. E. (2014). A decade of graphene research: Production, applications and outlook. *Materials Today*, *17*(9), 426–432. DOI: 10.1016/j.mattod.2014.06.001

Rizo-Téllez, S. A., Sekheri, M., & Filep, J. G. (2022). Myeloperoxidase: Regulation of neutrophil function and target for therapy. *Antioxidants*, *11*(11), 2302. DOI: 10.3390/antiox11112302 PMID: 36421487

Rozhin, P., Abdel Monem Gamal, J., Giordani, S., & Marchesan, S. (2022). Carbon nanomaterials (CNMs) and enzymes: From nanozymes to cnm-enzyme conjugates and biodegradation. *Materials (Basel)*, *15*(3), 1037. DOI: 10.3390/ma15031037 PMID: 35160982

Sahu, S., & Tiwari, S. (2023). Carbon allotropes: Fundamental, synthesis, characterization, and properties functionalization. In *Carbon allotropes* (pp. 41–70). CRC Press. DOI: 10.1201/9781003323976-5

Sajjadi, M., Nasrollahzadeh, M., Jaleh, B., Soufi, G. J., & Iravani, S. (2021). Carbon-based nanomaterials for targeted cancer nanotherapy: Recent trends and future prospects. *Journal of Drug Targeting*, *29*(7), 716–741. DOI: 10.1080/1061186X.2021.1886301 PMID: 33566719

Sanjeev, N. O., Vallabha, M. S., & Valsan, A. E. (2024). Azadirachta indica leaf extract based green-synthesized ZnO nanoparticles coated on spent tea waste activated carbon for pharmaceuticals and personal care products removal. *Environmental Research*, *252*, 119047. DOI: 10.1016/j.envres.2024.119047 PMID: 38704006

Sasidharan, S., Kumari, K., & Ramakrishnan, V. (2024). Aromatic π–π Interactions in Self-assembled Nanostructured Materials. Conference on Research and Industrial Conclave-Integration, Sergeeva, O. V., Luo, L., & Guiseppi-Elie, A. (2025). Cancer theragnostics: Closing the loop for advanced personalized cancer treatment through the platform integration of therapeutics and diagnostics. *Frontiers in Bioengineering and Biotechnology*, *12*, 1499474.

Sezer, E., Ulucan Karnak, F., & Akgöl, S. (2025). Green carbon dots in the era of AI: Sustainable synthesis, intelligent drug delivery, advanced diagnostics, and bioimaging. *Turkish Journal of Biology = Türk Biyoloji Dergisi*, *49*(5), 498–533. DOI: 10.55730/1300-0152.2762 PMID: 41246234

Shamsian, S., Siddique, A. B., Kordzadeh-Kermani, V., de la Vega Tejuca, L., Falcone, F., Ray, M., Ashrafizadeh, S. N., Chapa, S. O. M., Madou, M. J., & Madadelahi, M. (2025). Nanomaterials in PCR: Exploring light-to-heat conversion mechanisms and microfluidic integration. *Microsystems & Nanoengineering*, *11*(1), 127. DOI: 10.1038/s41378-025-00898-3 PMID: 40533488

Shinde, V. R., Revi, N., Murugappan, S., Singh, S. P., & Rengan, A. K. (2022). Enhanced permeability and retention effect: A key facilitator for solid tumor targeting by nanoparticles. *Photodiagnosis and Photodynamic Therapy*, *39*, 102915. DOI: 10.1016/j.pdpdt.2022.102915 PMID: 35597441

Souto, E. B., Silva, G. F., Dias-Ferreira, J., Zielinska, A., Ventura, F., Durazzo, A., Lucarini, M., Novellino, E., & Santini, A. (2020). Nanopharmaceutics: Part I—Clinical trials legislation and good manufacturing practices (GMP) of nanotherapeutics in the EU. *Pharmaceutics*, *12*(2), 146. DOI: 10.3390/pharmaceutics12020146 PMID: 32053962

Spada, A., & Gerber-Lemaire, S. (2025). Surface functionalization of nanocarriers with anti-EGFR ligands for cancer active targeting. *Nanomaterials (Basel, Switzerland)*, *15*(3), 158. DOI: 10.3390/nano15030158 PMID: 39940134

Su, Y., Wang, N., Liu, B., Du, Y., Li, R., Meng, Y., Feng, Y., Shan, Z., & Meng, S. (2020). A phototheranostic nanoparticle for cancer therapy fabricated by BODIPY and graphene to realize photo-chemo synergistic therapy and fluorescence/photothermal imaging. *Dyes and Pigments*, *177*, 108262. DOI: 10.1016/j.dyepig.2020.108262

Sun, B., Wang, W., & Sain, M. (2022). Carbonaceous nanocomposites for biomedical applications as high-drug loading nanocarriers for sustained delivery: A Review. *Journal of Composites Science*, *6*(12), 379. DOI: 10.3390/jcs6120379

Tang, L., Li, J., Pan, T., Yin, Y., Mei, Y., Xiao, Q., Wang, R., Yan, Z., & Wang, W. (2022). Versatile carbon nanoplatforms for cancer treatment and diagnosis: Strategies, applications and future perspectives. *Theranostics*, *12*(5), 2290–2321. DOI: 10.7150/thno.69628 PMID: 35265211

Ullah, N., Khan, A., Maqbool, H., Ali, M., Munawar, K. S., Fallatah, A. M., Mohamed, H. E. A., Hkiri, K., Maaza, M., & Shinwari, Z. K. (2026). Green synthesis and biomedical evaluation of graphene oxide nanoparticles using Monotheca buxifolia extract. *3 Biotech, 16*(4), 144. DOI: 10.1007/s13205-026-04758-3

Varma, S., Dey, S., & Palanisamy, D. (2022). Cellular uptake pathways of nanoparticles: Process of endocytosis and factors affecting their fate. *Current Pharmaceutical Biotechnology*, *23*(5), 679–706. DOI: 10.2174/1389201022666210714145356 PMID: 34264182

Venkataraman, S., Rajendran, D. S., Kumar, P. S., Rangasamy, G., & Vaidyanathan, V. K. (2024). A comprehensive review on the refinery of citrus peel towards the production of bioenergy, biochemical and biobased value-added products: Present insights and futuristic challenges. *Waste and Biomass Valorization*, *15*(11), 6491–6512. DOI: 10.1007/s12649-024-02557-6

Voss, M., Valle, C., Calcio Gaudino, E., Tabasso, S., Forte, C., & Cravotto, G. (2024). Unlocking the potential of agrifood waste for sustainable innovation in agriculture. *Recycling*, *9*(2), 25. DOI: 10.3390/recycling9020025

Wu, J., Zhou, X., Tsang, C. Y., Mei, Q., & Zhang, Y. (2025). Bioengineered nanomaterials for dynamic diagnostics in vivo. *Chemical Society Reviews*, *54*(11), 5470–5515. DOI: 10.1039/D5CS00136F PMID: 40289891

Yang, S., Liu, J., Ping, Y., Wang, Z., Zhang, J., Zhang, L., Cui, L., Xiao, Y., & Qu, L. (2022). Multi-functionalized single-walled carbon nanotubes as delivery carriers: Promote the targeting uptake and antitumor efficacy of doxorubicin. *Journal of Inclusion Phenomena and Macrocyclic Chemistry*, *102*(9), 801–817. DOI: 10.1007/s10847-022-01163-0

Ye, W., Han, H., Li, H., Jin, Q., Wu, Y., Chakrabortty, S., Weil, T., & Ji, J. (2020). Polymer coated nanodiamonds as gemcitabine prodrug with enzymatic sensitivity for pancreatic cancer treatment. *Progress in Natural Science*, *30*(5), 711–717. DOI: 10.1016/j.pnsc.2020.10.011

Ying, H., Zhang, K., Chen, Y., Deng, Q., Mao, C., & Li, J. (2026). Recent development on hyperthermia: An effective cotreatment improving radiotherapy outcome. *Oncology Letters*, *31*(3), 112. DOI: 10.3892/ol.2026.15465 PMID: 41613581

Yu, Y., Song, M., Chen, C., Du, Y., Li, C., Han, Y., Yan, F., Shi, Z., & Feng, S. (2020). Bortezomib-encapsulated CuS/carbon dot nanocomposites for enhanced photothermal therapy via stabilization of polyubiquitinated substrates in the proteasomal degradation pathway. *ACS Nano*, *14*(8), 10688–10703. DOI: 10.1021/acsnano.0c05332 PMID: 32790339

Zhang, G., Chang, H., Amatore, C., Chen, Y., Jiang, H., & Wang, X. (2013). Apoptosis induction and inhibition of drug resistant tumor growth in vivo involving daunorubicin-loaded graphene–gold composites. *Journal of Materials Chemistry. B*, *1*(4), 493–499. DOI: 10.1039/C2TB00378C PMID: 32260820

Zhang, H., Liu, Y., & Qu, S. (2024). Recent advances in photo-responsive carbon dots for tumor therapy. *Responsive Materials*, *2*(2), e20240012. DOI: 10.1002/rpm.20240012

Zhang, X., Wang, S., Cheng, G., Yu, P., Chang, J., & Chen, X. (2021). Cascade Drug-Release Strategy for Enhanced Anticancer Therapy. *Matter*, *4*(1), 26–53. DOI: 10.1016/j.matt.2020.10.002 PMID: 33718863

Zhu, S., Xu, R., Xu, H., Wang, Y., Wang, L., He, N., Wang, H.-H., & Ren, X. (2026). The emerging role of aptamers in targeted cancer immunotherapy. *Molecular Therapy. Oncology*, *34*(1), 201110. DOI: 10.1016/j.omton.2025.201110 PMID: 41659919

# Chapter 3
# Green Synthesis of Metallic Nanoparticles (Ag, Au, Se)


**Mujeeb ur Rehman**
*Shandong University, China*

**Mamoona Uzma Kalsoom**
*University of Mianwali, Pakistan*

**Hansa Gul**
http://orcid.org/0009-0001-5554-617X
*University of Mianwali, Pakistan*

**Zahida Nasreen**
http://orcid.org/0009-0002-0995-5041
*University of Mianwali, Pakistan*

**Hira Larab**
http://orcid.org/0009-0009-7224-4454
*Government Associate College for Women, Kundian, Pakistan*



## ABSTRACT

*This chapter provides a detailed analysis of the green synthesis of metallic nanoparticles (NPs) i.e. silver (Ag), gold (Au) and selenium (Se) using medicinal plant extracts and microbial systems in the framework of precision oncology. The green biosynthetic method utilizes bioactive phytochemicals including flavonoids, terpenoids and polyphenols, as natural reductants and stabilizers/capping agents, and, thus, enables the environmentally friendly synthesis of NPs. The characterization protocols, include UV- Vis spectroscopy, energy-dispersive X-ray spectroscopy,*



DOI: 10.4018/979-8-3373-9848-8.ch003

*X-ray diffraction, scanning electron microscopy and fourier-transform infrared spectroscopy. Biological activities indicate their antimicrobial, antioxidant, and anticancer effects, and some can selectively kill brain, prostate and colon cancer cell lines, causing ROS-mediated apoptosis of tumor cells with sparing the effect of normal cell populations. In conclusion, this chapter outlines sustainable, and biocompatible nano-synthesis bridging the gap between laboratory work and clinical practice in precise oncology.*

## 1. INTRODUCTION

Nanotechnology is the field of materials science that deals with materials between 1 – 100 nm or so. It has transformed various industry sectors like medicine, agriculture, environmental science and electronics by allowing nanotechnology innovations such as targeted drug delivery, diagnostic imaging, and sustainable agriculture (Malik et al., 2023; N. A. Singh, 2017; J. Wang, 2025). In medicine applications, these are employed in drug delivery, biosensing, gene therapy, biomedical implant devices and anticancer medicine, and in agriculture applications, they are utilized in controlled release nanopesticides and nanosensors for improved food security (Haleem et al., 2023; Malik et al., 2023; Ramos et al., 2017). Nanomaterials, including carbon nanotubes, liposomes, metal oxides, and ceramic nanoparticles, exhibit exceptional physicochemical properties, which are proving to have a beneficial effect on biomedical applications like bone repair and bioimaging, but there are still some concerns about toxicity and oxidative stress (Ramos et al., 2017; Soni & Jha, 2024). Moreover, nanotechnology can be used in flexible electronics by the way of conductive networks applicable in wearable technologies (Kanaoujiya et al., 2023; J. Wang, 2025)

Due to their high surface area to volume ratio, quantum effects and tunable surface chemistry, nanomaterials can be used to target and deliver drugs in a targeted manner as well as penetrate biological barriers like blood brain barrier (Chandrakala et al., 2022; Khursheed et al., 2022). The three most widely studied types of metallic nanoparticles are, namely, AgNPs, AuNPs and SeNPs, which are used in imaging, biosensing, and in cancer therapy, respectively (Chandrakala et al., 2022; Khursheed et al., 2022; Vijayaram et al., 2024). The size-, shape- and surface-chemistry dependent electronic, optical and surface plasmon resonance properties are expected to be useful for imaging, diagnosis and biosensing applications (Khursheed et al., 2022; Truong et al., 2024).

The traditional synthesis routes often use toxic reducing agents, harsh reaction conditions and excessive energy consumptions, thus, are not sustainable and suitable for biomedical applications. In green synthesis, the use of plants, microorganisms and biomolecules as natural reducing agents and stabilizing agents to prepare biocompatible nanoparticles under mild aqueous conditions (Chopra et al., 2022; Dikshit et al., 2021; Szczyglewska et al., 2023; Vijayaram et al., 2024). Initial research showed that fungi like *Fusarium oxysporum,*(Mukherjee et al., 2001) (Mukherjee et al., 2001) and plants like alfalfa (*Medicago sativa*) (Gardea-Torresdey et al., 2003), could minimize metal ions to stable metallic nanoparticles. Since then, various biomolecules, such as polyphenols, sugars, chlorophylls, carotenoids, bacterial nitrate reductase, and fungal laccase, have been employed in green synthesis, resulting in greater sustainability and scalability (Huston et al., 2021; Jeevanandam et al., 2022; Rónavári et al., 2021). Although these steps forward have been taken, toxicity, biological interactions, and regulatory approval have been significant hurdles to clinical translation (Truong et al., 2024; N. Zhang et al., 2022). Some promising new methods are combining synthetic biology with artificial intelligence, which can be used to fine-tune biosynthetic pathways, fine-tune process control, and scale up the production of nanoparticles (Iram et al., 2024; Song & Prather, 2024; Jeong et al., 2025).

This chapter presents a systematic analysis and detailed mechanisms of green biosynthesis of Ag, Au and Se nanoparticles via plant phytochemicals and microbial systems, physicochemical characterization models and reports the biological activities of each type of nanoparticles in the context of precision oncology. It also explores the use of microbial platforms engineered with CRISPR, as well as green chemistry indices to achieve reproducible, scalable and regulatory compliant synthesis of nanoparticles.

## 2. PRINCIPLES OF GREEN NANOTECHNOLOGY

The concept of green chemistry was established in the early 1990s by Paul Anastas and John C. Warner at the U.S. EPA. In 2001, Neil Winterton extended the framework by proposing the new principles to address emerging sustainability challenges. In the early twenty-first century, green chemistry combined with nanotechnology led to the development of green nanotechnology (Green Nano) which emphasizes on green synthesis of nanomaterials and sustainable nanoprocesses (Saratale et al., 2018). Plant, bacteria, fungi, algae, and other microorganisms are used in green synthesis as natural reducing and stabilizing agents in mild reaction conditions. The metal ion reducing capacities, stabilizing properties of nanoparticles, and an increase in biocompatibility of biomolecules like polyphenols, flavonoids, proteins, enzymes, and polysaccharides enable their use in biomedicine, agriculture, environmental remediation, food systems, and energy (Chopra et al., 2022; Khan et al., 2022; Osman et al., 2024; Salem & Fouda, 2021). The metal-reduction capacity of medicinal plants mainly depends on secondary metabolites like flavonoids, terpenoids, proteins, and alkaloids which stabilize the nanoparticles by the functional group interaction, resulting in antioxidant, antimicrobial, and anticancer properties (El-Seedi et al., 2019; Mittal et al., 2013). Apart from the hydrogenase, cytochrome, metallothionein, bacteriocin and extracellular polymeric substances (EPS), in microbial synthesis, metal ion reduction and stabilization of nanoparticles is accomplished by NADH-dependent enzymes like nitrate reductase which are more stable and biologically active (Hulkoti & Taranath, 2014; R. Singh et al., 2015; Ghosh et al., 2022). Bacterial extracellular polymeric substances (EPS) which are polysaccharides and protein products are also natural capping agents that stabilize nanoparticles following reduction (Tsekhmistrenko et al., 2020). The diverse classes of phytochemicals and microbial biomolecules involved in nanoparticle biosynthesis, along with their mechanistic roles, are systematically summarized as given in Table 1.

*Table 1. Mechanistic roles of plant-derived phytochemicals and bacterial biomolecules in the green synthesis of MNPs, highlighting their potential role as reducing and stabilizing (capping) agents*

| Compound Class | Source | Primary Role | Mechanism | References |
|---|---|---|---|---|
| Polyphenols | Plants | Reducing + capping | –OH → –C=O oxidation; electrostatic stabilization | (Deka et al., 2022) |
| Flavonoids | Plants | Reducing + capping | Enol–keto tautomerism; electron donation | (Deka et al., 2022; Olędzka & Czerwińska, 2023; Tumilaar et al., 2024) |
| Terpenoids | Plants | Reducing + capping | Hydroxyl group oxidation to carbonyl | (Deka et al., 2022; Olędzka & Czerwińska, 2023; Tumilaar et al., 2024) |
| Alkaloids | Plants | Reducing + capping | Electron donation via N-heterocycles; partially characterized | (Deka et al., 2022; Olędzka & Czerwińska, 2023; Tumilaar et al., 2024) |
| NADH-dependent reductases | Bacteria | Primary reducing | NADH → $NAD^+$ oxidation; electron transfer to metal ions | (Boros et al., 2020) |
| Nitrate reductase | Bacteria | Primary reducing + stabilizing | $NO_3^-$ → $NO_2^-$ electron shuttle; $Ag^+/Au^{3+} \rightarrow M^0$ | (Boros et al., 2020) |
| Hydrogenases / Cytochromes | Bacteria | Secondary reducing | Redox-active electron carrier proteins under metal stress | (Boros et al., 2020) |

continued on following page

*Table 1. Continued*

| Compound Class | Source | Primary Role | Mechanism | References |
|---|---|---|---|---|
| Bacteriocins | Bacteria | Capping + stabilizing; nano-conjugate formation | Surface adsorption post-nucleation; SPR peak shift confirmed at 428–453 nm; adds antimicrobial functionality | (Darbandi et al., 2022; Sugrue et al., 2024) |
| EPS / Secreted proteins | Bacteria | Capping + stabilizing | Surface adsorption; prevents aggregation | (Shleeva et al., 2023) |

## 3. BIOLOGICAL PATHWAYS FOR NANOPARTICLE SYNTHESIS

Utilizing the plants/microorganisms as natural reducing and stabilizing agents has paved the way for MNPs synthesis as an alternative, to facilitate the ecofriendly production of MNPs. The major pathways of synthesis, mediated by plants and microbes, are dsicussed in section 3, along with recent developments in the engineering of microbes using the CRISPR system

### 3.1. Plant-Mediated Nanoparticle Synthesis

Plant-mediated nanoparticles synthesis is green and environmentally friendly where the plant extracts act as reducing and stabilizing agents and no additional chemicals are required. The synthesis process is usually divided into three steps: activation, during which phytochemicals such as phenolic acids reduce metal ions to their zero-valent state in order to initiate nucleation; growth, which involves the addition of atoms and LaMer burst nucleation as well as Ostwald ripening that depends on the concentration of extracts; and termination, when capping biomolecules stabilize nanoparticles by preventing further growth (Khan et al., 2022; Monica Ahmad et al., 2024; H. Singh et al., 2023). The most important phytochemicals are polyphenols, flavonoids, and terpenoids that have a critical impact on the size of nanoparticles, their stability, and biological activity, with the chemical structures (e.g., ortho-substituted hydroxyl groups) associated with small and stable particles (Monica Ahmad et al., 2024; H. Singh et al., 2023). The morphology and uniformity of nanoparticles are highly sensitive to the process parameters, including pH, temperatures, concentration of reactants, and reaction duration; in other words, higher temperatures favor rapid reduction and smaller sizes whereas lower temperatures lead to polydispersity (Khan et al., 2022; Monica Ahmad et al., 2024). Although the benefits include low costs and eco-friendliness, reproducibility is a problem as phytochemical composition is not always consistent across species of plants, seasons, extraction procedures, and location (Adeyemi et al., 2022; Khan et al., 2022; Monica Ahmad et al., 2024). This is a challenge to standardization and mass-scale manufacture of nanoparticles with homogenous properties (Adeyemi et al., 2022; Khan et al., 2022). A clear difference between plant-mediated nanoparticle synthesis and microbial enzymatic synthesis routes has been elaborated in Figure 1.

*Figure 1. Plant and microbial based green synthesis routes for metallic nanoparticles. (a) Phytochemicals in plants are responsible for reduction of metal ions, nucleation and growth of nanoparticles, and capping with biomolecules to stabilize nanoparticles in plant-mediated synthesis. (b) Microbial mediated synthesis is carried out by the enzymes which are produced outside the cell and inside the cell and reduce metal ions to nanoparticles*

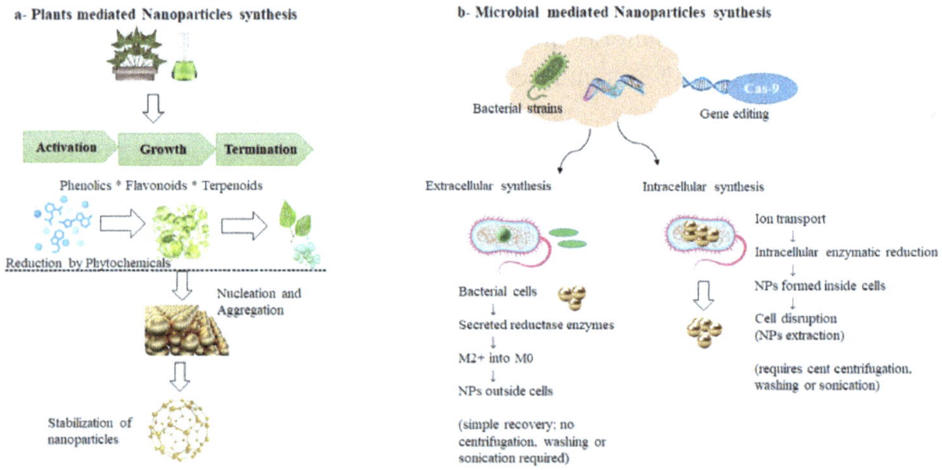


## 3.2. Microbial Systems for Nanoparticle Biosynthesis

### 3.2.1. Bacterial Synthesis Platforms

There are two essentially different pathways by which bacteria generate MNPs. Extracellular synthesis is more widely used since the reaction is easy, quick, and nanoparticle extraction does not involve other downstream processing measures like centrifugation, washing, or sonication, rendering intracellular synthesis relatively time-consuming and tedious (Yuksekdag et al., 2025) (figure 1). The movement of metal ions across bacterial cells in the presence of intracellular enzymes is called intracellular synthesis, and is therefore slower than extracellular routes; it has been shown in a variety of species, such as *Bacillus subtilis, Acinetobacter calcoaceticus,*

*Lactobacillus rhamnosus* and *Pseudomonas aeruginosa* (Alfryyan et al., 2022). *Bacillus licheniformis, Pseudomonas aeruginosa* and *Lactobacillus* spp. are some of the most widely characterized bacterial platforms, and *P. aeruginosa* can survive at high levels of heavy metals, and Lactobacillus strains have been shown to biosynthesize Ag and AuNPs extracellularly through postbiotic metabolite activity (Al-asbahi et al., 2024; Kalimuthu et al., 2008; S. Singh et al., 2014; Z. Wang et al., 2024).

### 3.2.2. CRISPR-Based Microbial Systems

Cas9 and Cas12a, the CRISPR-Cas systems have gained popularity in bacterial metabolic engineering owing to their accuracy in genome editing, gene regulation, and multiplex pathway coordination. These are accomplish for specific deletions, insertions, and promoter engineering to redirect metabolic fluxes to increase biosynthetic productivity and product yields in industrially relevant bacteria including B*acillus, Escherichia coli, Lactobacillus, Pseudomonas,* and *Corynebacterium* (Cho et al., 2018; Liu et al., 2020; Mougiakos et al., 2018). CRISPR-Cas9 has been applied in MNPs biosynthesis to upregulate reductase genes which enhance electron transfer capacity and eliminate competing pathways. Metal-binding protein-expressing engineered *E. coli* strains demonstrate much greater metal sorption and reduction performances in comparison to wild-type strains (Cho et al., 2018). For instance, Ijaz and Javed (2026) were able to show that enhanced antifungal activity of the CRISPR modified *Aspergillus niger* biosynthesis AgNPs with size (8 nm) much smaller than parent strain (14 nm). The engineered strain biosynthesized AgNPs having lower MIC (18.75 vs. 37.2 μg/mL), which is an indication of the potential of CRISPR engineering in optimizing the yield and activity of MNPs (Ijaz & Javed, 2026). The CRISPR-AuNP platform showed high gene-editing efficiency and cell viability, outperforming traditional non-viral delivery methods (Gottimukkala et al., 2026). Similarly, the development of synthetic biology has recent evidence of genetically modified microorganisms to enhance the uptake and metabolism of selenium to form SeNPs, which could facilitate more effective, reproducible, and scalable synthesis of nanoparticles (Muhammad et al., 2025). Recent innovations in CRISPR technologies encompass transcriptional activation systems (CRISPRa) to enable the simultaneous regulation of multiple genes in bacterial cells such as *Pseudomonas putida*, broadening the metabolic engineering approaches (Kiattisewee et al., 2021). To ensure the reproducibility and compliance with regulatory standards, whole-genome resequencing and metabolomics are essential to verify engineered strains by identifying on-target modifications and detecting modifications in the metabolism (Ding et al., 2020; Yao et al., 2018). A wide variety of plants and bacterial species have been employed for nanoparticle biosynthesis with diverse functional outcomes as given in table 2.

*Table 2. Representative examples of plant and bacterial systems used for green synthesis of MNPs, including associated biomolecular reducing agents and biological activities*

| Source Type | Organism | MNPs | Reducing Agents | Activity | Reference |
|---|---|---|---|---|---|
| Plants | *Ocimum sanctum* | AgNPs | Quercetin, Eugenol | Antimicrobial, anticancer | (N. Ahmad et al., 2024; Jain & Mehata, 2017) |
| | *Camellia sinensis* | AuNPs | Catechins, Polyphenols | Antioxidant | (Sliem, 2024; Zhu et al., 2021) |
| | *Tridax procumbens* | AgNPs | Terpenoids | Antimicrobial | (Gul et al., 2021) |
| | *Azadirachta indica* | AgNPs | Flavonoids | Antimicrobial | (I. Ahmad et al., 2018; Ahmed et al., 2016) |
| | *Moringa oleifera* | AgNPs | Phenolics | Antioxidant | (I. Ahmad et al., 2018) |
| | *Aloe vera* | AuNPs | Polysaccharides | Wound healing | (Chandran et al., 2006) |
| | *Eucalyptus globulus* | AgNPs | Polyphenols | Antimicrobial | (K. Ali et al., 2015) |
| | *Citrus sinensis* | AgNPs | Ascorbic acid | Antioxidant | (Sliem, 2024) |
| | *Zingiber officinale* | AuNPs | Gingerols | Anticancer | (Mittal et al., 2013) |
| | *Curcuma longa* | AgNPs | Curcumin | Antimicrobial | (D. Singh et al., 2014, 2016) |
| Bacteria | *Bacillus licheniformis, Lactobacillus spp.* | AgNPs | Nitrate reductase | Antimicrobial | (Hamouda et al., 2019) |
| | *Pseudomonas aeruginosa* | AuNPs | NADH reductase, EPS | Anticancer | (Mohammed & Al-Megrin, 2021) |
| | *Streptomyces spp.* | SeNPs | Reductases | Antioxidant | (Hemlata et al., 2020) |
| | *Escherichia coli* | AgNPs | Enzymatic | Antimicrobial | (Kalishwaralal et al., 2010) |
| | *Klebsiella pneumonia* | AgNPs | Reductase | Antimicrobial | (Mohammed Fayaz et al., 2009) |
| | *Bacillus subtilis* | AuNPs | Nitrate reductase | Catalytic | (Yang et al., 2021; X. Zhang et al., 2024) |
| | *Rhodococcus spp.* | AuNPs | Enzymes | Bioremediation | (A. Ahmad et al., 2003) |
| | *Enterobacter cloacae* | AgNPs | Nitrate reductase | Antimicrobial | (Shahverdi et al., 2007) |
| | *Acinetobacter spp.* | AgNPs | Enzymatic | Antimicrobial | (Mukherjee et al., 2013) |

# 4. PHYSICOCHEMICAL CHARACTERIZATION OF BIOGENIC NANOPARTICLES

## 4.1. UV-Visible Spectroscopy

Surface plasmon resonance (SPR) monitoring is one of the most commonly adopted methods to monitor the formation and stability of biogenic nanoparticles by UV–Visible (UV–Vis) spectroscopy. It offers quick information of the optical properties, relative particle size distribution, concentration, aggregation, and surface modification (Mobarak et al., 2026). The characteristic SPR peaks were found at 410–450 nm, indicating the successful synthesis and colloidal stability of the AgNPs in both plant- and bacteria-mediated methods (Alsamhary, 2020; Flieger et al., 2021). AuNPs exhibit SPR absorption between 520-550 nm, where peak shifts signify changes in AuNP size or aggregation state (Serdar, 2024). The optical properties of SeNPs are more dictated by allotropic form and structuring than by size itself, as they do not have a defined SPR peak, but rather a broad absorption at 265–300 nm from the electron transitions of Se–Se bond, characteristic of amorphous selenium (Lian et al., 2019). UV–Vis is not directly able to provide structural information of particles, such as particle size distribution, morphology, and surface chemistry, and should be complemented with XRD, TEM, or FTIR to obtain a complete structural characterization, especially for SeNPs (Gui & Jin, 2025). Although UV-Vis spectroscopy is a powerful tool to track the course of nanoparticles synthesis and assess the stability of the colloidal suspension, it does not give direct data on the particle size distribution, morphology, and surface chemistry, and thus, complementary methods, like TEM or FTIR, are required to fully characterize nanoparticles (Alsamhary, 2020; Lian et al., 2019).

## 4.2. X-Ray Diffraction (XRD)

XRD is used to characterize the crystalline structure, phase composition and average crystallite size by diffraction patterns (Mobarak et al., 2026). Biogenic AgNPs typically show peaks at 38.17°, 44.31°, 64.44°, and 77.34° (2θ), corresponding to the (111), (200), (220), and (311) planes of a face-centered cubic structure, consistent across synthesis methods (Al-Mahmud et al., 2024). Both the Williamson-Hall and the Size-Strain Plot methods are used to separate the crystallite size from the lattice strain (Khorsand Zak & Hashim, 2025; Mustapha et al., 2019); while the Rietveld refinement method confirms high crystallinity of AgNPs with an average size of ~32 nm (Al-Mahmud et al., 2024; Uvarov & Popov, 2013). XRD also distinguishes between amorphous (red) and crystalline trigonal (grey) phases of SeNPs: amorphous SeNPs exhibit a broad diffuse peak at 20-30° (2θ) with no Bragg reflections (no long-range atomic order), while trigonal SeNPs show sharp diffraction peaks, which are an indication of higher crystallinity, larger crystallite domains and lower lattice strain. Biosynthesized SeNPs are commonly found in their amorphous red form, which can transform to the more thermodynamically stable trigonal grey form after thermal treatment, ageing and post-synthesis processing, thus affecting crystallinity, optical properties and stability (Naveenkumar et al., 2024; Ferroudj et al., 2026).

## 4.3. Fourier-Transform Infrared Spectroscopy (FTIR)

FTIR can be used for the identification of functional groups and biomolecules on the surface of nanoparticles that thus confirm the role of phytochemicals or other agents in the reduction and capping of nanoparticles (Mobarak et al., 2026). The functional capping groups of biogenic nanoparticles (AuNPs, AgNPs and SeNPs) are identified by FTIR playing important role in formation and stabilization of the nanoparticle. The most common FTIR bands include broad peaks at 3385 to 3432 $cm^{-1}$ due to either O-H or N-H bond stretching, 1603 to 1650 $cm^{-1}$ due to C-O carbonyl stretching of carboxylic acids or amide bonds, and 1416 to 1441 $cm^{-1}$ due to C-N stretching of aromatic amines, which all confirm the presence of phytochemicals or proteins as capping agents (Bindhu & Umadevi, 2014; Gul et al., 2021; Pasieczna-Patkowska et al., 2025). The changes in the plant extracts and nanoparticle spectra of the peaks indicate that polyphenols, flavonoids, and proteins are involved in the reduction and binding of metal ions during biosynthesis, the amine and carboxylate groups are important reducing and stabilizing agents (Gul et al., 2021; Hulkoti & Taranath, 2014; Pasieczna-Patkowska et al., 2025). However, polyphenols, flavonoids, and proteins, especially the amine and carboxylate groups, are the major reducing and stabilizing agents involved in plant- and bacteria-mediated synthesis as indicated by peak shift in both the plant extract and the nanoparticles spectra (Hulkoti & Taranath, 2014).

## 4.4. Scanning/ Transmission Electron Microscopy (SEM/ TEM)

Scanning electron microscopy and transmission electron microscopy are important techniques for the study of the morphology and size of nanoparticles. TEM offers high resolution visualization of internal structure, crystallinity, particle size and shape, as well as the ability to examine lattice fringes and interplanar spacing (Mourdikoudis et al., 2018). SEM yields detailed images of surface morphology, texture, aggregation and particle distribution (Goldstein et al., 2018). Unlike the UV–Vis and XRD, electron microscopy images the nanoparticles itself, which means that structural properties are not deduced. TEM is well suited to nanoparticles<50 nm and suitable to surface characterization, while SEM is suitable to larger particles. SEM, in combination with energy-dispersive X-ray spectroscopy (EDX), also provides elemental composition. TEM and SEM are used in conjunction with XRD and DLS for full characterization of nanoparticles (Eaton et al., 2017; Huber et al., 2023; Jagadeesh et al., 2024; Mourdikoudis et al., 2018; Patil et al., 2022; Smirnov et al., 2023), since each of these methods has its own advantages.

## 4.5. Energy-Dispersive X-Ray Spectroscopy (EDX)

Elemental composition and purity of nanoparticles are usually analyzed by the energy-dispersive X-ray spectroscopy method, or EDX. Specific peaks of characteristic X-ray emission are used to identify the elements present including Ag Lα (~2.98 keV), Au Mα (~2.12 keV) and Se Lα (~1.37 keV). Combined with SEM or TEM, EDX can be used to map the distribution of elements at the spatial resolution of the imaging electron beam, thus revealing the elemental heterogeneity of the particles and their aggregation that is not possible by bulk analytical techniques. Carbon and oxygen signals tend to be signals from biomolecular capping layers, whether from plants or microbial metabolites, rather than contamination. Advanced STEM–EDX tomography is another tool which enables three dimensional elemental mapping of bimetallic nanoparticles with an eyesight into the internal composition and structure of the nanoparticles. The qualitative and semi-quantitative elemental analysis is possible by EDX with a spatial resolution of about 10 nm, but it is best used in combination with high-resolution imaging techniques like SEM or TEM (Md. H. Ali et al., 2023; Slater et al., 2016; Ferroudj et al., 2026).

## 4.6. Dynamic Light Scattering (DLS)

Dynamic light scattering (DLS) is a fast, non-destructive method for measuring nanoparticle hydrodynamic diameter, particle size distribution and polydispersity index (PDI), as well as colloidal stability in suspension. DLS is different from TEM as it measures the hydrodynamic diameter that encompasses the core of the particle and the solvent layer and surface-associated biomolecules. Thus, the particle size reported in DLS is usually larger than that reported by TEM. DLS is commonly employed to assess the stability of the suspensions, aggregation state and dispersion of green-synthesised AgNPs, AuNPs and SeNPs and is used in conjunction with zeta potential analysis. Controlled-synthesized Optimized SeNPs exhibited hydrodynamic diameter of 137–172 nm and low PDI values of 0.049–0.306, which suggested decent colloidal stability. The current use of DLS for data analysis of nanoparticles is, however, limited to the assumption of sphericity of the particles and is highly sensitive toward aggregation, and thus should not be used as a sole tool for assessing the morphology and size distribution of nanoparticles (Joudeh & Linke, 2022; Mourdikoudis et al., 2018; Oliveira et al., 2026).

As no single analytical methods can completely characterize biogenic nanoparticles, therefore, an integrating method including UV-Vis, SEM/TEM, EDX, FRTIR and DLS, can evaluate the optical, morphological, elemental, structural, and colloidal properties of biosynthesized MNPs. The multi-technique approach is consistent with ISO/TC 229 requirements and FDA guidance, which suggest orthogonal characterization of biosynthesized MNPs prior to biomedical or industrial applications. Complementary analytical approaches are also needed to include the study of protein corona formation and biological identity for green-synthesized nanoparticles, thus reinforcing the necessity of complementary bioanalytical approaches (Suárez Priede et al., 2026).

## 5. BIOLOGICAL ACTIVITIES AND ONCOLOGICAL APPLICATIONS

The green synthesized metallic nanoparticles such as AgNPs, AuNPs and SeNPs are found to have wide spectrum antimicrobial, antioxidant and anticancer activity and are good candidates for precision oncology. The antimicrobial activity of AgNPs is mainly attributed to the release of Ag+ ions, which disrupt bacterial membranes and increase their permeability, the production of Reactive Oxygen Species (ROS), interference with bacterial DNA replication, and by enhancing the activity of conventional antibiotics against resistant pathogens (Dakal et al., 2016; More et al., 2023; Rodrigues et al., 2024). The antioxidant properties are thought to be due to the presence of phytochemicals on the surface, while the anticancer properties are attributed to the ability to induce ROS-mediated mitochondrial dysfunction, cell-cycle arrest and apoptosis through p53- and caspase-dependent pathways (Lee & Jun, 2019; Xu et al., 2020). AuNPs are selective anticancer agents inducing apoptosis, ROS, and angiogenesis inhibition; under near-infrared radiation (~808 nm), AuNPs change light to heat to selectively ablate tumors, and PEG functionalization and targeting ligands enhance the circulation time and receptor-mediated endocytosis of the nanoparticle (Panzarini et al., 2018). SeNPs have a dual action in the treatment and antioxidant defense of cancer based on their doses. The higher doses of SeNPs cause cell death in variant cancer cell lines, i.e. breast cancer (MCF-7), lung adenocarcinoma (A549), and melanoma via production of ROS, caspase pathway, mitochondrial dysfunction, and cell cycle arrest, which causes intrinsic apoptosis (B. A. Ali et al., 2024; Ameen et al., 2025; Chen et al., 2008). As shown in Table 3, biosynthesized AgNPs, AuNPs, and SeNPs have shown promising in vitro anticancer activity against various human cancer cell lines. $IC_{50}$ values reported depend on the type of nanoparticles, their method of preparation, their physicochemical properties, the type of cancer cells, and the experimental conditions.

*Table 3. Reported in vitro $IC_{50}$ values of biosynthesized AgNPs, AuNPs, and SeNPs against cancer and normal cell lines, with corresponding selectivity indices (SI). (- = not reported)*

| Nanoparticle | Cancer cell line | $IC_{50}$ (Cancer) | $IC_{50}$ (Normal) | Selectivity Index (SI) | Reference |
|---|---|---|---|---|---|
| **AgNPs** | MCF-7 | 3.04 µg/mL | - | - | (Firdhouse, J., & Lalith, 2015) |
| | Hep-2 | 3.42 µg/mL | - | - | (Satyavani et al., 2011) |
| | HepG2 | 10.02 µg/mL | - | - | (Patra et al., 2015) |
| **AuNPs** | HeLa | 100 µg/mL | - | - | (Baharara et al., 2016) |
| | MCF-7 | >80 nM | < 60 nM (HEK-293) | - | (Arunkumar et al., 2013) |
| | U87 | 2 ng/mL | - | - | (Mishra et al., 2016) |
| **SeNPs** | HepG2 | 70.79 µg/mL | 165.5 ± 5.4 µg/mL (WI−38) | ~2.34 | (Fouda et al., 2022) |
| | A549 | 80 µg/mL (24h) | NR | - | (Menon & Shanmugam, 2020) |
| | HeLa | 5.5 µg/mL | - | - | (S. Pandey et al., 2021) |

In addition to their natural bioactivity, green synthesized nanoparticles have proved to be multifunctional nanoplatforms for targeted cancer therapy and theranostics. By using antibodies, peptides, aptamers, hyaluronic acid, folic acid or other targeting ligands, receptor-mediated delivery has been achieved that facilitates the accumulation in tumors, internalization by tumor cells and therapeutic efficacy, while reducing the toxicity to non-target cells (Deng et al., 2025). AuNPs have shown potential as drug delivery vectors, molecular imaging agents, tumor-targeting, and photothermal therapy (PGT) agents, while AgNPs exhibit selective cytotoxicity, ROS mediated apoptosis, and promise as biomarker-guided drug delivery systems (Ali Dheyab et al., 2025; Hamzah, 2026). The functionalized SeNPs also hold great promise for precision oncology by delivering targeted cytotoxicity, drug delivery, modulation of oxidative stress, apoptosis/autophagy, miRNA-mediated signaling, and increased chemosensitivity (Anjum et al., 2025). Multifunctional theranostic nanoplatforms, combining detection and therapeutic delivery with efficacy monitoring, have emerged as a result of the combination of molecular targeting with diagnostic imaging and therapy, highlighting the potential of green-synthesized MNPs in precision oncology, which can deliver minimal systemic toxicity. These mechanisms of ROS generation, disruption of membranes, mitochondrial apoptosis, photothermal effects, and targeted drug delivery are integrated in the three different AgNPs, AuNPs, and SeNPs shown in Figure 2.

*Figure 2. Biological activities and therapeutic mechanisms of the AgNPs, AuNPs and SeNPs synthesized by using green methods. (a) AgNPs exhibit antimicrobial activity via the release of $Ag^+$ ions, disruption of membranes, generation of ROS, damage to the DNA and caspase-mediated apoptosis, and polyphenol capping impart antioxidant activity. (b) Functionalized AuNPs could enable selective delivery, ROS-induced cell apoptosis, anti-angiogenesis and NIR-induced photothermal therapy of tumor destruction. (c) SeNPs show dose-dependent effects with antioxidant and cytoprotective effects at low doses, and ROS overproduction, mitochondrial dys-function, cell-cycle arrest and apoptosis at high doses, in cancer cells*

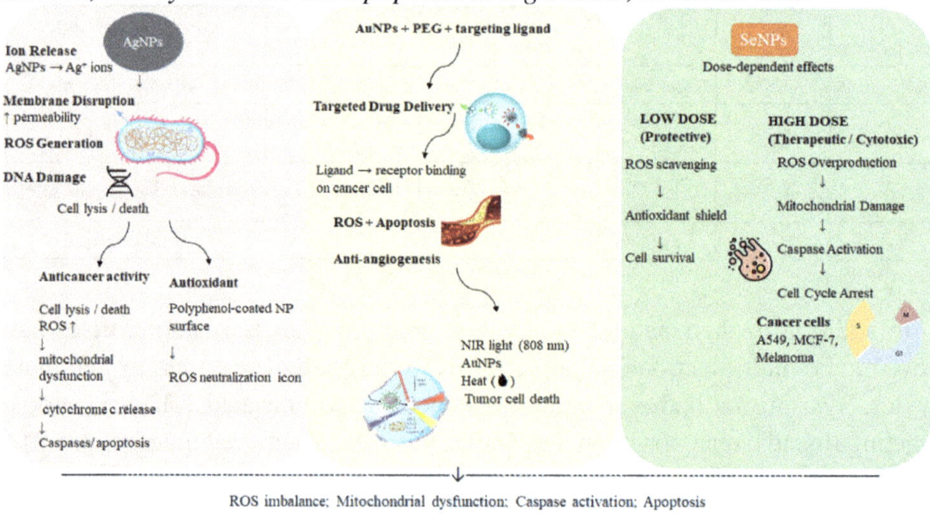


## 6. GREEN CHEMISTRY COMPLIANCE, SCALABILITY AND SUSTAINABILITY

Green synthesis of nanoparticles (NPs) with plants and microorganisms is an eco-friendly alternative to conventional chemical route, as it minimizes the use of hazardous chemicals, non-recyclable wastes and energy consumption, and promotes the circular economy. For industrial application, however, standardized sustainability indicators, integration of biorefinery elements and scalable production systems must be considered. Typical tools for assessing the sustainability of nanoparticle synthesis include the E-factor, atom economy (AE), process mass intensity (PMI), and life cycle assessment (LCA). The E-factor is a way to quantify the amount of waste that is produced for each unit of product, while AE is a measure of the theoretical efficiency of the conversion of the reactants into the desired product. Generally, the

E-factors produced in green synthesis are lower as the biomolecules from plants/microbes serve as both the reducing agents and capping agents, eliminating the need for toxic chemicals (García-Quintero & Palencia, 2021; Gupta et al., 2023; Saxena et al., 2025). While process improvement is beneficial through the use of PMI, it does not encompass wider environmental considerations. Thus, LCA is crucial to evaluate resource utilization, energy consumption, toxicity and environmental performance. However, the production of biomass and energy sources play a key role in sustainability, therefore, an integrated assessment is needed (García-Quintero & Palencia, 2021; Patiño-Ruiz et al., 2021; Rodríguez-Rojas et al., 2024).

Moreover, the integration of biorefinery enables the use of agricultural wastes like fruit peels, leaves and bagasse as renewable resources for polyphenols and flavonoids production for nanoparticle synthesis, further increasing the sustainability of the processes. The use of co-production of nanoparticles, biofuels, pectin and other value added products by the zero-waste biorefineries, and the replacement of organic solvents with aqueous systems, further reduces environment impacts (Dheyab et al., 2025; Saratale et al., 2025). While these benefits are true, there are still hurdles for commercialization, including species, seasonal, and extraction variability from plants; and the need for optimized fed-batch or continuous bioreactors and contamination control for microbial production. While energy savings can be estimated to be around 30% and costs 40%, these are process-specific and need to be confirmed by detailed LCAs. The most critical challenges for large-scale production are reproducibility, process control, raw material availability and environmental variability (Gacem & Abd-Elsalam, 2022; Ortiz-Tirado et al., 2025).

## 7. CLINICAL TRANSLATION AND REGULATORY ISSUES

Globally, regulatory frameworks of nano-medicines are still fragmented and no dedicated pathways are currently available to the regulation of nanoparticle-based products, but they are instead regulated through pre-existing routes, including drugs, biologics, or combination products. The critical quality attribute (CQA) such as the particle size, morphology, and drug release profiles should be well controlled, yet it is difficult to standardize them, particularly in the case of biogenic nanoparticles because of their variability in biological sources and production techniques, which requires genomic validation and certified reference materials under GMP/GLP schemes. The majority of translational breakages are due to the use of in vitro information with minimal in vivo pharmacokinetics, bio distribution, toxicity and long-term safety data, which makes clinical translation more challenging. The GMP-scale production is challenging as even a little batch-to-batch change in nanoparticle size (~10 nm) can have a considerable influence on the biological behavior and clinical outcome.

The absence of coherent regulatory advice and standardized manufacturing can only slow down the clinical implementation of nano-medicines despite the therapeutic potential. Researchers, industry participants, and regulatory bodies should closely work together to come up with consistent protocols that guarantee reproducibility, safety, and efficacy to achieve successful clinical translation (Desai et al., 2025; Đorđević et al., 2022; Dri et al., 2023).

## Conclusion and Future Prospective

The use of green nanotechnology as a sustainable alternative to the traditional synthesis of nanoparticles in a way that reduces the use of harmful chemicals, energy consumption and toxicity. Eco-friendly reducing and stabilizing agents such as plants and microorganisms synthesize biocompatible AgNPs, AuNPs and SeNPs with antimicrobial, antioxidant and anticancer properties by inducing ROS generation, membrane disruptive, inducing apoptosis, and targeted drug delivery properties. In addition to their bioactivity, these nanoparticles are also promising for drug delivery, biosensing, diagnosis imaging, and theranostic applications. But their in vivo validation, pharmacokinetic and biodistribution studies, and long-term safety studies are still not sufficiently performed. Another limitation is the biological variation in the sources of the biological precursors and the synthesis conditions, which result in varying properties of nanoparticles and poor reproducibility. Different characterization protocols and lack of comprehensive life cycle assessments (LCAs), also hinder cross-study comparison and validation of sustainability claims. Future studies should focus on standardized synthesis and characterization procedure, optimization of the procedure with the aid of AI, metabolic engineering with the aid of CRISPR and scaling up the production in bioreactors, in order to improve the reproducibility and industrial feasibility. The safe, sustainable, and successful translation of green nanotechnology will require systematic in vivo studies, long-term in vivo toxicological evaluations, pharmacokinetic studies, and nanotechnology-specific regulations, among other things.

## REFERENCES


Adeyemi, J. O., Oriola, A. O., Onwudiwe, D. C., & Oyedeji, A. O. (2022). Plant Extracts Mediated Metal-Based Nanoparticles: Synthesis and Biological Applications. *Biomolecules, 12*(5), 627. DOI: 10.3390/biom12050627 PMID: 35625555

Ahmad, A., Mukherjee, P., Senapati, S., Mandal, D., Khan, M. I., Kumar, R., & Sastry, M. (2003). Extracellular biosynthesis of silver nanoparticles using the fungus Fusarium oxysporum. *Colloids and Surfaces. B, Biointerfaces*, *28*(4), 313–318. DOI: 10.1016/S0927-7765(02)00174-1

Ahmad, I., Mobeen, M. F., Sheraz, M. A., Ahmed, S., Anwar, Z., Shaikh, R. S., Hussain, I., & Ali, S. M. (2018). Photochemical interaction of ascorbic acid and nicotinamide in aqueous solution: A kinetic study. *Journal of Photochemistry and Photobiology. B, Biology*, *182*, 115–121. DOI: 10.1016/j.jphotobiol.2018.04.011 PMID: 29660587

Ahmad, N., Ansari, M. A., Al-Mahmeed, A., Joji, R. M., Saeed, N. K., & Shahid, M. (2024). Biogenic silver nanomaterials synthesized from Ocimum sanctum leaf extract exhibiting robust antimicrobial and anticancer activities: Exploring the therapeutic potential. *Heliyon*, *10*(15), e35486. DOI: 10.1016/j.heliyon.2024.e35486 PMID: 39170333

Ahmed, S., Ahmad, M., Swami, B. L., & Ikram, S. (2016). A review on plants extract mediated synthesis of silver nanoparticles for antimicrobial applications: A green expertise. *Journal of Advanced Research*, *7*(1), 17–28. DOI: 10.1016/j.jare.2015.02.007 PMID: 26843966

Al-asbahi, M. G. S. S., Al-Ofiry, B. A., Saad, F. A. A., Alnehia, A., & Al-Gunaid, M. Q. A. (2024). Silver nanoparticles biosynthesis using mixture of Lactobacillus sp. And Bacillus sp. Growth and their antibacterial activity. *Scientific Reports*, *14*(1), 10224. DOI: 10.1038/s41598-024-59936-1 PMID: 38702368

Al-Mahmud, M., Hossain Shishir, M. K., Ahmed, S., Tabassum, S., Islam Sadia, S., Haque Sachchu, M. M., Tanbin Tama, R., Rahim Miah, A., & Ashraful Alam, M. (2024). Stoichiometry crystallographic phase analysis and crystallinity integration of silver nanoparticles: A Rietveld refinement study. *Journal of Crystal Growth*, *643*, 127815. DOI: 10.1016/j.jcrysgro.2024.127815

Alfryyan, N., Kordy, M. G. M., Abdel-Gabbar, M., Soliman, H. A., & Shaban, M. (2022). Characterization of the biosynthesized intracellular and extracellular plasmonic silver nanoparticles using Bacillus cereus and their catalytic reduction of methylene blue. *Scientific Reports*, *12*(1), 12495. DOI: 10.1038/s41598-022-16029-1 PMID: 35864132

Ali, B. A., Allam, R. M., Hasanin, M. S., & Hassabo, A. A. (2024). Biosynthesis of selenium nanoparticles as a potential therapeutic agent in breast cancer: G2/M arrest and apoptosis induction. *Toxicology Reports*, *13*, 101792. DOI: 10.1016/j.toxrep.2024.101792 PMID: 39554610

Ali, K., Ahmed, B., Dwivedi, S., Saquib, Q., Al-Khedhairy, A. A., & Musarrat, J. (2015). Microwave Accelerated Green Synthesis of Stable Silver Nanoparticles with Eucalyptus globulus Leaf Extract and Their Antibacterial and Antibiofilm Activity on Clinical Isolates. *PLoS One*, *10*(7), e0131178. DOI: 10.1371/journal.pone.0131178 PMID: 26132199

Ali, M., Azad, M. A. K., Khan, K. A., Rahman, M. O., Chakma, U., & Kumer, A. (2023). Analysis of Crystallographic Structures and Properties of Silver Nanoparticles Synthesized Using PKL Extract and Nanoscale Characterization Techniques. *ACS Omega*, *8*(31), 28133–28142. DOI: 10.1021/acsomega.3c01261 PMID: 37576647

Ali Dheyab, M., Tang, J. H., Abdul Aziz, A., Hussein Nowfal, S., Jameel, M. S., Alrosan, M., Oladzadabbasabadi, N., & Ghasemlou, M. (2025). Green synthesis of gold nanoparticles and their emerging applications in cancer imaging and therapy: A review. *Reviews in Inorganic Chemistry*, *45*(4), 663–685. DOI: 10.1515/revic-2024-0048

Alsamhary, K. I. (2020). Eco-friendly synthesis of silver nanoparticles by Bacillus subtilis and their antibacterial activity. *Saudi Journal of Biological Sciences*, *27*(8), 2185–2191. DOI: 10.1016/j.sjbs.2020.04.026 PMID: 32714045

Ameen, F., Almalki, N. S., Alshalan, R., & Sakayanathan, P. (2025). Green Synthesis of Selenium Nanoparticles Utilizing *Drimia indica*: Insights Into Anticancer and Antimicrobial Activities. *Microscopy Research and Technique*, *88*(3), 749–760. DOI: 10.1002/jemt.24726 PMID: 39558605

Anjum, S., Hashim, M., Imran, M., Babur, S., Adnan, S., Hano, C., & Ibrahim, W. N. (2025). Selenium Nanoparticles in Cancer Therapy: Unveiling Cytotoxic Mechanisms and Therapeutic Potential. *Cancer Reports*, *8*(6), e70210. DOI: 10.1002/cnr2.70210 PMID: 40452566

Arunkumar, P., Vedagiri, H., & Premkumar, K. (2013). Rapid bioreduction of trivalent aurum using banana stem powder and its cytotoxicity against MCF-7 and HEK-293 cell lines. *Journal of Nanoparticle Research*, *15*(3), 1481. DOI: 10.1007/s11051-013-1481-0

Baharara, J., Ramezani, T., Divsalar, A., Mousavi, M., & Seyedarabi, A. (2016). *Induction of Apoptosis by Green Synthesized Gold Nanoparticles Through Activation of Caspase-3 and 9 in Human Cervical Cancer Cells.*

Bindhu, M. R., & Umadevi, M. (2014). Silver and gold nanoparticles for sensor and antibacterial applications. *Spectrochimica Acta. Part A, Molecular and Biomolecular Spectroscopy*, *128*, 37–45. DOI: 10.1016/j.saa.2014.02.119 PMID: 24657466

Boros, E., Dyson, P. J., & Gasser, G. (2020). Classification of Metal-Based Drugs according to Their Mechanisms of Action. *Chem*, *6*(1), 41–60. DOI: 10.1016/j.chempr.2019.10.013 PMID: 32864503

Chandrakala, V., Aruna, V., & Angajala, G. (2022). Review on metal nanoparticles as nanocarriers: Current challenges and perspectives in drug delivery systems. *Emergent Materials*, *5*(6), 1593–1615. DOI: 10.1007/s42247-021-00335-x PMID: 35005431

Chandran, S. P., Chaudhary, M., Pasricha, R., Ahmad, A., & Sastry, M. (2006). Synthesis of Gold Nanotriangles and Silver Nanoparticles Using Aloe vera Plant Extract. *Biotechnology Progress*, *22*(2), 577–583. DOI: 10.1021/bp0501423 PMID: 16599579

Chen, T., Wong, Y.-S., Zheng, W., Bai, Y., & Huang, L. (2008). Selenium nanoparticles fabricated in Undaria pinnatifida polysaccharide solutions induce mitochondria-mediated apoptosis in A375 human melanoma cells. *Colloids and Surfaces. B, Biointerfaces*, *67*(1), 26–31. DOI: 10.1016/j.colsurfb.2008.07.010 PMID: 18805679

Cho, S., Shin, J., & Cho, B.-K. (2018). Applications of CRISPR/Cas System to Bacterial Metabolic Engineering. *International Journal of Molecular Sciences*, *19*(4), 1089. DOI: 10.3390/ijms19041089 PMID: 29621180

Chopra, H., Bibi, S., Singh, I., Hasan, M. M., Khan, M. S., Yousafi, Q., Baig, A. A., Rahman, M., Islam, F., Emran, T. B., & Cavalu, S. (2022). Green Metallic Nanoparticles: Biosynthesis to Applications. *Frontiers in Bioengineering and Biotechnology*, *10*, 874742. DOI: 10.3389/fbioe.2022.874742 PMID: 35464722

Dakal, T. C., Kumar, A., Majumdar, R. S., & Yadav, V. (2016). Mechanistic Basis of Antimicrobial Actions of Silver Nanoparticles. *Frontiers in Microbiology*, *7*. Advance online publication. DOI: 10.3389/fmicb.2016.01831 PMID: 27899918

Darbandi, A., Asadi, A., Mahdizade Ari, M., Ohadi, E., Talebi, M., Halaj Zadeh, M., Darb Emamie, A., Ghanavati, R., & Kakanj, M. (2022). Bacteriocins: Properties and potential use as antimicrobials. *Journal of Clinical Laboratory Analysis*, *36*(1), e24093. DOI: 10.1002/jcla.24093 PMID: 34851542

Deka, H., Choudhury, A., & Dey, B. K. (2022). An Overview on Plant Derived Phenolic Compounds and Their Role in Treatment and Management of Diabetes. *Journal of Pharmacopuncture*, *25*(3), 199–208. DOI: 10.3831/KPI.2022.25.3.199 PMID: 36186092

Deng, X., Zhang, Z., Ren, T., & Chen, L. (2025). Regulation of oxidative stress and inflammation caused by drug accumulation in the TME based on EPR-passive strategy and active targeting. *Cancer Nanotechnology*, *16*(1), 40. DOI: 10.1186/s12645-025-00342-1

Desai, N., Rana, D., Patel, M., Bajwa, N., Prasad, R., & Vora, L. K. (2025). Nanoparticle Therapeutics in Clinical Perspective: Classification, Marketed Products, and Regulatory Landscape. *Small*, *21*(29), 2502315. DOI: 10.1002/smll.202502315 PMID: 40454890

Dheyab, M. A., Abdullah, W., Aziz, A. A., Alanezi, S. T., Oladzadabbasabadi, N., Khaniabadi, P. M., Jameel, M. S., Braim, F. S., Al-Jawarneh, M. A., Mohammad, A. I. J. M., Rahman, A. A., & Mehrdel, B. (2025). Mechanistic insights and advances in polyphenol- and flavonoid-mediated sustainable synthesis of gold nanoparticles from agricultural waste: A review. *International Journal of Biological Macromolecules*, *320*, 145978. DOI: 10.1016/j.ijbiomac.2025.145978 PMID: 40669619

Dikshit, P., Kumar, J., Das, A., Sadhu, S., Sharma, S., Singh, S., Gupta, P., & Kim, B. (2021). Green Synthesis of Metallic Nanoparticles: Applications and Limitations. *Catalysts*, *11*(8), 902. DOI: 10.3390/catal11080902

Ding, W., Zhang, Y., & Shi, S. (2020). Development and Application of CRISPR/Cas in Microbial Biotechnology. *Frontiers in Bioengineering and Biotechnology*, *8*, 711. DOI: 10.3389/fbioe.2020.00711 PMID: 32695770

Đorđević, S., Gonzalez, M. M., Conejos-Sánchez, I., Carreira, B., Pozzi, S., Acúrcio, R. C., Satchi-Fainaro, R., Florindo, H. F., & Vicent, M. J. (2022). Current hurdles to the translation of nanomedicines from bench to the clinic. *Drug Delivery and Translational Research*, *12*(3), 500–525. DOI: 10.1007/s13346-021-01024-2 PMID: 34302274

Dri, D. A., Rinaldi, F., Carafa, M., & Marianecci, C. (2023). Nanomedicines and nanocarriers in clinical trials: Surfing through regulatory requirements and physico-chemical critical quality attributes. *Drug Delivery and Translational Research*, *13*(3), 757–769. DOI: 10.1007/s13346-022-01262-y PMID: 36450964

Eaton, P., Quaresma, P., Soares, C., Neves, C., De Almeida, M. P., Pereira, E., & West, P. (2017). A direct comparison of experimental methods to measure dimensions of synthetic nanoparticles. *Ultramicroscopy*, *182*, 179–190. DOI: 10.1016/j.ultramic.2017.07.001 PMID: 28692935

El-Seedi, H. R., El-Shabasy, R. M., Khalifa, S. A. M., Saeed, A., Shah, A., Shah, R., Iftikhar, F. J., Abdel-Daim, M. M., Omri, A., Hajrahand, N. H., Sabir, J. S. M., Zou, X., Halabi, M. F., Sarhan, W., & Guo, W. (2019). Metal nanoparticles fabricated by green chemistry using natural extracts: Biosynthesis, mechanisms, and applications. *RSC Advances*, *9*(42), 24539–24559. DOI: 10.1039/C9RA02225B PMID: 35527869

Ferroudj, A., Csík, A., & Prokisch, J. (2026). Allotrope-dependent physicochemical and optical properties of red and grey selenium nanoparticles. *RSC Advances*, *16*(22), 20391–20398. DOI: 10.1039/D6RA01409G PMID: 42017156

Firdhouse, J., & Lalitha, P. (2015). Apoptotic efficacy of biogenic silver nanoparticles on human breast cancer MCF-7 cell lines. *Progress in Biomaterials*, *4*(2–4), 113–121. DOI: 10.1007/s40204-015-0042-2 PMID: 26566469

Flieger, J., Franus, W., Panek, R., Szymańska-Chargot, M., Flieger, W., Flieger, M., & Kołodziej, P. (2021). Green Synthesis of Silver Nanoparticles Using Natural Extracts with Proven Antioxidant Activity. *Molecules (Basel, Switzerland)*, *26*(16), 4986. DOI: 10.3390/molecules26164986 PMID: 34443574

Fouda, A., Al-Otaibi, W. A., Saber, T., AlMotwaa, S. M., Alshallash, K. S., Elhady, M., Badr, N. F., & Abdel-Rahman, M. A. (2022). Antimicrobial, Antiviral, and In-Vitro Cytotoxicity and Mosquitocidal Activities of Portulaca oleracea-Based Green Synthesis of Selenium Nanoparticles. *Journal of Functional Biomaterials*, *13*(3), 157. DOI: 10.3390/jfb13030157 PMID: 36135592

Gacem, M. A., & Abd-Elsalam, K. A. (2022). Strategies for scaling up of green-synthesized nanomaterials: Challenges and future trends. In *Green Synthesis of Silver Nanomaterials* (pp. 669–698). Elsevier., DOI: 10.1016/B978-0-12-824508-8.00008-3

García-Quintero, A., & Palencia, M. (2021). A critical analysis of environmental sustainability metrics applied to green synthesis of nanomaterials and the assessment of environmental risks associated with the nanotechnology. *The Science of the Total Environment*, *793*, 148524. DOI: 10.1016/j.scitotenv.2021.148524 PMID: 34182452

Gardea-Torresdey, J. L., Gomez, E., Peralta-Videa, J. R., Parsons, J. G., Troiani, H., & Jose-Yacaman, M. (2003). Alfalfa Sprouts: A Natural Source for the Synthesis of Silver Nanoparticles. *Langmuir : The ACS Journal of Surfaces and Colloids, 19*(4), 1357–1361. DOI: 10.1021/la020835i

Ghosh, S., Sarkar, B., Kaushik, A., & Mostafavi, E. (2022). Nanobiotechnological prospects of probiotic microflora: Synthesis, mechanism, and applications. *The Science of the Total Environment, 838*, 156212. DOI: 10.1016/j.scitotenv.2022.156212 PMID: 35623529

Goldstein, J. I., Newbury, D. E., Michael, J. R., Ritchie, N. W. M., Scott, J. H. J., & Joy, D. C. (2018). *Scanning Electron Microscopy and X-Ray Microanalysis*. Springer New York., DOI: 10.1007/978-1-4939-6676-9

Gottimukkala, K. S. V., Lane, D. D., Cunningham, R., Malik, H. S., Jwa, Y., Cassidy, M. E., Castelli, J. M. P., Enstrom, M. R., Poljakov, K., Gastelum, G., Ho, S. H., Tassa, C., & Adair, J. E. (2026). CRISPR-AuNP: Physicochemical optimization of a gold nanoparticle platform for cost-effective and modular non-viral gene editing in HSPCs. *Gene Therapy, 33*(2), 188–202. DOI: 10.1038/s41434-025-00591-0 PMID: 41535609

Gui, R., & Jin, H. (2025). Synthesis, characterizations, properties, modifications and applications of zero-dimensional monoelemental selenium nanomaterials. *Coordination Chemistry Reviews, 535*, 216626. DOI: 10.1016/j.ccr.2025.216626

Gul, A., Fozia, , Shaheen, A., Ahmad, I., Khattak, B., Ahmad, M., Ullah, R., Bari, A., Ali, S. S., Alobaid, A., Asmari, M. M., & Mahmood, H. M. (2021). Green Synthesis, Characterization, Enzyme Inhibition, Antimicrobial Potential, and Cytotoxic Activity of Plant Mediated Silver Nanoparticle Using Ricinus communis Leaf and Root Extracts. *Biomolecules, 11*(2), 206. DOI: 10.3390/biom11020206 PMID: 33540690

Gupta, D., Boora, A., Thakur, A., & Gupta, T. K. (2023). Green and sustainable synthesis of nanomaterials: Recent advancements and limitations. *Environmental Research, 231*, 116316. DOI: 10.1016/j.envres.2023.116316 PMID: 37270084

Haleem, A., Javaid, M., Singh, R. P., Rab, S., & Suman, R. (2023). Applications of nanotechnology in medical field: A brief review. *Global Health Journal (Amsterdam, Netherlands), 7*(2), 70–77. DOI: 10.1016/j.glohj.2023.02.008

Hamouda, R. A., Hussein, M. H., Abo-elmagd, R. A., & Bawazir, S. S. (2019). Synthesis and biological characterization of silver nanoparticles derived from the cyanobacterium Oscillatoria limnetica. *Scientific Reports, 9*(1), 13071. DOI: 10.1038/s41598-019-49444-y PMID: 31506473

Hamzah, H. (2026). Targeted therapy and biomarker-guided applications of ecofriendly silver nanoparticles in precision oncology. *Journal of Pharmacy & Pharmaceutical Sciences*, *28*, 15403. DOI: 10.3389/jpps.2025.15403 PMID: 41608666

Hemlata, M., Meena, P. R., Singh, A. P., & Tejavath, K. K. (2020). Biosynthesis of Silver Nanoparticles Using *Cucumis prophetarum* Aqueous Leaf Extract and Their Antibacterial and Antiproliferative Activity Against Cancer Cell Lines. *ACS Omega*, *5*(10), 5520–5528. DOI: 10.1021/acsomega.0c00155 PMID: 32201844

Huber, M. J., Ivleva, N. P., Booth, A. M., Beer, I., Bianchi, I., Drexel, R., Geiss, O., Mehn, D., Meier, F., Molska, A., Parot, J., Sørensen, L., Vella, G., Prina-Mello, A., Vogel, R., & Caputo, F. (2023). Physicochemical characterization and quantification of nanoplastics: Applicability, limitations and complementarity of batch and fractionation methods. *Analytical and Bioanalytical Chemistry*, *415*(15), 3007–3031. DOI: 10.1007/s00216-023-04689-5 PMID: 37106123

Hulkoti, N. I., & Taranath, T. C. (2014a). Biosynthesis of nanoparticles using microbes—A review. *Colloids and Surfaces. B, Biointerfaces*, *121*, 474–483. DOI: 10.1016/j.colsurfb.2014.05.027 PMID: 25001188

Hulkoti, N. I., & Taranath, T. C. (2014b). Biosynthesis of nanoparticles using microbes—A review. *Colloids and Surfaces. B, Biointerfaces*, *121*, 474–483. DOI: 10.1016/j.colsurfb.2014.05.027 PMID: 25001188

Huston, M., DeBella, M., DiBella, M., & Gupta, A. (2021). Green Synthesis of Nanomaterials. *Nanomaterials (Basel, Switzerland)*, *11*(8), 2130. DOI: 10.3390/nano11082130 PMID: 34443960

Ijaz, A., & Javed, M. R. (2026). Synthesis, characterization, antifungal and SERS analysis of mycosynthesized silver nanoparticles against Aspergillus parasiticus. *Process Biochemistry (Barking, London, England)*, *162*, 170–182. DOI: 10.1016/j.procbio.2026.01.017

Iram, A., Dong, Y., & Ignea, C. (2024). Synthetic biology advances towards a bio-based society in the era of artificial intelligence. *Current Opinion in Biotechnology*, *87*, 103143. DOI: 10.1016/j.copbio.2024.103143 PMID: 38781699

Jagadeesh, P., Rangappa, S. M., & Siengchin, S. (2024). Advanced characterization techniques for nanostructured materials in biomedical applications. *Advanced Industrial and Engineering Polymer Research*, *7*(1), 122–143. DOI: 10.1016/j.aiepr.2023.03.002

Jain, S., & Mehata, M. S. (2017). Medicinal Plant Leaf Extract and Pure Flavonoid Mediated Green Synthesis of Silver Nanoparticles and their Enhanced Antibacterial Property. *Scientific Reports*, *7*(1), 15867. DOI: 10.1038/s41598-017-15724-8 PMID: 29158537

Jeevanandam, J., Kiew, S. F., Boakye-Ansah, S., Lau, S. Y., Barhoum, A., Danquah, M. K., & Rodrigues, J. (2022). Green approaches for the synthesis of metal and metal oxide nanoparticles using microbial and plant extracts. *Nanoscale*, *14*(7), 2534–2571. DOI: 10.1039/D1NR08144F PMID: 35133391

Jeong, W., Yu, U., & Lee, S. H. (2025). Global decomposition of networks into multiple cores formed by local hubs. *Physical Review. E*, *111*(5), 054302. DOI: 10.1103/PhysRevE.111.054302 PMID: 40534039

Joudeh, N., & Linke, D. (2022). Nanoparticle classification, physicochemical properties, characterization, and applications: A comprehensive review for biologists. *Journal of Nanobiotechnology*, *20*(1), 262. DOI: 10.1186/s12951-022-01477-8 PMID: 35672712

Kalimuthu, K., Suresh Babu, R., Venkataraman, D., Bilal, M., & Gurunathan, S. (2008). Biosynthesis of silver nanocrystals by Bacillus licheniformis. *Colloids and Surfaces. B, Biointerfaces*, *65*(1), 150–153. DOI: 10.1016/j.colsurfb.2008.02.018 PMID: 18406112

Kalishwaralal, K., Deepak, V., Ram Kumar Pandian, S., Kottaisamy, M., BarathManiKanth, S., Kartikeyan, B., & Gurunathan, S. (2010). Biosynthesis of silver and gold nanoparticles using Brevibacterium casei. *Colloids and Surfaces. B, Biointerfaces*, *77*(2), 257–262. DOI: 10.1016/j.colsurfb.2010.02.007 PMID: 20197229

Kanaoujiya, R., Saroj, S. K., Rajput, V. D., Alimuddin, , Srivastava, S., Minkina, T., Igwegbe, C. A., Singh, M., & Kumar, A. (2023). Emerging application of nanotechnology for mankind. *Emergent Materials*, *6*(2), 439–452. DOI: 10.1007/s42247-023-00461-8 PMID: 36743193

Khan, F., Shariq, M., Asif, M., Siddiqui, M. A., Malan, P., & Ahmad, F. (2022). Green Nanotechnology: Plant-Mediated Nanoparticle Synthesis and Application. *Nanomaterials (Basel, Switzerland)*, *12*(4), 673. DOI: 10.3390/nano12040673 PMID: 35215000

Khorsand Zak, A., & Hashim, A. M. (2025). Advanced XRD peak broadening analysis of gallium-doped ZnO nanoparticles for crystallite size evaluation. *Scientific Reports*, *16*(1), 1717. DOI: 10.1038/s41598-025-31317-2 PMID: 41372362

Khursheed, R., Dua, K., Vishwas, S., Gulati, M., Jha, N. K., Aldhafeeri, G. M., Alanazi, F. G., Goh, B. H., Gupta, G., Paudel, K. R., Hansbro, P. M., Chellappan, D. K., & Singh, S. K. (2022). Biomedical applications of metallic nanoparticles in cancer: Current status and future perspectives. *Biomedicine & Pharmacotherapy = Biomédecine & Pharmacothérapie*, *150*, 112951. DOI: 10.1016/j.biopha.2022.112951 PMID: 35447546

Kiattisewee, C., Dong, C., Fontana, J., Sugianto, W., Peralta-Yahya, P., Carothers, J. M., & Zalatan, J. G. (2021). Portable bacterial CRISPR transcriptional activation enables metabolic engineering in Pseudomonas putida. *Metabolic Engineering*, *66*, 283–295. DOI: 10.1016/j.ymben.2021.04.002 PMID: 33930546

Lee, S. H., & Jun, B.-H. (2019). Silver Nanoparticles: Synthesis and Application for Nanomedicine. *International Journal of Molecular Sciences*, *20*(4), 865. DOI: 10.3390/ijms20040865 PMID: 30781560

Lian, S., Diko, C. S., Yan, Y., Li, Z., Zhang, H., Ma, Q., & Qu, Y. (2019). Characterization of biogenic selenium nanoparticles derived from cell-free extracts of a novel yeast Magnusiomyces ingens. *3 Biotech, 9*(6), 221. DOI: 10.1007/s13205-019-1748-y

Liu, Z., Dong, H., Cui, Y., Cong, L., & Zhang, D. (2020). Application of different types of CRISPR/Cas-based systems in bacteria. *Microbial Cell Factories*, *19*(1), 172. DOI: 10.1186/s12934-020-01431-z PMID: 32883277

Malik, S., Muhammad, K., & Waheed, Y. (2023). Nanotechnology: A Revolution in Modern Industry. *Molecules (Basel, Switzerland)*, *28*(2), 661. DOI: 10.3390/molecules28020661 PMID: 36677717

Menon, S., & Shanmugam, V. (2020). Cytotoxicity Analysis of Biosynthesized Selenium Nanoparticles Towards A549 Lung Cancer Cell Line. *Journal of Inorganic and Organometallic Polymers and Materials*, *30*(5), 1852–1864. DOI: 10.1007/s10904-019-01409-4

Mishra, P., Ray, S., Sinha, S., Das, B., Khan, M., Behera, S. K., Yun, S.-I., Tripathy, S. K., & Mishra, A. (2016). Facile bio-synthesis of gold nanoparticles by using extract of Hibiscus sabdariffa and evaluation of its cytotoxicity against U87 glioblastoma cells under hyperglycemic condition. *Biochemical Engineering Journal*, *105*, 264–272. DOI: 10.1016/j.bej.2015.09.021

Mittal, A. K., Chisti, Y., & Banerjee, U. C. (2013). Synthesis of metallic nanoparticles using plant extracts. *Biotechnology Advances*, *31*(2), 346–356. DOI: 10.1016/j.biotechadv.2013.01.003 PMID: 23318667

Mobarak, M. H., & Al Mahmud, M., Hossain, A., Abir, Md. A. H., Hossain, N., & Chowdhury, M. A. (2026). Advances of Green-Synthesized Nanoparticles for Biomedical Applications. *BioMed Research International*, *9207147*(1), •••. DOI: 10.1155/bmri/9207147 PMID: 41925124

Mohammed, A. E., & Al-Megrin, W. A. (2021). Biological Potential of Silver Nanoparticles Mediated by Leucophyllum frutescens and Russelia equisetiformis Extracts. *Nanomaterials (Basel, Switzerland)*, *11*(8), 2098. DOI: 10.3390/nano11082098 PMID: 34443930

Mohammed Fayaz, A., Balaji, K., Kalaichelvan, P. T., & Venkatesan, R. (2009). Fungal based synthesis of silver nanoparticles—An effect of temperature on the size of particles. *Colloids and Surfaces. B, Biointerfaces*, *74*(1), 123–126. DOI: 10.1016/j.colsurfb.2009.07.002 PMID: 19674875

Monica Ahmad, N., Husaini Mohamed, A., Hasan, N., Zainal- Abidin, N., Zaini Nawahwi, M., & Mohamad Azzeme, A. (2024). Effect of optimisation variable and the role of plant extract in the synthesis of nanoparticles using plant-mediated synthesis approaches. *Inorganic Chemistry Communications*, *161*, 111839. DOI: 10.1016/j.inoche.2023.111839

More, P. R., Pandit, S., Filippis, A. D., Franci, G., Mijakovic, I., & Galdiero, M. (2023). Silver Nanoparticles: Bactericidal and Mechanistic Approach against Drug Resistant Pathogens. *Microorganisms*, *11*(2), 369. DOI: 10.3390/microorganisms11020369 PMID: 36838334

Mougiakos, I., Bosma, E. F., Ganguly, J., Van Der Oost, J., & Van Kranenburg, R. (2018). Hijacking CRISPR-Cas for high-throughput bacterial metabolic engineering: Advances and prospects. *Current Opinion in Biotechnology*, *50*, 146–157. DOI: 10.1016/j.copbio.2018.01.002 PMID: 29414054

Mourdikoudis, S., Pallares, R. M., & Thanh, N. T. K. (2018). Characterization techniques for nanoparticles: Comparison and complementarity upon studying nanoparticle properties. *Nanoscale*, *10*(27), 12871–12934. DOI: 10.1039/C8NR02278J PMID: 29926865

Muhammad, A. I., Dalia, A. A. M., Hemly, N. I. M., Zainudin, N. N., & Samsudin, A. A. (2025). Biofortified Bacteria: The Role of Selenium-Enriched Microorganisms in Enhancing Animal Selenium Uptake—A Review. *Journal of Animal Physiology and Animal Nutrition*, *109*(6), 1298–1320. DOI: 10.1111/jpn.70001 PMID: 40674621

Mukherjee, P., Ahmad, A., Mandal, D., Senapati, S., Sainkar, S. R., Khan, M. I., Parishcha, R., Ajaykumar, P. V., Alam, M., Kumar, R., & Sastry, M. (2001). Fungus-Mediated Synthesis of Silver Nanoparticles and Their Immobilization in the Mycelial Matrix: A Novel Biological Approach to Nanoparticle Synthesis. *Nano Letters, 1*(10), 515–519. DOI: 10.1021/nl0155274

Mukherjee, P., Bhattacharya, R., Bone, N., Lee, Y. K., Patra, C., Wang, S., Lu, L., Secreto, C., Banerjee, P. C., Yaszemski, M. J., Kay, N. E., & Mukhopadhyay, D. (2013). Correction: Potential therapeutic application of gold nanoparticles in B-chronic lymphocytic leukemia (BCLL): enhancing apoptosis. *Journal of Nanobiotechnology, 11*(1), 23. DOI: 10.1186/1477-3155-11-23

Mustapha, S., Ndamitso, M. M., Abdulkareem, A. S., Tijani, J. O., Shuaib, D. T., Mohammed, A. K., & Sumaila, A. (2019). Comparative study of crystallite size using Williamson-Hall and Debye-Scherrer plots for ZnO nanoparticles. *Advances in Natural Sciences: Nanoscience and Nanotechnology, 10*(4), 045013. DOI: 10.1088/2043-6254/ab52f7

Naveenkumar, S., Alagumanikumaran, N., Kaviyarasu, K., & Muthukumaran, A. (2024). Influence of encapsulated sodium alginates and pectin on selenium nanoparticles and efficient cardioprotective effect in C2C12 cell line. *Journal of Nanoparticle Research, 26*(3), 52. DOI: 10.1007/s11051-024-05956-x

Olędzka, A. J., & Czerwińska, M. E. (2023). Role of Plant-Derived Compounds in the Molecular Pathways Related to Inflammation. *International Journal of Molecular Sciences, 24*(5), 4666. DOI: 10.3390/ijms24054666 PMID: 36902097

Oliveira, T. P. D. S., Lima, A. K. O., Gonçalves, T. P., Florêncio, I., Báo, S. N., Silva, N. O. D., Albuquerque, P., Silva-Pereira, I., & Muehlmann, L. A. (2026). Synthesis of Selenium Nanoparticles: Influence of Reaction Parameters on Physicochemical, Morphological, and Biological Properties. *Drugs and Drug Candidates, 5*(1), 22. DOI: 10.3390/ddc5010022

Ortiz-Tirado, A., Medina-Ganem, L. F., Bandala, E. R., Conejo-Davila, A. S., Vega-Rios, A., Goonetilleke, A., & Rodriguez-Narvaez, O. M. (2025). Biogenic nanoparticles: Synthesis, characterization, applications and scaling up limitations in water treatment. *Journal of Environmental Chemical Engineering, 13*(5), 117730. DOI: 10.1016/j.jece.2025.117730

Osman, A. I., Zhang, Y., Farghali, M., Rashwan, A. K., Eltaweil, A. S., Abd El-Monaem, E. M., Mohamed, I. M. A., Badr, M. M., Ihara, I., Rooney, D. W., & Yap, P.-S. (2024). Synthesis of green nanoparticles for energy, biomedical, environmental, agricultural, and food applications: A review. *Environmental Chemistry Letters*, *22*(2), 841–887. DOI: 10.1007/s10311-023-01682-3

Pandey, S., Awasthee, N., Shekher, A., Rai, L. C., Gupta, S. C., & Dubey, S. K. (2021). Biogenic synthesis and characterization of selenium nanoparticles and their applications with special reference to antibacterial, antioxidant, anticancer and photocatalytic activity. *Bioprocess and Biosystems Engineering*, *44*(12), 2679–2696. DOI: 10.1007/s00449-021-02637-0 PMID: 34599397

Panzarini, E., Mariano, S., Carata, E., Mura, F., Rossi, M., & Dini, L. (2018). Intracellular Transport of Silver and Gold Nanoparticles and Biological Responses: An Update. *International Journal of Molecular Sciences*, *19*(5), 1305. DOI: 10.3390/ijms19051305 PMID: 29702561

Pasieczna-Patkowska, S., Cichy, M., & Flieger, J. (2025). Application of Fourier Transform Infrared (FTIR) Spectroscopy in Characterization of Green Synthesized Nanoparticles. *Molecules (Basel, Switzerland)*, *30*(3), 684. DOI: 10.3390/molecules30030684 PMID: 39942788

Patil, R. M., Deshpande, P. P., Aalhate, M., Gananadhamu, S., & Singh, P. K. (2022). An Update on Sophisticated and Advanced Analytical Tools for Surface Characterization of Nanoparticles. *Surfaces and Interfaces*, *33*, 102165. DOI: 10.1016/j.surfin.2022.102165

Patiño-Ruiz, D. A., Meramo-Hurtado, S. I., González-Delgado, Á. D., & Herrera, A. (2021). Environmental Sustainability Evaluation of Iron Oxide Nanoparticles Synthesized via Green Synthesis and the Coprecipitation Method: A Comparative Life Cycle Assessment Study. *ACS Omega*, *6*(19), 12410–12423. DOI: 10.1021/acsomega.0c05246 PMID: 34056392

Patra, S., Mukherjee, S., Barui, A. K., Ganguly, A., Sreedhar, B., & Patra, C. R. (2015). Green synthesis, characterization of gold and silver nanoparticles and their potential application for cancer therapeutics. *Materials Science and Engineering C*, *53*, 298–309. DOI: 10.1016/j.msec.2015.04.048 PMID: 26042718

Ramos, A. P., Cruz, M. A. E., Tovani, C. B., & Ciancaglini, P. (2017). Biomedical applications of nanotechnology. *Biophysical Reviews*, *9*(2), 79–89. DOI: 10.1007/s12551-016-0246-2 PMID: 28510082

Rodrigues, A. S., Batista, J. G. S., Rodrigues, M. Á. V., Thipe, V. C., Minarini, L. A. R., Lopes, P. S., & Lugão, A. B. (2024). Advances in silver nanoparticles: A comprehensive review on their potential as antimicrobial agents and their mechanisms of action elucidated by proteomics. *Frontiers in Microbiology*, *15*, 1440065. DOI: 10.3389/fmicb.2024.1440065 PMID: 39149204

Rodríguez-Rojas, M. D. P., Bustos-Terrones, V., Díaz-Cárdenas, M. Y., Vázquez-Vélez, E., & Martínez, H. (2024). Life Cycle Assessment of Green Synthesis of TiO2 Nanoparticles vs. Chemical Synthesis. *Sustainability (Basel)*, *16*(17), 7751. DOI: 10.3390/su16177751

Rónavári, A., Igaz, N., Adamecz, D. I., Szerencsés, B., Molnar, C., Kónya, Z., Pfeiffer, I., & Kiricsi, M. (2021). Green Silver and Gold Nanoparticles: Biological Synthesis Approaches and Potentials for Biomedical Applications. *Molecules (Basel, Switzerland)*, *26*(4), 844. DOI: 10.3390/molecules26040844 PMID: 33562781

Salem, S. S., & Fouda, A. (2021). Green Synthesis of Metallic Nanoparticles and Their Prospective Biotechnological Applications: An Overview. *Biological Trace Element Research*, *199*(1), 344–370. DOI: 10.1007/s12011-020-02138-3 PMID: 32377944

Saratale, R. G., Cho, S. K., Bharagava, R. N., Vivekanand, V., Bhatia, S. K., Romanholo Ferreira, L. F., Shin, H. S., Jeon, B.-H., Kumar, R., & Saratale, G. D. (2025). Biorefinery of avocado seed waste for sustainable production of bioplastics and functionalized silver nanocomposites. *International Journal of Biological Macromolecules*, *323*, 147042. DOI: 10.1016/j.ijbiomac.2025.147042 PMID: 40848789

Saratale, R. G., Karuppusamy, I., Saratale, G. D., Pugazhendhi, A., Kumar, G., Park, Y., Ghodake, G. S., Bharagava, R. N., Banu, J. R., & Shin, H. S. (2018). A comprehensive review on green nanomaterials using biological systems: Recent perception and their future applications. *Colloids and Surfaces. B, Biointerfaces*, *170*, 20–35. DOI: 10.1016/j.colsurfb.2018.05.045 PMID: 29860217

Satyavani, K., Gurudeeban, S., Ramanathan, T., & Balasubramanian, T. (2011). Biomedical potential of silver nanoparticles synthesized from calli cells of Citrullus colocynthis (L.) Schrad. *Journal of Nanobiotechnology*, *9*(1), 43. DOI: 10.1186/1477-3155-9-43 PMID: 21943321

Saxena, R., Kotnala, S., Bhatt, S. C., Uniyal, M., Rawat, B. S., Negi, P., & Riyal, M. K. (2025). A review on green synthesis of nanoparticles toward sustainable environment. *Sustainable Chemistry for Climate Action*, *6*, 100071. DOI: 10.1016/j.scca.2025.100071

Serdar, G. (2024). Biogenic Synthesis and Characterization of Gold Nanoparticles from Morus alba L. Leaves by Microwave Extraction Method. *Journal of the Turkish Chemical Society Section A: Chemistry, 11*(2), 615–622. DOI: 10.18596/jotcsa.1372302

Shahverdi, A. R., Minaeian, S., Shahverdi, H. R., Jamalifar, H., & Nohi, A.-A. (2007). Rapid synthesis of silver nanoparticles using culture supernatants of Enterobacteria: A novel biological approach. *Process Biochemistry (Barking, London, England), 42*(5), 919–923. DOI: 10.1016/j.procbio.2007.02.005

Shleeva, M. O., Kondratieva, D. A., & Kaprelyants, A. S. (2023). Bacillus licheniformis: A Producer of Antimicrobial Substances, including Antimycobacterials, Which Are Feasible for Medical Applications. *Pharmaceutics, 15*(7), 1893. DOI: 10.3390/pharmaceutics15071893 PMID: 37514078

Singh, D., Rathod, V., Ninganagouda, S., Hiremath, J., Singh, A. K., & Mathew, J. (2014). Optimization and Characterization of Silver Nanoparticle by Endophytic Fungi *Penicillium* sp. Isolated from *Curcuma longa* (Turmeric) and Application Studies against MDR *E. coli* and *S. aureus. Bioinorganic Chemistry and Applications, 2014,* 1–8. DOI: 10.1155/2014/408021 PMID: 24639625

Singh, D., Rathod, V., Singh, A. K., Ul Haq, M., Mathew, J., & Kulkarni, P. (2016). A Study on Extracellular Synthesis of Silver Nanoparticles from Endophytic Fungi, Isolated from Ethanomedicinal Plants *Curcuma longa* and *Catharanthus roseus. International Letters of Natural Sciences, 57,* 58–66. DOI: 10.56431/p-sx5071

Singh, H., Desimone, M. F., Pandya, S., Jasani, S., George, N., Adnan, M., Aldarhami, A., Bazaid, A. S., & Alderhami, S. A. (2023). Revisiting the Green Synthesis of Nanoparticles: Uncovering Influences of Plant Extracts as Reducing Agents for Enhanced Synthesis Efficiency and Its Biomedical Applications. *International Journal of Nanomedicine, 18,* 4727–4750. DOI: 10.2147/IJN.S419369 PMID: 37621852

Singh, N. A. (2017). Nanotechnology innovations, industrial applications and patents. *Environmental Chemistry Letters, 15*(2), 185–191. DOI: 10.1007/s10311-017-0612-8

Singh, R., Shedbalkar, U. U., Wadhwani, S. A., & Chopade, B. A. (2015). Bacteriagenic silver nanoparticles: Synthesis, mechanism, and applications. *Applied Microbiology and Biotechnology, 99*(11), 4579–4593. DOI: 10.1007/s00253-015-6622-1 PMID: 25952110

Singh, S., Vidyarthi, A. S., Nigam, V. K., & Dev, A. (2014). Extracellular facile biosynthesis, characterization and stability of gold nanoparticles by *Bacillus licheniformis. Artificial Cells, Nanomedicine, and Biotechnology, 42*(1), 6–12. DOI: 10.3109/21691401.2012.759122 PMID: 23438180

Slater, T. J. A., Janssen, A., Camargo, P. H. C., Burke, M. G., Zaluzec, N. J., & Haigh, S. J. (2016). STEM-EDX tomography of bimetallic nanoparticles: A methodological investigation. *Ultramicroscopy, 162*, 61–73. DOI: 10.1016/j.ultramic.2015.10.007 PMID: 26780684

Sliem, A. R. (2024). Green Synthesis of Antimicrobial Silver Nanoparticles: Exploring Plant-Derived Reducing Agents from Camellia Sinensis, Syzygium aromaticum, and Citrus Sinensis. *Al-Azhar Journal of Agricultural Research, 49*(1), 81–94. DOI: 10.21608/ajar.2024.269482.1329

Smirnov, I., Kaszkur, Z., & Hoell, A. (2023). Development of nanoparticle bulk morphology analysis: A multidomain XRD approach. *Nanoscale, 15*(19), 8633–8642. DOI: 10.1039/D3NR00456B PMID: 37073880

Song, Y., & Prather, K. L. J. (2024). Strategies in engineering sustainable biochemical synthesis through microbial systems. *Current Opinion in Chemical Biology, 81*, 102493. DOI: 10.1016/j.cbpa.2024.102493 PMID: 38971129

Soni, A. K., & Jha, R. K. (2024). Nanotechnology's Applications and Potential in Various Fields. *Cureus*. Advance online publication. DOI: 10.7759/cureus.59234 PMID: 38813303

Suárez Priede, A., Gómez-Sánchez, M., García-Cancela, P., Bettmer, J., & Díez, P. (2026). Deciphering the formation of biogenic nanoparticles and their protein corona: State-of-the-art and analytical challenges. *Analytical and Bioanalytical Chemistry, 418*(2), 415–435. DOI: 10.1007/s00216-025-06144-z PMID: 41076454

Sugrue, I., Ross, R. P., & Hill, C. (2024). Bacteriocin diversity, function, discovery and application as antimicrobials. *Nature Reviews. Microbiology, 22*(9), 556–571. DOI: 10.1038/s41579-024-01045-x PMID: 38730101

Szczyglewska, P., Feliczak-Guzik, A., & Nowak, I. (2023). Nanotechnology–General Aspects: A Chemical Reduction Approach to the Synthesis of Nanoparticles. *Molecules (Basel, Switzerland), 28*(13), 4932. DOI: 10.3390/molecules28134932 PMID: 37446593

Truong, T. T., Mondal, S., Doan, V. H. M., Tak, S., Choi, J., Oh, H., Nguyen, T. D., Misra, M., Lee, B., & Oh, J. (2024). Precision-engineered metal and metal-oxide nanoparticles for biomedical imaging and healthcare applications. *Advances in Colloid and Interface Science, 332*, 103263. DOI: 10.1016/j.cis.2024.103263 PMID: 39121830

Tsekhmistrenko, S. I., Bityutskyy, V. S., Tsekhmistrenko, O. S., Horalskyi, L. P., Tymoshok, N. O., & Spivak, M. Y. (2020). Bacterial synthesis of nanoparticles: A green approach. *Biosystems Diversity*, *28*(1), 9–17. DOI: 10.15421/012002

Tumilaar, S. G., Hardianto, A., Dohi, H., & Kurnia, D. (2024). A Comprehensive Review of Free Radicals, Oxidative Stress, and Antioxidants: Overview, Clinical Applications, Global Perspectives, Future Directions, and Mechanisms of Antioxidant Activity of Flavonoid Compounds. *Journal of Chemistry*, *2024*, 1–21. DOI: 10.1155/2024/5594386

Uvarov, V., & Popov, I. (2013). Metrological characterization of X-ray diffraction methods at different acquisition geometries for determination of crystallite size in nano-scale materials. *Materials Characterization*, *85*, 111–123. DOI: 10.1016/j.matchar.2013.09.002

Vijayaram, S., Razafindralambo, H., Sun, Y.-Z., Vasantharaj, S., Ghafarifarsani, H., Hoseinifar, S. H., & Raeeszadeh, M. (2024). Applications of Green Synthesized Metal Nanoparticles—A Review. *Biological Trace Element Research*, *202*(1), 360–386. DOI: 10.1007/s12011-023-03645-9 PMID: 37046039

Wang, J. (2025). The Application of Nanotechnology. *Applied and Computational Engineering*, *205*(1), 81–87. DOI: 10.54254/2755-2721/2025.TJ23325

Wang, Z., Li, N., Zhou, X., Wei, S., Zhu, Y., Li, M., Gong, J., He, Y., Dong, X., Gao, C., & Cheng, S. (2024). Optimization of fermentation parameters to improve the biosynthesis of selenium nanoparticles by Bacillus licheniformis F1 and its comprehensive application. *BMC Microbiology*, *24*(1), 271. DOI: 10.1186/s12866-024-03410-5 PMID: 39033096

Xu, L., Wang, Y.-Y., Huang, J., Chen, C.-Y., Wang, Z.-X., & Xie, H. (2020). Silver nanoparticles: Synthesis, medical applications and biosafety. *Theranostics*, *10*(20), 8996–9031. DOI: 10.7150/thno.45413 PMID: 32802176

Yang, T., Shi, Y., Yang, Q., Xin, Y., Gu, Z., & Zhang, L. (2021). A Novel Regulator Participating in Nitrogen Removal Process of Bacillus subtilis JD-014. *International Journal of Molecular Sciences*, *22*(12), 6543. DOI: 10.3390/ijms22126543 PMID: 34207153

Yao, R., Liu, D., Jia, X., Zheng, Y., Liu, W., & Xiao, Y. (2018). CRISPR-Cas9/Cas12a biotechnology and application in bacteria. *Synthetic and Systems Biotechnology*, *3*(3), 135–149. DOI: 10.1016/j.synbio.2018.09.004 PMID: 30345399

Yuksekdag, Z., Kilickaya, R., Kara, F., & Acar, B. C. (2025). Biogenic-Synthesized Silver Nanoparticles Using the Ligilactobacillus salivarius KC27L Postbiotic: Antimicrobial, Anti-Biofilm, and Antioxidant Activity and Cytotoxic Effects. *Probiotics and Antimicrobial Proteins*, *17*(5), 3685–3699. DOI: 10.1007/s12602-025-10481-x PMID: 40011382

Zhang, N., Xiong, G., & Liu, Z. (2022). Toxicity of metal-based nanoparticles: Challenges in the nano era. *Frontiers in Bioengineering and Biotechnology*, *10*, 1001572. DOI: 10.3389/fbioe.2022.1001572 PMID: 36619393

Zhang, X., Xu, P., Lou, Y., Liu, Y., Shan, Q., Xiong, Y., Wei, H., & Song, J. (2024). Characteristics of Novel Heterotrophic Nitrification–Aerobic Denitrification Bacteria Bacillus subtilis F4 and Alcaligenes faecalis P4 Isolated from Landfill Leachate Biochemical Treatment System. *Water (Basel)*, *16*(14), 1993. DOI: 10.3390/w16141993

Zhu, W., Fan, W., Zhang, X., & Gao, M. (2021). Sustained-Release Solid Dispersion of High-Melting-Point and Insoluble Resveratrol Prepared through Hot Melt Extrusion to Improve Its Solubility and Bioavailability. *Molecules (Basel, Switzerland)*, *26*(16), 4982. DOI: 10.3390/molecules26164982 PMID: 34443569

# Chapter 4

# Green Biopolymer-Based Nanocarriers for Precision Cancer Therapeutics:

## From Sustainable Synthesis to Clinical Translation


**Aiman Zahid**
*Department of Public Health, University of Sunderland, London, UK*

**Adnan Afzal**
http://orcid.org/0009-0001-4359-5695
*Government College University, Faisalabad, Pakistan*

**Abu Sulman**
http://orcid.org/0009-0008-7194-5219
*University of Agriculture, Faisalabad, Pakistan*

**Muhammad Saad Naeem**
*University of Agriculture, Faisalabad, Pakistan*

**Muhammad Yawar Saeed**
*University of Agriculture, Faisalabad, Pakistan*

**Faiza Mumtaz**
*Department of Pharmacy, The Superior University, Lahore, Pakistan*

**Aleeha Afzal**
*University of Agriculture, Faisalabad, Pakistan*

**Maria Shafaqat**
http://orcid.org/0009-0008-7399-2529
*School of Natural Sciences, University of Kent, UK*

**Hira Waheed**
http://orcid.org/0009-0000-2836-2335
*Department of Pharmaceutical Chemistry, Government College University, Faisalabad, Pakistan*

**Hina Nawab**
http://orcid.org/0009-0002-8250-5856
*Institute of Physiology and Pharmacology, University of Agriculture, Faisalabad, Pakistan*



DOI: 10.4018/979-8-3373-9848-8.ch004

## ABSTRACT

*Cancer remains a major global health challenge, with conventional chemotherapy limited by systemic toxicity, poor selectivity, multidrug resistance, and unfavorable pharmacokinetics. Nanomedicine offers advanced drug delivery systems that enhance solubility, biodistribution, and therapeutic efficacy. Recently, sustainable biodegradable biopolymers have gained attention as eco-friendly alternatives to synthetic materials for nanocarrier development. This chapter explores green biopolymer-based nanocarriers, including polysaccharide-, protein-, and microbial-derived systems, highlighting their functional properties, green fabrication methods, targeting strategies, and stimuli-responsive release mechanisms for precision and clinically translatable cancer therapy.*


## INTRODUCTION

Cancer is currently one of the major causes of death all around the world and it is a huge burden to health care systems and society across the world. In spite of incredible progress in early diagnosis, molecular diagnostics, and therapeutic interventions, the heterogeneity and complexity of cancer are still a challenge to both clinicians and researchers (Murray & Lopez, 1997). Conventional forms of treatment including surgery, radiotherapy and chemotherapy have contributed immensely to the survival of various types of cancer. Nevertheless, these methods tend not to be very precise and can be linked to significant side effects which reduce the quality of life of patients. The recent development of tumour biology, cellular signalling pathways and the tumour microenvironment has identified the necessity of more selective, efficient, and patient-friendly therapeutic approaches (Kaur et al., 2023). The interplay between materials science, molecular biology and pharmaceutical technology in recent decades has led to new methods of treating cancer. Nanomedicine has become one such revolution that can overcome most of the disadvantages of traditional therapies. Nanotechnology provides an opportunity to improve the therapeutic efficacy of drugs (reducing systemic toxicity) because nanotechnology facilitates controlled, targeted, and sustained drug delivery. At the same time, the growing interest in environmental sustainability as well as biocompatibility over the long term has stimulated the transition to biodegradable and naturally obtained materials in drug delivery systems (Niemeyer, 2001). Medicinal plants are also a source of bioactive compounds to treat cancer (Naeem et al., 2025). All these developments are revolutionary in the paradigm of oncology and are transforming the way anticancer therapies are designed, delivered and optimized to be safer and more effective in their clinical endpoints (Falzone et al., 2018).

## 1. ONCOLOGY: THE PARADIGM SHIFT

### The Limitations of Conventional Chemotherapy.

The treatment of cancer has been subjected to great evolution in the last hundred years, but traditional chemotherapy is one of the most popular therapeutic interventions in cancer treatment. Recent advancements in nanomedicine also show promise for treating neurobiological drug addiction, with targeted drug delivery systems offering new avenues for overcoming addiction-related challenges in the brain (Sulman et al., 2025). Although its application has a long history and has demonstrated effectiveness in the treatment of different malignancies, chemotherapy is starting to be understood as having severe limitations. Such limitations have led to a paradigm shift in the field of oncology that has led to the pursuit of safer, more effective, and more targeted treatment methods (Falzone et al., 2018).

The traditional chemotherapeutic agents act more by attacking cells that divide rapidly. Even though this method allows the elimination of malignant cells, it causes native damage to healthy tissues that grow with a fast pace, including bone marrow, gastrointestinal epithelium, hair follicles. Its effects are systemic toxicity, which presents itself in the form of myelosuppression, mucositis, alopecia, nausea, fatigue, and vulnerability to infections. One of the most essential disadvantages of conventional chemotherapy is its no selectivity in regard to malignant cells and normal cells (Kaur et al., 2023).

*Figure 1. The paradigm shift*

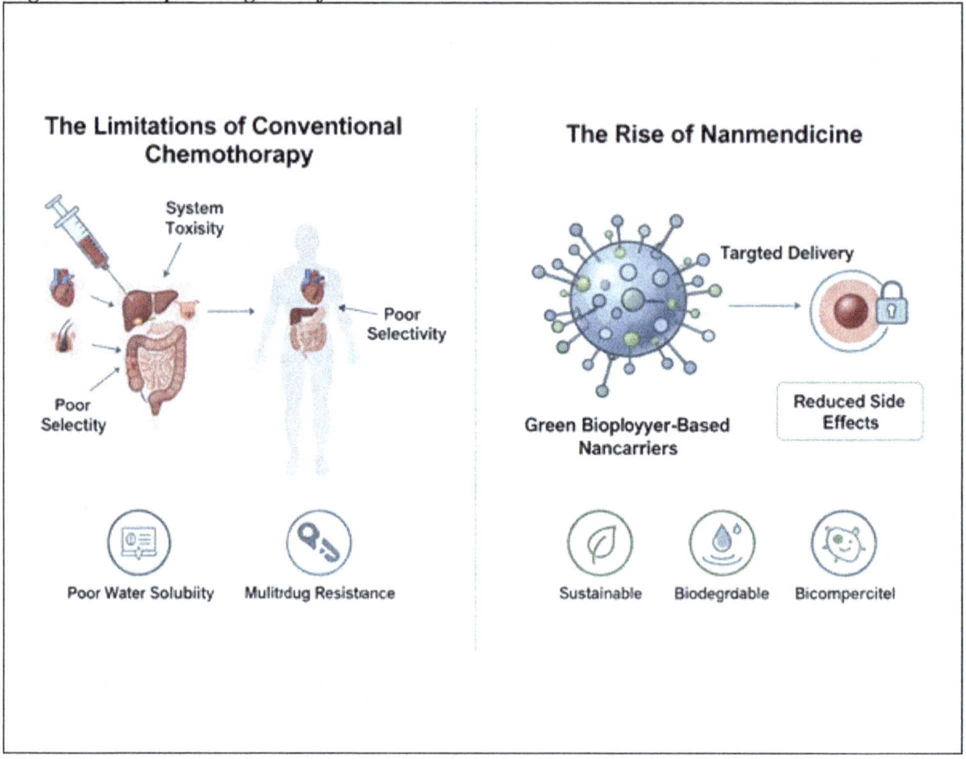


Poor solubility and bioavailability of most anticancer drugs is another significant problem. Many chemotherapeutic agents are hydrophobic meaning that they need toxic organic solvents or surfactants to be administered intravenously. The excipients themselves are capable of hypersensitivity reaction, nephrotoxicity, and other unfavourable events. Moreover, low aqueous solubility usually results in unpredictable pharmacokinetics, unpredictable responses to therapy and inadequate levels of drug within tumour foci (Bhalani et al., 2022).

Resistance to drugs is another major obstacle towards effective chemotherapy. Mechanisms of intrinsic or acquired resistance such as overexpression of efflux pumps, altered drug metabolism, increased DNA repair, and apoptosis evasion can be developed in tumour cells. Once resistance arises, an increased dose or combination therapy is frequently needed, and this will continue to raise the level of toxicity with no assurance of greater efficacy (Haider et al., 2020). The non-specificity of the delivery of chemotherapeutic agents in the body decreases the precision of therapeutics. Drugs are distributed indiscriminately throughout the body after systemic administration in both tumour and normal tissues. This does not only restrict the highest tolerable dose but also decreases the therapeutic index. It is very common that a very low percentage of the drug administered is adsorbed at the tumour site and the remaining counts to cause side effects to the whole-body systems (John et al., 2023). Treatment outcome is also complicated by pharmacokinetic limitations. Quick clearance, enzymatic degradation, and instability in the physiological environment are able to lower drug half-life and therapeutic efficacy. The high dosing schedule would be required frequently, which would cause more load to the patients and decrease compliance. Moreover, the cumulative toxicity may lead to permanent damage to the organs, such as cardiotoxicity and neurotoxicity (Masimirembwa et al., 2003). There is an extra complexity of tumour heterogeneity. Cancers do not represent a homogenous mass, but a heterogeneous group of cells with different genetic and phenotypic composition. The traditional chemotherapy fails to sufficiently handle this heterogeneity where sensitive cells will be killed, and the resistant clones will diversify. This is a factor that contributes to relapse and metastasis which are considered to be the major causes of mortality due to cancer (Marusyk et al., 2012). The microenvironment of the tumour also has an effect on the therapeutic effectiveness. The infiltration of solid tumours by drugs is inhibited by abnormal vasculature, hypoxia, elevated interstitial pressure, and dense extra-cellular matrices. The traditional small-molecule medications do not tend to accumulate in the tumour tissues to a high degree, which leads to the partial eradication (Li et al., 2021).

All these constraints highlight the immediate necessity to develop more advanced drug delivery systems that will be able to increase specificity, enhance solubility, lower toxicity rates, and overcome other biological obstacles. The need has opened up the way to new practices that restructure the treatment field of oncology (Ezike et al., 2023).

## THE RISE OF NANOMEDICINE

### These are the benefits of Nanocarriers in Targeted Delivery.

Nanomedicine has become an innovative technology in cancer treatment that has led to solutions to most of the limitations that come with the traditional chemotherapy. Nanoscale engineering of drug delivery systems has allowed researchers to create platforms that could enhance the solubility of the drug, targeting of the tumour and reduce the systemic toxicity (Rasool et al., 2022). The size of nanocarriers is usually developed in the range of 1 to 200 nanometres, which means that nanocarriers are able to interact with biological systems at a molecular and cellular level. They are small and thus increase their permeability and retention (EPR) in tumour tissues. Passive targeting is realized because nanoparticles are attracted preferentially in tumour tissue due to leaky vasculature and defective lymphatic drainage in the tumour (Torchilin, 2008). Among the best benefits of nanocarriers, there is enhanced solubility. Anticancer drugs that are hydrophobic can be incorporated into lipid-based, polymeric, or inorganic nanostructures, and thus there is no necessity of using solvents that are harmful to the human body. This increases the bioavailability and decreases the excipient-related toxicity. Encapsulation also helps to guard against premature degradation of drugs in circulation thus enhancing stability and half-life (Naderinezhad et al., 2017). Another characteristic of nanomedicine is the controlled and sustained release of drugs. Nanocarriers can be programmed to deliver their cargo in response to a given stimulus, e.g., pH, enzymes, temperature, or redox environment typical of tumour micro environs. This is a targeted release that reduces off-target effects, and it makes therapeutic concentrations attainable mostly in the malignant tissues (Natarajan et al., 2014).

Native cellulose exhibits excellent biocompatibility and mechanical strength; however, it is not readily enzymatically degraded in the human body because humans lack cellulase enzymes. Therefore, cellulose is generally used in modified forms such as nanocellulose or carboxymethyl cellulose, which can be engineered to achieve controlled degradation and drug-release behavior.

Active targeting also narrows down accuracy. Receptor receptors that are overexpressed on cancer cells can be selectively targeted by their surface modification with ligands, antibodies, peptides or small molecules on nanoparticles. Indicatively, they can be used to target receptors like HER2, folate receptors, or integrins to provide receptor-targeted endocytosis and increased drug delivery to the intracellular structures. This approach has a significant advantage of enhancing therapeutic effectiveness and limiting the damage of normal tissues (Kunjiappan et al., 2021). Multifunctionality is also available with nanocarriers. In addition to delivering drugs, they can use them to provide imaging agents to perform diagnostics, giving rise to theragnostic platforms, which are therapeutic and diagnostic solutions. This allows real time tracking of drug delivery and treatment response to aid individual medicine strategies (Fernandez-Fernandez et al., 2011). Another benefit is the capability to co-deliver several agents. The drug resistance may frequently need to be overcome by combination therapy and address more than one signalling pathway. Nanocarriers have the capacity to entrap two or more drugs that have synergistic action, which will be delivered in harmony at the best ratios. This minimizes the chance of developing resistance and improves the general treatment of the therapy results (Khan et al., 2017). One of the most significant advantages that have been witnessed in some clinically approved nanomedicines is reduced systemic toxicity. Liposomal formulations, polymeric nanoparticles, and albumin-bound drugs have also shown an increased safety profile over its traditional equivalents. Nanocarriers can provide therapeutic advantages by increasing the efficacy of drugs within toxicity limits by modifying both pharmacokinetics and biodistribution (Beltrán-Gracia et al., 2019). Nanotechnology also solves biological barriers impeding the delivery of drugs. Surface engineering methods may enhance the circulation time by avoiding the immune recognition process. Examples include polyethylene glycol (PEG) coated nanoparticles which form stealth nanoparticles that are resistant to opsonization and clearance by the mononuclear phagocyte system. Circulation prolongs the likelihood of tumour accretion (Fam et al., 2020).

The nanocarrier design allows flexibility in its design to target the tumour type, stage and patient specific factors. Nanoparticles made of lipids, polymers, dendrimers, and inorganics are useful in giving varying structural and functional characteristics. This flexibility is useful in the precision oncology model, whereby therapies are developed based on a personal molecular profile. Although these have been achieved, there are still obstacles. Problems tied to mass production, standardization, licensing by authorities, and safety over time should be resolved. However, nanomedicine is a major step towards the future of oncology that would provide a solution to the problem of the efficacy of drugs and safety of patients (Zhang et al., 2017).

## THE "GREEN" MANDATE

### The Changing of Synthetic Polymers to Sustainable Biopolymers.

Due to the continuous development of nanomedicine, more and more attention is paid to the materials to be employed in the fabrication of nanoparticles. In the past, a good number of nanocarriers have been based on synthetic polymers that are of petrochemical origin. As much as these materials provide tenable characteristics and structural flexibility, the issues of toxicity, environmental effects and long-term biocompatibility have led to the need to move to alternative materials that are more sustainable (Boulaiz et al., 2011). The green requirement in medical research is a consequent of the larger international concern towards environmental sustainability and healthier medical technologies. Synthetic polymers can also have harmful byproducts in the course of manufacturing and degradation. Incomplete degradation in certain scenarios may result into bioaccumulation or chronic inflammatory reactions. Such dangers have generated an interest in polymers that are biodegradable and naturally found (Zhang et al., 2024). A possible solution is biopolymers, which are obtained either through plants, animals or microorganisms. Chitosan, alginate, gelatine, cellulose, starch, and hyaluronic acid are naturally biodegradable and tend to have high levels of biocompatibility. Their products of degradation are usually not toxic and can be readily metabolized or excreted by the body (Tamariz & Rios-Ramírez, 2013). The next important benefit of drug delivery systems is biodegradability.

Once the nanoparticles have served the therapeutic purpose, the idea is that when the nanoparticles are broken down, they are destroyed into harmless elements and do not build up in the body over a long period of time. Biopolymers are biodegradable, which means that they break down via enzymatic or hydrolytic processes, and their breakdown is in line with physiological routes (Ghuttora, 2016). Biocompatibility minimises the chances of immune responses and chronic toxicity. Biopolymers are generally well tolerated since many of them have a structural similarity to components of the extracellular matrix. This compatibility increases patient safety and increases clinical applicability (Arif et al., 2019). Sustainability is also applied to the manufacturing practices. Biopolymers are renewable in nature and can be extracted using agricultural or marine wastes hence avoiding use of fossil fuels. The idea behind green synthesis processes is that very few hazardous solvents are used, less energy is used, and the environmental footprint is minimized (Sheldon, 2005).

The alternative of biopolymers is also supported by functional versatility. Natural polymers contain functional groups which enable chemical modification and therefore controlled drug loading, kinetics of release and functionalization of the surface. Their endogenous bioactivity is also capable of improving therapeutic results. As an example, certain biopolymers can be used as mucoadhesive, antimicrobials or wound healing, which are complementary to anticancer therapy (Fazal et al., 2023). Green chemistry principles incorporated in nanomedicine design encourage less harmful innovation. Scholars are moving towards using green crosslinking reagents, solvent-free methods of fabricating materials, and biodegradable stabilizers. This would guarantee that the development of cancer treatment does not jeopardize environmental health (Trombino et al., 2023). Sustainability is also being focused on by regulatory agencies and healthcare systems. With the increase in the level of environmental awareness among the population, the pharmaceutical industries are pressured to engage in responsible production. The transition to biodegradable nanocarriers corresponds to these expectations and increases the confidence of the patient (Veleva et al., 2003). The merging of green materials science and nanotechnology is the future in the field of oncology. Since they can deliver targeted therapy and preserve the environment, the researchers are looking forward to developing not only effective but also ethically and environmentally friendly therapies (Jaffee et al., 2017).

## 2. SOURCES AND CLASSIFICATION OF BIOPOLYMERS TO NANOTHERAPEUTICS

Biopolymers are also highly promising materials in the fabrication of nanotherapeutic systems because of biodegradability, bio compatibility, renewability and versatile functional chemistry. These polymers are natural derivatives obtained using plants, animals and microorganisms thus possessing distinct structural and physicochemical properties that render them helpful in drug encapsulation, controlled release and targeted delivery (Kučuk et al., 2023). Biopolymers in nanomedicine of cancer and other advanced therapeutic use are widely grouped in polysaccharide-based, protein-based and microbial/fungal-based polymers. Classical mechanical properties, degradation behaviour, and functional group, which have effects on nanoparticle fabrication and performance, are unique to each class (Wu et al., 2024).

## Polysaccharide-Based Biopolymers

Polysaccharides are long chains of carbohydrate polymers made up of monosaccharides connected with the help of glycosidic bonds. They are of particular interest because of their ease of chemical modification, low toxicity, and abundance as nanotherapeutics (Chaudhary et al., 2022). Chitosan is a cationic polysaccharide that is derived out of chitin, which is usually derived by crustacean shells. It has a positive charge, which allows it to exhibit strong interaction through electrostatic forces with negatively charged cell membranes and nucleic acids, and find application in gene delivery and mucoadhesive drug delivery. Chitosan nanoparticles are characterised by good film-forming capacity, antimicrobial property and pH-sensitive behaviour which is beneficial in controlled drug release in acidic tumour microenvironment (Mikušová & Mikuš, 2021).

Alginate is a polysaccharide anionic polysaccharide that is a combination of guluronic and mannuronic acid and is extracted mainly through brown seaweed. Hydrogels can be prepared by it through the presence of divalent cations like calcium ions, which enables mild and biocompatible preparation of nanoparticles. Alginate type systems have been highly utilized in the encapsulation of chemotherapeutic agents, proteins and cells because they have mild gelation process and can be controlled to swell (Lin et al., 2022). Hyaluronic acid (HA) is a naturally occurring glycosaminoglycan that is present in the extracellular skeleton of connective tissue. It is very biodegradable and biocompatible. HA is especially useful in specific nanotherapeutics as it attaches specifically to CD44 receptors that are over-expressed in most cancer cells. Targeting This targeting involves use of receptors to aid in the delivery of drugs selectively and to increase therapeutic efficacy (Marinho et al., 2021).

The most common natural polymer on earth is cellulose which is produced by the cell walls of plants. In nanomedicine, cellulose usually exists in its modified, nanocellulose or carboxymethyl cellulose, forms. It has great mechanical, chemical, and film-forming. Though it is not naturally biodegradable in the human body without modification its derivatives can be developed to become more or less biodegradable and have control over the release of drugs. Nanocarriers made of polysaccharides are especially appreciated because of their hydrophilicity, structural flexibility, or the availability of reactive hydroxyl, carboxyl, or amino functional groups that allow functionalization (Bhilare et al., 2024).

## Protein-Based Biopolymers

Protein-based biopolymers are created through the bands of amino acids that offer a wide range of functional groups and secondary structures. Their natural bioactivity and resemblance with natural biological constituents increase their suitability in medicine (Joyce et al., 2021). Albumin as a naturally occurring plasma protein has been of importance in nanotherapeutics as it is capable of binding hydrophobic drugs and enhancing their solubility. Albumin nanoparticles take advantage of natural routes of transportation, such as receptor-mediated transcytosis and increase tumour uptake. The albumin-based systems are also characterized by a long circulation time and lower immunogenicity (Tincu et al., 2023).

Collagen, which is the major structural protein in connective tissues, offers great biocompatibility and biodegradability. It facilitates cell adhesion and cell growth rendering it appropriate in tissue-engineered drug delivery systems. The collagen nanoparticles are enzymatically degradable such that therapeutic agents can be released on demand (Lotfi et al., 2013). Silk Fibroin is a product of silkworm cocoons that has great mechanical strength and structural stability. It could be made into nanoparticles or films or made into hydrogel that can be degraded on different rates depending on the crystallinity. Silk fibroin has reactive amino acid side chains which can be used in conjugating drugs and surface modification (Biju, 2014).

Gelatine, which is a denatured collagen, has found extensive application in pharmaceutical preparations because of its easy processing and biodegradation. It has positively and negatively charged groups which allows versatility in drug loading. Gelatine nanoparticles are mainly applicable in the encapsulation of proteins, peptides, and chemotherapeutic drugs. Polymers of protein are typically better in terms of biological recognition, enzymatic biodegradability, and functional diversity as they contain amino, carboxyl and thiol groups (Gorgieva & Kokol, 2011).

## Biopolymers Obtained by Fermentation.

Microorganisms and fungi are biopolymers that are sustainable with distinct physicochemical properties applicable in nanotherapeutics. Pullulan is an example of a water-soluble polysaccharide that is synthesized by fungi Aureo basidium pullulans. It is made up of 8 units; malt triose chains that are joined together by 2 alpha (1 6) bonds. Pullulan is non-toxic, biodegradable and can form stable films and nanoparticles. It has a high number of hydroxyl groups, which can easily be chemically modified to serve targeted drug delivery and release systems (Rai et al., 2021).

Xanthan Gum is an extracellular polysaccharide produced by bacterium Xanthomonas campestris and it is very high-viscosity and has a wide pH-temperature stability. It creates hydrogel and nanoparticulate systems that would be useful in sustained release of drugs. Xanthan gum also has carboxyl and hydroxyl groups which allow it to crosslink and functionalize. Microbial biopolymers are of particular interest since they can be obtained by means of controlled fermentation and therefore have scalability, uniformity and less environmental impact (Sharma et al., 2024).

## Comparative Analysis of Biopolymers for Nanotherapeutics

Below is the comparative analysis of biopolymers based on key properties relevant to nanotherapeutic applications: (Stevanović & Filipović, 2024)

*Table 1. Comparative analysis of biopolymers based on key properties relevant to nanotherapeutic applications*

| Biopolymer | Source Type | Mechanical Strength | Degradation Rate | Key Functional Groups | Notable Advantages |
|---|---|---|---|---|---|
| Chitosan | Polysaccharide | Moderate | Moderate (enzymatic) | Amino (-$NH_2$), Hydroxyl (-OH) | Mucoadhesive, pH-sensitive, gene delivery potential |
| Alginate | Polysaccharide | Moderate (gel-forming) | Slow to Moderate | Carboxyl (-COOH), Hydroxyl (-OH) | Mild gelation, biocompatible encapsulation |
| Hyaluronic Acid | Polysaccharide | Low to Moderate | Rapid (enzymatic) | Carboxyl (-COOH), Hydroxyl (-OH) | CD44 targeting capability |
| Cellulose (modified) | Polysaccharide | High | Slow (requires modification) | Hydroxyl (-OH) | Strong mechanical stability |
| Albumin | Protein | Moderate | Enzymatic, Moderate | Amino, Carboxyl, Thiol | Natural transport pathways |
| Collagen | Protein | Moderate | Enzymatic, Controlled | Amino, Carboxyl | Excellent cell compatibility |
| Silk Fibroin | Protein | High | Tenable (slow to moderate) | Amino, Carboxyl | High mechanical strength, tenable crystallinity |
| Gelatine | Protein | Low to Moderate | Rapid to Moderate | Amino, Carboxyl | Easy processing, versatile loading |
| Pullulan | Microbial | Moderate | Moderate | Hydroxyl (-OH) | Water-soluble, easily modified |
| Xanthan Gum | Microbial | Moderate (viscous gel) | Slow to Moderate | Carboxyl, Hydroxyl | Stable across pH and temperature |

## 3. ECO-FRIENDLY REACTION METHOD AND GREEN DISCOVERY PROCESSES

Nanotherapeutics have rapidly developed, which has increased the requirement of environmentally conscious and biologically safe production techniques. The transition to biodegradable and renewable biopolymers in the field should be accompanied by an equal focus on sustainable synthesis processes that would reduce ecological effects and preserve high levels of therapeutic activity (Nayak & Sahoo, 2025). Green fabrication incorporates the concepts of environmental stewardship, resource efficiency and human safety into the creation and manufacturing of nanoparticles. With the introduction of the concept of green chemistry and new low-energy techniques, scientists will be able to come up with nanocarriers that are, in addition to being clinically useful, also eco-friendly (Gilbertson et al., 2015).

### Green Chemistry Principles

Green chemistry offers a systematic basis on how chemical processes can be designed to limit or reduce toxic chemicals. With regard to nanotherapeutics, these principles are used to inform material and solvents selection, reaction conditions and energy inputs to allow safer and more sustainable manufacture (Beach et al., 2009). Green chemistry revolves around atom economy. It focuses on the extent of using all the initial materials into the end product and therefore minimizing wastage. In the production of nanoparticles, high atom economy guarantees the generation of fewer byproducts and that raw materials are put into effective use. This minimizes the purification steps, cost of production and decreases environmental burden (Nalla et al., 2025). Another important principle is the use of non-toxic solvents especially aqueous systems. The method of fabrication of traditional nanoparticles is usually dependent on organic solvents like chloroform, dichloromethane or acetone, all of which are dangerous to human health and the environment. Green nanofabrication is taking into consideration water, ethanol, or other bio-based solvents as less harmful solvents. Not only do aqueous systems decrease toxicity, but they also increase the ability of aqueous systems to interact with sensitive biomolecules like proteins and nucleic acids (Majhi et al., 2019). Green fabrication also is characterized by energy efficiency. In some cases, traditional chemical techniques can be energy intensive by operating at high temperatures, long reaction times or during intensive mechanical stirring of the reaction mixture. Greener synthesis techniques seek to minimize thermal and mechanical energy use with improved reaction kinetics, catalytic systems and other more energetically efficient technologies such as the use of microwave or ultrasound energy (Gronnow et al., 2005).One more principle is that biodegradable products are designed, derivatives are reduced and that renewable feedstocks are utilized. Using this on nanotherapeutics, the materials, as well as the fabrication processes, are ensured to comply with environmental and biomedical safety standards (Chand et al., 2025).

Recent studies (2023–2025) have reported the use of natural deep eutectic solvents (NADES), enzyme-assisted nanoparticle synthesis, and plant-extract-mediated fabrication techniques as promising green approaches for producing biopolymer-based nanocarriers with reduced environmental impact and improved sustainability.

Recent targeting strategies have expanded beyond conventional folate and HER2 ligands to include aptamers, transferrin, epidermal growth factor receptor (EGFR)-targeting peptides, and immune-cell-directed ligands, improving tumor selectivity and therapeutic precision.

*Figure 2. Sustainable synthesis and green fabrication technique*

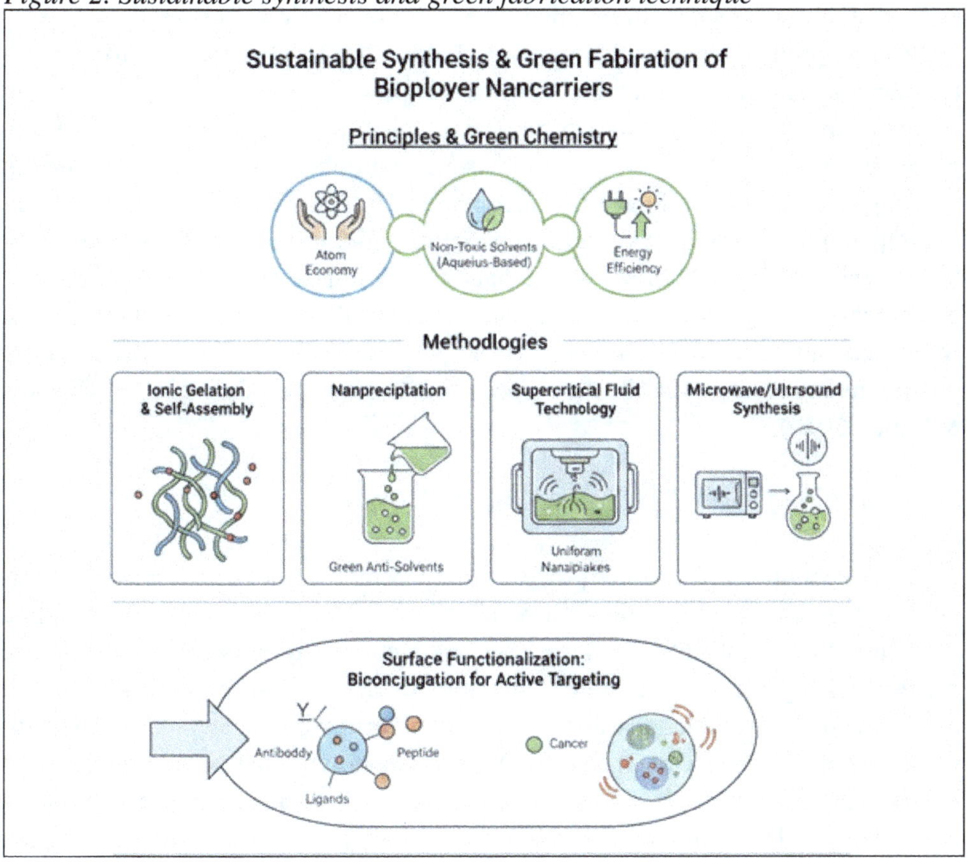

## Technologies in Green Nanofabrication:

### Ionic Gelation and Self- Assembly.

One of the most popular green fabrication methods of biopolymer-based nanoparticles is the ionic gelation. This process is based on the electrostatic interactions of oppositely charged polymers with crosslinking agents usually in aqueous conditions. As an illustration, multivalent anions can be associated with positively charged polymers to establish stable nanostructures under the absence of organic solvents and extreme conditions (Joye & McClements, 2014). It is nonaggressive, gentle and very reproducible. It usually takes place at room temperature and atmospheric pressure, which saves on energy. Since the procedure does not use toxic crosslinkers and harsh reaction conditions, the biological activity of drugs, peptides, or nucleic acids that are encapsulated remains intact in ionic gelation. This renders it especially appropriate to delicate therapeutics (Tsiklis, 2012). The other environmental-friendly strategy is self-assembly. During this process, amphiphilic polymers are spontaneously assembled into nanoscale, i.e., micelles, vesicles or nanogels when placed in aqueous conditions. These forces are hydrophobic interactions, hydrogen bonding, van der Waals forces and electrostatic interactions. Self-assembly does not require more complicated chemical reactions, and instead, it uses intrinsic molecular characteristics to create stable nanoparticles (Luo et al., 2015). These two techniques, ionic gelation and self-assembly, are scalable, do not produce a lot of waste and can be used with biodegradable polymers. They are very mild in their operations and therefore can be developed as sustainable nanotherapeutics (Okesola & Smith, 2016).

### Nanoprecipitation With Green Anti-Solvents.

Nanoprecipitation is a simple process through which the polymeric nano nanoparticles can be manufactured through precipitation of a polymer in a solvent into a non-solvent (anti-solvent). Conventional nanoprecipitation procedures tend to use detrimental organic solvents. Green nanoprecipitation substitutes them with less harmful sources like ethanol or solvents of plant origin and uses water as the anti-solvent (Truong-Dinh Tran et al., 2016). Here, the polymer and drug are dissolved in biodegradable solution, and then in drops added to an aqueous phase as the solution is stirred. The solution is swiftly diffused into water resulting in the creation of nanoparticles through supersaturation. Evaporation of the solvent under reduced pressure or gentle heating to eliminate the solvent is also done (Herrmann, 2003). Green nanoprecipitation has the benefits of being simple, requiring low energy and lower toxicity. Due to the fact that the method does not imply any high temperatures or complicated equipment, it is inexpensive and can be used in mass production. Also, the safety profile of the final nanotherapeutic product is improved because of the lack of residual hazards (Hussain et al., 2009). The aim of the optimization strategies is to reduce the volume of solvent, recycle the solvents where necessary and adjust the size of the particles in terms of stirring speed and concentration. Such improvements also make nanoprecipitation closer to green chemistry (Pinto et al., 2004).

Despite significant progress, large-scale manufacturing, batch-to-batch reproducibility, regulatory approval pathways, and long-term safety validation remain major challenges for clinical translation. Harmonized regulatory guidelines, standardized characterization methods, and scalable green manufacturing technologies will be essential for the successful commercialization of biopolymer-based nanocarriers.

## Supercritical Fluid Technology.

The technology of supercritical fluid is a very innovative and environmentally friendly method of the synthesis of nanoparticles. A supercritical fluid is a fluid in existence above its critical pressure and temperature and has properties between the properties of liquids and gases. The most frequently made use of supercritical fluids is carbon dioxide because it is a non-toxic, non-flammable, and cheap fluid (Sheth et al., 2012). Supercritical carbon dioxide has a great solvent capability of some compounds and can also be removed by the simple use of depressurization process and leaves no compounds behind. This avoids the use of organic solvents and minimizes the need of purification. Supercritical methods used in nanoparticle production are associated with a small size distribution and purity (Zhang et al., 2014). The technique is especially beneficial in wrapping hydrophobic drugs. The researchers can control the morphology and size of particles carefully by controlling pressure and temperature. In addition, supercritical fluid technology has the advantage of being carried out at relatively moderate temperatures, and therefore, thermolabile bioactive compounds are retained. Though, the initial equipment cost might be strong, the environmental and safety benefits are significant in the long term. This technology is the foundation of sustainable nanomanufacturing because the waste of the solvents is absent and the carbon dioxide can be recycled (Şengül et al., 2008).

## Microwave Assisted and Ultrasound Mediated Synthesis.

More energy-assisted techniques like the use of microwave irradiation and ultrasound are also being considered due to their effectiveness and less impact on the environment. The use of Microwave-assisted synthesis increases the rate of reactions in chemicals by directly interacting the polar molecules with the energy of microwaves. This causes fast and even heating, which contributes to a vast reaction time and use of energy. Microwave techniques usually enhance yield and particle homogeneity and reduce solvent need than traditional heating processes (Meraj et al., 2023).

Polymers Microwave processing seems to be especially appropriate in cross-linking of polymers, hydrogel formation, and stabilization of nanoparticles. The reduced reaction time has the benefit of reducing thermal deterioration of sensitive compounds and improving reproducibility. The method also allows aqueous-based reactions or solvent-free reactions, which is further in conformity to the principles of green chemistry (Mavila et al., 2016).

Sono chemical synthesis Sono chemical synthesis is also called ultrasound-mediated synthesis and involves the use of acoustic cavitation to produce localized high-energy microenvironments. When microscopic bubbles burst, a lot of heat and pressure is generated at a very tiny scale, which facilitates quick growth of the nucleation process and generation of particles. Although these are local effects, the general operation of the process under mild bulk conditions takes place (Huang et al., 2024). Increased mixing efficiency, shorter reaction time and even smaller and homogenous nanoparticles can be formed with ultrasound. It is also greatly applicable in emulsification, dispersion, and surface modification procedures. Notably, ultrasound techniques usually need lesser temperatures and reduced chemical additives in comparison to traditional ones. Microwave and ultrasound methods can help to make the processes more energy-efficient, shorten reactions time, and enhance the material performance (Bang & Suslick, 2010).

## Surface Functionalization: Bioconjugation Strategies of Active Targeting.

Sustainable nanotherapeutic fabrication goes further than the formation of core particles to environmentally friendly surface functionalisation. Active targeting involves the binding of ligands to nanoparticle surfaces including antibodies, peptides, aptamers, or small molecules. The green bioconjugation strategies are meant to ensure this is done using minimum toxic reagents and wastes. Covalent conjugation technique usually takes advantage of naturally occurring functional groups including amino, carboxyl groups or hydroxyl groups that are found in biopolymers. An example is the use of carbodiimide chemistry in aqueous systems, whereby an amide bond can be easily formed without too many byproducts. Another alternative that is eco-friendly and provides a high specificity is the enzyme-mediated conjugation, which can be carried out in mild conditions (Spada & Gerber-Lemaire, 2025). Surface modification can be efficiently and selectively achieved by click chemistry methods, especially those that avoid using copper catalysts and that do not result in toxicity. These reactions are normally done in water and produce only little wastes. Physical adsorption is a less complicated strategy, which depends on electrostatic interactions or hydrogen bonding to conjugate targeting ligands. Adsorption does not require chemical crosslinkers or deplorable reaction conditions as compared to covalent bonds, though they are less stable. Emphasis in green surface functionalization is the preservation of ligand bioactivity, minimization of excess reagents and the use of biodegradable linkers. Through sustainable incorporation of targeting moieties, nanocarriers will be capable of high cellular uptake, receptor-selective binding and increased therapeutic effectiveness without jeopardizing the environment (Lombardi et al., 2025).

## 4. PASSIVE AND ACTIVE PRECISION CANCER THERAPEUTIC TARGETING

Precision cancer therapeutics aim to target the tumour tissue with a high degree of drug delivery and reduce effects on the normal cells. Passive and active targeting strategies are the main bases of this selectivity. Passive targeting uses Enhanced Permeability and Retention (EPR) effect which is an effect found in most solid tumours. Tumours proliferate at a very fast pace and form abnormal and leaky blood vessels with large fenestrations that enable nanoscale drug carriers to get through with greater ease as compared to normal tissues. Also, tumours do not have effective lymphatic drainage and, as a result, nanoparticles remain in the tumour microenvironment. This leads to a higher accumulation of nanocarriers in tumour tissues than that of traditional small-molecule drugs that are not targeted (Chen, 2010). The EPR effect increases the drug concentration in the locality, decreases the systemic toxicity, and improves the therapeutic index. Its effectiveness, however, differs when it comes to the tumour type, size, vascularization, and patient-specific factors. Deep penetration of nanoparticles can be inhibited by heterogeneous blood flow and high interstitial pressure in tumours. Thus, the passive targeting increases the tumour accumulation but does not guarantee the effective uptake by the cancer cells (Bazak et al., 2014). Active targeting is based on passive accumulation in the sense that it adds molecular recognition processes. Under this method, nanocarriers are targeted with ligands (antibodies, peptides, or some small molecules) that bind to receptors overexpressed by cancer cells. These ligand-receptors interactions facilitate receptor-mediated endocytosis which helps in the more effective delivery of drugs into the cell. Active targeting enhances the cellular specificity, uptake, and off-target interactions are minimized. Advanced nanotherapeutics are more precise and effective in their therapy by combining passive accumulation with the EPR effect and receptor-mediated internalization (Deng et al., 2025).

*Figure 3. Mechanisms of precision cancer therapeutics*

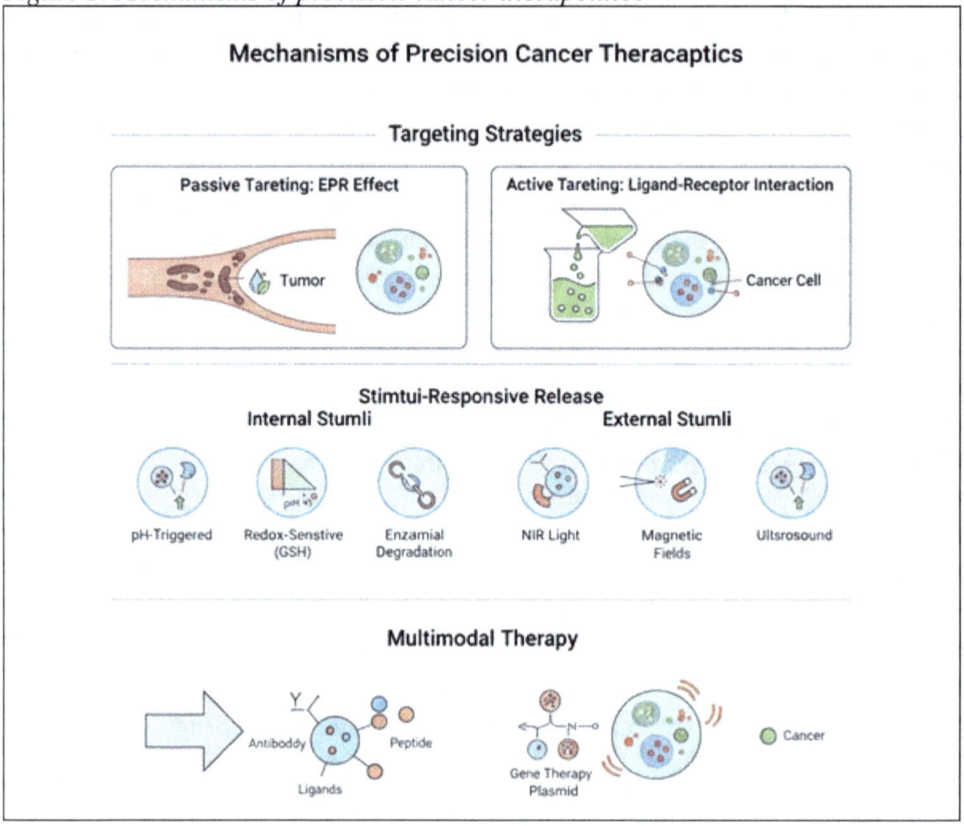


## Stimulus-Releases and Multimodal Therapy.

In addition to selective delivery, stimuli-responsive drug delivery systems have gained popularity as the leading method of precision cancer therapy. These systems are designed to deliver therapeutic agents based on certain internal or external stimuli and therefore, localized activation of drugs at the tumour site is achieved. Biological differences between tumour and normal tissues are used in internal stimuli-responsive systems. Acidic pH is one of the triggers. Nanocarriers are also used to deliver drugs in selective acidic conditions since the tumour microenvironment is slightly acidic because of the changes in metabolic activity. Redox systems exploit high intracellular glutathione of cancer cells. Nanocarriers with redox-cleavable bonds are stable in circulation yet they degrade in tumour cells, guaranteeing intracellular delivery of drugs. Tumour-overexpressed enzymes are also used in enzyme-responsive systems that cause site-specific degradation and the release of drugs (Zhao et al., 2024).

There are external stimuli that provide further regulation of drug activation. The near-infrared (NIR) light is capable of passing through tissues and promoting photosensitive nanocarriers that release drugs or produce reactive oxygen species that are cytotoxic. Targeting Magnetic field-sensitive systems allow guided targeting or localized hyperthermia, which increases the effectiveness of the treatment. Mechanical disruption can be induced with ultrasound and permeability of cell membranes can be improved, which allows the release of drugs into the cell in a controlled way, and penetration of tumours more deeply (Wen et al., 2023).

Multimodal therapy also improves the therapeutic outcomes brought about by the combination of various therapeutic modalities into a single nanocarrier. As an illustration, chemotherapy may be combined with gene therapy to silence the genes of resistance as cytotoxic agents are transferred. On the same note, chemotherapy with photodynamic therapy (PDT) has synergistic effects of destroying tumour cells by both light and chemical activations. Such combined systems overcome the idea of heterogeneity of tumours, minimize resistance, and enhance the overall therapeutic effect. Combined, targeted delivery, stimuli-responsive release, and multimodal strategies are more sophisticated mechanisms in precision cancer therapeutics, which have the benefits of being safer and more effective in treatment (Mohamed et al., 2025).

## 5. BIOCOMPATIBILITY, BIODISTRIBUTION AND TOXICOLOGY

The success of a green nanocarrier in the clinical application is not only assessed by their therapeutic efficacy, but also the safety profile, biodistribution, and metabolism. A material that can be used without causing any harmful local or systemic effects is said to be biocompatible so that it can fulfil its intended purpose. Natural-based biopolymer nanocarriers in general are less toxic than synthetic nanocarriers because it is structurally close to natural biomolecules. Nonetheless, there in vivo activity should be appropriately assessed to provide predictable pharmacokinetics and low adverse responses (Khan et al., 2025).

After their administration, nanoparticles face a biological barrier and the circulation period and the distribution of the particles in the organs determine the remaining period. One significant difficulty is the Reticuloendothelial System (RES) or mononuclear phagocytosis system that comprises macrophages in the liver, the spleen, and the lymph nodes. The RES detects and removes foreign particles in circulation, which restricts the half-life of nanoparticles. The green nanocarriers should be optimized in size, surface charge, and hydrophilic surface engineering to partly avoid the uptake of the nanocarrier by the RES. The aggregates with the size in the range of 10 to 200 nm tend to have a longer circulation than the larger aggregates. The neutral or slightly negative charges on the surface slow down the process of rapid opsonization, which is the antigens attaching plasma proteins to particles to be cleared. The natural polymers used to produce hydrophilic coating provide a steric barrier that reduces the adsorption of proteins and recognition by macrophages. Such techniques promote tumour deposition at the expense of hepatic and splenic sequestration (Shafaei et al., 2023).

The other crucial parameter affecting clinical translation is immunogenicity. Though the concept of natural polymers as biocompatible is widely agreed upon, a polysaccharide, nevertheless, may cause an immune reaction depending on its origin, purity, and structural alteration. The impurities could be endotoxins, residual proteins, crosslinking agents, etc that would cause an inflammatory or hypersensitivity reaction. Thus, strict purification and characterization is necessary. The immune assessment normally entails assessment of cytokine release, complement activation, antibody formation, as well as cellular immune response in preclinical models. Biopolymers which are correctly processed tend to show low immunogenicity as they have a similarity to naturally produced extracellular matrix components or plasma proteins. However, even in repeated dosage, the adaptive immune response can still occur and thus the long-term safety trials are to be conducted (Esfahani et al., 2020).

Unlike previous reviews that focus primarily on nanocarrier design or cancer-targeting mechanisms, this chapter integrates green biopolymer sources, sustainable fabrication approaches, precision-targeting strategies, and safety considerations within a single framework, providing a comprehensive perspective on environmentally sustainable precision oncology.

The fact that they have metabolic pathways that are conducive to their toxicological profile is another factor that makes biopolymeric nanocarriers a good option. In comparison to non-degradable synthetic materials, biodegradable biopolymers can be decomposed by enzymes or hydrolytic mechanisms into small and non-toxic metabolites. Carriers made of polysaccharides break down to simple sugars or ironic acid, which may be transferred to normal metabolic processes or may be cleared through renal excretion. Carriers based on proteins are broken down by proteolytic enzymes to amino acids which are either used again in the production of proteins or used to provide energy. Degradation rate is determined by the structure of the polymer, crosslinking density and environmental factors including pH and availability of enzymes. Efficient biodegradation minimizes accumulation in the tissues over the long term as well as decreasing the chances of chronic toxicity. Finally, biodistribution, immune compatibility and metabolic clearance should be mastered to make sure that green nanocarriers have therapeutic effect without jeopardizing the life of a patient (Jain & Bhise, 2025).

## 6. FUTURE PERSPECTIVES

The future of the green nanocarriers is at the cross-section (sustainability, technological innovation, and personalized medicine). Circular economy in medicine is one of the novel ideas that encourages the utilization of renewable and waste-resources to extract biopolymers. The agricultural byproducts like crustacean shells, plant fibres and microbial fermentation residues can be used as the raw materials of high-value biomedical polymers. By converting waste into therapeutic substances researchers maximize waste reduction to the environment and develop cost effective supply chains. This strategy will place nanomedicine at the centre of the sustainability agenda and sustainable use of the environment (Falsini et al., 2018).

The innovation will be further accelerated with the use of artificial intelligence (AI) and machine learning. Polymer-drug interactions can be investigated using predictive modelling to optimize nanoparticles, predict biodistribution trends, or optimize nanoparticles. The algorithms developed with AI can be used to discover the best polymer compositions, ligand densities, and release kinetics based on a patient profile. Such tools enable to design nanocarriers and tumour-specific biomarkers and genetic signatures, facilitating personalized oncology. Optimization of manufacturing and quality control can also be optimized by machine learning, which makes reproducibility and compliance with the regulations more effective (Albertoni et al., 2023).

## CONCLUSION

A shift in the traditional chemotherapy to precision nanomedicine, is a big step in the field of oncology. Green bio-based biopolymer nanocarriers, according to the information presented in this chapter, are a combination of therapeutic effectiveness and environmental friendliness. Polymers that are formed naturally are biodegradable, biocompatible and versatile in their functions, thus are applicable in targeted and controlled delivery of drugs. Green chemistry based eco-friendly fabrication methods further boost the clinical prospects of these systems because they reduce toxic reagents, use of less energy and decreases environmental degradation. Passive and active targeting mechanisms combined with stimuli-responsive and multimodal strategies enhance specificity of the targets in an improved tumor and reduce systemic toxicity. Even with this promising developments, scalability, regulatory harmonization, and long-term biosafety assessment are the issues that are of concern to successful clinical translation. Sustainable manufacturing and the need to continue interdisciplinary research will be necessary. Finally, green nanocarriers have a great potential to transform cancer therapeutics to match precision medicine with sustainability and patient-centered care.

## REFERENCES


Albertoni, R., Colantonio, S., & Stefanowski, J. (2023). Reproducibility of machine learning: Terminology, recommendations and open issues. *arXiv preprint arXiv:2302.12691*.

Arif, U., Haider, S., Haider, A., Khan, N., Alghyamah, A. A., Jamila, N., Khan, M. I., Almasry, W. A., & Kang, I.-K. (2019). Biocompatible polymers and their potential biomedical applications: A review. *Current Pharmaceutical Design*, *25*(34), 3608–3619. DOI: 10.2174/1381612825999191011105148 PMID: 31604409

Bang, J. H., & Suslick, K. S. (2010). Applications of ultrasound to the synthesis of nanostructured materials. *Advanced Materials*, *22*(10), 1039–1059. DOI: 10.1002/adma.200904093 PMID: 20401929

Bazak, R., Houri, M., El Achy, S., Hussein, W., & Refaat, T. (2014). Passive targeting of nanoparticles to cancer: A comprehensive review of the literature. *Molecular and Clinical Oncology*, *2*(6), 904–908. DOI: 10.3892/mco.2014.356 PMID: 25279172

Beach, E. S., Cui, Z., & Anastas, P. T. (2009). Green Chemistry: A design framework for sustainability. *Energy & Environmental Science*, *2*(10), 1038–1049. DOI: 10.1039/b904997p

Beltrán-Gracia, E., López-Camacho, A., Higuera-Ciapara, I., Velázquez-Fernández, J. B., & Vallejo-Cardona, A. A. (2019). Nanomedicine review: Clinical developments in liposomal applications. *Cancer Nanotechnology*, *10*(1), 11. DOI: 10.1186/s12645-019-0055-y

Bhalani, D. V., Nutan, B., Kumar, A., & Singh Chandel, A. K. (2022). Bioavailability enhancement techniques for poorly aqueous soluble drugs and therapeutics. *Biomedicines*, *10*(9), 2055. DOI: 10.3390/biomedicines10092055 PMID: 36140156

Bhilare, N. V., Marulkar, V. S., Chatap, V. K., Kulkarni, A. S., Dombe, S. A., & Shirote, P. J. (2024). Biodegradability and Biodegradable Polymers: An Overview. *Handbook of Biodegradable Polymers*, 1-38.

Biju, V. (2014). Chemical modifications and bioconjugate reactions of nanomaterials for sensing, imaging, drug delivery and therapy. *Chemical Society Reviews*, *43*(3), 744–764. DOI: 10.1039/C3CS60273G PMID: 24220322

Boulaiz, H., Alvarez, P. J., Ramirez, A., Marchal, J. A., Prados, J., Rodríguez-Serrano, F., Perán, M., Melguizo, C., & Aranega, A. (2011). Nanomedicine: Application areas and development prospects. *International Journal of Molecular Sciences*, *12*(5), 3303–3321. DOI: 10.3390/ijms12053303 PMID: 21686186

Chand, B., & Phogat, P. Shreya, Singh, S., & Jha, R. (2025). Regulatory Framework and Standards on Bionanocomposites Manufacturing and Applications: Social Impact and Ethical Considerations. In *Advanced Bionanocomposite Materials: Innovations for Sustainable Development* (pp. 625-667). Springer.

Chaudhary, S., Jain, V. P., & Jaiswar, G. (2022). The composition of polysaccharides: monosaccharides and binding, group decorating, polysaccharides chains. In *Innovation in nano-polysaccharides for eco-sustainability* (pp. 83–118). Elsevier. DOI: 10.1016/B978-0-12-823439-6.00005-2

Chen, Z. (2010). Small-molecule delivery by nanoparticles for anticancer therapy. *Trends in Molecular Medicine*, *16*(12), 594–602. DOI: 10.1016/j.molmed.2010.08.001 PMID: 20846905

Deng, X., Zhang, Z., Ren, T., & Chen, L. (2025). Regulation of oxidative stress and inflammation caused by drug accumulation in the TME based on EPR-passive strategy and active targeting. *Cancer Nanotechnology*, *16*(1), 40. DOI: 10.1186/s12645-025-00342-1

Esfahani, K., Elkrief, A., Calabrese, C., Lapointe, R., Hudson, M., Routy, B., Miller, W. H.Jr, & Calabrese, L. (2020). Moving towards personalized treatments of immune-related adverse events. *Nature Reviews. Clinical Oncology*, *17*(8), 504–515. DOI: 10.1038/s41571-020-0352-8 PMID: 32246128

Ezike, T. C., Okpala, U. S., Onoja, U. L., Nwike, C. P., Ezeako, E. C., Okpara, O. J., Okoroafor, C. C., Eze, S. C., Kalu, O. L., Odoh, E. C., Nwadike, U. G., Ogbodo, J. O., Umeh, B. U., Ossai, E. C., & Nwanguma, B. C. (2023). Advances in drug delivery systems, challenges and future directions. *Heliyon*, *9*(6), e17488. DOI: 10.1016/j.heliyon.2023.e17488 PMID: 37416680

Falsini, S., Bardi, U., Abou-Hassan, A., & Ristori, S. (2018). Sustainable strategies for large-scale nanotechnology manufacturing in the biomedical field. *Green Chemistry*, *20*(17), 3897–3907. DOI: 10.1039/C8GC01248B

Falzone, L., Salomone, S., & Libra, M. (2018). Evolution of cancer pharmacological treatments at the turn of the third millennium. *Frontiers in Pharmacology*, *9*, 421926. DOI: 10.3389/fphar.2018.01300 PMID: 30483135

Fam, S. Y., Chee, C. F., Yong, C. Y., Ho, K. L., Mariatulqabtiah, A. R., & Tan, W. S. (2020). Stealth coating of nanoparticles in drug-delivery systems. *Nanomaterials (Basel, Switzerland)*, *10*(4), 787. DOI: 10.3390/nano10040787 PMID: 32325941

Fazal, T., Murtaza, B. N., Shah, M., Iqbal, S., Rehman, M., Jaber, F., Dera, A. A., Awwad, N. S., & Ibrahium, H. A. (2023). Recent developments in natural biopolymer based drug delivery systems. *RSC Advances, 13*(33), 23087–23121. DOI: 10.1039/D3RA03369D PMID: 37529365

Fernandez-Fernandez, A., Manchanda, R., & McGoron, A. J. (2011). Theranostic applications of nanomaterials in cancer: Drug delivery, image-guided therapy, and multifunctional platforms. *Applied Biochemistry and Biotechnology, 165*(7), 1628–1651. DOI: 10.1007/s12010-011-9383-z PMID: 21947761

Ghuttora, N. (2016). Increase the usage of biopolymers and biodegradable polymers for sustainable environment.

Gilbertson, L. M., Zimmerman, J. B., Plata, D. L., Hutchison, J. E., & Anastas, P. T. (2015). Designing nanomaterials to maximize performance and minimize undesirable implications guided by the Principles of Green Chemistry. *Chemical Society Reviews, 44*(16), 5758–5777. DOI: 10.1039/C4CS00445K PMID: 25955514

Gorgieva, S., & Kokol, V. (2011). Collagen-vs. gelatine-based biomaterials and their biocompatibility: review and perspectives. *Biomaterials applications for nanomedicine, 2*(2011), 17-52.

Gronnow, M. J., White, R. J., Clark, J. H., & Macquarrie, D. J. (2005). Energy efficiency in chemical reactions: A comparative study of different reaction techniques. *Organic Process Research & Development, 9*(4), 516–518. DOI: 10.1021/op0498060

Haider, T., Pandey, V., Banjare, N., Gupta, P. N., & Soni, V. (2020). Drug resistance in cancer: Mechanisms and tackling strategies. *Pharmacological Reports, 72*(5), 1125–1151. DOI: 10.1007/s43440-020-00138-7 PMID: 32700248

Herrmann, H. (2003). Kinetics of aqueous phase reactions relevant for atmospheric chemistry. *Chemical Reviews, 103*(12), 4691–4716. DOI: 10.1021/cr020658q PMID: 14664629

Huang, H., Zheng, Y., Chang, M., Song, J., Xia, L., Wu, C., Jia, W., Ren, H., Feng, W., & Chen, Y. (2024). Ultrasound-based micro-/nanosystems for biomedical applications. *Chemical Reviews, 124*(13), 8307–8472. DOI: 10.1021/acs.chemrev.4c00009 PMID: 38924776

Hussain, S. M., Braydich-Stolle, L. K., Schrand, A. M., Murdock, R. C., Yu, K. O., Mattie, D. M., Schlager, J. J., & Terrones, M. (2009). Toxicity evaluation for safe use of nanomaterials: Recent achievements and technical challenges. *Advanced Materials, 21*(16), 1549–1559. DOI: 10.1002/adma.200801395

Jaffee, E. M., Van Dang, C., Agus, D. B., Alexander, B. M., Anderson, K. C., Ashworth, A., Barker, A. D., Bastani, R., Bhatia, S., & Bluestone, J. A. (2017). Future cancer research priorities in the USA: A Lancet Oncology Commission. *The Lancet. Oncology*, *18*(11), e653–e706. DOI: 10.1016/S1470-2045(17)30698-8 PMID: 29208398

Jain, A., & Bhise, K. (2025). Nano-pharmacokinetics and pharmacodynamics of green-synthesized ZnO nanoparticles: A pathway to safer therapeutic applications. *Xenobiotica; The Fate of Foreign Compounds in Biological Systems*, *55*(4), 265–276. DOI: 10.1080/00498254.2025.2505062 PMID: 40380770

John, P. M., Emmanuel, M., Beegum, J., John, F., & George, J. (2023). Nano-drug delivery systems for tumour-targeting: overcoming the limitations of chemotherapy. In *Targeted Cancer Therapy in Biomedical Engineering* (pp. 487–521). Springer. DOI: 10.1007/978-981-19-9786-0_14

Joyce, K., Fabra, G. T., Bozkurt, Y., & Pandit, A. (2021). Bioactive potential of natural biomaterials: Identification, retention and assessment of biological properties. *Signal Transduction and Targeted Therapy*, *6*(1), 122. DOI: 10.1038/s41392-021-00512-8 PMID: 33737507

Joye, I. J., & McClements, D. J. (2014). Biopolymer-based nanoparticles and microparticles: Fabrication, characterization, and application. *Current Opinion in Colloid & Interface Science*, *19*(5), 417–427. DOI: 10.1016/j.cocis.2014.07.002

Kaur, R., Bhardwaj, A., & Gupta, S. (2023). Cancer treatment therapies: Traditional to modern approaches to combat cancers. *Molecular Biology Reports*, *50*(11), 9663–9676. DOI: 10.1007/s11033-023-08809-3 PMID: 37828275

Khan, I. A., Yu, T., Yang, M., Liu, J., & Chen, Z. (2025). A systematic review of toxicity, biodistribution, and biosafety in upconversion nanomaterials: Critical insights into toxicity mitigation strategies and future directions for safe applications. *BME frontiers, 6*, 0120.

Khan, I. U., Khan, R. U., Asif, H., Khalid, S. H., Asghar, S., Saleem, M., Shah, K. U., Shah, S. U., Rizvi, S. A., & Shahzad, Y. (2017). Co-delivery strategies to overcome multidrug resistance in ovarian cancer. *International Journal of Pharmaceutics*, *533*(1), 111–124. DOI: 10.1016/j.ijpharm.2017.09.060 PMID: 28947245

Kučuk, N., Primožič, M., Knez, Ž., & Leitgeb, M. (2023). Sustainable biodegradable biopolymer-based nanoparticles for healthcare applications. *International Journal of Molecular Sciences*, *24*(4), 3188. DOI: 10.3390/ijms24043188 PMID: 36834596

Kunjiappan, S., Pavadai, P., Vellaichamy, S., Ram Kumar Pandian, S., Ravishankar, V., Palanisamy, P., Govindaraj, S., Srinivasan, G., Premanand, A., Sankaranarayanan, M., & Theivendren, P. (2021). Surface receptor-mediated targeted drug delivery systems for enhanced cancer treatment: A state-of-the-art review. *Drug Development Research, 82*(3), 309–340. DOI: 10.1002/ddr.21758 PMID: 33170541

Li, Z., Ning, F., Wang, C., Yu, H., Ma, Q., & Sun, Y. (2021). Normalization of the tumor microvasculature based on targeting and modulation of the tumor microenvironment. *Nanoscale, 13*(41), 17254–17271. DOI: 10.1039/D1NR03387E PMID: 34651623

Lin, J., Jiao, G., & Kermanshahi-Pour, A. (2022). Algal polysaccharides-based hydrogels: Extraction, synthesis, characterization, and applications. *Marine Drugs, 20*(5), 306. DOI: 10.3390/md20050306 PMID: 35621958

Lombardi, L., Genio, V. D., Albericio, F., & Williams, D. R. (2025). Advances in peptidomimetics for next-generation therapeutics: Strategies, modifications, and applications. *Chemical Reviews, 125*(15), 7099–7166. DOI: 10.1021/acs.chemrev.4c00989 PMID: 40698392

Lotfi, M., Nejib, M., & Naceur, M. (2013). Cell adhesion to biomaterials: Concept of biocompatibility. *Advances in biomaterials science and biomedical applications, 8*, 208-240.

Luo, D., Yan, C., & Wang, T. (2015). Interparticle forces underlying nanoparticle self-assemblies. *Small, 11*(45), 5984–6008. DOI: 10.1002/smll.201501783 PMID: 26436692

Majhi, K. C., Karfa, P., Kumar, S., & Madhuri, R. (2019). Water as the green solvent in organic synthesis. *Ind. Appl. Green Solvents, 2*, 182–201.

Marinho, A., Nunes, C., & Reis, S. (2021). Hyaluronic acid: A key ingredient in the therapy of inflammation. *Biomolecules, 11*(10), 1518. DOI: 10.3390/biom11101518 PMID: 34680150

Marusyk, A., Almendro, V., & Polyak, K. (2012). Intra-tumour heterogeneity: A looking glass for cancer? *Nature Reviews. Cancer, 12*(5), 323–334. DOI: 10.1038/nrc3261 PMID: 22513401

Masimirembwa, C. M., Bredberg, U., & Andersson, T. B. (2003). Metabolic stability for drug discovery and development: Pharmacokinetic and biochemical challenges. *Clinical Pharmacokinetics, 42*(6), 515–528. DOI: 10.2165/00003088-200342060-00002 PMID: 12793837

Mavila, S., Eivgi, O., Berkovich, I., & Lemcoff, N. G. (2016). Intramolecular cross-linking methodologies for the synthesis of polymer nanoparticles. *Chemical Reviews*, *116*(3), 878–961. DOI: 10.1021/acs.chemrev.5b00290 PMID: 26270435

Meraj, A., Singh, S. P., Jawaid, M., Nasef, M. M., Alomar, T. S., & AlMasoud, N. (2023). A review on eco-friendly isolation of lignin by natural deep eutectic solvents from agricultural wastes. *Journal of Polymers and the Environment*, *31*(8), 3283–3316. DOI: 10.1007/s10924-023-02817-x

Mikušová, V., & Mikuš, P. (2021). Advances in chitosan-based nanoparticles for drug delivery. *International Journal of Molecular Sciences*, *22*(17), 9652. DOI: 10.3390/ijms22179652 PMID: 34502560

Mohamed, R. G. A., Ali, S. M., Ahmed, I. S., Rawas-Qalaji, M., & Hussain, Z. (2025). Next-generation nanocarriers for colorectal cancer: Passive, active, and stimuli-responsive strategies for precision therapy. *Biomaterials Science*, *13*(20), 5626–5664. DOI: 10.1039/D5BM01176K PMID: 40928395

Murray, C. J., & Lopez, A. D. (1997). Mortality by cause for eight regions of the world: Global Burden of Disease Study. *Lancet*, *349*(9061), 1269–1276. DOI: 10.1016/S0140-6736(96)07493-4 PMID: 9142060

Naderinezhad, S., Amoabediny, G., & Haghiralsadat, F. (2017). Co-delivery of hydrophilic and hydrophobic anticancer drugs using biocompatible pH-sensitive lipid-based nano-carriers for multidrug-resistant cancers. *RSC Advances*, *7*(48), 30008–30019. DOI: 10.1039/C7RA01736G

Naeem, M. S., Ahmed, S., Sulman, A., Usman, S., Farooq, M. A., Talha, M., Awan, A., Afzal, Z., Khan, M. Z., & Hanif, A. (2025). Role of Medicinal Plants in the Management of Metabolic Disorders: Mechanisms, Efficacy, and Therapeutic Prospects.

Nalla, B. T., Choudhary, B. K., Sunil Kumar, M., Ramachandra, C. G., Vaghela, D. R., Pradhan, P., Smerat, A., & Kamakshi Priya, K. (2025). Nanocatalyst-enabled waste-to-energy systems: from material innovation to techno-economic and sustainability pathways. *Reviews in Inorganic Chemistry*(0).

Natarajan, J. V., Darwitan, A., Barathi, V. A., Ang, M., Htoon, H. M., Boey, F., Tam, K. C., Wong, T. T., & Venkatraman, S. S. (2014). Sustained drug release in nanomedicine: A long-acting nanocarrier-based formulation for glaucoma. *ACS Nano*, *8*(1), 419–429. DOI: 10.1021/nn4046024 PMID: 24392729

Nayak, N., & Sahoo, T. R. (2025). Sustainable Nanomaterials Production Using Bio-based Polymers. *Green Chemistry Methods for Producing Biopolymeric Nanomaterials*, 133-160.

Niemeyer, C. M. (2001). Nanoparticles, proteins, and nucleic acids: Biotechnology meets materials science. *Angewandte Chemie International Edition*, *40*(22), 4128–4158. DOI: 10.1002/1521-3773(20011119)40:22<4128::AID-ANIE4128>3.0.CO;2-S PMID: 29712109

Okesola, B. O., & Smith, D. K. (2016). Applying low-molecular weight supramolecular gelators in an environmental setting–self-assembled gels as smart materials for pollutant removal. *Chemical Society Reviews*, *45*(15), 4226–4251. DOI: 10.1039/C6CS00124F PMID: 27241027

Pinto, G. A., Durao, F. O., Fiúza, A. M., Guimaraes, M. M., & Madureira, C. N. (2004). Design optimisation study of solvent extraction: Chemical reaction, mass transfer and mixer–settler hydrodynamics. *Hydrometallurgy*, *74*(1-2), 131–147. DOI: 10.1016/j.hydromet.2004.02.002

Rai, M., Wypij, M., Ingle, A. P., Trzcińska-Wencel, J., & Golińska, P. (2021). Emerging trends in pullulan-based antimicrobial systems for various applications. *International Journal of Molecular Sciences*, *22*(24), 13596. DOI: 10.3390/ijms222413596 PMID: 34948392

Rasool, M., Malik, A., Waquar, S., Arooj, M., Zahid, S., Asif, M., Shaheen, S., Hussain, A., Ullah, H., & Gan, S. H. (2022). New challenges in the use of nanomedicine in cancer therapy. *Bioengineered*, *13*(1), 759–773. DOI: 10.1080/21655979.2021.2012907 PMID: 34856849

Şengül, H., Theis, T. L., & Ghosh, S. (2008). Toward sustainable nanoproducts: An overview of nanomanufacturing methods. *Journal of Industrial Ecology*, *12*(3), 329–359. DOI: 10.1111/j.1530-9290.2008.00046.x

Shafaei, N., Khorshidi, S., & Karkhaneh, A. (2023). The immune-stealth polymeric coating on drug delivery nanocarriers: In vitro engineering and in vivo fate. *Journal of Biomaterials Applications*, *38*(2), 159–178. DOI: 10.1177/08853282231185352 PMID: 37480331

Sharma, M., Tellili, N., Kacem, I., & Rouissi, T. (2024). Microbial biopolymers: From production to environmental applications—a review. *Applied Sciences (Basel, Switzerland)*, *14*(12), 5081. DOI: 10.3390/app14125081

Sheldon, R. A. (2005). Green solvents for sustainable organic synthesis: State of the art. *Green Chemistry*, *7*(5), 267–278. DOI: 10.1039/b418069k

Sheth, P., Sandhu, H., Singhal, D., Malick, W., Shah, N., & Serpil Kislalioglu, M. (2012). Nanoparticles in the pharmaceutical industry and the use of supercritical fluid technologies for nanoparticle production. *Current Drug Delivery*, *9*(3), 269–284. DOI: 10.2174/156720112800389052 PMID: 22283656

Spada, A., & Gerber-Lemaire, S. (2025). Surface functionalization of nanocarriers with anti-EGFR ligands for cancer active targeting. *Nanomaterials (Basel, Switzerland)*, *15*(3), 158. DOI: 10.3390/nano15030158 PMID: 39940134

Stevanović, M., & Filipović, N. (2024). A review of recent developments in biopolymer nano-based drug delivery systems with antioxidative properties: Insights into the last five years. *Pharmaceutics*, *16*(5), 670. DOI: 10.3390/pharmaceutics16050670 PMID: 38794332

Sulman, A., Farooq, M. A., Naseer, R. D., Hanif, A., Afzal, Z., Khan, M. Z., Naeem, M. S., Ahmad, H. M., Mahmood, S., & Kharl, H. A. A. (2025). Neurobiology and Management of Drug Addiction and Dependence: A Multidisciplinary Review. *Journal of Pharma and Biomedics*, *3*(1), 191–207.

Tamariz, E., & Rios-Ramírez, A. (2013). Biodegradation of medical purpose polymeric materials and their impact on biocompatibility. *Biodegradation-Life of Science Croatia: Intech*, 1-29.

Tincu, C.-E., Andri oiu, C. V., Popa, M., & Ochiuz, L. (2023). Recent advancements and strategies for overcoming the blood–brain barrier using albumin-based drug delivery systems to treat brain cancer, with a focus on glioblastoma. *Polymers*, *15*(19), 3969. DOI: 10.3390/polym15193969 PMID: 37836018

Torchilin, V. (2008). Multifunctional pharmaceutical nanocarriers: development of the concept. In *Multifunctional pharmaceutical nanocarriers* (pp. 1–32). Springer. DOI: 10.1007/978-0-387-76554-9_1

Trombino, S., Sole, R., Di Gioia, M. L., Procopio, D., Curcio, F., & Cassano, R. (2023). Green chemistry principles for nano-and micro-sized hydrogel synthesis. *Molecules (Basel, Switzerland)*, *28*(5), 2107. DOI: 10.3390/molecules28052107 PMID: 36903352

Truong-Dinh Tran, T., Ha-Lien Tran, P., Tu Nguyen, K., & Tran, V.-T. (2016). Nano-precipitation: Preparation and application in the field of pharmacy. *Current Pharmaceutical Design*, *22*(20), 2997–3006. DOI: 10.2174/1381612822666160408151702 PMID: 27055935

Tsiklis, D. S. (2012). *Handbook of techniques in high-pressure research and engineering*. Springer Science & Business Media.

Veleva, V., Hart, M., Greiner, T., & Crumbley, C. (2003). Indicators for measuring environmental sustainability: A case study of the pharmaceutical industry. *Benchmarking*, *10*(2), 107–119. DOI: 10.1108/14635770310469644

Wen, Z., Liu, C., Teng, Z., Jin, Q., Liao, Z., Zhu, X., & Huo, S. (2023). Ultrasound meets the cell membrane: For enhanced endocytosis and drug delivery. *Nanoscale*, *15*(33), 13532–13545. DOI: 10.1039/D3NR02562D PMID: 37548587

Wu, X., Xin, Y., Zhang, H., Quan, L., & Ao, Q. (2024). Biopolymer-based nanomedicine for cancer therapy: Opportunities and challenges. *International Journal of Nanomedicine*, *19*, 7415–7471. DOI: 10.2147/IJN.S460047 PMID: 39071502

Zhang, R. X., Ahmed, T., Li, L. Y., Li, J., Abbasi, A. Z., & Wu, X. Y. (2017). Design of nanocarriers for nanoscale drug delivery to enhance cancer treatment using hybrid polymer and lipid building blocks. *Nanoscale*, *9*(4), 1334–1355. DOI: 10.1039/C6NR08486A PMID: 27973629

Zhang, X., Heinonen, S., & Levänen, E. (2014). Applications of supercritical carbon dioxide in materials processing and synthesis. *RSC Advances*, *4*(105), 61137–61152. DOI: 10.1039/C4RA10662H

Zhang, X., Yin, Z., Xiang, S., Yan, H., & Tian, H. (2024). Degradation of polymer materials in the environment and its impact on the health of experimental animals: A review. *Polymers*, *16*(19), 2807. DOI: 10.3390/polym16192807 PMID: 39408516

Zhao, J., Li, X., Ma, T., Chang, B., Zhang, B., & Fang, J. (2024). Glutathione-triggered prodrugs: Design strategies, potential applications, and perspectives. *Medicinal Research Reviews*, *44*(3), 1013–1054. DOI: 10.1002/med.22007 PMID: 38140851

# Chapter 5
# Polymeric Nanocarriers for Controlled Drug Release

**Zeeshan Ahmad**
http://orcid.org/0009-0009-9459-659X
*COMSATS University Islamabad, Pakistan*

**Mujtaba ul Hassan**
*COMSATS University Islamabad, Pakistan*

**Hira Maryam**
*COMSATS University Islamabad, Pakistan*

**Nabeel ur Rehman**
http://orcid.org/0009-0009-6632-6218
*COMSATS University Islamabad, Pakistan*

**Kasim Sakran Abass**
*University of Kirkuk, Iraq*

**Ayesha Fazal**
http://orcid.org/0009-0007-5328-267X
*COMSATS University Islamabad, Pakistan*

**Muhammad Ilyas**
http://orcid.org/0009-0008-2573-367X
*COMSATS University Islamabad, Pakistan*

**Humaira Yasmin**
*COMSATS University Islamabad, Pakistan*

## ABSTRACT

*Polymeric nanocarriers are promising tools in cancer nanomedicine, offering better control over drug delivery and tumor targeting. Conventional therapies often suffer from poor selectivity, systemic toxicity, and multidrug resistance, creating a need for safer delivery systems. Built from natural or synthetic polymers, these carriers provide biodegradability, biocompatibility, high drug loading, and tunable release. They can encapsulate or conjugate therapeutic agents and release them through diffusion, degradation, swelling, or stimuli-responsive mechanisms. Surface ligands*

DOI: 10.4018/979-8-3373-9848-8.ch005

*support passive, active, and biomimetic tumor targeting, reducing off-target effects. Beyond chemotherapy, they are useful for gene delivery, combination therapy, and theranostics. However, biological barriers, instability, scale-up problems, and regulation still limit clinical translation. Smart design, AI-guided optimization, and hybrid nanomaterials may help address these limits and support precision oncology.*

## 1. INTRODUCTION

Despite all advances towards disease complexity and challenge, cancer in modern medicine is still one of the most complex and challenging diseases, as it is characterized by uncontrolled cell proliferation, metastatic spread and high genetic and phenotypic heterogeneity (Brown et al., 2023) .Although cancer has seen a huge improvement in early detection and treatment, the disease remains one of the world's top killers and reasons for early death. Although traditional treatment methods such as chemotherapy, radiotherapy, and surgery have shown their efficacy, they are limited in terms of the distribution of drugs in a very specific manner with minimal systemic toxicity and rapid development of drug resistance (Zafar et al., 2025). These constraints highlight the need for improved, selective, and less damaging therapeutic methods that are either less invasive or don't harm any cells to help treat the growing number of cancers (Colonna, 2025).

With the emergence of nanotechnology, the approach of its application to the delivery of drugs has inflicted a paradigm shift in oncology, and now it is possible to design multifunctional platforms for controlled and targeted therapy (Sell et al., 2023). Unique physicochemical properties including large surface area-to-volume ratio, tunable surface chemistry, and molecular/cellular interaction are featured by nanoscale systems. These features enable Nano carriers to enhance the solubility of the drugs, improve their pharmacokinetics, extend their systemic circulation time and promote preferential accumulation in tumor tissues via enhanced permeability and retention (EPR) effect. Thus, nanotechnology offers a promising approach to address some intrinsic challenges to traditional cancer therapy (Dalbanjan et al., 2025).

Poly-synthetic Nano carriers are increasingly being recognized as one of the most promising and clinically translatable platforms for the delivery of cancer drugs among all the different nanomaterial platforms (Bozzer et al., 2022). These carriers are based on natural, synthetic or hybrid polymers and have a tremendous versatility with respect to structural design, functionalization and drug loading capacities. Their properties of loading therapeutic agents and releasing them in a controlled way make them very suitable for precision oncology applications (Ranjit De et al., 2022). Moreover, developments in polymer chemistry have allowed engineering of Nano carriers responsive to stimuli or smart Nano carriers which are responsive to specific biological stimuli, including pH, redox gradients, enzymatic activity, or external stimuli, e.g., light and magnetic field (Mosa Alsehli, 2020).

The purpose of this chapter is to gain a broad perspective of polymeric Nano carriers for controlled drug delivery in cancer therapy. It attempts to discuss them systematically with respect to their design principle, classification, fabrication strategies, and mechanism that dictates the release of drugs. In addition, the problems and techniques associated with targeting, biomedical applications, and main challenges in clinical translation are explored. Lastly, emerging trends and future perspectives are discussed to give insights on the development of polymeric Nano carriers toward next generation Nano-oncology and precision cancer therapeutics.

## 2. FUNDAMENTALS OF POLYMERIC NANO CARRIERS

Polymeric Nano carriers are a very flexible type of Nano sized drug vehicles designed to enhance drug therapeutic index. They depend on structural architecture, surface chemistry and physicochemical behavior for their performance in biological systems. Grasping these basic features are crucial for the rational design and optimization of polymer–based Nano medicines in cancer therapy (Tripathi et al., 2025).

### 2.1. Definition and Structural Features

Polymeric Nano carriers are Nano sized delivery platforms, typically < 200 nm in diameter, that are made of a natural, synthetic, or semi-synthetic polymer. Their structure might be in the form of a solid matrix (Nano spheres), hollow core–shell particles (Nano capsules), amphiphilic self-assembled structures (micelles) or three-dimensional, cross-linked network (nanogels and dendrimers) (Haider et al., 2022).

Most of the time they are made from a hydrophobic core, a hydrophilic shell, or a polymer-drug conjugated backbone, which is determined by the design strategy and therapeutic needs. This modular architecture facilitates the efficient loading of the hydrophobic chemotherapeutics in the core or the loading of the hydrophilic drugs on the polymer backbone, improving its solubility, stability and bioavailability. Further, it can have its surface modified by polyethylene glycol (PEG) and targeting ligands to further prolong circulation time, and deliver it to tumor targets (Blagojevic & Kamaly, 2025).

## 2.2. Physicochemical Properties Relevant to Drug Delivery

Physicochemical features enable Nano carriers to either reside in biological fluids or to be taken up by cells, and influence their bio distribution and clearance, making these parameters crucial for their therapeutic performance (Ewii et al., 2025).

### 2.2.1. Particle Size and Surface Area

One of the most determining particle size in the Nano carrier definition is particle size. The optimal accumulation in tumor areas occurs for Nano carriers in the size range of 20–150 nm, where the cells are sensitive to the enhanced permeability and retention (EPR) effect. Smaller particles have shown better penetrating ability into the tissues, and larger particles might have a good capability for drug loading. Moreover, higher surface area leads to better drug interaction, loading and drug release (Batool et al., 2023).

### 2.2.2. Surface Charge

The stability of the nanoparticles and interactions with cells is mediated by the zeta potential. Positively charged Nano carriers are likely to have stronger electrostatic interactions with the negatively charged cell membranes, which results in improved cellular uptake. But high levels of positive charge can trigger a cell's toxicity and rapid clearance. On the other hand, a nominal to slightly negative surface charge promotes better systemic circulation and minimizes non-specific protein adsorption, which results in longer blood half-life (Eljack et al., 2022).

### 2.2.3. Biodegradability and Biocompatibility

After the drug is delivered, polymeric carriers are biodegradable so the tissue accumulation is kept to a minimum because they break down into metabolites that are not harmful. Commonly used biodegradable polymers, e.g. PLGA, PLA and PCL, are degraded by hydrolysis into the naturally metabolized lactic acid and glycolic acid. Equally important is biocompatibility, with minimal immunogenicity, hemolysis and inflammatory response resulting in safe clinical translation (Gagliardi et al., 2021).

## 2.3. Drug Loading Strategies

For efficient therapeutic drug incorporation, there is a need for the controlled and sustained release. Multiple strategies for loading polymeric Nano carriers can be used, depending on drug chemistry and polymer composition (Beach et al., 2024).

### 2.3.1. Physical Encapsulation

In this method, drugs are simply contained in the polymer matrix or in the core of the nanoparticles. This approach is very common for hydrophobic drugs which will provide diffusion controlled release, no alteration of drug structure (Hawthorne et al., 2022).

### 2.3.2. Chemical Conjugation

In chemical conjugation, polymer chains are covalently linked to drug molecules via cleavable spacers. These linkers can be designed to be sensitive to different stimuli such as pH or enzyme activity to achieve site-specific drug release in tumor microenvironment (M. Alsehli, 2020).

### 2.3.3. Adsorption and Entrapment

Adsorption is a binding of the drug to the surface of the nanoparticle, through electrostatic or hydrogen bonding or by van der Waals forces. One way of achieving moderate stability and tunable release profiles is to achieve the incorporation of certain drugs during the polymer solidification process, an effect called entrapment (Islam et al., 2025).

## 2.4. Advantages in Cancer Therapeutics

However, polymeric Nano carriers have multiple different benefits which render them highly appropriate for cancer applications. Some are used to improve the solubility and stability of chemotherapeutic agents that do not dissolve easily in water, and therefore are less bioavailable. Their small size permits preferential accumulation in tumor tissues by EPR effect and surface functionalization permits active targeting to the receptors expressed on the cancer cells (Mangu et al., 2025).

Moreover, platform of the controlled and sustained drug release reduces systemic toxicity and enhances dosing frequency which increases the patient's compliance. Polymeric system also offers an advantage of shielding the encapsulated drug from enzyme degradation and early clearance. Notably, the capacity to deliver more than one therapeutic agent is useful for implementing combination therapy, especially to overcome multidrug resistance (MDR) in cancer cells (Thang et al., 2023).

## 3. CLASSIFICATION OF POLYMERIC NANOCARRIERS

The amazing diversity in the structure of the polymeric nanocarriers in both structure and functionality allows for the customization of the nanocarriers' structures for a wide range of purposes in cancer therapy. They are usually classed on the basis of polymer origin, structure, drug delivery function and behaviour. This multi-dimensional classification framework gives a systematic knowledge of their design principle and application specific performance in nano-oncology (Figure 1) (Jayaraj et al., 2026).

*Figure 1. Structural architectures of major polymeric nanocarriers and their corresponding drug-loading configurations for cancer therapy*

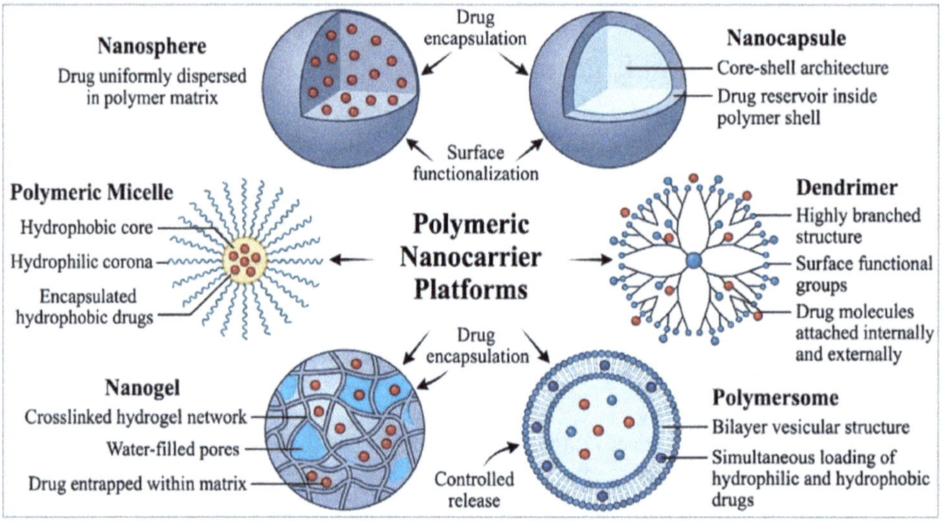


### 3.1.1. Natural Polymers

Natural polymers are those that can be obtained from biological sources and highly biocompatible with low immunogenicity. These are examples such as chitosan, alginate, gelatin and albumin. Chitosan is mucoadhesive and the inherent antimicrobial activity, alginate has mild gelation for encapsulation. Protein-based carriers, such as gelatin and albumins, increase drug stability as well as facilitating efficient cellular uptake. However, natural polymers have the disadvantage of batch to batch variation and low mechanical strength (Zhao et al., 2023).

### 3.1.2. Synthetic Polymers

Synthetic polymers are scientifically designed and chemically controlled to ensure consistent, accurate MW, degradation rate and mechanical properties. The most commonly used synthetic polymers are PLGA (poly lactic-co-glycolic acid), PLA (polylactic acid), PCL (polycaprolactone), and PEG (polyethylene glycol). PLGA, because of its FDA approval, is widely used because of its predictable biodegradation into lactic and glycolic acids. PEG has been widely used for surface modification, to help increase circulation time and decrease immune recognition. These polymers possess a number of advantages such as better reproducibility, and tunable drug release kinetics (Jumadi et al., 2025).

## 3.2. Based on Structural Architecture

The structure of polymeric nanocarriers affects both the amount of drug it can carry and the release rate of the drug, as well as the interaction of the nanocarriers with biological systems (R. De et al., 2022).

### 3.2.1. Nanospheres

Nanospheres are nano sized systems composed of a solid matrix in which the drug molecules are uniformly loaded within the polymeric matrix. They are achieved by diffusion and degradation of the matrix (Almontasser et al., 2025).

### 3.2.2. Nanocapsules

Nanocapsules are a core–shell structure: The drug is kept inside a central cavity and surrounded by a polymeric membrane which allows reservoir-type controlled release (Abdelhamid et al., 2021).

### 3.2.3. Polymeric Micelles

Polymeric micelles are polymers assembled in a self-organized manner that have a hydrophobic core and a hydrophilic shell, which is ideal for solubilizing hydrophobic anticancer drugs and for the purpose of enhancing systemic circulation (Guzmán Rodríguez et al., 2022).

### 3.2.4. Dendrimers

Dendrimers are monodisperse, highly branched, nanoparticles possessing multiple numbers of functional surface groups which are useful to impart high drug payloads and exact surface modifications for targeted therapy (Fadaei et al., 2025).

### 3.2.5. Nanogels

These crosslinked hydrophilic polymer networks are nanogels with a very good water uptake, suitable for stimuli-responsive and controlled drug release (Delgado-Pujol et al., 2025).

### 3.2.6. Polymersomes

Vesicle-like structures composed of amphiphilic block copolymers, e.g., polymersomes are more stable than lipid vesicles and simultaneously allows co-encapsulation of hydrophilic and hydrophobic drugs (Triantafyllopoulou et al., 2026).

## 3.3. Based on Drug Delivery Function

The space distribution and release mechanism of therapeutic agents depends on the functional design of polymeric nanocarriers (Yetisgin et al., 2020).

### 3.3.1. Matrix-Type Systems

With matrix systems, the drug molecules are distributed in a polymer continuous matrix and the drug is released mainly by diffusion and degradation of the polymer. These systems with predictable release profiles are capable of sustained drug delivery (Herdiana et al., 2022).

### 3.3.2. Reservoir-Type Systems

Unlike matrix systems, reservoir systems feature a drug core and polymeric membrane which provides more precisely controlled and controlled drug release rates (Li et al., 2022).

### 3.3.3. Conjugated Polymer–Drug Systems

In the case of conjugated systems, drugs are connected through a chemical(linker) to the polymer chains. Tumors have physiological or pathological properties that can be used to trigger drug release for specific tumor microenvironment(s) including acid and enzyme-activated (Parashar et al., 2024).

## 3.4. Based on Functional Behavior

The functional classification involves the interaction of nanocarriers with the biological system and their response in the tumor site (Kaushik et al., 2022).

### 3.4.1. Passive Targeting Systems

Passive targeting exploits the enhanced permeability and retention (EPR) effect, which is based on leaky tumor blood vessels and on the absence of lymphatic drainage that causes the preferential accumulation of nanocarriers in tumor tissue (Subhan et al., 2021).

### 3.4.2. Active Targeting Systems

For improving cellular uptake and specificity, the surfaces were modified by active targeting using monoclonal antibodies, peptides and/or small molecules that specifically bind to receptors which are usually discovered in receptors' up-regulation of expression on cancer cells (Hong et al., 2023).

### 3.4.3. Stimuli-Responsive (Smart) Nanocarriers

Stimuli-responsive systems are systems that are designed to release a drug in response to a stimuli, that is either internal, such as a pH, redox gradient, or enzyme level, or external, such as light, ultrasound, or magnetic fields, allowing a more highly selective level of therapeutic precision achieved through releasing the drug at the appropriate time and in the right place (Liu et al., 2020).

# 4. SYNTHESIS AND FABRICATION STRATEGIES OF POLYMERIC NANOCARRIERS

The fabrication of polymeric nanocarriers is a crucial step to determine their physicochemical properties, drug loading efficiency, and therapeutic effectiveness in cancer nanomedicine. Several synthesis methods have been designed to ensure accurate control of the particle architecture, stability and release properties. These methods may generally be divided into three broad categories, depending on the method's principles and fabricating mechanism (R. De et al., 2022).

## 4.1. Emulsion, Solvent-Based, and Self-Assembly Techniques

The most commonly and widely used and scalable approaches for polymeric nanocarrier fabrication will fall into this category. The emulsion-based techniques are based on the dispersion of one immiscible phase into another with the assistance of surfactants which create nanoscale droplets that are solidified to nanoparticles (Bufalini & Campardelli, 2025). Double emulsion (W/O/W) systems can be used for entrapping hydrophilic drugs in the internal aqueous compartment, whereas the hydrophobic drugs can be efficiently encapsulated in polymer matrices in single emulsion (O/W) systems. The solvent displacement and nanoprecipitation processes are based on the rapid diffusion of a water-miscible organic solvent containing polymers into an aqueous phase, where spontaneous nanoparticle formation occurs with controlled, narrow size distribution and high reproducibility (Jacob et al., 2025). Moreover, amphiphilic block copolymers (block copolymers composed of both hydrophobic and hydrophilic segments) are capable of self-assembling in the aqueous milieu to form micelles or vesicles containing hydrophobic drug molecules – the self-assembling nature of these polymers enables higher efficiency in the delivery of these drug molecules; the hydrophobic block form the core while the hydrophilic block envelops and encloses the core assembly to give better systemic stability to the micelle/vesicle (Wei et al., 2025).

## 4.2. Gelation and Polymerization-Based Approaches

Gelation and polymerization-based methods offer structurally stable and chemically tunable nanocarriers, which have better control over the drug release behavior. Ionic gelation is especially interesting for the natural polymers, like chitosan and alginate, which form nanoparticles by electrostatic interaction between oppositely charged polymers and the crosslinkers used, thus avoiding denaturing drug substances which could be sensitive to harsh conditions (Blagojevic & Kamaly, 2025). Polymerizations, such as in situ polymerization and controlled radical polymerization, rely on chemical methods that synthesize polymer chains within a media or template, allowing for exquisite control of the functional group incorporated, architecture and molecular weight. These approaches boost reproducibility of production and enable the design of rather customized nanocarriers for cancer therapy for precisely designed drug release profiles (Alsubaie et al., 2026).

## 4.3. Advanced Microfluidic Fabrication Techniques

Microfluidic fabrication is an emerging technology that can be used to prepare polymeric nanocarriers in a next-generation fashion with high precision and uniformity. This method allows for precise control of fluid mixing at a small scale into which the ability to control flow rates, channel geometry, and solvent diffusion rates can be used to fine tune the size, polydispersity, and composition of nanoparticles produced (Nunziata et al., 2025). Microfluidics can provide better batches to batches variability, scalability and higher reproducibilities than conventional batch methods. Therefore, it is gaining increasing traction for the development of clinically relevant polymeric nanocarriers, ultimately facilitating the translation of nanomedicine from the laboratory to large-scale pharmaceutical manufacturing (Gimondi et al., 2023).

## 5. MECHANISMS OF CONTROLLED AND SUSTAINED DRUG RELEASE

### 5.1. Diffusion- and Swelling-Mediated Release Mechanisms

One of the most basic mechanisms of the polymeric nanocarrier system is drug diffusion controlled release, in which the drug molecules move from the polymer matrix to the biological environment according to the concentration gradient (Figure 2A). These polymer porosity, drug-polymer interaction and matrix density greatly affect the release rate (Herdiana et al., 2022). In many hydrophilic polymer systems diffusion is related to swelling behaviour: i.e. water uptake causes relaxation and increased chain mobility of the polymer. Swelling-induced controlled release is achieved as the polymer network expands to allow for a more gradual diffusion of the drug, allowing for more prolonged therapeutic effects. The interplay of diffusion and swelling can be tuned by changing the polymer composition, crosslinking density, and hydrophilicity and lead to a tunable release profile (Kowalski et al., 2024) .

### 5.2. Polymer Degradation-Driven Release Mechanisms

The degradation of polymers is important especially for the biodegradable nanocarriers like PLGA, PLA and PCL for long-term drug release. In bulk degradation systems, the water will invade the total polymer matrix, resulting in uniform hydrolysis of polymer chains and in a gradual release of the drug from the volume of the nanoparticles (Figure 2B) (Visan et al., 2021). This often leads to a characteristic biphasic release profile; first, a burst release and then a sustained release. Surface erosion systems, on the other hand, degrade the nanocarrier from the exterior, with gradual release of the loaded drug while preserving the structural integrity of the nanocarrier. Surface erosion yields more linear and predictable release kinetics, which is very beneficial for precise dosing in cancer therapy. Polymer hydrophobicity, crystallinity and molecular weight determine the balance between bulk and surface erosion (Herdiana et al., 2022).

## 5.3. Stimuli-Responsive and Triggered Release Mechanisms

Stimuli responsive polymeric nanocarriers are specifically designed to respond to internal/external stimuli for allowing site specific/on-demand targeting of drug release in the context of tumor microenvironment. pH-responsive systems takes advantage of the acidic environment of tumors and intracellular endosomal compartments, that is: protonation of a polymer, cleavage of bonds, or destabilization of a structure that allows for the release of drugs (Figure 2C) (M. Alsehli, 2020). Redox-responsive systems take advantage of the high levels of glutathione found within cancer cells to break up the disulfide bond and cause rapid release of the drug. Temperature-responsive is based on thermosensitive polymers which undergo phase transitions at hyperthermic conditions either by physiological or external means, increasing drug diffusion. Enzyme responsive systems are developed for the aim of degradation in presence of tumor overexpressed enzymes like matrix metalloproteinases for the purpose of localised release (Abed et al., 2022). Furthermore, the externally stimulated systems such as light, ultrasound, or magnetic field responsive systems allow for controlled drug release not only potentially at the tumor site but also at spatiotemporal coordinates (La Monica et al., 2026).

*Figure 2. Diffusion-, degradation-, and stimuli-responsive mechanisms governing controlled drug release from polymeric nanocarriers*

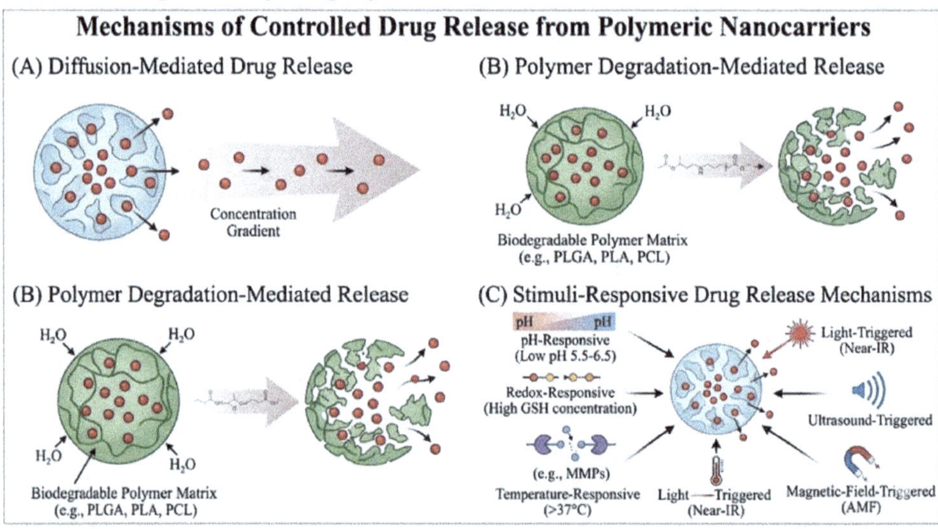

## 6. TARGETING STRATEGIES OF POLYMERIC NANOCARRIERS

Efficiency of targeting – such as the selective enrichment of tumour site without affecting the off-target toxicity – is a key factor for the success of polymeric nanocarrier-based cancer drug delivery. In the last decades, targeting mechanisms have progressed from rather simple passive accumulation strategies towards truly advanced active and biomimetic approaches, taking the advantages of molecular recognition and biological camouflage to optimize tumor specificity and to promote intracellular uptake (Xia et al., 2021).

*Figure 3. Mechanistic overview of passive, active, and intracellular targeting processes employed by polymeric nanocarriers for tumor-specific drug delivery*

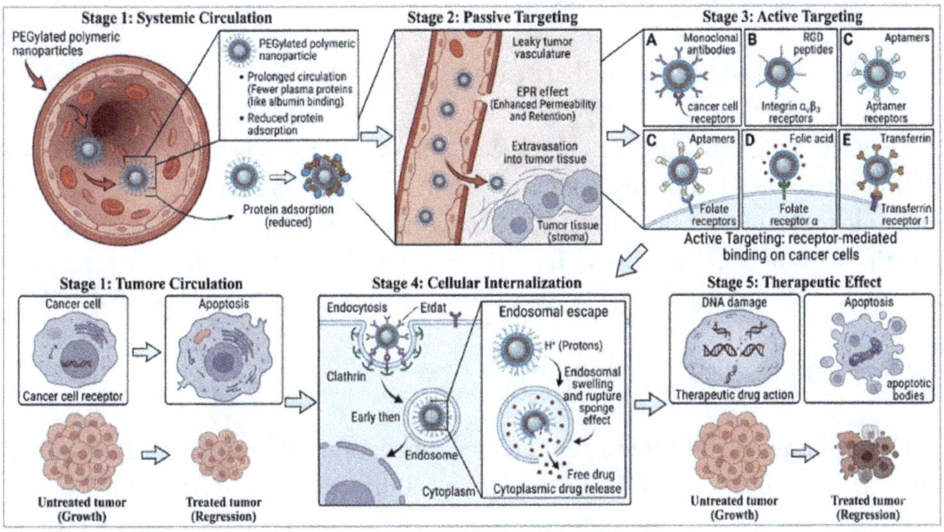


### 6.1. Passive Targeting: Enhanced Permeability and Retention (EPR) Effect

The key feature of passive targeting is the intrinsic pathophysiological features of tumor blood vessels, including the increased permeability and retention (EPR) effect (Figure 3). Tumor tissues are characterized by overproduction of blood vessels, with highly permeable and disorganized vessels with large inter-endothelial gaps, which is why nanosized polymeric carriers are able to preferentially extravasate into the tumor interstitial (Subhan et al., 2021). Further, the compromised lymphatic system in tumor microenvironments leads to enhanced accumulation of nanocarriers in the vicinity of a tumor. Efficiency of passive targeting is highly dependent on the size, surface charge and hydrophilicity of nanoparticles, and in many cases, nanoparticles are PEGylated to promote prolonged systemic circulation and to minimize particle opsonization (Vagena et al., 2025).

## 6.2. Active Targeting Strategies

Active targeting involves the incorporation of an appropriate ligand which is capable of recognizing and binding to the receptor that is overexpressed on tumour associated endothelial cells and/or cancer cells for enhancing the selectivity of the polymeric nanocarriers. Cellular internalization and the efficiency of intracellular drug delivery are greatly enhanced by this receptor-mediated recognition relative to passive systems (Figure 3) (Yoo et al., 2019).

### 6.2.1. Antibody-Mediated Targeting

Monoclonal antibodies are very specific targeting ligands that selectively bind to tumor-associated antigens, including HER2, EGFR and CD20. Antibodies, as targeting molecules, when attached to polymersomes, can enable very accurate targeting of cancer cells and also facilitate receptor-mediated endocytosis resulting in increased accumulation of therapeutic dose in the cells (Chehelgerdi et al., 2023).

### 6.2.2. Peptide-Based Targeting

High binding affinity & low immunogenicity peptides as an alternate to antibody. Targeting peptides like RGD with a strong ability to home to tumorous regions with high expression of the integrin receptor help nanocarriers achieve efficient penetration and internalization within tumor cells (Javid et al., 2024).

### 6.2.3. Aptamer-Based Targeting

Aptamers are ssDNA of short length with modified 3D structures capable of high affinity binding of molecular targets of interest. As ligands for precisely targeting tumors in polymeric nanocarrier systems they are very promising because of their chemical stability, simplicity of synthesis, and low immunogenicity (Egbe Vydaline et al., 2025).

### 6.2.4. Ligand-Mediated Targeting (Folic Acid, Transferrin)

Typical receptor-mediated targeting ligands are small molecule compounds such as folic acid and transferrin because the receptors of both compounds are overexpressed in many types of cancer. These ligands allow to improve cell uptake via endocytosis, and are simple and cost-effective in structure (Yan et al., 2024).

## 6.3. Biomimetic and Cell Membrane-Coated Systems

Biomimetic targeting strategies include the use of polymeric nanocarriers disguised with natural cell membranes of erythrocytes, platelets, leukocytes or cancerous cells. This cell membrane disguise facilitates immune resistance, long-term persistence in the circulation and targeting to the corresponding homotypic cells (Li et al., 2021). Cancer cell membrane coated nanoparticles in particular take advantage of the self-recognition mechanisms, which result in a selective targeting of the primary and metastatic tumor formation and decrease clearance in the mononuclear phagocyte system (Zhao et al., 2025).

## 6.4. Dual and Multi-Targeting Approaches

Dual/targeting and multi-targeting approaches involve using various targeting ligands attached to one nanocarrier and/or simultaneously using passive, active and biomimetic targeting mechanism within one nanocarrier platform to provide the synergy in targeting efficiency (Jurczyk et al., 2021). The design of these systems aims to defeat the heterogeneity of the tumor and improve the accumulation of the drug in heterogeneous tumor microenvironments. The multi-functional nanocarriers - combination of targeting different receptors or biological pathways can enhance cell uptake, tumor penetration, and therapeutic accuracy; reflecting next generation precision nano-oncology (Yan et al., 2024).

# 7. APPLICATIONS IN CANCER THERAPY

Nowadays, polymeric nanocarriers have completely revolutionized the perspectives of cancer treatment thanks to their capacity to enable the delivery on demand of a therapeutic agent targeting the tumor site with efficacy and without damaging the surrounding healthy tissue. They can be engineered to deliver a variety of therapeutic agents, such as small molecule drugs, nucleic acids and imaging probes making them very versatile for precision oncology (Shanahan et al., 2025). These applications include, and are not limited to, a number of non-traditional chemotherapy based applications such as gene therapy, combination applications and integrated diagnostic and therapeutic (theranostic) applications (Haider et al., 2022).

## Delivery of Conventional Chemotherapeutics

The most well-developed uses of polymeric nanocarriers are in the delivery of conventional anticancer drugs where the potential benefit is enhanced solubility, stability, pharmacokinetics, and tumor selectivity. Polymer-based systems are widely used to reduce systemic toxicity and improve the therapeutic effect of drugs like doxorubicin, paclitaxel and cisplatin (Avramović et al., 2020). Controlled and sustained release can be achieved through encapsulation in any polymeric matrix or in a micelle, ensuring optimal concentrations of the drug at the tumor sites with little harmful effects off-target. Furthermore, surface modification of these carriers further enhances the accumulation of the nanocarrier in the tumor and its cellular uptake (Bai et al., 2022).

## 7.1. Gene and Nucleic Acid Delivery Systems

Furthermore, polymeric nanocarriers are useful as efficient non-viral vectors for delivering genes and nucleic acids, since these are also hampered by the problems of instability, enzymatic degradation, and low cellular uptake. These systems can be used to deliver siRNA for gene silencing, mRNA for protein expression and even DNA plasmids to facilitate long term genetic modulation (Khan, 2024). They possess both in addition to great ability to protect nucleic acids from the degradation by nucleases and to induce the endosomal escape and have high potential for targeted genetic interventions in cancer therapy. In addition, polymeric systems are available that are designed to deliver in a sequence-specific manner, thus enhancing that therapeutic accuracy at the molecular level (Zhang et al., 2023).

## 7.2. Combinatory and synergistic therapeutic approaches

Evolving polymeric nanomedicines have proven to be an effective solution to address the problems of drug resistance and tumor cell diversity through combination therapy. These systems allow for the simultaneous delivery of different therapeutic strategies, including synergistic combination of chemotherapy and photothermal therapy, immunotherapy or various chemotherapeutic drugs from a single nano platform (Kurul et al., 2025). These synergistic approaches improve killing of tumour cells with complementary effects of the drugs and lower the dose of each one needed, reducing systemic effects. Polymeric carriers also provide controlled release of drugs and ensure synchronicity in drug delivery, a factor vital to maximize synergies in the therapy (Wagh & Bhattacharya, 2025).

## 7.3. Theranostic Polymeric Nanocarriers

Theranostic nanocarriers combine therapeutic and diagnostic roles in a single platform that allow monitoring of drug delivery and treatment outcomes in real time. Polymeric systems can be designed to include imaging agents that enable targeted visualization of tumor localization and drug delivery through either fluorescence imaging, magnetic resonance imaging (MRI or computed tomography (CT). Fluorescence polymeric nanocarriers enable optical monitoring at the cellular level, whereas MRI/CT-observable systems are used to monitor at the deep tissue level for clinical applications. This dual purpose increases the accuracy of treatment, facilitates individual treatment and allows for adaptive treatment based on real-time feedback (Hosseini et al., 2023) (Shanahan et al., 2025).

*Table 1. Comprehensive overview of polymeric nanocarriers for controlled drug delivery in cancer therapy*

| Category | Type / Strategy | Key Materials / Examples | Mechanism / Principle | Therapeutic Advantage in Cancer | References |
|---|---|---|---|---|---|
| **Polymer Origin** | Natural Polymers | Chitosan, Alginate, Gelatin, Albumin | Biodegradable protein/polysaccharide matrices | High biocompatibility, low toxicity | (Ishola et al., 2026) |
| | Synthetic Polymers | PLGA, PLA, PCL, PEG | Controlled hydrolysis and structural tuning | Predictable degradation and drug release | |
| **Nanocarrier Architecture** | Nanospheres / Nanocapsules | PLGA, PLA systems | Matrix or core–shell encapsulation | Sustained and protected drug delivery | (Gressler et al., 2025) |
| | Micelles / Polymersomes | Block copolymers (PEG-b-PLA) | Self-assembly in aqueous media | Enhanced solubility of hydrophobic drugs | |
| | Dendrimers / Nanogels | PAMAM, crosslinked polymers | Branched / 3D networks | High drug loading and multifunctionality | |
| **Fabrication Methods** | Emulsion Techniques | O/W, W/O/W systems | Droplet formation and solvent evaporation | Scalable nanoparticle synthesis | (Procopio et al., 2022) |
| | Nanoprecipitation | PLGA, PEG systems | Solvent diffusion-driven formation | Uniform size distribution | |

continued on following page

*Table 1. Continued*

| Category | Type / Strategy | Key Materials / Examples | Mechanism / Principle | Therapeutic Advantage in Cancer | References |
|---|---|---|---|---|---|
| | Ionic Gelation | Chitosan, Alginate | Electrostatic crosslinking | Mild, drug-friendly synthesis | |
| | Microfluidics | All polymer systems | Controlled micro-mixing | High reproducibility and scalability | |
| **Drug Release Mechanism** | Diffusion-Controlled | Matrix systems | Concentration gradient diffusion | Sustained release | (Adepu & Ramakrishna, 2021) |
| | Degradation-Controlled | PLGA, PLA | Bulk/surface erosion | Time-dependent release | |
| | Stimuli-Responsive | Smart polymers | pH, redox, enzyme, light triggers | Site-specific drug activation | |
| **Targeting Strategy** | Passive Targeting | PEGylated nanoparticles | EPR effect | Tumor accumulation | (Beh et al., 2021) |
| | Active Targeting | Antibodies, folic acid, peptides | Ligand-receptor binding | High cellular specificity | |
| | Biomimetic Targeting | Cell membrane-coated systems | Immune evasion + homotypic binding | Enhanced circulation & tumor homing | |
| **Therapeutic Applications** | Chemotherapy Delivery | Doxorubicin, Paclitaxel | Encapsulation or conjugation | Reduced systemic toxicity | (Jayaraj et al., 2026) |
| | Gene Delivery | siRNA, mRNA, DNA | Nucleic acid protection & delivery | Gene regulation in tumors | |
| | Theranostics | Imaging + drug agents | Dual diagnostic-therapeutic systems | Real-time treatment monitoring | |

## 8. FUTURE PERSPECTIVES, CHALLENGES AND LIMITATIONS OF POLYMERIC NANOCARRIER SYSTEMS

Although polymeric nanocarrier-based drug delivery systems have made notable progress, their implementation at the clinical level is still hindered by various problems related to several intricate interactions between biological, physicochemical, technological and regulatory factors. The tumour microenvironment has several challenges to overcome in terms of efficient nanocarrier delivery and therapeutic potential. These involve fast clearance by the mononuclear phagocyte system, generation of a protein corona that alters the 'nature' and behavior of the nanoparticles, and biologic variation in tumor structure that will create differing drug penetration and response. All these together greatly lower the efficiency of the targeting and predictability of therapeutic action in vivo.

The problem from a physic-chemical point-of-view is to keep nanoparticles stable in complex biological fluids. However, polymeric nanocarriers can aggregate, dissemble too early and release the drugs inappropriately before reaching the tumour, which decreases the therapeutic efficiency. Further, as well as size distribution variations, surface morphology and batch to batch reproducibility make their clinical reliability and standardization more problematic.

In addition to the limitations imposed by the biological and physicochemical properties, size scale up and reproducibility are major constraints hampering commercial value. A number of fabrication methods are still in the laboratory and have limited scalability with high production costs. In addition, the long-term effects of polymer accumulation, breakdown products, and potential off-target effects remain safety issues to be thoroughly investigated.

Other issues associated with polymeric nanocarriers in oncology include regulatory and clinical translation issues. As the lack of standardized regulations for the development of complex nanomedicines contributes, as well as limited long-term clinical data, this delays approval procedures and speed of integration in the clinical arena.

In the future of the polymeric nanocarrier systems is the creation of smart multi-stimuli responsive nanocarriers to achieve highly selective drug release, the application of artificial intelligence to develop rational nanocarriers and the step-wise progression to personalized precision oncology. Multifunctionality and efficiency of the therapy is expected to be improved using hybrid systems composed of a polymer matrix with lipidic or inorganic components. Furthermore, sustainable and green approach in synthesis is becoming even more relevant in order to minimize the negative impact to the environment. Polymeric nanocarriers are anticipated to move from bench to bedside with further innovations and harmonization with regulations to usher in a new era of cancer nanomedicine.

## 9. CONCLUSION

Polymeric nanocarriers can be very easily rendered versatile and highly tunable, making them a revolutionary carrier platform in cancer nanomedicine for controlled and targeted drug delivery. They are unique in their physicochemical characteristics such as nanoscale dimensions, structural diversity, and surface modifiability that allows efficient encapsulation of a variety of therapeutic agents to improve pharmacokinetics, biodistribution, and tumor selectivity. These systems enable sustained and site-specific drug delivery, often via processes of diffusion, degradation, swelling, and stimuli-responsive release, which result in improvements of therapeutic efficacy with a decreased systemic toxicity.

The combination of passive and active targeting, plus sophisticated biomimetic methods, has further enhanced the specificity of tumor tissue drug accumulation. Moreover, the combination with "multifunctionality" of polymeric nanocarriers paves the way to combination therapies, gene delivery or theranostic applications, filling the fore of next-generation strategies for cancer treatment. There are, however, still issues associated with biological barriers, biochemical instabilities, large-scale production, safety and regulatory issues which hinder their complete clinical translation.

However, the current progress in the development of smart nanocarriers, artificial intelligence (AI) driven engineering for nanocarriers, hybrid nanomaterials systems, and green synthesis of nanocarriers is quickly overcoming these disadvantages. The next step of polymeric nanocarriers is to combine them with personalized and precision oncology platforms where patient-specific treatment options are possible with utmost precision. Yet, polymeric nanocarriers are an emerging and very promising family of nanomedicines that may arguably provide the path for revolutionizing cancer therapy by overcoming the current challenges of conventional treatment approaches and introducing new precision based healthcare solutions.

## REFERENCES


Abdelhamid, A., Zayed, D., Heikal, L., Khattab, S., Mady, O., El-Gizawy, S., & Elzoghby, A. (2021). Recent Advances in Polymer Shell Oily-Core Nanocapsules for Drug-Delivery Applications. *Nanomedicine (London), 16*(18), 1613–1625. Advance online publication. DOI: 10.2217/nnm-2021-0037 PMID: 34189946

Abed, H. F., Abuwatfa, W. H., & Husseini, G. A. (2022). Redox-Responsive Drug Delivery Systems: A Chemical Perspective. *Nanomaterials (Basel, Switzerland), 12*(18), 3183. Advance online publication. DOI: 10.3390/nano12183183 PMID: 36144971

Adepu, S., & Ramakrishna, S. (2021). Controlled Drug Delivery Systems: Current Status and Future Directions. *Molecules (Basel, Switzerland), 26*(19), 5905. Advance online publication. DOI: 10.3390/molecules26195905 PMID: 34641447

Almontasser, A., Al-Mufti, S. M. S., & Arya, R. K. (2025). Nanofillers in Drug Delivery Industry: Applications. In Mallakpour, S., & Hussain, C. M. (Eds.), *Handbook of Nanofillers* (pp. 1479–1519). Springer Nature Singapore., DOI: 10.1007/978-981-96-2407-2_78

Alsehli, M. (2020). Polymeric nanocarriers as stimuli-responsive systems for targeted tumor (cancer) therapy: Recent advances in drug delivery. *Saudi Pharmaceutical Journal : SPJ : The Official Publication of the Saudi Pharmaceutical Society, 28*(3), 255–265. https://doi.org/https://doi.org/10.1016/j.jsps.2020.01.004. DOI: 10.1016/j.jsps.2020.01.004 PMID: 32194326

Alsehli, M. (2020). Polymeric nanocarriers as stimuli-responsive systems for targeted tumor (cancer) therapy: Recent advances in drug delivery. *Saudi Pharmaceutical Journal : SPJ : The Official Publication of the Saudi Pharmaceutical Society, 28*(3), 255–265. DOI: 10.1016/j.jsps.2020.01.004 PMID: 32194326

Alsubaie, F., Alshammari, B., Grira, S., Mohammad Sayem Mozumder, A. S., & Mourad, A.-H. I. (2026). Recent advances, methods, and applications of copper-mediated Atom Transfer Radical Polymerization (ATRP) for precise block copolymers synthesis. *Materials Today. Communications, 53*, 115385. https://doi.org/https://doi.org/10.1016/j.mtcomm.2026.115385. DOI: 10.1016/j.mtcomm.2026.115385

Avramović, N., Mandić, B., Savić-Radojević, A., & Simić, T. (2020). Polymeric Nanocarriers of Drug Delivery Systems in Cancer Therapy. *Pharmaceutics, 12*(4), 298. Advance online publication. DOI: 10.3390/pharmaceutics12040298 PMID: 32218326

Bai, X., Smith, Z. L., Wang, Y., Butterworth, S., & Tirella, A. (2022). Sustained Drug Release from Smart Nanoparticles in Cancer Therapy: A Comprehensive Review. *Micromachines*, *13*(10), 1623. Advance online publication. DOI: 10.3390/mi13101623 PMID: 36295976

Batool, S., Sohail, S., ud Din, F., Alamri, A. H., Alqahtani, A. S., Alshahrani, M. A., Alshehri, M. A., & Choi, H. G. (2023). A detailed insight of the tumor targeting using nanocarrier drug delivery system. *Drug Delivery*, *30*(1), 2183815. DOI: 10.1080/10717544.2023.2183815 PMID: 36866455

Beach, M. A., Nayanathara, U., Gao, Y., Zhang, C., Xiong, Y., Wang, Y., & Such, G. K. (2024). Polymeric Nanoparticles for Drug Delivery. *Chemical Reviews*, *124*(9), 5505–5616. DOI: 10.1021/acs.chemrev.3c00705 PMID: 38626459

Beh, C. Y., Prajnamitra, R. P., Chen, L. L., & Hsieh, P. C. (2021). Advances in Biomimetic Nanoparticles for Targeted Cancer Therapy and Diagnosis. *Molecules (Basel, Switzerland)*, *26*(16), 5052. Advance online publication. DOI: 10.3390/molecules26165052 PMID: 34443638

Blagojevic, L., & Kamaly, N. (2025). Nanogels: A chemically versatile drug delivery platform. *Nano Today*, *61*, 102645. https://doi.org/https://doi.org/10.1016/j.nantod.2025.102645. DOI: 10.1016/j.nantod.2025.102645

Bozzer, S., Dal Bo, M., Grimaldi, M. C., Toffoli, G., & Macor, P. (2022). Nanocarriers as a Delivery Platform for Anticancer Treatment: Biological Limits and Perspectives in B-Cell Malignancies. *Pharmaceutics*, *14*(9), 1965. https://www.mdpi.com/1999-4923/14/9/1965. DOI: 10.3390/pharmaceutics14091965 PMID: 36145713

Brown, J. S., Amend, S. R., Austin, R. H., Gatenby, R. A., Hammarlund, E. U., & Pienta, K. J. (2023). Updating the Definition of Cancer. *Molecular Cancer Research*, *21*(11), 1142–1147. DOI: 10.1158/1541-7786.MCR-23-0411 PMID: 37409952

Bufalini, C., & Campardelli, R. (2025). Versatile Emulsion-Based Encapsulation System Production Processes: A Review. *Processes (Basel, Switzerland)*, *13*(5), 1409. DOI: 10.3390/pr13051409

Chehelgerdi, M., Chehelgerdi, M., Allela, O. Q. B., Pecho, R. D. C., Jayasankar, N., Rao, D. P., Thamaraikani, T., Vasanthan, M., Viktor, P., Lakshmaiya, N., Saadh, M. J., Amajd, A., Abo-Zaid, M. A., Castillo-Acobo, R. Y., Ismail, A. H., Amin, A. H., & Akhavan-Sigari, R. (2023). Progressing nanotechnology to improve targeted cancer treatment: Overcoming hurdles in its clinical implementation. *Molecular Cancer*, *22*(1), 169. DOI: 10.1186/s12943-023-01865-0 PMID: 37814270

Colonna, G. (2025). Overcoming Barriers in Cancer Biology Research: Current Limitations and Solutions. *Cancers (Basel)*, *17*(13), 2102. https://www.mdpi.com/2072-6694/17/13/2102. DOI: 10.3390/cancers17132102 PMID: 40647400

Dalbanjan, N. P., Korgaonkar, K., Eelager, M. P., Gonal, B. N., Kadapure, A. J., Arakera, S. B., & Kumar, S. K. P. (2025). In-silico strategies in nano-drug design: Bridging nanomaterials and pharmacological applications. *Nano TransMed, 4*, 100091. https://doi.org/https://doi.org/10.1016/j.ntm.2025.100091

De, R., Mahata, M. K., & Kim, K.-T. (2022). Structure-Based Varieties of Polymeric Nanocarriers and Influences of Their Physicochemical Properties on Drug Delivery Profiles. *Advanced Science (Weinheim, Baden-Württemberg, Germany)*, *9*(10), 2105373. https://doi.org/https://doi.org/10.1002/advs.202105373. DOI: 10.1002/advs.202105373 PMID: 35112798

De, R., Mahata, M. K., & Kim, K. T. (2022). Structure-Based Varieties of Polymeric Nanocarriers and Influences of Their Physicochemical Properties on Drug Delivery Profiles. *Advanced Science (Weinheim, Baden-Württemberg, Germany)*, *9*(10), e2105373. DOI: 10.1002/advs.202105373 PMID: 35112798

Delgado-Pujol, E. J., Martínez, G., Casado-Jurado, D., Vázquez, J., León-Barberena, J., Rodríguez-Lucena, D., Torres, Y., Alcudia, A., & Begines, B. (2025). Hydrogels and Nanogels: Pioneering the Future of Advanced Drug Delivery Systems. *Pharmaceutics*, *17*(2), 215. Advance online publication. DOI: 10.3390/pharmaceutics17020215 PMID: 40006582

Egbe Vydaline, A. O., Rozhkov, S., Sosa, G., & Mallikaratchy, P. (2025). Aptamers as target-specific recognition elements in drug delivery. *Advanced Drug Delivery Reviews*, *226*, 115685. https://doi.org/https://doi.org/10.1016/j.addr.2025.115685. DOI: 10.1016/j.addr.2025.115685 PMID: 40925520

Eljack, S., David, S., Faggad, A., Chourpa, I., & Allard-Vannier, E. (2022). Nanoparticles design considerations to co-deliver nucleic acids and anti-cancer drugs for chemoresistance reversal. *International Journal of Pharmaceutics: X*, *4*, 100126. DOI: 10.1016/j.ijpx.2022.100126 PMID: 36147518

Ewii, U. E., Attama, A. A., Olorunsola, E. O., Onugwu, A. L., Nwakpa, F. U., Anyiam, C., Chijioke, C., & Ogbulie, T. (2025). Nanoparticles for drug delivery: Insight into in vitro and in vivo drug release from nanomedicines. *Nano TransMed*, *4*, 100083. https://doi.org/https://doi.org/10.1016/j.ntm.2025.100083

Fadaei, M. R., Fadaei, M. S., Kheirieh, A. E., Hatami, H., Rahmanian-Devin, P., Tayebi-Khorrami, V., Fathabadi, M. F., Baradaran Rahimi, V., & Askari, V. R. (2025). Overview of dendrimers as promising drug delivery systems with insight into anticancer and anti-microbial applications. *International Journal of Pharmaceutics: X, 10*, 100390. https://doi.org/https://doi.org/10.1016/j.ijpx.2025.100390. DOI: 10.1016/j.ijpx.2025.100390 PMID: 41497304

Gagliardi, A., Giuliano, E., Venkateswararao, E., Fresta, M., Bulotta, S., Awasthi, V., & Cosco, D. (2021). Biodegradable Polymeric Nanoparticles for Drug Delivery to Solid Tumors. *Frontiers in Pharmacology*, *12*, 601626. DOI: 10.3389/fphar.2021.601626 PMID: 33613290

Gimondi, S., Ferreira, H., Reis, R. L., & Neves, N. M. (2023). Microfluidic Devices: A Tool for Nanoparticle Synthesis and Performance Evaluation. *ACS Nano*, *17*(15), 14205–14228. DOI: 10.1021/acsnano.3c01117 PMID: 37498731

Gressler, S., Hipfinger, C., Part, F., Pavlicek, A., Zafiu, C., & Giese, B. (2025). A systematic review of nanocarriers used in medicine and beyond - definition and categorization framework. *Journal of Nanobiotechnology*, *23*(1), 90. DOI: 10.1186/s12951-025-03113-7 PMID: 39920688

Guzmán Rodríguez, A., Sablón Carrazana, M., Rodríguez Tanty, C., Malessy, M. J. A., Fuentes, G., & Cruz, L. J. (2022). Smart Polymeric Micelles for Anticancer Hydrophobic Drugs. *Cancers (Basel)*, *15*(1), 4. Advance online publication. DOI: 10.3390/cancers15010004 PMID: 36612002

Haider, M., Zaki, K. Z., El Hamshary, M. R., Hussain, Z., Orive, G., & Ibrahim, H. O. (2022). Polymeric nanocarriers: A promising tool for early diagnosis and efficient treatment of colorectal cancer. *Journal of Advanced Research*, *39*, 237–255. https://doi.org/https://doi.org/10.1016/j.jare.2021.11.008. DOI: 10.1016/j.jare.2021.11.008 PMID: 35777911

Hawthorne, D., Pannala, A., Sandeman, S., & Lloyd, A. (2022). Sustained and targeted delivery of hydrophilic drug compounds: A review of existing and novel technologies from bench to bedside. *Journal of Drug Delivery Science and Technology*, *78*, 103936. https://doi.org/https://doi.org/10.1016/j.jddst.2022.103936. DOI: 10.1016/j.jddst.2022.103936

Herdiana, Y., Wathoni, N., Shamsuddin, S., & Muchtaridi, M. (2022). Drug release study of the chitosan-based nanoparticles. *Heliyon*, *8*(1), e08674. https://doi.org/https://doi.org/10.1016/j.heliyon.2021.e08674. DOI: 10.1016/j.heliyon.2021.e08674 PMID: 35028457

Hong, L., Li, W., Li, Y., & Yin, S. (2023). Nanoparticle-based drug delivery systems targeting cancer cell surfaces. *RSC Advances*, *13*(31), 21365–21382. DOI: 10.1039/D3RA02969G PMID: 37465582

Hosseini, S. M., Mohammadnejad, J., Salamat, S., Beiram Zadeh, Z., Tanhaei, M., & Ramakrishna, S. (2023). Theranostic polymeric nanoparticles as a new approach in cancer therapy and diagnosis: A review. *Materials Today. Chemistry*, *29*, 101400. https://doi.org/https://doi.org/10.1016/j.mtchem.2023.101400. DOI: 10.1016/j.mtchem.2023.101400

Ishola, B. O., Rahaman, K. A., Razzak, S. A., Rumon, M. M. H., Shakil, M. S., & Uddin, S. (2026). Advanced mechanisms of polymer-based drug delivery systems for clinical applications. *RSC Pharmaceutics*.

Islam, S., Ahmed, M. M., Islam, M., Hossain, N., & Chowdhury, M. (2025). Advances in Nanoparticles in Targeted Drug Delivery- A Review. *Results in Surfaces and Interfaces*, *19*, 100529. DOI: 10.1016/j.rsurfi.2025.100529

Jacob, S., Varkey, N. R., Boddu, S. H. S., Gorain, B., Rao, R., & Nair, A. B. (2025). Advances in Lipid-Polymer Hybrid Nanoparticles: Design Strategies, Functionalization, Oncological and Non-Oncological Clinical Prospects. *Pharmaceuticals (Basel, Switzerland)*, *18*(12), 1772. https://www.mdpi.com/1424-8247/18/12/1772. DOI: 10.3390/ph18121772 PMID: 41471261

Javid, H., Oryani, M. A., Rezagholinejad, N., Esparham, A., Tajaldini, M., & Karimi-Shahri, M. (2024). RGD peptide in cancer targeting: Benefits, challenges, solutions, and possible integrin-RGD interactions. *Cancer Medicine*, *13*(2), e6800. DOI: 10.1002/cam4.6800 PMID: 38349028

Jayaraj, S., Jayaprakash, N., Devaraji, M., Murugavel, S., Robinson, A. X., & Murugesan, L. (2026). Nanocarrier-based strategies for advanced cancer drug delivery and therapeutics: Design principles, biological interactions, and clinical potential. *Nano Trends, 14*, 100203. https://doi.org/https://doi.org/10.1016/j.nwnano.2026.100203

Jumadi, J., Harun, W. S. W., Kadirgama, K., Samylingam, L., Aslfattahi, N., S, C. K., Kok, C. K., & Kiai, M. S. (2025). A Comparative Review of Polylactic Acid and Poly(Lactic-co-Glycolic Acid) Biomaterials: Optimizing Drug Delivery. *BioNanoScience*, *16*(1), 19. DOI: 10.1007/s12668-025-02233-6

Jurczyk, M., Jelonek, K., Musiał-Kulik, M., Beberok, A., Wrześniok, D., & Kasperczyk, J. (2021). Single- versus Dual-Targeted Nanoparticles with Folic Acid and Biotin for Anticancer Drug Delivery. *Pharmaceutics*, *13*(3), 326. Advance online publication. DOI: 10.3390/pharmaceutics13030326 PMID: 33802531

Kaushik, N., Borkar, S. B., Nandanwar, S. K., Panda, P. K., Choi, E. H., & Kaushik, N. K. (2022). Nanocarrier cancer therapeutics with functional stimuli-responsive mechanisms. *Journal of Nanobiotechnology*, *20*(1), 152. DOI: 10.1186/s12951-022-01364-2 PMID: 35331246

Khan, M. (2024). Polymers as Efficient Non-Viral Gene Delivery Vectors: The Role of the Chemical and Physical Architecture of Macromolecules. *Polymers*, *16*(18), 2629. Advance online publication. DOI: 10.3390/polym16182629 PMID: 39339093

Kowalski, G., Witczak, M., & Kuterasiński, Ł. (2024). Structure Effects on Swelling Properties of Hydrogels Based on Sodium Alginate and Acrylic Polymers. *Molecules (Basel, Switzerland)*, *29*(9), 1937. Advance online publication. DOI: 10.3390/molecules29091937 PMID: 38731429

Kurul, F., Turkmen, H., Cetin, A. E., & Topkaya, S. N. (2025). Nanomedicine: How nanomaterials are transforming drug delivery, bio-imaging, and diagnosis. *Next Nanotechnology, 7*, 100129. https://doi.org/https://doi.org/10.1016/j.nxnano.2024.100129

La Monica, G., Bono, A., Alamia, F., Lauria, A., & Martorana, A. (2026). Spatiotemporal control of prodrug activation through external stimuli for effective and safe on-site release in solid tumors: From current advances to future perspectives. *Drug Discovery Today*, *31*(1), 104595. https://doi.org/https://doi.org/10.1016/j.drudis.2025.104595. DOI: 10.1016/j.drudis.2025.104595 PMID: 41494612

Li, A., Zhao, Y., Li, Y., Jiang, L., Gu, Y., & Liu, J. (2021). Cell-derived biomimetic nanocarriers for targeted cancer therapy: Cell membranes and extracellular vesicles. *Drug Delivery*, *28*(1), 1237–1255. DOI: 10.1080/10717544.2021.1938757 PMID: 34142930

Li, L., Lee, C., Cruz, D. F., Krovi, S. A., Hudgens, M. G., Cottrell, M. L., & Johnson, L. M. (2022). Reservoir-Style Polymeric Drug Delivery Systems: Empirical and Predictive Models for Implant Design. *Pharmaceuticals (Basel, Switzerland)*, *15*(10), 1226. Advance online publication. DOI: 10.3390/ph15101226 PMID: 36297338

Liu, G., Lovell, J. F., Zhang, L., & Zhang, Y. (2020). Stimulus-Responsive Nanomedicines for Disease Diagnosis and Treatment. *International Journal of Molecular Sciences*, *21*(17), 6380. Advance online publication. DOI: 10.3390/ijms21176380 PMID: 32887466

Mangu, K., Gudeli, R., & Rizwanullah, M. (2025). Recent advancement in polymeric nanoparticles for oral chemotherapy: Transforming cancer treatment. *BioImpacts*, *15*, 31117. DOI: 10.34172/bi.31117 PMID: 41185868

Nunziata, G., Borroni, A., & Rossi, F. (2025). Advanced microfluidic strategies for core-shell nanoparticles: The next-generation of polymeric and lipid-based drug nanocarriers. *Chemical Engineering Journal Advances*, *22*, 100759. https://doi.org/https://doi.org/10.1016/j.ceja.2025.100759. DOI: 10.1016/j.ceja.2025.100759

Parashar, A. K., Saraogi, G. K., Jain, P. K., Kurmi, B., Shrivastava, V., & Arora, V. (2024). Polymer-drug conjugates: Revolutionizing nanotheranostic agents for diagnosis and therapy. *Discover. Oncology*, *15*(1), 641. DOI: 10.1007/s12672-024-01509-9 PMID: 39527173

Procopio, A., Lagreca, E., Jamaledin, R., La Manna, S., Corrado, B., Di Natale, C., & Onesto, V. (2022). Recent Fabrication Methods to Produce Polymer-Based Drug Delivery Matrices (Experimental and In Silico Approaches). *Pharmaceutics*, *14*(4), 872. Advance online publication. DOI: 10.3390/pharmaceutics14040872 PMID: 35456704

Sell, M., Lopes, A. R., Escudeiro, M., Esteves, B., Monteiro, A. R., Trindade, T., & Cruz-Lopes, L. (2023). Application of Nanoparticles in Cancer Treatment: A Concise Review. *Nanomaterials (Basel, Switzerland)*, *13*(21), 2887. https://www.mdpi.com/2079-4991/13/21/2887. DOI: 10.3390/nano13212887 PMID: 37947732

Shanahan, K., Coen, D., & Nafo, W. (2025). Polymer-based nanoparticles for cancer theranostics: Advances, challenges, and future perspectives. *Exploration of Biomat-X*, *2*, 101342. DOI: 10.37349/ebmx.2025.101342

Subhan, M. A., Yalamarty, S. S. K., Filipczak, N., Parveen, F., & Torchilin, V. P. (2021). Recent Advances in Tumor Targeting via EPR Effect for Cancer Treatment. *Journal of Personalized Medicine*, *11*(6), 571. Advance online publication. DOI: 10.3390/jpm11060571 PMID: 34207137

Thang, N. H., Chien, T. B., & Cuong, D. X. (2023). Polymer-Based Hydrogels Applied in Drug Delivery: An Overview. *Gels (Basel, Switzerland)*, *9*(7), 523. Advance online publication. DOI: 10.3390/gels9070523 PMID: 37504402

Triantafyllopoulou, E., Selianitis, D., Chountoulesi, M., Karayianni, M., Valsami, G., Pispas, S., & Pippa, N. (2026). Lipid/copolymer hybrid drug delivery systems: From design to performance. *Advances in Colloid and Interface Science*, *353*, 103884. https://doi.org/https://doi.org/10.1016/j.cis.2026.103884. DOI: 10.1016/j.cis.2026.103884 PMID: 41923680

Tripathi, D., Pandey, P., Sharma, S., Rai, A. K., & Prabhu, B. H. M. (2025). Advances in nanomaterials for precision drug delivery: Insights into pharmacokinetics and toxicity. *BioImpacts*, *15*(1), 30573–30573. DOI: 10.34172/bi.30573 PMID: 40256227

Vagena, I.-A., Malapani, C., Gatou, M.-A., Lagopati, N., & Pavlatou, E. A. (2025). Enhancement of EPR Effect for Passive Tumor Targeting: Current Status and Future Perspectives. *Applied Sciences (Basel, Switzerland)*, *15*(6), 3189. https://www.mdpi.com/2076-3417/15/6/3189. DOI: 10.3390/app15063189

Visan, A. I., Popescu-Pelin, G., & Socol, G. (2021). Degradation Behavior of Polymers Used as Coating Materials for Drug Delivery—A Basic Review. *Polymers*, *13*(8), 1272. https://www.mdpi.com/2073-4360/13/8/1272. DOI: 10.3390/polym13081272 PMID: 33919820

Wagh, H., & Bhattacharya, S. (2025). Therapeutic potential, formulation strategies and cancer treatment applications of ripe papaya pectin (Carica papaya L.): A comprehensive review. *Carbohydrate Polymer Technologies and Applications*, *11*, 100942. https://doi.org/https://doi.org/10.1016/j.carpta.2025.100942. DOI: 10.1016/j.carpta.2025.100942

Wei, H.-X., Liu, M.-H., Wang, T.-Y., Shih, M.-H., Yu, J., & Yeh, Y.-C. (2025). Fabrication of pH- and Ultrasound-Responsive Polymeric Micelles: The Effect of Amphiphilic Block Copolymers with Different Hydrophilic/Hydrophobic Block Ratios for Self-Assembly and Controlled Drug Release. *Biomacromolecules*, *26*(4), 2116–2130. Advance online publication. DOI: 10.1021/acs.biomac.4c01202 PMID: 40067950

Xia, W., Tao, Z., Zhu, B., Zhang, W., Liu, C., Chen, S., & Song, M. (2021). Targeted Delivery of Drugs and Genes Using Polymer Nanocarriers for Cancer Therapy. *International Journal of Molecular Sciences*, *22*(17), 9118. Advance online publication. DOI: 10.3390/ijms22179118 PMID: 34502028

Yan, S., Na, J., Liu, X., & Wu, P. (2024). Different Targeting Ligands-Mediated Drug Delivery Systems for Tumor Therapy. *Pharmaceutics*, *16*(2), 248. Advance online publication. DOI: 10.3390/pharmaceutics16020248 PMID: 38399302

Yetisgin, A. A., Cetinel, S., Zuvin, M., Kosar, A., & Kutlu, O. (2020). Therapeutic Nanoparticles and Their Targeted Delivery Applications. *Molecules (Basel, Switzerland)*, *25*(9), 2193. Advance online publication. DOI: 10.3390/molecules25092193 PMID: 32397080

Yoo, J., Park, C., Yi, G., Lee, D., & Koo, H. (2019). Active Targeting Strategies Using Biological Ligands for Nanoparticle Drug Delivery Systems. *Cancers (Basel)*, *11*(5), 640. Advance online publication. DOI: 10.3390/cancers11050640 PMID: 31072061

Zafar, A., Khatoon, S., Khan, M. J., Abu, J., & Naeem, A. (2025). Advancements and limitations in traditional anti-cancer therapies: A comprehensive review of surgery, chemotherapy, radiation therapy, and hormonal therapy. *Discover. Oncology*, *16*(1), 607. DOI: 10.1007/s12672-025-02198-8 PMID: 40272602

Zhang, Y. Q., Guo, R. R., Chen, Y. H., Li, T. C., Du, W. Z., Xiang, R. W., Guan, J. B., Li, Y. P., Huang, Y. Y., Yu, Z. Q., Cai, Y., Zhang, P., & Ling, G. X. (2023). Ionizable drug delivery systems for efficient and selective gene therapy. *Military Medical Research*, *10*(1), 9. DOI: 10.1186/s40779-023-00445-z PMID: 36843103

Zhao, L., Zhou, Y., Zhang, J., Liang, H., Chen, X., & Tan, H. (2023). Natural Polymer-Based Hydrogels: From Polymer to Biomedical Applications. *Pharmaceutics*, *15*(10), 2514. Advance online publication. DOI: 10.3390/pharmaceutics15102514 PMID: 37896274

Zhao, X., Chen, W., Wu, J., Shen, Y., Xu, B., Chen, Z., & Sun, Y. (2025). Application of Biomimetic Cell Membrane-Coated Nanocarriers in Cardiovascular Diseases. *International Journal of Nanomedicine*, *20*, 8249–8289. DOI: 10.2147/IJN.S531558 PMID: 40589723

# Chapter 6
# Stimuli-Responsive "Smart" Nanomaterials in Oncology

**Ihsan Ullah**
http://orcid.org/0009-0005-6630-9486
*Kyungpook National University, South Korea*

**Hamza Raheem**
*Department of Zoology, Faculty of Biological Sciences, Quaid-i-Azam University, Pakistan*

**Iram Bashir**
*Department of Zoology, Faculty of Biological Sciences, Quaid-i-Azam University, Pakistan*

**Muhammad Ali**
http://orcid.org/0009-0002-7593-0395
*Department of Zoology, Faculty of Biological Sciences, Quaid-i-Azam University, Pakistan*

**Jawad Elahi**
*Asfandyar Bukhari District Headquarters Hospital, Attock, Pakistan*

**Fahim Ullah Khan**
*Shanghai Ocean University, China*

**Dil Muhammad**
*MGRI, Russian State Geological Prospecting University, Russia*

## ABSTRACT

*This chapter examines the design, mechanisms, and biomedical applications of stimuli-responsive smart nanomaterials for precision cancer theranostics. It highlights nanocarriers that respond to endogenous stimuli, including pH, redox potential, enzymes, reactive oxygen species, and hypoxia, as well as exogenous triggers such as light, temperature, magnetic fields, and ultrasound, enabling controlled drug release and targeted tumor therapy. The chapter also discusses multifunctional nanoplatforms for image-guided treatment, combination therapies, and real-time*

DOI: 10.4018/979-8-3373-9848-8.ch006

*monitoring. Emerging advances in green nanotechnology and artificial intelligence for sustainable nanoparticle fabrication and personalized nanomedicine are reviewed, together with key challenges related to tumor heterogeneity, biosafety, standardization, large-scale manufacturing, and clinical translation. Future perspectives on multi-stimuli-responsive and AI-assisted nanotherapeutic systems are also presented.*

## INTRODUCTION

Cancer is a major cause of morbidity and mortality globally causing millions of deaths and incident cases every year. In spite of the progress made with regards to surgery as well as chemotherapy, radiotherapy, immunotherapy and targeted cancer therapies, treatment options available for numerous cancers are limited by low drug selectivity, systemic toxicity, multidrug resistance (MDR) and tumor heterogeneity. As a result, many conventional chemotherapeutic agents are unable to distinguish between healthy tissues and tumor cells therefore impairing their therapeutic window upon systemic administration. Recently, attempts have been made to explore new drug delivery strategies that possess the ability to diagnose disease conditions while limiting their deleterious side effects (Wang et al., 2024; Abdullah et al., 2025).

Nanomedicine, the application of nanotechnology in disease diagnosis, monitoring and therapy has gained significant attention owing to its promising role in cancer therapeutics.

During the early stages of nanomedicine, attempts were centered on developing nanocarriers that could increase the aqueous solubility, stability and pharmacokinetics-associated half-life of drugs. For example, early nanocarriers like liposomes and polymeric nanoparticles were developed to passively accumulate in tumor tissue through the enhanced permeability and retention (EPR) effect. In addition to EPR-mediated tumor accumulation, targeted ligands were conjugated onto nanocarrier surfaces to actively recognize tumor associated receptors (Afzal et al., 2022). In the past two decades however, there has been significant research put into evolving nanoparticles from passive drug carriers into versatile platforms capable of combining treatment, molecular imaging and even therapeutic response monitoring into a single platform (Puri et al., 2023; Kong et al., 2023; Gull et al., 2025).

Stimuli-responsive nanomaterials also referred to as "smart" nanomaterials have gained increasing interest owing to their tunable therapeutic response under internal or externally applied stimuli. Unlike traditional nanocarriers, smart nanomaterials are specifically designed to respond to certain stimuli by altering their physicochemical properties for controlled release of their cargo. Stimuli used for responsive nanomaterials are often associated with tumor microenvironment such as low pH, high levels of reactive oxygen species (ROS), hypoxia, overexpressed enzymes and reductive glutathione (GSH) as well as external triggers such as light irradiation, magnetic fields, ultrasound and hyperthermia (Mi, 2020; Chen et al., 2024; Sun et al., 2023). Exploiting these stimuli enables drugs to remain stable during systemic circulation but responsive to changes in tumor tissue for selective drug release at the target site.

Precision oncology which refers to the customization of cancer treatment to the biological and molecular profiles of the patient and their disease shares key features with smart nanomaterials. Precision oncology mainly involves targeting, controlled drug exposure and therapy accompanied with diagnostic evaluations to improve patient outcomes. Stimuli responsive nanoplatforms allow for selective drug release at tumor sites, allows combination of multimodal therapeutic approaches as well as integrating molecular imaging probes for visualizing therapeutic response during treatment. Other rapidly evolving nanotechnologies such as biomimetic nanocarriers, hybrid nanoparticles and artificial intelligence guided nanomedicine design have also shown promise for advancing precision oncology management strategies (Fang et al., 2022; Younas et al., 2025; Abdullah et al., 2025).

Although smart nanomaterials have been widely studied over the years, numerous challenges still need to be overcome before they can be clinically translated including biological barriers, ease of manufacture, scalability, safety assessments and current regulatory restrictions. Despite these setbacks, continuous advancements being made in stimuli-responsive nanoplatforms and their ability to be combined with various therapeutic platforms show great promise in cancer precision therapy.

In this chapter, we review stimuli-responsive smart nanomaterials for oncology applications. We begin by highlighting the principles behind stimuli-responsive smart nanocarriers and the various internal and external stimuli that are used to provoke a responsive therapeutic release. We then highlight various classes of stimuli responsive nanomaterials along with their applications in targeted drug delivery, cancer imaging and combination therapies as well as sustainable approaches used to synthesize these nanomaterials. We also discuss current challenges and translational relevance of stimuli-responsive nanomaterials along future perspectives.

## 1 DESIGN PRINCIPLES AND MECHANISMS

Stimuli-responsive smart nanomaterials behavior is mainly correlated with the intelligent incorporation of design variables, stimuli-responsive moieties, and triggered drug release mechanism(s). Traditional nanocarriers deliver their payload passively by simple diffusion, whereas smart nanomaterials should be stable under physiological conditions during systemic circulation until they experience physical and/or chemical changes under internal or external stimuli response. The response to stimuli alters the configuration of the carrier or the surface functional groups of the carrier triggering spatiotemporal controlled drug release behaviors. Rational combination of carrier morphology, responsive moieties, and drug release profile are the critical principles in designing smart nanoplatforms for smart-on-target cancer therapy (Chen et al., 2024; Sun & Davis, 2021). The general design principles and mechanisms underlying stimuli-responsive smart nanomaterials are illustrated in **Figure 1**.

## 1.1 Structural Characteristics

In particular, smart nanomaterials can take advantage of their nanostructured architecture to improve biological efficacy and responsivity. Liposomes, polymeric micelles, dendrimers, mesoporous silica nanoparticles (MSNs), metal-organic frameworks (MOFs), hydrogels, and nanogels are examples of nanoplatforms that have been fabricated and optimized to enhance drug loading capacity, colloidal stability, and stimuli-responsive release properties. Core-shell nanostructures are especially popular due to their versatility; these nanoparticles allow loading of the therapeutic payload into the core, while leaving room in the shell region for modification with targeting ligands and/or responsive polymers (Singh et al., 2022; Santhamoorthy et al., 2025).

MSNs have also become one of the most popular types of smart delivery vehicles studied in the literature due to their large surface area and high pore volume, ease of tunability, and readily functionalizable surfaces. Typically, the nanopores of MSNs are plugged with polymeric “gatekeepers” that trap the drug payload and prevent premature release while the nanoparticle travels through circulation. The gatekeepers then experience conformational changes or degrade upon stimulus exposure which opens the pores and initiates drug release (Santhamoorthy et al., 2025).

Amphiphilic polymeric assemblies and micelles have been commonly used due to their ability to self-assemble through hydrophobic and hydrophilic forces. Their stability and responsiveness can be tuned by parameters such as molecular packing, crosslink density, hydrophobic/hydrophilic balance, etc. When exposed to the appropriate stimuli, these assemblies can reversibly disassemble to control drug release kinetics (Liu et al., 2023; Maboudi et al., 2024). Hydrogels and nanogels are similar in that they both consist of three-dimensional crosslinked polymer networks. Ionizable or degradable linkages are often incorporated into hydrogel architectures which dictate when the hydrogel will swell, shrink, or degrade in response to external stimuli. These features make hydrogels promising carriers that allow sustained and localized drug release into tissues of interest (Weng et al., 2022; Zhang & Wu, 2023).

In addition to these platforms, there has been recent work developing peptide-based nanoarchitectures, carbon nanomaterials, transition metal dichalcogenides, cellulose-based nanoparticles, and crystalline nanoparticles based on supramolecular assembly as smart nanomaterials. Each of these material systems provide unique physiochemical properties that can be leveraged and tuned for enhanced responsivity, biocompatibility, and multifunctionality (Sun & Davis, 2021; Torres- Cavanillas & Forment- Aliaga, 2024; Zhao et al., 2023; Zhu et al., 2020).

## 1.2 Trigger-Responsive Mechanisms

Smart nanomaterials are designed to respond to specific internal or external stimuli. While internal triggers are based on tumor microenvironmental properties, external triggers involve externally applied physical cues. Common internal triggers include pH, redox state, enzymes, hypoxia, and reactive oxygen species (ROS). Due to increased metabolic activity within tumors extracellular pH is often lower than that observed in normal tissues making pH one of the most exploited triggers. pH-responsive carriers are often designed with ionizable groups or acid-cleavable chemical linkers that undergo protonation or cleavage at acidic pHes leading to carrier disassembly and payload release (Zhou et al., 2022; Santhamoorthy et al., 2025). Redox-responsive carriers often contain disulfide or diselenide linkages that are reduced by high concentrations of glutathione present within cancer cells (Wang et al., 2023). Enzyme-responsive carriers are triggered by proteases, matrix metalloproteinases (MMPs), and other tumor-specific enzymes that can cleave coatings or linker molecules (Zhao et al., 2023).

Another type of enzyme-responsive carrier is antibodies or antigen-functionalized carriers that will bind to antigens overexpressed on tumor cells leading to drug release. Other internal triggers that have been developed include hypoxia-responsive and ROS-responsive carriers. Hypoxia often presents in solid tumors and can trigger hypoxia-sensitive chemical moieties. Similar to pH-responsive carriers, some ROS-responsive carriers are also designed with chemical linkers that can be cleaved under high ROS conditions found within tumors leading to drug release (Santhamoorthy et al., 2025; Wang et al., 2023).

Exogenous stimuli allow spatiotemporal control of drug release. Light-responsive carriers often contain photoresponsive molecules that change structurally upon irradiation with light. Thermal-responsive carriers change their physiochemical properties at higher temperatures. Magnetic nanoparticles can also be triggered externally with alternating magnetic fields. Ultrasound-responsive carriers use acoustic forces to cause cavitations and mechanical disruptions of the nanocarrier (Devnarain et al., 2020; Chen et al., 2024). Stimuli-responsive carriers can be combined within a single nanoplatform to gain better targeting specificity.

## 1.3 Drug Release Pathways

Controlled drug release is the ultimate functional readout of smart nanomaterial design. Carrier systems can facilitate drug release by a variety of means, depending on both their architecture and their underlying responsive mechanism.

Arguably the simplest strategy is based on barrier opening/pore uncapping mechanisms. Protective polymeric coronas or attached gatekeeper molecules undergo volume changes or conformational rearrangements that permit encapsulated drug cargo to leak out. This strategy has been widely used in mesoporous silica nanoparticles as well as gated nanocarriers. (Santhamoorthy et al., 2025)

Bond cleavage is another widespread release mechanism. Drugs can be tethered to the carrier backbone via stimuli-responsive linkages. Common chemical bonds that are sensitive to acidic pH, reducing conditions, or tumor-associated proteases include hydrazone, disulfide, orthoesters, and others. Triggered cleavage of these linkages thus allows for release of active drug payloads. These strategies often, but not always, induce carrier disassembly, further facilitating drug availability. (Wang et al., 2023; Zhang & Wu, 2023)

Structural assembly-disassembly of carrier nanosystems represents another approach towards achieving triggered drug release. Polymeric micelles, peptide nanofibers/sheets, and hydrogels are some examples of smart materials that reversibly assemble/dissociate under different conditions, thereby coupling structural transitions on the nanoscale to modulated release. (Sun & Davis, 2021; Maboudi et al., 2024)

Combination approaches make use of two or more independent release mechanisms/ stimuli (such as pH/redox, pH/temperature, pH/enzymes, etc.). Multistimuli-responsive drug delivery platforms have become increasingly popular in recent years because they allow for sequential/hierarchical release of cargo. This additional layer of control may offer some benefit in terms of achieving more precise targeting within the spatially and temporally heterogeneous TME. (Basety et al., 2026; Zhou et al., 2022)

In summary, smart nanomaterials can be tailored using design principles centered around architectural framework, responsive functional groups, and programmable release mechanisms. Coupling targeted stimuli to predictable structural changes on the nanoscale provides tremendous utility for drug delivery applications.

*Figure 1. Overview and principles of stimuli-responsive smart nanomaterials*

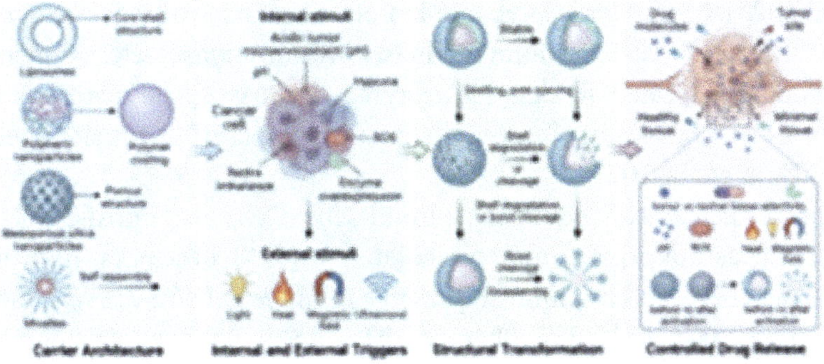

## 2 TUMOR MICROENVIRONMENT AS A TARGET

The tumor microenvironment (TME) consists of tumor cells, stromal cells, infiltrating immune cells, extracellular matrix (ECM), and aberrant vasculatures, forming a dynamic, complex, and heterogeneous microenvironment. Distinct from normal tissue microenvironments, TME usually possesses physicochemical and biochemical characteristics that favor tumor growth, making stimuli-responsive nanomedicine become smarter by taking advantage of these pathological features for tumor-specific drug activation/release and decreasing systemic side effects (Mi, 2020; Wang et al., 2024).

Currently, the common pathophysiological features of TME being widely applied in smart nanomaterial design include acidic pH, hypoxia, abnormal redox level, elevated enzyme expression, etc. Tumor cells exhibit enhanced glycolysis metabolism than normal cells, producing excessive acidic metabolites such as lactic acid and leading to mild extracellular acidosis in TME. Therefore, carrier's sensitive to the acidic pH changes have been developed to enable stability in normal physiological microenvironments and trigger drug release specifically at tumor sites (Li & Burgess, 2020; Liu et al., 2025). Similarly, the rapid growth of solid tumors is always accompanied by inefficient vasculatures, leading to local hypoxic regions that contribute to tumor metastasis and drug resistance. Thus, hypoxia-responsive nanoplatforms have also been designed for triggered drug activation in hypoxic tumor regions (Meng et al., 2024).

On the other hand, cancer cells tend to maintain an imbalanced redox homeostasis with high intracellular glutathione (GSH) levels and excessive reactive oxygen species (ROS) production. Thus, nanocarriers responsive to redox and ROS have been constructed, which can achieve structures' transition or trigger bond breakage in tumor cells, allowing controlled intracellular drug release (Gaddimath et al., 2024; Zou et al., 2025). Moreover, a large number of tumors have upregulated expressions of various enzymes (matrix metalloproteinases, hyaluronidases, etc.). Cancer-specific enzymes can be applied to degrade enzyme-sensitive coats/materials or cleave enzyme-sensitive linkers conjugated on smart nanomaterials for site-specific drug delivery (Zhou et al., 2022).

Except for these physicochemical stimuli, the TME also consists of biological barriers, such as abnormal tumor vasculatures, ECM networks, high interstitial fluid pressure (IFP), and immunosuppressive immune cells/tissue structures. These biological barriers in TME might hamper the nanocarriers' penetration and impede the effectiveness of passive targeting. As a result, the latest generation of smart nanocarriers is being developed towards TME-responsive systems combined with either active targeting, biomimetic modifications, or multifunctional carrier systems to overcome these biological barriers (Ejigah et al., 2022; Zhou et al., 2020).

In summary, the intrinsic features of TME endow stimuli-responsive nanomaterials with an opportunity to realize precision medicine, laying a biological foundation for guided cancer therapy. A summary of the key TME features and the associated stimuli-responsive nanomaterial strategies is provided in **Table 1**.

*Table 1. Major tumor microenvironment characteristics exploited by stimuli-responsive smart nanomaterials for targeted cancer therapy*

| TME Feature | Biological Basis | Smart Nanomaterial Strategy | Reference |
|---|---|---|---|
| Acidic pH | Enhanced aerobic glycolysis (Warburg effect) and lactate accumulation reduce extracellular pH (≈6.0–7.0). | pH-sensitive polymers, acid-labile bonds, and pH-responsive inorganic nanocarriers. | Li & Burgess 2020; Liu et al. 2025; Wang et al. 2024 |
| Hypoxia | Rapid tumor growth and abnormal vasculature create oxygen-deficient regions. | Hypoxia-responsive prodrugs, oxygen-generating nanoparticles, and hypoxia-sensitive carriers. | Meng et al. 2024; Sun et al. 2025; Wang et al. 2024 |
| Redox Imbalance | Elevated intracellular glutathione (GSH) concentrations create a reducing environment. | Disulfide/diselenide-linked nanocarriers for intracellular drug release. | Gaddimath et al. 2024; Meng et al. 2024; Zhou et al. 2022 |
| Reactive Oxygen Species (ROS) | Cancer cells exhibit increased ROS production compared with normal tissues. | ROS-responsive linkers and ROS-amplifying nanoplatforms for controlled release. | Zou et al. 2025; Karimi et al. 2025; Wang et al. 2024 |

continued on following page

*Table 1. Continued*

| TME Feature | Biological Basis | Smart Nanomaterial Strategy | Reference |
|---|---|---|---|
| Enzyme Overexpression | Matrix metalloproteinases (MMPs), hyaluronidases, and other enzymes are highly expressed in tumors. | Enzyme-cleavable coatings, peptide linkers, and degradable polymer shells. | Zhou et al. 2022; Sun et al. 2025; Li & Burgess 2020 |
| Biological Barriers | Dense extracellular matrix, abnormal vasculature, elevated interstitial pressure, and stromal cells hinder drug penetration. | Biomimetic nanocarriers, active targeting ligands, and matrix-modulating systems. | Ejigah et al. 2022; Zhou et al. 2020; Mi 2020 |

## 3 CLASSIFICATION OF STIMULI-RESPONSIVE NANOMATERIALS IN ONCOLOGY

Stimuli-responsive nanomaterials can usually be classified based on the stimulus's origin, which is typically categorized as internal (endogenous) or external (exogenous). This classification system is useful for oncologic applications since tumor tissues offer unique microenvironmental stimuli and exogenous triggers allow precise spatial-temporal control of drug release by clinicians (Chen et al., 2024; Mi, 2020; Pham et al., 2020).

### 3.1 Internal Stimuli

Endogenous/Internal stimuli are derived from TME and involve biochemical and physiochemical differences existing between tumor tissues and normal ones. Guided by these intrinsic signals, nanocarriers can perform stimulus-responsive behavior without external manipulation, leading to enhanced selectivity and reduced off-target toxicity. A summary of the internal stimuli-responsive nanoplatforms is presented in **Table 2**.

#### 3.1.1 pH-responsive Systems

The acidic tumor microenvironment is probably the most investigated internal stimuli. The physiological pH of human tissues/organs is tightly regulated between pH 7.4 and 7.5, while pH in the extracellular tumor space ranges from pH 6.5 to 6.8 due to the high glycolytic activity and resultant accumulation of lactic acid. pH-responsive nanomaterials are designed based on pH sensitive acid-labile linkers, pH-triggered conformation changes via protonation, and charge-reversal mechanisms to ensure tumor-selective drug release (Alsawaftah et al., 2022; Kumar et al., 2024; Zhao et al., 2023).

### 3.1.2 Redox-responsive Systems

Redox-responsive nanocarriers are designed to sense elevated intracellular GSH concentrations compared to normal tissues. They are engineered with disulfide groups or redox-recognizable linkers that are stable in circulation but cleaved quickly in intracellular reductive conditions to allow controlled release of encapsulated drugs (Chen et al., 2024; Mi, 2020; Wei et al., 2023).

### 3.1.3 Enzyme-responsive Systems

Enzyme-responsive nanomaterials are designed to sense overexpressed enzymes such as matrix metalloproteinases, phosphatases, or esterases in tumor tissues. They can be tuned with peptide sequence tags or enzymatic-cleavable polymeric backbones for enzyme-triggered cleavage and selective release/degradation in diseased tissues (Kumar et al., 2024; Zhang et al., 2022; Zhou et al., 2023).

### 3.1.4 ROS-responsive Systems

ROS levels are significantly higher in most cancer cells due to mitochondrial damage and oxidative stress. ROS-responsive nanomaterials are often designed with oxidation-sensitive linkages/cross-links that can be cleaved upon exposure to high levels of ROS for site-selective drug release or nanostructure degradation (Das et al., 2020; Sun et al., 2022; Zhang et al., 2023).

### 3.1.5 Hypoxia-responsive Systems

Hypoxia refers to the low oxygen levels in solid tumors due to poor perfusion rates in growing tumor bulks. Hypoxia-responsive nanocarriers are designed using bioreductive moieties or oxygen-sensitive linkers that can be selectively activated under hypoxic conditions. The response can be combined with other stimuli including pH and ROS sensitivity for better response control and therapeutic efficacy (Alsawaftah et al., 2022; Sun et al., 2022; Zhao et al., 2023).

*Table 2. Internal stimuli-based nanoplatforms*

| Nanoplatform | Stimuli | Functional Role | Reference |
|---|---|---|---|
| Polymeric nanoparticles & nanogels | pH, redox, enzymes, ROS, hypoxia | Controlled and multifunctional drug delivery systems | Chen et al. 2024; Wei et al. 2023 |
| Peptide-based self-assemblies | Enzymes, redox, pH | Biocompatible and programmable nanostructures | Zhou et al. 2023 |
| TME-responsive systems | Multi-stimuli integration | Enhanced specificity and cascade activation | Mi 2020; Kumar et al. 2024 |

## 3.2 External Stimuli

Systems reliant on external stimuli permit clinician-applied spatiotemporal control of therapeutic activation, thus allowing exquisite precision over dose modulation. Such systems are especially useful in modalities like image-guided therapy or therapies administered via minimally invasive routes as part of treatment for cancer (Chen et al., 2024; Pham et al., 2020). A summary of external stimuli-responsive is presented in **Table 3.**

### 3.2.1 Temperature-Responsive Systems

Thermo-responsive systems are based on nanomaterials that undergo a change in physicochemical properties with temperature. Most polymers responsive to mild changes in temperature are polymeric materials that have an LCST. Mild hyperthermia has also been used to trigger phase change/release or structural changes of nanocarriers to enable controlled drug release as well as increase tumor sensitivity to drugs or radiation (Kumar et al., 2024; Wei et al., 2023).

### 3.2.2 Light-Responsive Systems

Light-responsive nanomaterials are triggered using ultraviolet, visible, or near-infrared (NIR) light irradiation to achieve photochemical bond cleavage, photothermal conversions, or lightmediated production of ROS via photodynamic therapy. NIR light has been favored over UV and visible wavelengths due to its ability to penetrate tissue further with minimal photoxicity (Armenia et al., 2022; Mi, 2020; Zhang et al., 2021).

### 3.2.3 Magnetic Field-Responsive Systems

Magnetically-responsive systems are predominately based on iron oxide nanomaterials that can be directed and localized to a region of interest using an external magnetic field. In addition to allowing improved localization of therapeutics, these nanoparticles can be made to heat under alternating magnetic fields for hyperthermia-mediated drug release or ablation of tumors (Armenia et al., 2022; Wei et al., 2023).

### 3.2.4 Ultrasound-Responsive Systems

Ultrasound-responsive carriers use sound waves to trigger mechanical disruption, cavitation, and localized heating. The mechanical indices can also improve tissue permeability and promote release of drug from nanocarriers. Ultrasound is non-invasive and easily modifiable during clinical applications, making it an ideal external stimulus (Chen et al., 2024; Zhang et al.,2021; Nascimento et al., 2025).

### 3.2.5 Electric/Radiofrequency (RF)-Responsive Systems

Electric and RF-responsive nanomaterials are most commonly derived from conductive polymers or electrically conductive carbon nanomaterials that change properties in response to electromagnetic stimuli. Through local heating or changes in structural conformation these materials allow for remote-controlled drug release and hold promise for implantable devices or therapeutics with external control mechanisms (Thomas et al., 2020; Zhang et al., 2021).

*Table 3. External stimuli-based nanoplatforms*

| Nanoplatform | Stimuli | Functional Role | References |
|---|---|---|---|
| Gold, iron oxide, polymeric nanoparticles | Light, magnetic field | Targeted delivery and imaging-guided therapy | Armenia et al. 2022 |
| Carbon-based nanomaterials | Electric/RF fields | Controlled heating and on-demand release | Thomas et al. 2020 |
| Lipid and polymeric carriers | Ultrasound | Deep tissue penetration and non-invasive activation | Zhang et al. 2021 |

# 4 SMART NANOMATERIALS FOR PRECISION DRUG DELIVERY IN CANCER

Smart nanomaterials are poised to offer the greatest potential toward spatiotemporally precise drug delivery by synergistically incorporating controlled release kinetics, tumor targeting, payload diversification, and overcoming multidrug resistance (MDR).

Controlled Release Stimuli-responsive nanocarriers enable both sustained or triggered drug release in response to internal stimuli (pH, enzymes, redox) or external stimuli (light, heat, ultrasound) allowing for preferential release of therapeutics at the tumor site with improved pharmacokinetic profiles (Basety et al., 2026; Wang et al., 2022; Agiba et al., 2024).

## 4.1 Controlled Release

Nanoformulations such as polymeric nanoparticles, hydrogels, and layered nanostructures provide tunable release kinetics, and computational models have been developed to help predict the dynamic delivery profiles of nanocarriers across heterogeneous tumor microenvironments. Multi-responsive nanoplatforms that respond to two or more stimuli (e.g., pH/redox, light/enzymatic) allow for cascade trigger activation of nanocarriers for improved precision in response to complex tumor microenvironments (Park et al., 2025).

## 4.2 Tumor Targeting

Smart nanocarriers also enable improved tumor targeting by actively or passively homing in on tumor tissues and cancer cells. Active targeting approaches functionalize the surface of nanocarriers with tumor-homing ligands including antibodies, peptides, folate, and hyaluronic acid for receptor-mediated internalization by cancer cells (drug-resistant cancer cells included) (Sabit et al., 2025; Wang et al., 2024).

Likewise, tumor microenvironment-responsive nanocarriers leverage cues such as acidic pH, low oxygen levels, high glutathione levels, and overexpressed enzymes for improved site-specific release and activation. Some stimuli-responsive nanoplatforms have switchable properties (size, charge) allowing for improved tumor penetration and cellular uptake (Sun et al., 2023; Agiba et al., 2024).

## 4.3 Combination Payloads

Smart nanoplatforms permit loading and delivery of combination payloads for synergistic cancer treatment. Examples include co-delivery of small molecule chemotherapeutics with nucleic acids (siRNA, miRNA, antisense oligonucleotides), proteins, and immunomodulators to target multiple disease-onucing pathways simultaneously (Chen et al., 2023).

Remarkably, therapeutic combinations like chemo–photodynamic therapy (chemo–PDT), chemo–photothermal therapy (chemo–PTT), and chemo-immunotherapy are also being integrated into single nanoformulations. Innovations such as DNA dendrimers and anchorable protein "nanococktails" afford precise control of therapeutic ratios as well as temporal synchronization of drug release for improved coordination of payloads at tumor sites (Tian et al., 2025; Zou et al., 2024).

### 4.4 Overcoming Multidrug Resistance

Chemotherapy is often limited by multidrug resistance (MDR), a cancer cell property that promotes tolerance to chemotherapeutics through mechanisms like drug efflux, altered drug metabolism, and evasion of apoptosis. Multiple strategies enabled by smart nanomaterials can be leveraged to overcome MDR. Nano-delivery improves drug accumulation within cancer cells and protects therapeutic cargo from efflux pumps such as glycoprotein. Multicomponent delivery can also be used to deliver gene-silencing molecules such as siRNA, miRNA, and CRISPR-based gene editing molecules to downregulate genes and signaling pathways associated with chemoresistance (Wei et al., 2021; Zou et al., 2024).

Smart nanocarriers can also deliver specific chemical MDR modulators including inhibitors of P- gp, nitric oxide, metabolism modulators, or target organelles such as mitochondria to trigger apoptosis downstream of efflux transporter activity. Strategies for overcoming MDR continue to emerge such as induction of cuproptosis and manipulation of intracellular redox balance (Maimaitijiang et al., 2024; Tian et al., 2025).

## 5 THERANOSTIC AND COMBINATION APPLICATIONS

Theranostic nanomaterials allow for imaging guided therapy or concurrent imaging and therapeutic functions. Applications include all modalities discussed previously with theranostic nanoplatforms allowing for real-time imaging feedback of drug distribution/tumor regression during treatment.

### 5.1 Imaging-Guided Therapy / Real-Time Monitoring

Dual imaging guided therapy allows for precise tumor targeting as well as real time monitoring of therapeutic response. Activatable fluorescent probes allow for real time imaging of drug distribution and tumor regression. Carbon dot based nanotheranostics as well as gold nanoparticle platforms serve similar purposes (Siddique & Chow, 2022). Deep-tissue imaging can be achieved with biomimetic optotheranostic nanovesicles or photoacoustic biosensors which can be used during photothermal therapy or chemotherapy for real time tumor imaging/response monitoring while reducing off-target toxicity (Prasad et al., 2024; Taha et al., 2025).

## 5.2 Combination Therapy Modalities

Smart nanoplatforms are often used in combination with multiple therapy modalities for synergistic treatment efficacy:

- **Chemotherapy:** Smart nanocarriers can enhance solubility, stability, and passive accumulation of chemotherapeutics.
- **Phototherapy (PTT/PDT):** Near infrared (NIR) light responsive nanomaterials can convert light energy to heat or reactive oxygen species for localized tumor kill.
- **Gene Therapy:** Nanocarriers can deliver siRNA, miRNA, or gene editing complexes to regulate gene expression.
- **Immunotherapy:** Nanomaterials can either target immune checkpoints or modulate tumor antigen presentation to activate an immune response. Immunotherapy can be used alongside PTT or chemotherapy for enhanced treatment responses (Liu et al., 2022; Arif et al., 2023; Karnwal et al., 2024).

Multi-function systems can allow for spatiotemporally controlled therapy depending on the treatment needed.

# 6 GREEN AND SUSTAINABLE FABRICATION APPROACHES

Environmentally benign or green nanomedicine utilizes sustainable fabrication approaches aiming to reduce energy consumption, toxic byproducts, and overall ecological footprint during synthesis while maintaining the therapeutic efficacy of nanoparticulate systems.

## 6.1 Plant-Based Synthesis and Biomaterial Platforms

Plant extracts, microbes, or other natural compounds/organisms can function as reducing/stabilizing agents to produce nanomaterials. Utilizing biological synthesis can reduce the need for harmful chemical synthesis while imparting inherent biocompatibility and biological function to the nanoparticles (Chopra et al., 2022; Karnwal et al., 2024). Plant derived nanoparticles or phyto-functionalized nanomaterials show promise in cancer therapy, antimicrobial applications, and tissue regeneration therapy (Zanbili & Marjani, 2025).

Biopolymers such as chitosan represent sustainable drug delivery platforms. Biopolymers are biocompatible and biodegradable polymers that have shown promise in cancer therapy due to their positive surface charge allowing for interaction with negatively charged cell membranes. (El- Saadony et al., 2025). Similar biomaterial platforms include biopolymer-based nanosponges as well as biopolymer scaffolds for 3D-printed implants (Iravani & Varma, 2022; Hossain et al., 2025).

### 6.2 Eco-Friendly Functionalization and Manufacturing Challenges

Major drawbacks of green synthesis include lack of control over particle size, poor reproducibility, scalability, and lack of understanding of biosynthetic mechanisms. (Taha et al., 2025). Scale-up for clinical use will require stringent protocols and quality-by-design approaches along with controlled parameters to ensure consistent manufacturing. Biological synthesis of nanoparticles also faces challenges with functional performance reproducibility which must be addressed prior to clinical translation.

## 7 ARTIFICIAL INTELLIGENCE IN SMART NANOMEDICINE

Application of Artificial intelligence (AI) and; Machine learning in medicine is revolutionizing medicine as we know it, allowing for tailor made treatments of even the most complex diseases. AI has a wide range of applications in nanomedicine as well. AI can be used to design nanoparticles and help create predictive models that can allow for optimization of nanoparticle design.

### 7.1 AI-Driven Design and Optimization of Nanomedicine

Machine learning can be used to predict physiochemical properties of nanoparticles as well as their synthesis parameters. Modelling has been used to optimize chitosan and polymer-based nanoparticle formulations as well as predict drug loading efficiency, nanoparticle stability, and drug release profiles (Duan et al., 2023). Machine learning models can be trained to understand the structure-activity relationships of nanoparticles which can allow for faster rational design of nanomedicines.

## 7.2 AI-Enhanced Precision Oncology

Tumor heterogeneity can be analyzed by AI algorithms and used to design patient specific treatments by allowing for analysis of medical imaging, tumor genetics, and patient histories to design personalized medicine treatments. AI can allow for drug and dosage selection to allow for optimal treatment and scheduling based on the patient's needs. (El- Saadony et al., 2025). Photoacoustic imaging and biosensing can be monitored by AI for smart treatment adjustments in real-time based on how the tumor is responding to treatment. (Taha et al., 2025). AI will revolutionize smart nanomedicine and will push the field even further along with green nanofabrication techniques and multifunctional theranostics.

# 8 CLINICAL TRANSLATION AND CURRENT PROGRESS

Applications of smart nanomedicines into clinical cancer care has made promising progress in recent years with ample preclinical studies and clinical trials. To date, few nano enabled drugs have been approved by the FDA and integrated into standard cancer treatment. (Liu et al., 2023; Mundekkad & Cho, 2022).

## 8.1 Preclinical Advances

Preclinical evidence unanimously shows superior tumor accumulation, increased drug bioavailability, and multimodal imaging–therapy possibilities with smart nanocarriers. Stimuli-responsive drug delivery vehicles and TME-modulating systems demonstrated superior penetration and bioeffect across biological barriers, showing great potency even in aggressive and metastatic disease models in preclinical systems. (Gavas et al., 2021; Zhang et al., 2025; Abdullah et al., 2025). Combination therapies were also enabled through these smart nanoplatforms. (Wang et al., 2022)

## 8.2 Clinical Trials and Approved Systems

Approved drug delivery systems (primarily liposomal and albumin-bound drugs) have shown improved pharmacokinetics and favorable safety profiles compared to standard chemotherapeutics. (Rodríguez et al., 2022; Wang et al., 2024) Many clinical trials also exist testing polymeric nanoparticles, ligand-targeted drug delivery vehicles, stimuli-responsive nanomedicines, and more across a multitude of cancers. Early clinical-trial candidates have also shown promising translational potential with manageable toxicity and improved tumor accumulation. (Liu et al., 2023) An example includes CRLX101, a stimuli-responsive nanocarrier (Figure 1B).

## 8.3 Regulatory Considerations and Scale-Up

Despite myriad successes, few nanoparticle drug delivery platforms have received regulatory approval from the FDA. This is in part due to nanoparticle drug delivery systems lacking standardized characterization guidelines that address their size, long-term stability, surface chemistry, drug dissociation behavior, and more. Without established guidelines, nanoparticle batch-to-batch reproducibility is low and poses challenges for regulatory bodies. (Zhang et al., 2025; Mundekkad & Cho, 2022)

Scale-up for clinical translation of nanoparticle formulations is also hindered by their limited commercial scalability, poor batch-to-batch reproducibility, and elevated production costs. (Souri et al., 2022; Liu et al., 2024) Approaches currently being developed to address the need for scalable and manufacturable nanoparticles in clinical settings include microfluidic synthesis of nanoparticles, modular scalable manufacturing units, and quality-by-design (QbD) nanoparticle manufacturing paradigms.

# 9 CHALLENGES AND FUTURE PERSPECTIVES

Despite remarkable advances, the clinical success of smart nanomedicine is still constrained by biological complexity, safety concerns, and manufacturing limitations.

## 9.1 Tumor Heterogeneity

While nanoparticles have demonstrated superior permeation across biological barriers, tumor heterogeneity poses a large challenge to their clinical success. Specifically, heterogeneity within and between tumors result in variable nanoparticle accumulation and therapeutic response between patients. As aforementioned, the EPR effect is drastically reduced in human tumors compared to early preclinical studies, greatly limiting the efficiency of passive targeting nanoparticles. (Sabit et al., 2025; Liu et al., 2023) Therefore, research is shifting toward developing personalized nanomedicine and biomarker-driven nanotherapeutics.

## 9.2 Safety and Toxicity

Immune-system interactions, long-term biocompatibility, and organ toxicology of nanoparticles are not yet well characterized for all nanomaterials. While preclinical testing has shown excellent progress, reproducible and standardized toxicological profiling of nanoparticles needs to be developed to help with translational safety studies. (Thapa & Kim, 2022; Rodríguez et al., 2022) In particular, reproductive toxicology, hepatic toxicity, and renal toxicity are key considerations for future clinical translation.

## 9.3 Standardization Issues

A lack of uniform standards of nanoparticle characterization, in vitro cancer model efficacy testing, and in vivo toxicology methods exist. Standardized and uniform guidelines are needed to help develop globally-applicable testing methods to accelerate translational success of nanoparticle drug delivery. (Mundekkad & Cho, 2022; Wang et al., 2024) Since many nanoparticle studies utilize different protocols and methodologies that may not be reproduced in other laboratories, this presents a large problem for reproducibility and scale-up.

## 9.4 Multi-Stimuli Nanoplatforms

Designing multifunctional nanoplatforms that can respond to two or more different stimuli within the TME is an oncoming trend in future nanomedicine. By synthesizing nanoparticles that can respond to pH, redox, enzymatic, thermal, and optical stimuli, researchers can design hierarchical treatment strategies with improved tumor specificity. (Venturini et al., 2025; Younas et al., 2025) Combining different forms of cancer therapy is also made possible with these smart nanoplatforms. For example, researchers have developed nanomedicine that combines chemo- and phototherapy with immunotherapy to provoke a stronger anti-tumor immune response.

## 9.5 Autonomous Smart Nanomedicine

Advancements in AI will help drive future nanoparticle development and personalized medicine. Developing self-learning and intelligent nanoparticle drug delivery platforms may one day allow us to create autonomous nanomedicine systems. Future possibilities include AI-assisted nanoparticle synthesis and characterization (Liu et al., 2024) computer-generated predictive models that determine optimal treatment schedules and dosages, (Wang et al., 2021) and closed-loop nanoparticle systems that can modulate drug release based on tumor biofeedback. (Sabit et al., 2025) While these applications may seem far-off, the field of intelligent nanomedicine is rapidly developing.

## CONCLUSION

The development of stimuli-responsive nanomaterials is one of the most important progress in modern oncology, which allows to realize the precision, controlled and multifunctional approaches to the treatment of cancer. These smart systems provide unprecedented control of drug release, targeting efficiency and therapeutics activation in tumor microenvironment by Internal/external applied stimuli. They can be integrated with imaging modalities, further enhancing their significance in theranostics, enabling simultaneous diagnosis, treatment, and monitoring of cancer progression.

Furthermore, the use of smart nanoplatforms has allowed co-delivery of molecules like chemotherapeutic agents, gene therapies, and immunomodulatory drugs, which has broadened the possibilities of combination therapy and helped overcome the tumor heterogeneity and therapeutic resistance. Incorporating green synthesis techniques ensures sustainable and environmentally friendly methods for creating nanomaterials, while AI provides powerful tools for designing new materials rationally, modeling outcomes, and tailoring treatment strategies to individuals.

But despite the excellent preclinical results and early clinical data, there are still a number of obstacles to widespread clinical translation. These encompass rare long-term safety and toxicity issues, inconsistent evaluation procedures and problems with manufacturing on a large scale. All of these challenges will need to be solved through collaboration between disciplines, harmonization of regulations and sophisticated engineering solutions.

The development of multi-stimuli-responsive systems, autonomous systems with AI integration, and highly personalized therapeutic platforms are expected to help drive progress in this field in the future. These innovations will help to revolutionize the way cancer is treated from traditional treatment paradigms to precise, adaptive and patient-specific nanotherapeutic approaches.

## REFERENCES


Abdullah, K. M., Sharma, G., Singh, A. P., & Siddiqui, J. A. (2025). Nanomedicine in Cancer Therapeutics: Current Perspectives from Bench to Bedside. *Molecular Cancer, 24*(1), 169. Advance online publication. DOI: 10.1186/s12943-025-02368-w PMID: 40490771

Afzal, O., Altamimi, A. S., Nadeem, M., Alzarea, S. I., Almalki, W., Tariq, A., Mubeen, B., Murtaza, B. N., Iftikhar, S., Riaz, N., & Kazmi, I. (2022). Nanoparticles in Drug Delivery: From History to Therapeutic Applications. *Nanomaterials (Basel, Switzerland), 12*(24), 4494. Advance online publication. DOI: 10.3390/nano12244494 PMID: 36558344

Agiba, A. M., Elsayyad, N., ElShagea, H. N., Metwalli, M., Mahmoudsalehi, A. O., Beigi-Boroujeni, S., Lozano, O., Aguirre-Soto, A., Arreola-Ramírez, J. L., Segura-Medina, P., & Hamed, R. R. (2024). Advances in Light-Responsive Smart Multifunctional Nanofibers: Implications for Targeted Drug Delivery and Cancer Therapy. *Pharmaceutics, 16*(8), 1017. Advance online publication. DOI: 10.3390/pharmaceutics16081017 PMID: 39204362

Alsawaftah, N., Awad, N. S., Pitt, W., & Husseini, G. (2022). pH-Responsive Nanocarriers in Cancer Therapy. *Polymers, 14*(5), 936. Advance online publication. DOI: 10.3390/polym14050936 PMID: 35267759

Arif, Z. U., Khalid, M., Noroozi, R., Hossain, M., Shi, H., Tariq, A., Ramakrishna, S., & Umer, R. (2023). Additive manufacturing of sustainable biomaterials for biomedical applications. *Asian Journal of Pharmaceutical Sciences, 18*(3), 100812. Advance online publication. DOI: 10.1016/j.ajps.2023.100812 PMID: 37274921

Armenia, I., Ayllón, C. C., Herrero, B. T., Bussolari, F., Alfranca, G., Grazú, V., & De La Fuente, M. J. (2022). Photonic and magnetic materials for on-demand local drug delivery. *Advanced Drug Delivery Reviews, 114584*, 114584. Advance online publication. DOI: 10.1016/j.addr.2022.114584 PMID: 36273514

Basety, S., Gudepu, R., & Velidandi, A. (2026). Smart Nanoformulations for Oncology: A Review on Overcoming Biological Barriers with Active Targeting, Stimuli-Responsive, and Controlled Release for Effective Drug Delivery. *Pharmaceutics, 18*(2), 196. Advance online publication. DOI: 10.3390/pharmaceutics18020196 PMID: 41754938

Chen, D., Liu, X. W., Lu, X., & Tian, J. (2023). Nanoparticle drug delivery systems for synergistic delivery of tumor therapy. *Frontiers in Pharmacology, 14*, 1111991. Advance online publication. DOI: 10.3389/fphar.2023.1111991 PMID: 36874010

Chen, X., Wu, D., & Chen, Z. (2024). Biomedical applications of stimuli-responsive nanomaterials. *MedComm*, *5*(8), e643. Advance online publication. DOI: 10.1002/mco2.643 PMID: 39036340

Chopra, H., Bibi, S., Singh, I., Hasan, M. M., Khan, M. S., Yousafi, Q., Baig, A., Rahman, M., Islam, F., Emran, T., & Cavalu, S. (2022). Green Metallic Nanoparticles: Biosynthesis to Applications. *Frontiers in Bioengineering and Biotechnology*, *10*, 874742. Advance online publication. DOI: 10.3389/fbioe.2022.874742 PMID: 35464722

Das, S. S., Bharadwaj, P., Bilal, M., Barani, M., Rahdar, A., Taboada, P., Bungău, S., & Kyzas, G. (2020). Stimuli-Responsive Polymeric Nanocarriers for Drug Delivery, Imaging, and Theragnosis. *Polymers*, *12*(6), 1397. Advance online publication. DOI: 10.3390/polym12061397 PMID: 32580366

Devnarain, N., Osman, N., Fasiku, V., Makhathini, S., Salih, M., Ibrahim, U. H., & Govender, T. (2020). Intrinsic stimuli-responsive nanocarriers for smart drug delivery of antibacterial agents-An in-depth review of the last two decades. *Wiley Interdisciplinary Reviews. Nanomedicine and Nanobiotechnology*, *1664*. Advance online publication. DOI: 10.1002/wnan.1664 PMID: 32808486

Duan, Q., Zhao, Z., Zhang, Y.-J., Fu, L., Yuan, Y., Du, J.-Z., & Wang, J. (2023). Activatable fluorescent probes for real-time imaging-guided tumor therapy. *Advanced Drug Delivery Reviews*, *114793*, 114793. Advance online publication. DOI: 10.1016/j.addr.2023.114793 PMID: 36963569

Ejigah, V., Owoseni, O., Bataille-Backer, P., Ogundipe, O. D., Fisusi, F., & Adesina, S. (2022). Approaches to Improve Macromolecule and Nanoparticle Accumulation in the Tumor Microenvironment by the Enhanced Permeability and Retention Effect. *Polymers*, *14*(13), 2601. Advance online publication. DOI: 10.3390/polym14132601 PMID: 35808648

El-Saadony, M., Saad, A., Sitohy, M., Alkafaas, S. S., Dladla, M., Ghosh, S., Mohammed, D. M., Soliman, T. N., Ibrahim, E. H., Fahmy, M. A., Qamar, J. A. S., & Tarabily, K. E. A. (2025). Chitosan nanoparticles: Green synthesis, biological activities, and sustainable frontiers in targeted drug delivery and cancer nanomedicine – A comprehensive review. *Materials Today. Bio*, *35*, 102358. Advance online publication. DOI: 10.1016/j.mtbio.2025.102358 PMID: 41215768

Fang, R. H., Gao, W., & Zhang, L. (2022). Targeting drugs to tumours using cell membrane-coated nanoparticles. *Nature Reviews. Clinical Oncology*, *20*(1), 33–48. DOI: 10.1038/s41571-022-00699-x PMID: 36307534

Gaddimath, S., Payamalle, S., Puttaningaiah, K. P. C. H., & Hur, J.-K. (2024). Recent Advances in pH and Redox Responsive Polymer Nanocomposites for Cancer Therapy. *Journal of Composites Science*, *8*(1), 28. Advance online publication. DOI: 10.3390/jcs8010028

Gavas, S., Quazi, S., & Karpiński, T. (2021). Nanoparticles for Cancer Therapy: Current Progress and Challenges. *Nanoscale Research Letters*, *16*(1), 173. Advance online publication. DOI: 10.1186/s11671-021-03628-6 PMID: 34866166

Gull, S. A., Khan[1], M. A., Hashmi, S., & Noor, T. (2025). A COMPREHENSIVE REVIEW OF TREATMENT APPROACHES FOR CONGENITAL AND HEREDITARY ANOMALIES. survival, 3(2). https://doi.org/DOI: 10.5281/zenodo.15286622

Hossain, A., Manik, M. H., Rakib, S., Mahmud, N., Khan, S., Ahsan, Z., Islam, M. S., Hossain, N., & Akter, M. A. (2025). Green Nanotechnology for Implantable Biosensors: Biocompatibility and Functional Integration in Medical Applications. *Biosensors & Bioelectronics: X*, *27*, 100678. Advance online publication. DOI: 10.1016/j.biosx.2025.100678

Iravani, S., & Varma, R. (2022). Nanosponges for Drug Delivery and Cancer Therapy: Recent Advances. *Nanomaterials (Basel, Switzerland)*, *12*(14), 2440. Advance online publication. DOI: 10.3390/nano12142440 PMID: 35889665

Karimi, S., Bakhshali, R., Bolandi, S., Zahed, Z., Zadeh, S. S. M., Zenjanab, M. K., & Esfahlan, R. J. (2025). For and against tumor microenvironment: Nanoparticle-based strategies for active cancer therapy. *Materials Today. Bio*, *31*, 101626. Advance online publication. DOI: 10.1016/j.mtbio.2025.101626 PMID: 40124335

Karnwal, A., Jassim, A. Y., Mohammed, A. A., Sharma, V., Al-Tawaha, A. R., & Sivanesan, I. (2024). Nanotechnology for Healthcare: Plant-Derived Nanoparticles in Disease Treatment and Regenerative Medicine. *Pharmaceuticals (Basel, Switzerland)*, *17*(12), 1711. Advance online publication. DOI: 10.3390/ph17121711 PMID: 39770553

Kong, X., Gao, P., Wang, J., Fang, Y., & Hwang, K. (2023). Advances of medical nanorobots for future cancer treatments. *Journal of Hematology & Oncology*, *16*(1), 74. Advance online publication. DOI: 10.1186/s13045-023-01463-z PMID: 37452423

Kumar, N., Singh, S., Sharma, P., Kumar, B., & Kumar, A. (2024). Single-, Dual-, and Multi-Stimuli-Responsive Nanogels for Biomedical Applications. *Gels (Basel, Switzerland)*, *10*(1), 61. Advance online publication. DOI: 10.3390/gels10010061 PMID: 38247784

Li, J., & Burgess, D. (2020). Nanomedicine-based drug delivery towards tumor biological and immunological microenvironment. *Acta Pharmaceutica Sinica. B*, *10*(11), 2110–2124. DOI: 10.1016/j.apsb.2020.05.008 PMID: 33304781

Liu, H., Lu, H.-H., Alp, Y., Wu, R., & Thayumanavan, S. (2023). Structural Determinants of Stimuli-Responsiveness in Amphiphilic Macromolecular Nano-assemblies. *Progress in Polymer Science*, *148*, 101765. Advance online publication. DOI: 10.1016/j.progpolymsci.2023.101765 PMID: 38476148

Liu, J., Zhao, C., Chen, W. R., & Zhou, B. (2022). Recent progress in two-dimensional nanomaterials for cancer theranostics. *Coordination Chemistry Reviews*, *469*, 214654. Advance online publication. DOI: 10.1016/j.ccr.2022.214654

Liu, Y., Li, C., Yang, X., Yang, B., & Fu, Q. (2024). Stimuli-responsive polymer-based nanosystems for cardiovascular disease theranostics. *Biomaterials Science*, *12*(15), 3805–3825. Advance online publication. DOI: 10.1039/D4BM00415A PMID: 38967109

Liu, Y., Si, L., Jiang, Y., Jiang, S., Zhang, X., Li, S., Chen, J., & Hu, J. (2025). Design of pH-Responsive Nanomaterials Based on the Tumor Microenvironment. *International Journal of Nanomedicine*, *20*, 705–721. DOI: 10.2147/IJN.S504629 PMID: 39845771

Maboudi, A. H., Lotfipour, M. H., Rasouli, M., Azhdari, M., Macloughlin, R., Bekeschus, S., & Doroudian, M. (2024). Micelle-based nanoparticles with stimuli-responsive properties for drug delivery. *Nanotechnology Reviews*, *13*(1), 20230218. Advance online publication. DOI: 10.1515/ntrev-2023-0218

Maimaitijiang, A., He, D., Li, D., Li, W., Su, Z., Fan, Z., & Li, J. (2024). Progress in Research of Nanotherapeutics for Overcoming Multidrug Resistance in Cancer. *International Journal of Molecular Sciences*, *25*(18), 9973. Advance online publication. DOI: 10.3390/ijms25189973 PMID: 39337463

Meng, W., Huang, L., Guo, J., Xin, Q., Liu, J., & Hu, Y. (2024). Innovative Nanomedicine Delivery: Targeting Tumor Microenvironment to Defeat Drug Resistance. *Pharmaceutics*, *16*(12), 1549. Advance online publication. DOI: 10.3390/pharmaceutics16121549 PMID: 39771528

Mi, P. (2020). Stimuli-responsive nanocarriers for drug delivery, tumor imaging, therapy and theranostics. *Theranostics*, *10*(10), 4557–4588. DOI: 10.7150/thno.38069 PMID: 32292515

Mundekkad, D., & Cho, W. (2022). Nanoparticles in Clinical Translation for Cancer Therapy. *International Journal of Molecular Sciences*, *23*(3), 1685. Advance online publication. DOI: 10.3390/ijms23031685 PMID: 35163607

Nascimento, A. T. D., Stoddart, P. R., Goris, T., Kael, M., Manasseh, R., Alt, K., Tashkandi, J., Kim, B. C., & Moulton, S. E. (2025). Stimuli-Responsive Materials for Biomedical Applications. *Advanced Materials*, *37*(36), e07559. Advance online publication. DOI: 10.1002/adma.202507559 PMID: 40801175

Park, S., Lu, G., Zheng, Y., Davison, E. K., & Li, Y. (2025). Nanoparticle-Based Delivery Strategies for Combating Drug Resistance in Cancer Therapeutics. *Cancers (Basel)*, *17*(16), 2628. Advance online publication. DOI: 10.3390/cancers17162628 PMID: 40867257

Pham, S. H., Choi, Y., & Choi, J. (2020). Stimuli-Responsive Nanomaterials for Application in Antitumor Therapy and Drug Delivery. *Pharmaceutics*, *12*(7), 630. Advance online publication. DOI: 10.3390/pharmaceutics12070630 PMID: 32635539

Prasad, R., Peng, B., Mendes, B. B., Kilian, H. I., Gorain, M., Zhang, H., Kundu, G., Xia, J., Lovell, J. F., & Conde, J. (2024). Biomimetic bright optotheranostics for metastasis monitoring and multimodal image-guided breast cancer therapeutics. *Journal of Controlled Release : Official Journal of the Controlled Release Society*, *367*, 300–315. Advance online publication. DOI: 10.1016/j.jconrel.2024.01.056 PMID: 38281670

Puri, S., Mazza, M., Roy, G., England, R. M., Zhou, L., Nourian, S., & Subramony, A. J. (2023). Evolution of Nanomedicine Formulations for Targeted Delivery and Controlled Release. *Advanced Drug Delivery Reviews*, *114962*, 114962. Advance online publication. DOI: 10.1016/j.addr.2023.114962 PMID: 37321376

Rodríguez, F., Caruana, P., De La Fuente, N., Español, P., Gamez, M., Balart, J., Llurba, E., Rovira, R., Ruiz, R., Martín-Lorente, C., Corchero, J., & Céspedes, M. V. (2022). Nano-Based Approved Pharmaceuticals for Cancer Treatment: Present and Future Challenges. *Biomolecules*, *12*(6), 784. Advance online publication. DOI: 10.3390/biom12060784 PMID: 35740909

Sabit, H., Pawlik, T., Radwan, F. F. Y., Abdel-Hakeem, M. A., Abdel-Ghany, S., Wadan, A.-H. S., Elzawahri, M., El-Hashash, A., & Arneth, B. (2025). Precision nanomedicine: Navigating the tumor microenvironment for enhanced cancer immunotherapy and targeted drug delivery. *Molecular Cancer*, *24*(1), 160. Advance online publication. DOI: 10.1186/s12943-025-02357-z PMID: 40457437

Santhamoorthy, M., Asaithambi, P., Ramkumar, V., Elangovan, N., Perumal, I., & Kim, S.-C. (2025). A Review on the Recent Advancements of Polymer-Modified Mesoporous Silica Nanoparticles for Drug Delivery Under Stimuli-Trigger. *Polymers, 17*(12), 1640. Advance online publication. DOI: 10.3390/polym17121640 PMID: 40574168

Siddique, S., & Chow, J. (2022). Recent Advances in Functionalized Nanoparticles in Cancer Theranostics. *Nanomaterials (Basel, Switzerland), 12*(16), 2826. Advance online publication. DOI: 10.3390/nano12162826 PMID: 36014691

Singh, R., Sharma, A., Saji, J., Umapathi, A., Kumar, S., & Daima, H. (2022). Smart nanomaterials for cancer diagnosis and treatment. *Nano Convergence, 9*(1), 21. Advance online publication. DOI: 10.1186/s40580-022-00313-x PMID: 35569081

Souri, M., Soltani, M., Kashkooli, F. M., Shahvandi, M. K., Chiani, M., Shariati, F. S., Mehrabi, M., & Munn, L. (2022). Towards principled design of cancer nanomedicine to accelerate clinical translation. *Materials Today. Bio, 13*, 100208. Advance online publication. DOI: 10.1016/j.mtbio.2022.100208 PMID: 35198957

Sun, H., Li, Y., Xue, M., & Feng, D. (2025). Tumor Microenvironment-Responsive Nanoparticles: Promising Cancer PTT Carriers. *International Journal of Nanomedicine, 20*, 7987–8001. DOI: 10.2147/IJN.S526497 PMID: 40584786

Sun, L., Liu, H., Ye, Y., Lei, Y., Islam, R., Tan, S., Tong, R., Miao, Y., & Cai, L. (2023). Smart nanoparticles for cancer therapy. *Signal Transduction and Targeted Therapy, 8*(1), 418. Advance online publication. DOI: 10.1038/s41392-023-01642-x PMID: 37919282

Sun, Q., Wang, Z., Liu, B., He, F., Gai, S., Yang, P., Yang, D., Li, C., & Lin, J. (2022). Recent advances on endogenous/exogenous stimuli-triggered nanoplatforms for enhanced chemodynamic therapy. *Coordination Chemistry Reviews, 451*, 214267. Advance online publication. DOI: 10.1016/j.ccr.2021.214267

Sun, Y., & Davis, E. W. (2021). Nanoplatforms for Targeted Stimuli-Responsive Drug Delivery: A Review of Platform Materials and Stimuli-Responsive Release and Targeting Mechanisms. *Nanomaterials (Basel, Switzerland), 11*(3), 746. Advance online publication. DOI: 10.3390/nano11030746 PMID: 33809633

Taha, B., Sulaiman, G. M., Addie, A. J., Khalil, K. A., Ahmed, E. M., Chaudhary, V., & Arsad, N. (2025). Multifunctional nanophotonic photoacoustic biosensors: A new era in molecular imaging-guided deep-tissue cancer monitoring. *Methods : A Companion to Methods in Enzymology, 242*, 1–23. Advance online publication. DOI: 10.1016/j.ymeth.2025.06.005 PMID: 40516638

Thapa, R., & Kim, J. O. (2022). Nanomedicine-based commercial formulations: Current developments and future prospects. *Journal of Pharmaceutical Investigation*, *53*(1), 19–33. DOI: 10.1007/s40005-022-00607-6 PMID: 36568502

Thomas, R., Surendran, S. P., & Jeong, Y. (2020). Tumor Microenvironment-Stimuli Responsive Nanoparticles for Anticancer Therapy. *Frontiers in Molecular Biosciences*, *7*, 610533. Advance online publication. DOI: 10.3389/fmolb.2020.610533 PMID: 33392264

Tian, Y., Sun, M., Song, R., Yang, Z., & Zhang, H. (2025). DNA dendrimer-based nanocarriers for targeted Co-delivery and controlled release of multiple chemotherapeutic drugs. *RSC Advances*, *15*(4), 2981–2987. DOI: 10.1039/D4RA07839J PMID: 39882009

Torres-Cavanillas, R., & Forment-Aliaga, A. (2024). Design of stimuli-responsive transition metal dichalcogenides. *Communications Chemistry*, *7*(1), 241. Advance online publication. DOI: 10.1038/s42004-024-01322-z PMID: 39462088

Venturini, J., Chakraborty, A., Baysal, M. A., & Tsimberidou, A. M. (2025). Developments in nanotechnology approaches for the treatment of solid tumors. *Experimental Hematology & Oncology*, *14*(1), 76. Advance online publication. DOI: 10.1186/s40164-025-00656-1 PMID: 40390104

Wang, B., Hu, S., Teng, Y., Chen, J., Wang, H., Xu, Y., Wang, K., Xu, J., Cheng, Y., & Gao, X. (2024). Current advance of nanotechnology in diagnosis and treatment for malignant tumors. *Signal Transduction and Targeted Therapy*, *9*(1), 200. Advance online publication. DOI: 10.1038/s41392-024-01889-y PMID: 39128942

Wang, X., Li, C., Wang, Y., Chen, H., Zhang, X., Luo, C., Zhou, W., Li, L., Teng, L., Yu, H., & Wang, J. (2022). Smart drug delivery systems for precise cancer therapy. *Acta Pharmaceutica Sinica. B*, *12*(11), 4098–4121. DOI: 10.1016/j.apsb.2022.08.013 PMID: 36386470

Wang, X., Xiong, T., Cui, M., Li, N., Li, Q., Zhu, L., Duan, S., Wang, Y., & Guo, Y. (2021). A novel targeted co-delivery nanosystem for enhanced ovarian cancer treatment via multidrug resistance reversion and mTOR-mediated signaling pathway. *Journal of Nanobiotechnology*, *19*(1), 444. Advance online publication. DOI: 10.1186/s12951-021-01139-1 PMID: 34949180

Wang, Y.-H., Zhan, J., Huang, J., Wang, X., Chen, Z., Yang, Z., & Li, J. (2023). Dynamic responsiveness of self-assembling peptide-based nano-drug systems. *Interdisciplinary Medicine*, *1*(1), e20220005. Advance online publication. DOI: 10.1002/INMD.20220005

Wei, D., Sun, Y., Zhu, H., & Fu, Q. (2023). Stimuli-Responsive Polymer-Based Nanosystems for Cancer Theranostics. *ACS Nano*, *17*(23), 23223–23261. Advance online publication. DOI: 10.1021/acsnano.3c06019 PMID: 38041800

Wei, X., Song, M., Li, W., Huang, J., Yang, G., & Wang, Y. (2021). Multifunctional nanoplatforms co-delivering combinatorial dual-drug for eliminating cancer multidrug resistance. *Theranostics*, *11*(13), 6334–6354. DOI: 10.7150/thno.59342 PMID: 33995661

Weng, Z., Xu, Y., Gao, J., & Wang, X. (2022). Research progress of stimuli-responsive ZnO-based nanomaterials in biomedical applications. *Biomaterials Science*, *11*(1), 76–95. Advance online publication. DOI: 10.1039/D2BM01460B PMID: 36385188

Younas, A., Wang, S., Asad, M., Mamun, A. A., Majeed, S., Sharif, A., Zhou, Q., Liu, Y., Geng, P., Shao, C., & Xiao, J. (2025). Recent advances in cancer nanomedicine: From smart targeting to personalized therapeutics - pioneering a new era in precision oncology. *Materials Today. Bio*, *36*, 102660. Advance online publication. DOI: 10.1016/j.mtbio.2025.102660 PMID: 41560793

Zanbili, F., & Marjani, A. P. (2025). Innovative green and bio-based approaches for photosensitive nanoparticle synthesis: A review on methodologies, characterization, and applications. *Micro and Nano Systems Letters*, *13*(1), 3. Advance online publication. DOI: 10.1186/s40486-025-00223-7

Zhang, Y., Li, Y., Yang, K., Liu, W., Zhao, S., Tang, Y., & Chen, Z. (2025). Synergistic innovations of nanomedicine in lymphoma treatment: A systematic review. *Journal of Translational Medicine*, *23*(1), 1389. Advance online publication. DOI: 10.1186/s12967-025-07249-w PMID: 41366795

Zhang, Y., Wang, Q., Zhao, X. Y., Zhang, H., & Pan, G. (2023). Molecularly Imprinted Nanomaterials with Stimuli Responsiveness for Applications in Biomedicine. *Molecules (Basel, Switzerland)*, *28*(3), 918. Advance online publication. DOI: 10.3390/molecules28030918 PMID: 36770595

Zhang, Y., & Wu, B. M. (2023). Current Advances in Stimuli-Responsive Hydrogels as Smart Drug Delivery Carriers. *Gels (Basel, Switzerland)*, *9*(10), 838. Advance online publication. DOI: 10.3390/gels9100838 PMID: 37888411

Zhao, C., Kang, J., Wang, Y., Wang, Y., Tang, X., & Jiang, Z. (2023). Carbon-Based Stimuli-Responsive Nanomaterials: Classification and Application. *Cyborg and Bionic Systems (Washington, D.C.)*, *4*, 0022. Advance online publication. DOI: 10.34133/cbsystems.0022 PMID: 37223546

Zhou, W., Jia, Y., Liu, Y., Chen, Y., & Zhao, P. (2022). Tumor Microenvironment-Based Stimuli-Responsive Nanoparticles for Controlled Release of Drugs in Cancer Therapy. *Pharmaceutics*, *14*(11), 2346. Advance online publication. DOI: 10.3390/pharmaceutics14112346 PMID: 36365164

Zhou, Y., Chen, X., Cao, J., & Gao, H. (2020). Overcoming the biological barriers in the tumor microenvironment for improving drug delivery and efficacy. *Journal of Materials Chemistry. B*, *8*(31), 6765–6781. Advance online publication. DOI: 10.1039/D0TB00649A PMID: 32315375

Zhou, Y., Li, Q., Wu, Y., Li, X., Zhou, Y., Wang, Z., Liang, H., Ding, F., Hong, S., Steinmetz, N., & Cai, H. (2023). Molecularly Stimuli-Responsive Self-Assembled Peptide Nanoparticles for Targeted Imaging and Therapy. *ACS Nano*, *17*(9), 8004–8025. Advance online publication. DOI: 10.1021/acsnano.3c01452 PMID: 37079378

Zhu, Q., Liu, S., Sun, J., Liu, J., Kirubaharan, C. J., Chen, H., Xu, W.-H., & Wang, Q. (2020). Stimuli-responsive cellulose nanomaterials for smart applications. *Carbohydrate Polymers*, *235*, 115933. DOI: 10.1016/j.carbpol.2020.115933 PMID: 32122477

Zou, J., Jiang, C., Hu, Q., Jia, X., Wang, S., Wan, S., Mao, Y., Zhang, D., Zhang, P., Dai, B., & Li, Y. (2025). Tumor microenvironment-responsive engineered hybrid nanomedicine for photodynamic-immunotherapy via multi-pronged amplification of reactive oxygen species. *Nature Communications*, *16*(1), 424. Advance online publication. DOI: 10.1038/s41467-024-55658-0 PMID: 39762214

Zou, J., Xing, X., Teng, C., Zhao, Q., He, W., Wu, X., & Xia, Y. (2024). Cocrystal@protein-anchoring nanococktail for combinatorially treating multidrug-resistant cancer. *Acta Pharmaceutica Sinica. B*, *14*(10), 4509–4525. DOI: 10.1016/j.apsb.2024.08.014 PMID: 39525582

Chapter 7
# Surface Functionalization and Active Targeting in Nano-Oncology

**Umar Sajid Kayani**
http://orcid.org/0009-0009-6133-8156
*Yangzhou University, China*

**Ayesha Khan**
*COMSATS University Islamabad, Pakistan*

**Sakandar Khan**
*KBCMA College of Veterinary and Animal Sciences, Narowal, Pakistan*

**Muhammad Zakria Iqbal Khan**
*COMSATS University Islamabad, Pakistan*

**Shahid Hussain**
http://orcid.org/0000-0002-9820-7477
*COMSATS University Islamabad, Pakistan*

**Shab Niaz Khan**
*COMSATS University Islamabad, Pakistan*

**Muhammad Shimran Khan Afridi**
http://orcid.org/0009-0004-8300-6922
*COMSATS University Islamabad, Pakistan*

**Ayesha Fazal**
http://orcid.org/0009-0007-5328-267X
*COMSATS University Islamabad, Pakistan*

## ABSTRACT

*This chapter reviews surface functionalization and active targeting strategies in nano-oncology for improved cancer diagnosis, drug delivery, and therapeutic monitoring. It highlights polymer modifications, ligand-mediated targeting, biomimetic coatings, and stimuli-responsive systems that enhance stability, biocompatibility, circulation, tumor accumulation, and cellular uptake while reducing off-target effects. Theranostic applications and challenges related to biological barriers,*

DOI: 10.4018/979-8-3373-9848-8.ch007

*protein corona formation, scalability, toxicity, and regulation are discussed, along with emerging approaches such as AI-guided nanoparticle design, CRISPR-based delivery, and green nanotechnology for precision oncology.*

## 1. INTRODUCTION

Cancer has become one of the most prominent worldwide health problems and a cause of disability and mortality in the world (Deivayanai et al., 2025). Even after tremendous improvements in the initial diagnosis, surgery, radiotherapy, immunopathy and chemotherapy, the successful treatment of most malignancies is still hampered by heterogeneity, metastatic evolution, and the resistance produced by therapy. Traditional types of cancer therapy are usually less selective and, therefore, lead to inadvertent distribution of therapeutic elements to cancerous and non-cancerous tissues (Fan et al., 2023). This often leads to serious adverse effects such as systemic toxicity, immunosuppression, organ damage and poor quality of life in patients (Deivayanai et al., 2025). Moreover, emergence of multidrug resistance (MDR), low drug bioavailability, rapid clearance of systemic components and poor accumulation of therapeutic drugs in tumor locations play a key role in undermining the effectiveness of the treatment process and is a major cause of poor clinical outcomes (Fan et al., 2023).

Over the past few decades, nanotechnology has proven to be a revolution in the cancer diagnosis and treatment and has spawned the highly fast-growing nano-oncology discipline (Deivayanai et al., 2025). Using materials and structures in the nanoscale, nano-oncology provides new answers to a number of constraints of traditional cancer therapies. Nanoparticles have characteristic physicochemical features, such as large surface area-to-volume ratios, size and shape control, surface chemistry and ability to deliver a wide range of therapeutic and diagnostic molecules (Fan et al., 2023). The above characteristics make it possible to improve drug solubility, improve the pharmacokinetic profiles of the drugs, control drug release, and the accumulation of drugs in tumor tissues (Ciftci et al., 2025). Additionally, nanotechnology also helps in the creation of multifunctional platforms that can combine diagnosis, imaging, targeted therapy and real-time monitoring of therapy thus helping to enable development of precision cancer therapeutics (Pallares et al., 2025).

Surface functionalization and active targeting are some of the design solutions that have become critical in various factors that dictate the success of nanomedicine-based cancer interventions (Ciftci et al., 2025). Surface functionalization is the controlled alteration of the surface of nanoparticles with polymers, biomolecules, ligands, antibodies, peptides, aptamers and other functional ligands to enhance stability, biocompatibility, circulatory half-life and biological functions. These changes may lessen any inappropriate interactions with biological parts, decrease immunogenicity, and improve the persistence of nanoparticle in the circulation (Figure 1) (Ciftci et al., 2025). More significantly, nanoparticles that are surface-engineered can be engineered to actively sense and interact with a given receptor or biomarker that is either overexpressed on cancer cells or a part of the tumor microenvironment (Pallares et al., 2025).

Active targeting also increases the specificity of nanotherapeutics through facilitated interaction with receptors and uptake into the diseased site and reduced off-target effects (Sabir et al., 2025). Not only does this selective method of delivery enhance the therapeutic effect, but also decreases systemic toxicity, enhances biosafety and facilitates personalized treatment choices favoring specific tumor pathophysiology (Sabir et al., 2025). With the further development of precision oncology, combining active targeting and advanced surface engineering tools gain more significance in the design of most efficient, safe, and patient-specific cancer treatment (Pallares et al., 2025).

*Figure 1. Schematic illustration of surface-functionalized nanoparticles enhancing tumor targeting, therapeutic efficacy, and safety compared with conventional cancer therapies*

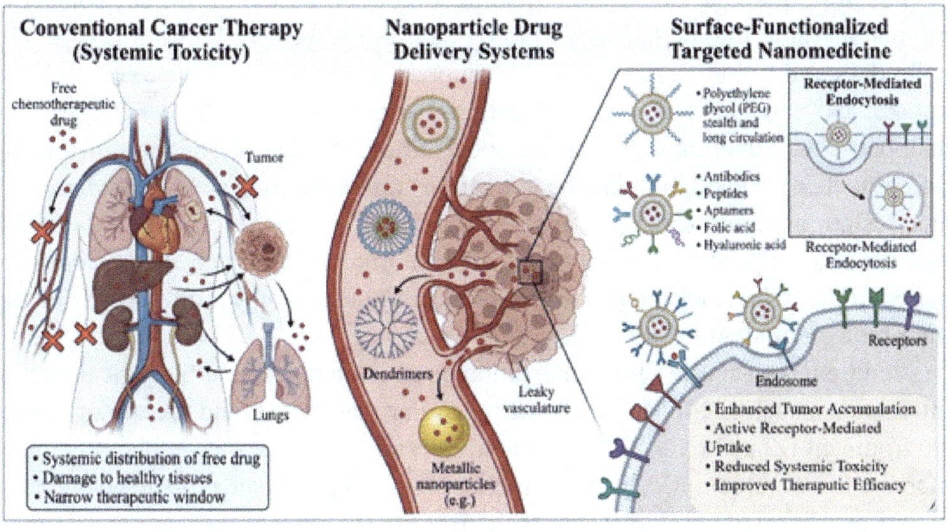


This chapter gives a detailed linkage of the surface functionalization and active targeting options of nano-oncology. It covers the basics of surface engineering nanoparticles, key functionalization strategies, biological barriers to nanoparticle delivery, the concept of targeted delivery, theranostics, translational issues and emerging trends that collectively shape the future of precision cancer therapeutics. The chapter seeks to bring to the fore the manner in which rationally engineered surface-modified nanoplatforms can enhance the selectivity, efficacy and clinical promise of the next generation cancer-therapies.

## 2. FUNDAMENTALS OF SURFACE FUNCTIONALIZATION IN NANOMEDICINE

### 2.1. Concept and Principles of Surface Functionalization

Surface functionalization is a general approach in nanomedicine, which comprises the targeted modification of the surfaces of nanoparticles with attaching chemical groups, polymer, biomolecule, ligand, antibody, peptides, aptamers or other functional groups (Al-Shadidi et al., 2024). The main aim of this process is to modify physicochemical and biological features of nanoparticles to improve performance of nanoparticles in biomedical processes. Surface functionalization has become a vital tool in nano-oncology in enhancing the specificity, stability and therapeutic effect of nanocarriers (Figure 2) (Ly et al., 2024).

The importance of surface functionalization is because it allows the control of the nanoparticle interactions with biology system (Al-Shadidi et al., 2024). The colloidal stability, biodistribution, cellular uptake and nanoparticles-protein, cellular uptake, and nanoparticle-tissue interactions are governed by surface chemistry of nanoparticles. Functional groups are added to the surface of nanoparticles using different chemical and biological conjugation strategies, such as covalent bonding, electrostatic adsorption and bio-orthogonal reactions (Al-Shadidi et al., 2024). Such changes affect the corona layer composed of biomolecules which adsorbs onto the nanoparticles when they interact with biological fluids, thereby profoundly affecting their biological identity (Ly et al., 2024).

Additionally, surface functionalization controls nanoparticle-biological barrier interactions, such as blood constituents, cellular membrane, and the tumor microenvironment (Zhao et al., 2025). Correctly designed surfaces have the potential to increase tumor retention, rely on interaction with receptors, and reduce the presence of nonspecific interactions that can result in optimal therapeutic efficacy and reduced systemic toxicity (Ly et al., 2024).

*Figure 2. Stepwise process of nanoparticle surface functionalization for improved stability, targeting, and biocompatibility*

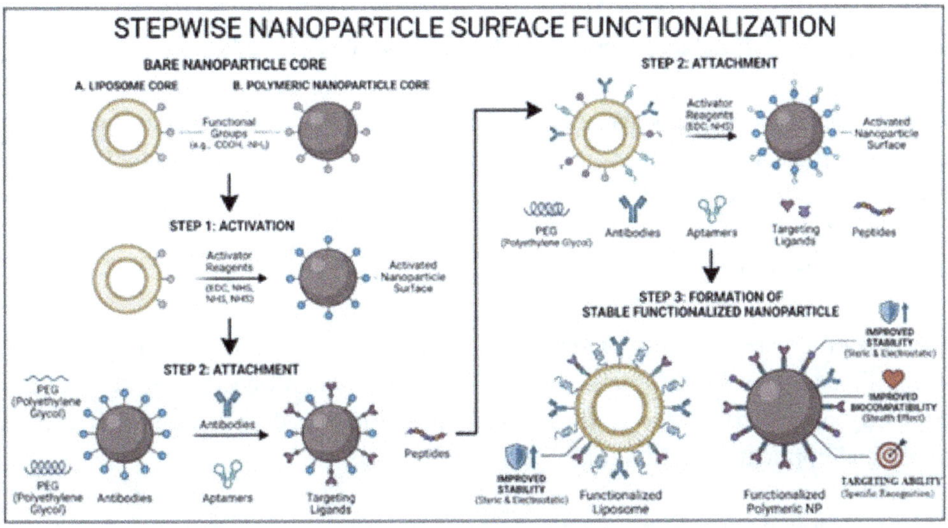


## 2.2. Objectives of Surface Engineering

Surface engineering has a variety of applications in nanoparticle-based cancer therapies (Zhao et al., 2025). Among the key aims is to improve the stability of nanoparticles, by eliminating aggregation, whilst preserving structural integrity during physiological conditions. Stable nanoparticles offer a better storage ability and predictable biological behavior (Zheng et al., 2025).

The biocompatibility is another significant objective (Zhao et al., 2025). Cytotoxicity is minimized by using surface coating made of polyethylene glycol (PEG), chitosan and biopolymer coatings which create improved compatibility with biological tissue. Surface engineering can also delay the systemic circulation time through reduced recognition and clearance of the mononuclear phagocyte system (MPS), as well as, enhancing the likelihood of accumulation of nanoparticle at tumor sites (Zheng et al., 2025).

The other significant benefit of surface modification is controlled drug release (Kumar et al., 2024). Nanoparticles functionalized can also be capable of responding to a specific stimulus e.g. pH/enzyme/temperature/reactive oxygen species (ROS) or external stimuli and release their therapeutic cargo to specifically target a specific area, thereby limiting destruction of non-cancerous tissues. Moreover, surface engineering minimizes immunogenicity by covering nanoparticle surfaces with an immune surveillance response, thus, diminishing inflammatory reactions and increasing therapeutic safety (Zheng et al., 2025).

## 2.3. Types of Nanocarriers Used in Cancer Therapy

Various types of nanocarriers have been developed to treat cancer and each of them possesses some unique properties, which can be further obtained by surface functionalization (Liu et al., 2025).

### 2.3.1. Liposomes

Liposomes are a type of phospholipid based vesicles that can entrap both hydrophilic drugs and hydrophobic drugs. They have been one of the most used in oncology nanocarriers due to their biocompatibility, biodegradation, and success in clinical applications (Liu et al., 2025).

### 2.3.2. Polymeric Nanoparticles

Nanoparticles made of natural or synthetic polymers are polymeric in nature and offer great control over loading of the drug being carried, discharge features, and surface modification. They are versatile such that they can be used in different therapies (Liu et al., 2025).

### 2.3.3. Dendrimers

Dendrimers are extremely branched macromolecules with a well-defined structure with a large number of functional groups on the surface. These characteristics allow effective loading of drugs and their delivery (Sun et al., 2023).

### 2.3.4. Micelles

Polymeric micelles have a core-shell architecture to allow an encapsulation of poorly soluble anticancer drugs in water. Their small sizes increase tumor penetration and delivery into cells (Sun et al., 2023).

### 2.3.5. Metallic Nanoparticles

Gold, silver, platinum and iron oxide nanoparticles are metallic nanoparticles that have interesting optical, magnetic, and photothermal characteristics. The latter properties are useful in imaging, drug delivery, photothermal therapy and theranostics (Sun et al., 2023).

### 2.3.6. Mesoporous Silica Nanoparticles

The mesoporous silica nanoparticles have large surface areas, sizeable pore sizes which can be tuned, and high drug-loading capacities (Kumar et al., 2024). The ease with which they can be modified and functionalized facilitates them being good candidates in targeted drug delivery system (Wen et al., 2025).

### 2.3.7. Carbon-Based Nanomaterials

Graphene oxide, carbon nanotubes, and graphene quantum dots, as well as fullerene derivatives, have outstanding mechanical, electrical, and thermal characteristics (Kumar et al., 2024). These nanomaterials have revealed a lot of potential in cancer imaging, drug delivery and photothermal therapy (Wen et al., 2025).

### 2.3.8. Biomimetic Nanoparticles

Another new type of nanocarrier is biomimetic nanoparticles that involve the use of biological membranes of red blood cells, immune cells, platelets, or cancer cells. The systems have increased immune evasion, extended circulability and targeting (Wen et al., 2025).

## 2.4. Physicochemical Factors Influencing Functionalization

Intrinsic physicochemical properties of nanoparticles have a large impact on the effectiveness of surface functionalization (Taghavimandi et al., 2025).

### 2.4.1. Particle Size

Particle size is important in the biodistribution, tumor deposition, cellular uptake, and clearance pathways. With a size of 10-200 nm, nanoparticles usually show good pharmacokinetic efficiencies and a better tumor penetration (Taghavimandi et al., 2025).

### 2.4.2. Shape

The shape of nanoparticles determines the processes of circulation, cellular internalization and retention by tumors (Pashkina et al., 2025). Spherical, rod-shaped, cubic and filamentous nanoparticles, do not have the same interactions with biological systems and can have different therapeutic performances (Taghavimandi et al., 2025).

### 2.4.3. Surface Charge

Surface charge influences cell membrane interactions, plasma protein interactions and immune cell interactions with nanoparticles (Pashkina et al., 2025). The positively charged nanoparticles tend to exhibit an improved cellular uptake because of the electrostatic attraction with negatively charged cellular membranes, and neutral or slightly negatively charged nanoparticles tend to be more biocompatible and able to have a longer circulation (Gutiérrez Coronado et al., 2025).

### 2.4.4. Hydrophobicity and Hydrophilicity

The ratio between hydrophobic and hydrophilic characteristics is known to have considerable impact on protein adsorption, colloidal stability and the biodistribution (Pashkina et al., 2025). Widespread use of hydrophilic surface coatings (PEGylation, in particular) is to minimize nonspecific contact and enhance systemic circulation (Gutiérrez Coronado et al., 2025).

### 2.4.5. Surface Roughness

Surface roughness influences both protein corona formation and cellular adhesion and endocytosis (Bajracharya et al., 2022). The topography of a surface can cause differences in the interactions of nanoparticles with cells and ultimately determine therapeutic efficacy and biodistribution (Gutiérrez Coronado et al., 2025).

## 3. SURFACE FUNCTIONALIZATION STRATEGIES FOR CANCER NANOTHERAPEUTICS

Strategies of surface functionalization are essential in improving the therapeutic efficacy, specificity and biocompatibility of the nanocarriers in cancer therapy (Jia et al., 2022). Nanoparticles contain many sites, whose modification with polymers, targeting ligands, biological membranes, and stimuli-responsive materials can help get past the biological barriers, enhance the efficiency of drug delivery, and achieve precision therapeutic effects by researchers in the future (Jia et al., 2022). The latest nanotechnological achievements resulted in the creation of advanced techniques in surface engineering allowing to control how nanotherapeutics and tumor tissues interact and reduce the off-target effect (Zalba and Garrido, 2022).

### 3.1. Polymer-Based Functionalization

The most common strategies used in nano-oncology which are based on polymer based functionalization are because it enhances the stability, circulatory time as well as the biocompatibility of nanoparticles (Jia et al., 2022). The polymers are used to coated the surface of nanoparticles to generate a protective layer that prevents aggregation, immune recognition and degradation of nanoparticles by enzymes (Zalba and Garrido, 2022).

#### 3.1.1. PEGylation

Polyethylene glycol (PEG) is among the best-used polymers to modify the surface of nanoparticles (Torres et al., 2025). PEGylation is the initial process that involves attaching PEG chains to surfaces of nanoparticles, to form a hydrophilic steric shell decreasing protein adsorption and minimizing mononuclear phagocytic system uptake. This stealth feature increases blood circulation time, improves tumor retention and accumulation due to increased permeability and retention of the tumors which enhances therapeutic effectiveness (Torres et al., 2025). PEGylation has already been effectively introduced into a number of clinically approved nanomedicines, and continues to serve as a best practice surface engineering approach in nanocarriers (Zalba and Garrido, 2022).

#### 3.1.2. Chitosan Coating

Chitosan is a polysaccharide derived out of chitin and is found naturally which has attracted a lot of attention because of its biodegradation ability, biocompatibility as well as its mucoadhesiveness (Torres et al., 2025). The nanoparticles are characterized by a chitosan-coated surface, which has a higher cellular uptake by the nanoparticles due to the presence of positively charged amino groups that give the particle the ability to form interactions with negatively charged cellular membranes (Chu et al., 2022). Additionally, chitosan can enhance the encapsulation of drugs, the protection of delicate therapeutic molecules, as well as, the ability of drugs to release in response to pH in acidic tumor environments (Miao et al., 2023).

#### 3.1.3. Hyaluronic Acid Modification

Hyaluronic acid (HA) is a natural glycosaminoglycan with a wide variety of applications in active targeting (Chu et al., 2022). HA has a high affinity towards CD44 receptors that are overexpressed in most types of cancer such as breast, ovarian, colorectal and pancreatic cancers, among others. HA modification of surfaces increases selectivity in tumor targeting, receptor-mediated endocytosis, and assists in intracellular delivery of drugs and has great biocompatibility (Dudhwala et al., 2026).

### 3.1.4. Polyethyleneimine (PEI)

Polyethyleneimine is a cationic polymer that is widely used in gene delivery system since it has great capacity of binding nucleic acids (Chu et al., 2022). PEI-functionalized nanoparticle assists in effective condensation of DNA and RNA and internalization of cells via electrostatic interactions. Also, PEI has a proton sponge effect to increase endosomal escape, thereby boosting cytoplasmic delivery of genetic therapeutics. Nonetheless, issues about cytotoxicity have motivated the creation of bioengineered and biodegradable derivatives of PEI (Miao et al., 2023).

## 3.2. Ligand-Based Functionalization

Ligand-based functionalization further increases nanoparticle specificity, through which it can recognize specific receptors or biomarkers, which are overexpressed on cancer cells and in the tumor microenvironment that are selectively exploited (Hong et al., 2023).

### 3.2.1. Small Molecules

Small-molecule ligands e.g. folic acid, transferrin and galactose are commonly used to localize particular cellular receptors (Bajracharya et al., 2022). One of them in particular, folic acid, has high affinity to folate receptors which are overexpressed in many malignancies. Small molecules due to their low molecular weight have excellent stability, are easy to synthesize, and immunogenicity is low (Zhang et al., 2025).

### 3.2.2. Peptides

Peptide ligands can deliver very specific recognition to the receptors and they also have relatively simple synthesis and modification methods (Bajracharya et al., 2022). iRGD-penetrating peptide, tumor-homing peptide and integrin-targeting peptide have been shown to hold promise in enhancing tumor accumulation and intracellular delivery of drugs (Sidorenko et al., 2025).

### 3.2.3. Proteins

The ligand can be based on protein, such as transferrin, albumin, and growth factors, and take advantage of the already existing cellular transportative methods to provide targeted delivery (Omidian et al., 2025). They inherently have biological functioning, which tends to promote cellular uptake and biodistribution of nanotherapeutic agents (Zhang et al., 2025).

### 3.2.4. Aptamers

Aptamers are brief single-stranded DNA or RNA aptamers which can identify particular molecules with high affinity and selectivity (Omidian et al., 2025). Aptamers are less immunogenic and more chemically stable and easier to make in comparison to antibodies. They have been useful targeting agents in cancer therapies and have been very appealing due to their versatility (Gao et al., 2026).

### 3.2.5. Antibodies

Achieving specific targeting is one of the primary objectives in nano-oncology, and monoclonal antibodies are one of the most helpful targeting ligands (Omidian et al., 2025). Nanoparticles that are functionalized with antibodies can be used to specifically target tumor-associated antigens like HER2, EGFR, and VEGF receptors, and increase therapeutic selectivity and decrease systemic toxicity. The use of targeting by antibodies has helped in the generation of personalized cancer therapies (Hong et al., 2023).

## 3.3. Biomimetic Surface Functionalization

Biomimetic functionalization aims at emulating functions found in nature onto enhance the performances of nano-particles and immunological evasion (Zhang et al., 2024). This new process has become an exciting solution method towards improving the time of circulation and achieving efficiency (Zhong et al., 2024).

### 3.3.1. Cell Membrane-Coated Nanoparticles

Nanoparticles produced by coating cell membranes offer a way to penetrate synthetic nanomaterials with cells, allowing nanoparticles to acquire the biological characteristics of cells of origin (Zhang et al., 2024). Such hybrid systems are able to escape immune clearance and remain in circulation longer as well as to being more biocompatible (Zhong et al., 2024).

### 3.3.2. Cancer Cell Membrane Camouflage

The targeting of cancer cell membrane-coated nanoparticles uses the homologous targeting phenomenon whereby nanoparticles designing in tumors are of the same type of cancer (Zhang et al., 2024). The approach increases delivery specificity to the tumors and becomes more effective in therapy (Zhong et al., 2024).

### 3.3.3. Red Blood Cell Membrane Coating

The erythrocytes have the long circulation life advantage that the nanoparticles that are coated with their red blood cell membranes have (Ijaz et al., 2024). The membrane proteins CD47 are a cause of a self-recognition signal that decreases phagocytic clearance and extends systemic circulation (Graván et al., 2025).

### 3.3.4. Immune Cell Membrane Functionalization

Macrophage, neutrophil, dendritic-cell and T-cell membranes can be used to build membrane-coated nanoparticles with enhanced targeting of inflamed tissues and the tumor microenvironment (Ijaz et al., 2024). These systems have the ability to take advantage of the natural immune trafficking to enhance drug delivery (Graván et al., 2025).

## 3.4. Stimuli-Responsive Surface Modification

The idea behind stimuli-responsive nanocarriers is to release the drug in a controlled and targeted manner in response to one or more of various external or internal triggers (Jin et al., 2024).

### 3.4.1. pH-Responsive Systems

Compared to normal tissues, the tumor tissues and intracellular compartments tend to have acidic microenvironments. pH-responsive nanoparticles capitalize on these variations to deliver therapeutic effects selectively in tumors, with minimum systemic exposure (Jin et al., 2024).

### 3.4.2. Enzyme-Responsive Systems

Some of the enzymes such as matrix metalloproteinases (MMPs), cathepsins and phospholipases are overexpressed in tumor tissues. When exposed to the enzymes, enzyme-sensitive surface coating in effect degrades and this initiates localized drug delivery (Jin et al., 2024).

### 3.4.3. ROS-Responsive Nanocarriers

Reactive oxygen species (ROS) may be excessively high into the cancer cells (Mousavi-Kiasary et al., 2025). ROS-responsive nanoparticles make use of oxidation-sensitive linkages which release therapeutic payloads when subjected to oxidative conditions with an increase in tumor specificity (AlSawaftah et al., 2022).

### 3.4.4. Temperature-Responsive Nanomaterials

Thermo-responsive nanocarriers are reactive to variations of temperatures (local) obtained by hyperthermia or external stimuli. These are able to offer controlled drug release and also increase efficacy of therapy (AlSawaftah et al., 2022).

### 3.4.5. Light-Responsive Systems

Light-responsive nanoparticle makes use of ultraviolet, visible or near-infrared light to deliver drug cargo or to facilitate a therapeutic action (Mousavi-Kiasary et al., 2025). These systems provide good spatial and temporal control which are desirable in photodynamic and photothermal cancer therapies (AlSawaftah et al., 2022).

## 3.5. Green and Sustainable Functionalization Approaches

The growing sustainability interest in nanomedicine has triggered creating ecofriendly surface functionalization schemes that minimize the utilization of hazardous chemicals and processes that consume a lot of energy (Amirinezhadfard et al., 2025).

### 3.5.1. Plant-Derived Biomolecules

Plant extracts are rich in various phytochemicals such as polyphenols, flavonoids, terpenoids, and alkaloids and they can be used as natural reducing, stabilizing, and functionalizing agents (Lee et al., 2025). These biomolecules facilitate the biocompatibility of nanoparticles and green biomethods of synthesis (Amirinezhadfard et al., 2025).

### 3.5.2. Microbial Metabolites

Microorganisms also generate a host of bioactive substances which can be used in the creation of nanoparticles or in the modification of the surfaces (Sabir et al., 2025). An alternative approach of producing bio-functionalized nanocarriers with high biological strengths involves the use of microbial metabolites as environmental-friendly alternatives (Amirinezhadfard et al., 2025).

### 3.5.3. Biopolymer Coatings

Alginate, cellulose, gelatin, dextran, starch as well as chitosan are natural biopolymers providing biodegradation and non-toxic alternative to synthetic polymers (Samathoti et al., 2025). They are an excellent material of sustainable nanomedicine due to their abundance, renewability and great biological compatibility (Lee et al., 2025).

## 4. NANOPARTICLE DELIVERY: BIOLOGICAL BARRIERS

Even with the impressive developments in nano-oncology, delivery of nanotherapeutics to tumor tissues is a major issue as a number of biological obstacles will stand in its way (Hoang et al., 2025). After being circulated systemically, nanoparticles can be faced by a set of physiological, cellular, and intracellular barriers that may restrict their circulation, biodistribution, accumulation in tumors, and therapeutic efficacy (Hoang et al., 2025). It is also necessary to understand these barriers to make rational design of surface-functionalized nanocarriers that can ensure efficient and targeted treatment of cancer (Bevilacqua et al., 2024).

### 4.1. Bloodstream Challenges

The first compartment that nanoparticles experience, upon administration, is the bloodstream (Chai et al., 2025). When nanoparticles enter the circulatory system they are instantly subjected to a dynamic biological environment, full of a variety of proteins, lipids and immune factors (Bevilacqua et al., 2024).

#### 4.1.1. Protein Corona Formation

The initial processes that take place after administration of the nanoparticle include adsorption of the plasma proteins to the nanoparticle surface, which leads to the development of protein corona (Chai et al., 2025). This interactive shell changes the physicochemical characteristics and biological identity of the nanoparticles impacting efforts of biodistribution, uptake into cells and efficacy of targeting. The formation of protein corona in most instances may obscure targeting ligands and diminish the efficacy of active targeting strategies (Bevilacqua et al., 2024).

#### 4.1.2. Opsonization

Opsonization is defined as the attachment of opsonin proteins, which are immunoglobulins and complement factors to the surface of nanoparticles (Chai et al., 2025). These proteins are molecular labels which give the immune system signals to identify and get rid of foreign objects. Based on large-scale opsonization, fast clearance and reduction in dose to tumor tissue are achieved (Wang et al., 2022).

#### 4.1.3. Clearance by the Mononuclear Phagocyte System

Opsonin-recognized nanoparticles are quickly taken out of circulation via mononuclear phagocyte system (MPS), also referred to as the reticuloendothelial system (RES) (Sun et al., 2024). This is done by macrophages found in the liver, in the spleen and at lymphatic tissues. Whether it is rapid phagocytic clearance, the duration of circulation and the effect of therapeutic efficacy is markedly reduced (Wang et al., 2022).

### 4.2. Tumor Microenvironment Barriers

Although nanoparticles are able to dodge the systemic clearance in some instances, they have to pass numerous challenges in the tumor microenvironment to attain successful delivery of drugs (Wang et al., 2022).

#### 4.2.1. Dense Extracellular Matrix

A solid tumor can have a very dense extra cellular matrix (ECM) consisting of collagen fibres, proteoglycans and other structural proteins (Mousavi-Kiasary et al., 2025). This complicated network is a physical impediment which prevents penetration of nanoparticle and constrains its diffusion through tumor tissues (Chou and Lin, 2024).

#### 4.2.2. Hypoxia

There are abnormal vascularization and insufficient oxygen supply to the tumors which result in tumor hypoxia (Samathoti et al., 2025). Hypoxic areas are frequently related with aggressive tumor growth, resistance to therapy, and changed cellular metabolism. All these circumstances may create barriers to the delivery of drugs and decrease the efficacy of traditional therapeutic interventions (Chou and Lin, 2024).

### 4.2.3. Elevated Interstitial Pressure

The increase in interstitial fluid pressure is common in tumors and is attributed to defective lymphatics and leaky vasculature (Samathoti et al., 2025). This gradient of abnormal pressure acts against entry of nanoparticles out of the blood vessels to tumor tissues and thus, decreases drug concentration and infiltration (Chou and Lin, 2024).

## 4.3. Cellular and Intracellular Barriers

### 4.3.1. Endosomal Entrapment

The vast majority of nanoparticles find their way into cells via endocytosis and can be trapped in endosomal vesicles (Murar et al., 2023). Therapeutic payloads can be endosomally entrapped, avoiding the delivery of the therapeutic payload to the intracellular site of action, which grossly impairs its efficacy in therapy (Sun et al., 2024).

### 4.3.2. Lysosomal Degradation

Endosomes can turn to lysosomes that may have enzymes with hydrolytic potential and acidic conditions that can destroy treatment molecules (Murar et al., 2023). This is because lysosomal degradation of drugs is a significant setback to intracellular protein, nucleic acid and other biomolecule delivery (Soukar and Gaharwar, 2025).

### 4.3.3. Limited Nuclear Penetration

Some treatment modalities, especially gene therapies and nucleic acid-based treatment modalities, demand entry into the cell nucleus (Murar et al., 2023). The nuclear membrane, however, is a very selective barrier, limiting the size of the nanoparticles and macromolecules that can be transported through it and consequently curbing therapeutic efficacy (Soukar and Gaharwar, 2025).

### 4.4. Role of Surface Functionalization in Overcoming Biological Barriers

Surface functionalization is another potent approach towards the limit of biological barriers when it comes to drug delivery of nanoparticles (Al Jayoush et al., 2025). PEG coatings can be applied to surface to prevent formation of a corona of proteins, reduce opsonization, and extend the systemic circulation by reducing immune recognition. Ligand-based targeting increases specificity to tumor specific receptors, which facilitated the uptake to the cells and localization to the tumors (Al Jayoush et al., 2025). Red blood cell, immune cell or cancer cell biomimetic coating is further used to aid immune evasion and extended circulation (Sun et al., 2024).

Also, surface modifications using stimuli responses allow the targeted release of drugs in response to tumor-related cues like acidic pH, increased reactive oxygen species, or overexpressed enzymes (Al Jayoush et al., 2025). Extensive studies on the development of polyethyleneimine (PEI), membrane-disrupting peptides, and various functional materials have aided the endosomal escape of drugs to increase intracellular drug delivery (Ghosh et al., 2025). Organelle-targeting ligands may also be used to deliver therapeutic agents to particularly the nucleus and mitochondria, facilitated by advanced targeting ligands (Soukar and Gaharwar, 2025).

## 5. ACTIVE TARGETING STRATEGIES IN NANO-ONCOLOGY

A key breakthrough in contemporary nano-oncology is the ability of nanocarriers to specifically identify, bind, and carry therapeutic agents to cancer cells without impacting healthy tissues (Ijaz et al., 2024). The difference between conventional drug delivery systems and actively targeted nanoparticles is the engineering of the former with specific ligands, which bind to receptors, biomarkers, or intracellular structures that are over-expressed in tumors (Graván et al., 2025). This will lead to improved drug delivery to diseased tissues, therapeutic efficacy and reduced systemic toxicity, which will allow the development of precision cancer therapeutics (Hong et al., 2023).

### 5.1. Passive versus Active Targeting

Passive and active targeting are both employed in tumor targeted nanomedicine (Figure 3) (Tang et al., 2025). Passive targeting is mostly based on the enhanced permeability and retention (EPR) effect, which refers to the preferential penetration of nanoparticles into tumor tissue because of the leaky blood capillaries and impaired lymphatic drainage (Tang et al., 2025). While the EPR effect enables more than therapeutic accumulation in tumor cells than with conventional therapeutics, the effectiveness of this effect can be limited by tumor heterogeneity, lack of vascularization, and patient-to-patient variability in tumor permeability (Sidorenko et al., 2025).

Active targeting, by contrast, uses surface-functionalized nanoparticles with specific ligands that are able to bind to a receptor that is overexpressed in cancer cells or in the microenvironment of the tumor (Tang et al., 2025). This receptor-specific interaction leads to increased cellular uptake and intracellular drug delivery and thus to better therapeutic results (Figure 3). Therefore, the active targeting is more selective and precise than passive accumulation (Sidorenko et al., 2025).

*Figure 3. Comparison of non-targeted systemic drug distribution versus tumor-specific targeted nanoparticle delivery*

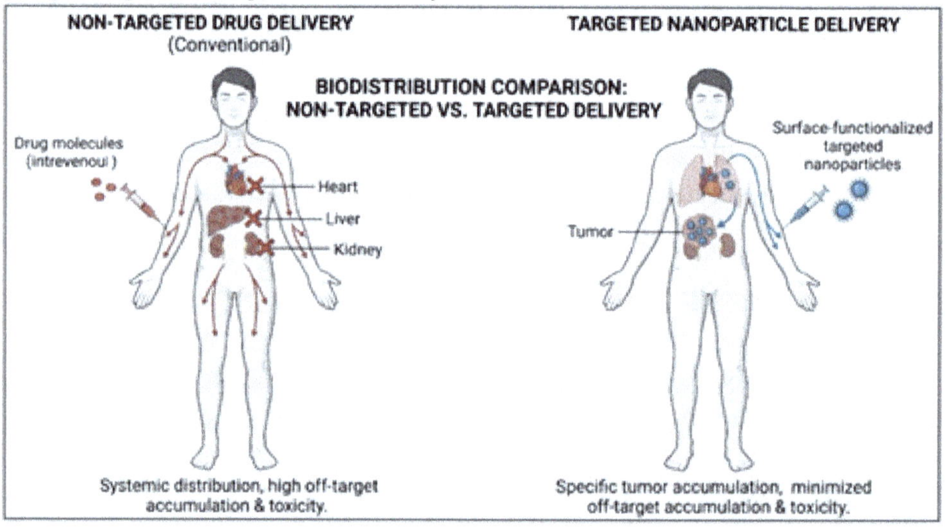


## 5.2. Receptor-Mediated Targeting

One of the most popular approaches in nano-oncology is receptor-mediated targeting (Shirzad et al., 2025). There are a number of cancer cells that overexpress certain receptors which are expressed at low levels in normal tissues, which offer unique molecular targets for selective drug delivery. Receptor-specific ligands can be attached to nanoparticles that will bind to the receptors and allow receptor-mediated endocytosis of the nanoparticle (Thongpon et al., 2025).

It's typically against folate receptors, transferrin receptors, epidermal growth factor receptor (EGFR), vascular endothelial growth factor receptor (VEGFR), and

CD44 receptors (Shirzad et al., 2025). These molecular signatures can be used to achieve receptor mediated targeting with a much improved tumor specificity and in-cell delivery of therapeutic agents (Zhang et al., 2025).

## 5.3. Antibody-Mediated Targeting

Monoclonal antibodies are extremely specific against tumor-associated antigens and therefore are very effective targeting agents for nanotherapeutics (Rauf et al., 2025). Biomarker-targeting antibody-coated nanoparticles can selectively bind to cancer cells with expression of markers like prostate-specific membrane antigen (PSMA), programmed death-ligand 1 (PD-L1), human epidermal growth factor receptor (EGFR), and human epidermal growth factor receptor 2 (HER2) (Thongpon et al., 2025).

After antigen recognition, the antibody-conjugated nanoparticles are internalized through receptor-mediated endocytosis, which will allow targeted delivery of drugs and thus decrease off-target toxicity (Rauf et al., 2025). In addition, antibody targeting can be combined with chemotherapy, immunotherapy and imaging methods, allowing for the creation of multifunctional therapeutic platforms. Although these have proven effective, issues including high production cost, immunogenicity, and instability remain to be overcome in developing alternative targeting agents (Thongpon et al., 2025).

## 5.4. Peptide-Mediated Targeting

The small size, simple synthetic route, low-immunogenic nature and high receptor specificity make peptides promising targeting ligands (Sharma et al., 2022). Tumor-homing peptides can selectively bind receptors which are over-expressed in tumor tissues and help to concentrate particles in the tumor (Venturini et al., 2025).

Arginine-glycine-aspartic acid (RGD) peptides are amongst the more widely studied peptide ligands, targeting integrin receptors which play a role in tumor angiogenesis and metastasis. Cell penetrating peptides and tumor penetrating peptides further enhance intracellular uptake and penetration. Peptide-functionalized nanoparticles are more likely to penetrate tumors than antibodies, and are typically easier to manufacture (Venturini et al., 2025).

## 5.5. Aptamer-Mediated Targeting

Aptamers are single-stranded DNA or RNA oligonucleotides with antibody-like affinity and specificity for a particular molecular target that have a short length. Aptamers are synthetic and have several merits compared to conventional antibodies, such as high stability, low immunogenicity, and a low production cost, because of their small size (Gao et al., 2026).

Targeting tumor biomarkers with aptamers can enable efficient cellular uptake and release of intracellular drug molecules by nanoparticles. Aptamers have been successfully developed to target a plethora of cancer-specific targets, such as nucleolin, EGFR, HER2, and prostate-specific membrane antigen. They are flexible and programmable, and thus are promising ingredients of precision nanomedicines of the future (Gao et al., 2026).

## 5.6. Small Molecule-Mediated Targeting

Small molecules are one of the simplest and lowest cost of the targeting ligands employed in nano-oncology. Because they are low molecular weight, they can easily be attached to the surface of nanoparticles without greatly impacting the nanoparticles' properties (Ghosh et al., 2025).

Folate receptors are frequently over-expressed in ovarian, breast, lung and colorectal cancers making folic acid one of the most widely used small molecule ligands. Likewise, the receptors linked with the increase in nutrition and cell growth in tumor tissue are targets of transferrin, lactoferrin and hyaluronic acid. The small molecule-mediated targeting approach is very stable, less immunogenic, and has the potential for scale up production, which is very attractive for clinical translation (Ghosh et al., 2025).

## 5.7. Organelle-Specific Targeting

Current targeting strategies mainly aim at targeting tumor cells, whereas organelle specific targeting targets therapeutic agents directly to the intracellular compartments where they can reach their maximum biological activity. This is especially useful in gene therapy, PDT and apoptosis inducing therapeutics (Tan et al., 2025).

Nuclear targeting is required for the delivery of nucleic acids, gene-editing systems and DNA-interacting drugs (Long et al., 2022). Many nuclear localization signals and specialized transport peptides are used to assist in transport through the nuclear membrane. This is called mitochondrial targeting and involves the use of lipophilic cations (e.g. triphenylphosphonium, TPP), which are used to transport drugs to the mitochondria where they may interfere with energy production and induce apoptosis (Yang et al., 2025). Lysosomal targeting has been the focus for those therapies based on the disruption of the lysosomal function and lysosomal degradation pathways. Delivery to organelles allows for greater targeting to the site of action and potentially greater treatment effectiveness (Tan et al., 2025).

## 5.8. Multi-Targeting and Dual-Ligand Approaches

However, because of the heterogeneity of the tumor, one-targeted approach is often less effective in cancer therapy. In order to overcome this drawback, multi-targeting and dual-ligand nanoplatforms that target multiple receptors or biological pathways have been developed (Wong et al., 2024).

Dual-ligand nanoparticles utilize more than one targeting ligand on a single nanoparticle for increased binding affinity and increased tumor recognition. For instance, folic acid and RGD peptides can be attached to the same nanoparticles, which allows for greater accumulation of the nanoparticles in the tumor and internalization by the cells. Likewise, combinations of antibody, peptides, aptamers, and small molecule can be used for improving targeting efficiency in a heterogenous tumor population (Venturini et al., 2025).

Multi-targeting strategies also enable sequential targeting in which nanoparticles target the tumor vasculature and then the cancer cells or organelles within cells. This hierarchical targeting method leads to better tissue penetration, cellular uptake, and therapeutic efficacy. With the development of precision oncology, multifunctional and dual-targeted nanoplatforms are anticipated to have a greater and greater impact on coping with tumor heterogeneity and improving clinical efficacy (Tan et al., 2025).

*Figure 4. Theranostic nanoplatforms integrate targeted drug delivery and diagnostic imaging to enable real-time monitoring and personalized cancer treatment*

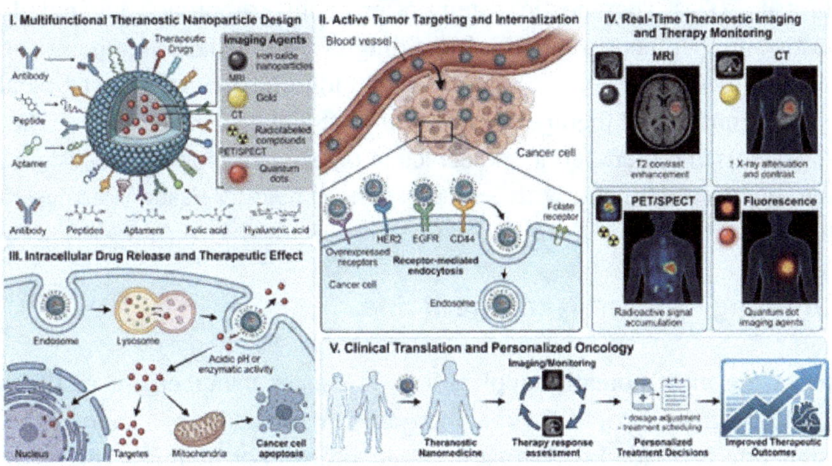

## 6. SURFACE FUNCTIONALIZATION FOR THERANOSTIC APPLICATIONS

Theranostics is a novel precision oncology strategy that has emerged due to the integration of diagnostic and therapeutic functions in a single nanoplatform (Shanahan et al., 2025). The concept of theranostic nanomedicine combines the disease diagnosis, targeted treatment, and monitoring of the therapy into a single system, allowing the disease to be monitored during therapy and therapeutic responses to be assessed on-the-fly (Gupta et al., 2024). To achieve these goals, improving the targeting ability, imaging ability, biocompatibility, and therapeutic efficiency of theranostic nanoparticles, surface functionalization is a key step in the development of these nanotechnology systems (Salgueiro et al., 2025). Rational surface engineering makes it possible to design nanoparticles with a targeted ability to concentrate in tumor tissue, with high resolution imaging signals and in controlled and efficient ways to deliver therapeutic payloads (Figure 4) (Salgueiro et al., 2025).

### 6.1. Concept of Theranostics

Theranostics is a combination of the words "therapy" and "diagnostics" and describes platforms that can diagnose and treat diseases at the same time (Azimizonuzi et al., 2025). Theranostic nanoplatforms provide immense potential in oncology as compared to conventional methods by enabling early detection of tumors, targeted delivery of therapeutic agents and real-time monitoring of therapeutic success (Gupta et al., 2024). Theranostic systems offer a holistic approach that includes both diagnostic and therapeutic capabilities, aiding in personalized treatment decisions and adaptive therapy strategies, as opposed to traditional therapies that typically use separate diagnostic and therapeutic procedures (Salgueiro et al., 2025). The use of imaging agents and therapeutic molecules in the same nanoparticle system allows for more precise treatment, with less exposure to healthy tissue (Parthasarathy et al., 2025).

### 6.2. Imaging-Guided Nanomedicine

One of the most potential applications in cancer management of the surface functionalized nanoparticles is imaging guided nanomedicine. Nanoparticles can be designed to be loaded with imaging agents for tumor imaging, disease staging and therapeutic monitoring. Through surface modification, these imaging agents can be made more stable, targeted and biodistributed, which consequently increases the diagnostic accuracy (Parthasarathy et al., 2025).

Incorporation of different imaging modalities into nanotherapeutic systems. Iron oxide nanoparticles are the most commonly used magnetic nanoparticles as contrast agents in magnetic resonance imaging (MRI). The high X-ray attenuation properties of gold nanoparticles have proven to be excellent for computed tomography (CT) imaging applications. In the same way, highly sensitive detection of cancer is achieved using radiolabeled nanoparticles with positron emission tomography (PET) and single photon emission computed tomography (SPECT) (Parthasarathy et al., 2025). Nanoparticles that are fluorescent and quantum dot based have also been created for optical imaging applications, which allow for real-time visualization of the distribution of drugs and tumor localization. These nanoplatforms are engineered to bring together both imaging and therapeutic capabilities, allowing for image-guided interventions and enhancing therapeutic outcomes (Najdian et al., 2024).

## 6.3. Multifunctional Surface Engineering

To achieve the goals of theranostic nanomedicine, multi-functional surface engineering strategies that are able to integrate multiple functions into a single nanoparticle system are essential (Shanahan et al., 2025). Surface functionalization enables the co-delivery of targeting ligands, imaging probes, therapeutic agents, and stimuli-responsive components all at once. This multifunctionality leads to increased specificity of tumor and increased therapeutic precision (Najdian et al., 2024).

For instance, nanoparticles equipped with antibodies, peptides or aptamers can be used for targeted delivery, carrying imaging agents for tumor visualization, and anticancer drugs for tumor therapy (Akpe et al., 2025). Stimuli-responsive coatings allow for the control of drug delivery, particularly when varying levels of acidity, reactive oxygen species, and/or enzymes are associated with a tumor. Further, multifunctional nanoplatforms can be used for the simultaneous delivery of chemotherapy, photothermal therapy, PDT, gene therapy, and/or immunotherapy, leading to a synergic anticancer effect (Kashyap et al., 2023). This “whole system” approach helps to optimize therapeutic effectiveness but limit systemic toxicity and adverse effects (Najdian et al., 2024).

## 6.4. Personalized Cancer Management through Theranostic Nanoplatforms

In personalized medicine, treatment will be adjusted based on the molecular and genetic features of individual patients. Theranostic nanoplatforms have an important role to play in this as they are able to deliver diagnostics and targeted therapeutics in a personalized manner. The surface functionalization of nanoparticles can be targeted against specific cancer subtypes to allow selective delivery of drugs and better treatment efficacy (Fernandes et al., 2023).

In addition, imaging-guided monitoring can assess the accumulation of the drug, response to treatment, and disease progression as it happens. This data can be utilized to refine treatment regimens, adjust treatments and detect non-responsive tumors early. The combination of molecular targeting, cutting-edge imaging and controlled drug delivery is thus significant to making precision oncology a reality (Fernandes et al., 2023).

With the ongoing advancement of nanotechnology, theranostic systems will become more complex and sophisticated, including the use of artificial intelligence, biosensors and personalized profiling of biomarkers (Dhanapal et al., 2026). The improvements will make it possible for doctors to treat cancer patients with even more precision and effectively, while reducing the risk of side effects (Younas et al., 2025). Overall, the surface-functionalized theranostic nanoplatforms are a significant step towards the present progress of cancer management, moving towards the personalized approach and linking diagnosis and treatment (Fernandes et al., 2023).

*Table 1. Clinical translation and commercial progress of surface-functionalized nanomedicines in oncology*

| Category | Key Developments | Representative Systems / Examples | Clinical/ Commercial Status | Key Challenges | References |
|---|---|---|---|---|---|
| **Clinically approved nanomedicines** | Surface-functionalized lipid and polymer-based drug carriers with improved pharmacokinetics and reduced toxicity | PEGylated liposomal doxorubicin, liposomal daunorubicin, albumin-bound paclitaxel | Approved and widely used in cancer therapy | Limited tumor specificity, cost of production | Zeinali et al., 2025 |
| **PEGylated formulations** | Stealth coatings to reduce opsonization and extend circulation time | PEG-liposomes, PEGylated protein-drug systems | Clinically established | Accelerated blood clearance (ABC phenomenon), immune response variability | Miao et al., 2023 |
| **Albumin-based nanocarriers** | Natural protein-based drug delivery with improved tumor accumulation | Albumin-bound paclitaxel (nab-paclitaxel) | FDA-approved and globally commercialized | Drug loading limitations, stability issues | Zeinali et al., 2025 |

continued on following page

*Table 1. Continued*

| Category | Key Developments | Representative Systems / Examples | Clinical/ Commercial Status | Key Challenges | References |
|---|---|---|---|---|---|
| **Ongoing clinical trials** | Actively targeted nanocarriers and theranostic systems under evaluation | Antibody-conjugated nanoparticles, ligand-targeted liposomes, siRNA nanocarriers | Phase I–III trials | Biological barriers, inconsistent EPR effect | Wong et al., 2024 |
| **Immuno-nanomedicines** | Nanoparticles integrated with immune checkpoint modulation | PD-L1 targeted nanocarriers, immunoliposomes | Early clinical evaluation | Immune toxicity, complex regulation | Wong et al., 2024 |
| **Theranostic nanoplatforms** | Combined imaging and therapeutic delivery systems | Iron oxide MRI agents, gold nanoparticle-based photothermal systems | Preclinical to early clinical trials | Regulatory uncertainty, imaging standardization | Zeinali et al., 2025 |
| **Regulatory landscape** | Evaluation of safety, pharmacokinetics, and biodistribution of nanomedicines | FDA/EMA-approved nanodrugs | Evolving regulatory framework | Lack of standardized nanotoxicology guidelines | EMA, 2025 |
| **Manufacturing & scalability** | Large-scale production of functionalized nanoparticles | Continuous synthesis platforms, GMP-based nanofabrication | Industrial development stage | Batch variability, high production cost, reproducibility issues | Abdullah et al., 2025 |

## 7. FUTURE DIRECTIONS AND CHALLENGES IN NANO-ONCOLOGY

Although remarkable progress has been made in surface-engineered nanotherapeutics, nano-oncology still has substantial hurdles that make the potential of this field for clinical and commercial applications difficult to realize (Abdullah et al., 2025). Surface modification is complex, and there is often batch-to-batch variation in the amount and consistency of the nanoparticles, which can result in variable physicochemical properties and biological function (Abdullah et al., 2025). Furthermore, the protein corona is not static and can change in physiological environments in a random manner, which may impact targeting efficiency and biodistribution. Further, therapeutic efficacy becomes complicated by the presence of tumor heterogeneity, as differences in receptor expression, vascular structure and microenvironmental conditions result in less consistent therapeutic response from the targeted nanocarriers (Hoang et al., 2025). However, long-term toxicity questions such as accumulation of nanoparticles in important organs and possible immune reactions are key obstacles to clinical use (European Medicines Agency, 2025). Furthermore, production costs remain high, the scale has not been proven, and there is no complete harmonization of the regulatory requirements, which still prevents the commercialization of the product (European Medicines Agency, 2025). In spite of these difficulties, the field is undergoing rapid change, with novel approaches such as the use of artificial intelligence to guide nanoparticle design, allowing for prediction of surface properties and targeting efficiencies (Shirzad et al., 2025). Precision and personalized nano-oncology is helping to enable personalized treatments that target molecular markers and tumor-specific biomarkers in a patient-specific manner (Dhanapal et al., 2026). The biomimetic, cell-derived nanocarriers are proving to be very efficient in escaping immune detection and improving biological compatibility, whereas the CRISPR-based gene-editing delivery systems are poised to provide new opportunities for the precise genetic manipulation of tumor cells in cancer treatment (Rauf et al., 2025). At the same time, the advent of smart multifunctional nanoplatforms potentially capable of integrated diagnosis, therapy and real-time monitoring is revolutionizing the framework of cancer treatment (Younas et al., 2025). In addition, sustainable and green nanotechnology methods are becoming increasingly popular, focusing on eco-friendly methods for producing nanotechnology and biocompatible materials. These developments together reflect that although many hurdles exist, new and emerging interdisciplinary innovations are steadily taking nano-oncology towards more efficient, safe, and clinically translatable cancer therapeutics (Lee et al., 2025).

## 8. CONCLUSION

Surface functionalization and active targeting have become key strategies for the development of nano-oncology, revolutionizing the field of cancer diagnosis and treatment. Surface engineering can achieve accurate regulation of the interaction between nanoparticles and biological systems, thereby improving the stability, circulation time in the body, immunogenicity, and tumor targeting accumulation of nanoparticles. This can be further improved through active targeting strategies which take advantage of molecular differences between cancerous and normal cells, allowing targeted uptake and organelle-specific delivery. Together, these advances address numerous challenges of traditional cancer treatments such as off-target effects, poor bioavailability and multidrug resistance. Furthermore, the introduction of biomimetic systems, stimuli-responsive materials and multifunctional nanoplatforms has introduced the scope of nanomedicine towards theranostic approaches, which allows the real-time monitoring of the disease state and personalized adjustments of the therapeutic approach.

Despite all the good news, clinical translation is a challenge as a result of biological barriers, tumor heterogeneity, regulatory requirements, and manufacturing scalability. But continuous advances in artificial intelligence-driven design, green nanotechnology, and precision medicine are beginning to overcome these downsides, enabling more efficient and clinically relevant nanotherapeutic systems. The combination of nanotechnology, molecular biology and biomedical engineering holds even more promise for the future in making cancer treatments more specific, safe, and effective.

Finally, surface-functionalized and targeted nanoplatforms are poised for revolutionizing the field of oncology and enabling precision cancer therapeutics with unprecedented opportunities. Ongoing interdisciplinary research and translation efforts will be key to close this gap between innovation in the lab and clinical implementation, leading to the development of next generation nanomedicine for efficient and sustainable cancer management.

## REFERENCES


Abdullah, K. M., Sharma, G., Singh, A. P., & Siddiqui, J. A. (2025). Nanomedicine in cancer therapeutics: Current perspectives from bench to bedside. *Molecular Cancer*, *24*(1), 169. DOI: 10.1186/s12943-025-02368-w PMID: 40490771

Akpe, V., Cock, I. E., Telukutla, S. R., & Shah, N. (2025). Advances in cancer treatment through nanotheranostics and emerging therapies. *Drugs and Drug Candidates*, *6*(4), 29. DOI: 10.3390/ddc6040029

Al Jayoush, A. R., Haider, M., Khan, S. A., & Hussain, Z. (2025). Hyaluronic acid-functionalized nanomedicines for CD44-receptors-mediated targeted cancer therapy: A review of selective targetability and biodistribution to the tumor microenvironment. *International Journal of Biological Macromolecules*, *308*(Pt 2), 142486. DOI: 10.1016/j.ijbiomac.2025.142486 PMID: 40139601

Al-Shadidi, J. R. M. H., Al-Shammari, S., Al-Mutairi, D., Alkhudhair, D., Thu, H. E., & Hussain, Z. (2024). Chitosan nanoparticles for targeted cancer therapy: A review of stimuli-responsive, passive, and active targeting strategies. *International Journal of Nanomedicine*, *19*, 8373–8400. DOI: 10.2147/IJN.S472433 PMID: 39161363

AlSawaftah, N. M., Awad, N. S., Paul, V., Kawak, P. S., Al-Sayah, M. H., & Husseini, G. A. (2022). pH-responsive nanocarriers in cancer therapy. *Polymers*, *14*(5), 936. DOI: 10.3390/polym14050936 PMID: 35267759

Amirinezhadfard, E., Niazi Tabar, A., Xia, L.-J., & Yang, W.-C. (2025). Plant-derived nanomaterials in cancer therapy: Advances from green synthesis to application of PANoptosis-mediated tumor suppression. *Industrial Crops and Products*, *236*, 121850. DOI: 10.1016/j.indcrop.2025.121850

Azimizonuzi, H., Ghayourvahdat, A., Ahmed, M. H., Kareem, R. A., Zrzor, A. J., Mansoor, A. S., Athab, Z. H., & Kalavi, S. (2025). A state-of-the-art review of the recent advances of theranostic liposome hybrid nanoparticles in cancer treatment and diagnosis. *Cancer Cell International*, *25*(1), 26. DOI: 10.1186/s12935-024-03610-z PMID: 39871316

Bajracharya, R., Song, J. G., Patil, B. R., Lee, S. H., Noh, H. M., Kim, D. H., Kim, G. L., Seo, S. H., Park, J. W., Jeong, S. H., Lee, C. H., & Han, H. K. (2022). Functional ligands for improving anticancer drug therapy: Current status and applications to drug delivery systems. *Drug Delivery*, *29*(1), 1959–1970. DOI: 10.1080/10717544.2022.2089296 PMID: 35762636

Bevilacqua, G., Corvino, R., Capriotti, A. L., Montone, C. M., Moriconi, M., Salciccia, S., Brunelli, V., Santarelli, V., Sciarra, B., Laganà, A., Santini, D., Sciarra, A., & Gentilucci, A. (2024). The protein corona on nanoparticles for tumor targeting in prostate cancer: A review of the literature and experimental trial protocol. *Biology (Basel), 13*(12), 1024. DOI: 10.3390/biology13121024 PMID: 39765691

Chai, M., Wang, S., Chen, Y., Pei, X., & Zhen, X. (2025). Targeted and intelligent nano-drug delivery systems for colorectal cancer treatment. *Frontiers in Bioengineering and Biotechnology, 13*, 1582659. DOI: 10.3389/fbioe.2025.1582659 PMID: 40352359

Chou, W.-C., & Lin, Z. (2024). Impact of protein coronas on nanoparticle interactions with tissues and targeted delivery. *Current Opinion in Biotechnology, 85*, 103046. DOI: 10.1016/j.copbio.2023.103046 PMID: 38103519

Chu, S., Shi, X., Tian, Y., & Gao, F. (2022). pH-responsive polymer nanomaterials for tumor therapy. *Frontiers in Oncology, 12*, 855019. DOI: 10.3389/fonc.2022.855019 PMID: 35392227

Ciftci, F., Özarslan, A. C., Kantarci, İ. C., Yelkenci, A., Tavukcuoglu, O., & Ghorbanpour, M. (2025). Advances in drug targeting, drug delivery, and nanotechnology applications: Therapeutic significance in cancer treatment. *Pharmaceutics, 17*(1), 121. DOI: 10.3390/pharmaceutics17010121 PMID: 39861768

Deivayanai, V. C., Thamarai, P., Karishma, S., Saravanan, A., Yaashikaa, P. R., Vickram, A. S., Hemavathy, R. V., Rohith Kumar, R., Rishikesavan, S., & Shruthi, S. (2025). Advances in nanoparticle-mediated cancer therapeutics: Current research and future perspectives. *Cancer Pathogenesis and Therapy, 3*(4), 293–308. DOI: 10.1016/j.cpt.2024.11.002 PMID: 40697517

Dhanapal, Y., Sridhar, D., Khute, S., Subash, P., & AnandaThangadurai, S. (2026). Artificial intelligence assisted design of nanomedicines for colorectal cancer diagnosis and therapy: Advances, challenges, and future directions. Intelligent Pharmacy. DOI: 10.1016/j.ipha.2026.01.001

Dudhwala, Y. D., Mehta, R. K., Kapoor, D. U., Shah, D. P., Shah, K. V., & Saini, V. (2026). Ligand-functionalized nanocarriers for targeted breast cancer therapy: Advances in receptor-mediated drug delivery. *Health Nanotechnology, 2*(1), 7. DOI: 10.1186/s44301-026-00029-5

European Medicines Agency. (2025). Nanotechnology-based medicinal products for human use: EU-IN Horizon Scanning Report. EMA/20989/2025/Rev.1. https://www.ema.europa.eu/

Fan, D., Cao, Y., Cao, M., Wang, Y., Cao, Y., & Gong, T. (2023). Nanomedicine in cancer therapy. *Signal Transduction and Targeted Therapy*, *8*(1), 293. DOI: 10.1038/s41392-023-01536-y PMID: 37544972

Fernandes, D. A.. (2023). Review on metal-based theranostic nanoparticles for cancer therapy and imaging. *Technology in Cancer Research & Treatment*, *22*, 15330338231191493. DOI: 10.1177/15330338231191493 PMID: 37642945

Gao, Z., Du, S., Song, J., Gao, Y., Peng, X., Lin, X., e, S., Zhao, Y., & Zhang, S. (2026). Aptamer-liposome targeted nanotherapeutics for cancer therapy: Bibliometric analysis, recent developments and future perspectives. *Materials Today. Bio*, *36*, 102766. DOI: 10.1016/j.mtbio.2026.102766 PMID: 41560835

Ghosh, R., Kumar, M., Kumar, S., Komal, K., Sharma, R., & Kurmi, B. D. (2025). Small molecule therapeutics for receptor-mediated targeting through liposomes in breast cancer treatment: A comprehensive review. *Bioorganic Chemistry*, *160*, 108442. DOI: 10.1016/j.bioorg.2025.108442 PMID: 40199009

Graván, P., Marchal, J. A., & Galisteo-González, F. (2025). Improving tumor treatment: Cell membrane-coated nanoparticles for targeted therapies. *Materials Today. Bio*, *32*, 101716. DOI: 10.1016/j.mtbio.2025.101716 PMID: 40391023

Gupta, D., Gupta, T. K., Roy, P., Sharma, R., Kasana, R., & Rathore, P. (2024). Recent nanotheranostic approaches in cancer research. *Clinical and Experimental Medicine*, *24*(1), 8. DOI: 10.1007/s10238-023-01262-3 PMID: 38240834

Gutiérrez Coronado, O., Sandoval Salazar, C., Muñoz Carrillo, J. L., Gutiérrez Villalobos, O. A., Miranda Beltrán, M. L., Soriano Hernández, A. D., Beltrán Campos, V., & Villalobos Gutiérrez, P. T. (2025). Functionalized nanomaterials in cancer treatment: A review. *International Journal of Molecular Sciences*, *26*(6), 2633. DOI: 10.3390/ijms26062633 PMID: 40141274

Hoang, C. N. M., Nguyen, S. H., & Tran, M. T. (2025). Nanoparticles in cancer therapy: Strategies to penetrate and modulate the tumor microenvironment - A review. *Smart Materials in Medicine*, *6*(2), 270–284. DOI: 10.1016/j.smaim.2025.07.004

Hong, L., Li, W., Li, Y., & Yin, S. (2023). Nanoparticle-based drug delivery systems targeting cancer cell surfaces. *RSC Advances*, *13*(31), 21365–21382. DOI: 10.1039/D3RA02969G PMID: 37465582

Ijaz, M., Aslam, B., Hasan, I., Ullah, Z., Roy, S., & Guo, B. (2024). Cell membrane-coated biomimetic nanomedicines: Productive cancer theranostic tools. *Biomaterials Science*, *12*(4), 863–891. DOI: 10.1039/D3BM01552A PMID: 38230669

Jia, Y., Chen, S., Wang, C., Sun, T., & Yang, L. (2022). Hyaluronic acid-based nano drug delivery systems for breast cancer. *Frontiers in Bioengineering and Biotechnology*, *10*, 990145. DOI: 10.3389/fbioe.2022.990145 PMID: 36091467

Jin, Z., & Amili, H.. (2024). Tumor microenvironment-responsive drug delivery based on micelles. *Biomedicines*, *12*(2), 417. DOI: 10.3390/biomedicines12020417 PMID: 38398021

Kashyap, B. K., Singh, V. V., Solanki, M. K., Kumar, A., Ruokolainen, J., & Kesari, K. K. (2023). Smart nanomaterials in cancer theranostics: Challenges and opportunities. *ACS Omega*, *8*(16), 14290–14320. DOI: 10.1021/acsomega.2c07840 PMID: 37125102

Kumar, N., Singh, S., Sharma, P., Kumar, B., & Kumar, A. (2024). Single-, dual-, and multi-stimuli-responsive nanogels for biomedical applications. *Gels (Basel, Switzerland)*, *10*(1), 61. DOI: 10.3390/gels10010061 PMID: 38247784

Lee, B. J., Kanaan, M. H. G., Abdullah, S. S., & Ghasemian, A. (2025). Plant-derived nanoparticles in cancer therapy: A comprehensive review of recent advances and future prospects. *OBM Genetics*, *9*(3), 308. DOI: 10.21926/obm.genet.2503308

Liu, Y., Zhang, Y., Li, H., & Hu, T. Y. (2025). Recent advances in the bench-to-bedside translation of cancer nanomedicines. *Acta Pharmaceutica Sinica. B*, *15*(1), 97–122. DOI: 10.1016/j.apsb.2024.12.007 PMID: 40041906

Long, X., Zhang, X., Chen, Q., Liu, M., Xiang, Y., Yang, Y., Xiao, Z., Huang, J., Wang, X., Liu, C., Nan, Y., & Huang, Q. (2022). Nucleus-targeting phototherapy nanodrugs for high-efficiency cancer treatment. *Frontiers in Pharmacology*, *13*, 905375. DOI: 10.3389/fphar.2022.905375 PMID: 35645841

Ly, P. D., Ly, K. N., Phan, H. L., Nguyen, H. H. T., Duong, V. A., & Nguyen, H. V. (2024). Recent advances in surface decoration of nanoparticles in drug delivery. *Frontiers in Nanotechnology*, *6*, 1456939. DOI: 10.3389/fnano.2024.1456939

Miao, G., Zhou, L., Zhu, Y., & Zhang, Y. (2023). Accelerated blood clearance of PEGylated nanoparticles in cancer therapy: Mechanisms and mitigation strategies. *Acta Biomaterialia*, *170*, 1–16. DOI: 10.1016/j.actbio.2023.08.040

Mousavi-Kiasary, S. M. S., Senabreh, A., Zandi, A., Pena, R., Cruz, F., Adibi, A., & Hooshmand, N. (2025). Synergistic cancer therapies enhanced by nanoparticles: Advancing nanomedicine through multimodal strategies. *Pharmaceutics*, *17*(6), 682. DOI: 10.3390/pharmaceutics17060682 PMID: 40573996

Murar, M., Pujals, S., & Albertazzi, L. (2023). Multivalent effect of peptide functionalized polymeric nanoparticles towards selective prostate cancer targeting. *Nanoscale Advances*, *5*(5), 1378–1385. DOI: 10.1039/D2NA00601D PMID: 36866255

Najdian, A., Beiki, D., Abbasi, M., Gholamrezanezhad, A., Ahmadzadehfar, H., Amani, A. M., Shafiee Ardestani, M., & Assadi, M. (2024). Exploring innovative strides in radiolabeled nanoparticle progress for multimodality cancer imaging and theranostic applications. *Cancer Imaging; the Official Publication of the International Cancer Imaging Society*, *24*(1), 127. DOI: 10.1186/s40644-024-00762-z PMID: 39304961

Omidian, H., Cubeddu, L. X., & Wilson, R. L. (2025). Peptide-functionalized nanomedicine: Advancements in drug delivery, diagnostics, and biomedical applications. *Molecules (Basel, Switzerland)*, *30*(7), 1572. DOI: 10.3390/molecules30071572 PMID: 40286158

Pallares, R. M., Barmin, R. A., Wang, A., Kiessling, F., & Lammers, T. (2025). Clinical cancer nanomedicines. *Journal of Controlled Release : Official Journal of the Controlled Release Society*, *385*, 113991. DOI: 10.1016/j.jconrel.2025.113991 PMID: 40582644

Parthasarathy, N.. (2025). Recent progress in multifunctional theranostic inorganic nanoparticles for cancer therapy. *Smart Materials in Medicine*, *6*, 801–825. DOI: 10.1016/j.smaim.2025.07.006

Pashkina, E., Bykova, M., Berishvili, M., Lazarev, Y., & Kozlov, V. (2025). Hyaluronic acid-based drug delivery systems for cancer therapy. *Cells*, *14*(2), 61. DOI: 10.3390/cells14020061 PMID: 39851489

Rauf, M. A., Rao, A., Sivasoorian, S. S., & Iyer, A. K. (2025). Nanotechnology-based delivery of CRISPR/Cas9 for cancer treatment: A comprehensive review. *Cells*, *14*(15), 1136. DOI: 10.3390/cells14151136 PMID: 40801569

Sabir, S., Thani, A. S. B., & Abbas, Q. (2025). Nanotechnology in cancer treatment: Revolutionizing strategies against drug resistance. *Frontiers in Bioengineering and Biotechnology*, *13*, 1548588. DOI: 10.3389/fbioe.2025.1548588 PMID: 40370595

Salgueiro, M. J., & Zubillaga, M. (2025). Theranostic nanoplatforms in nuclear medicine: Current advances, emerging trends, and perspectives for personalized oncology. *Journal of Nanotheranostics*, *6*(4), 27. DOI: 10.3390/jnt6040027

Samathoti, P., Kumarachari, R. K., Bukke, S. P. N., Rajasekhar, E. S. K., Jaiswal, A. A., & Eftekhari, Z. (2025). The role of nanomedicine and artificial intelligence in cancer health care: Individual applications and emerging integrations-a narrative review. *Discover. Oncology*, *16*(1), 697. DOI: 10.1007/s12672-025-02469-4 PMID: 40338421

Shanahan, K., Coen, D., & Nafo, W. (2025). Polymer-based nanoparticles for cancer theranostics: Advances, challenges, and future perspectives. *Exploration of Biomat-X*, *2*, 101342. DOI: 10.37349/ebmx.2025.101342

Sharma, R., Borah, S. J., Bhawna, , Kumar, S., Gupta, A., Singh, P., Goel, V. K., Kumar, R., & Kumar, V. (2022). Functionalized peptide-based nanoparticles for targeted cancer nanotherapeutics. *ACS Omega*, *7*(41), 37362–37389. DOI: 10.1021/acsomega.2c03974 PMID: 36278104

Shirzad, M.. (2025). Artificial intelligence-assisted design of nanomedicines for cancer diagnosis and therapy. *Drug Delivery and Translational Research*, *15*, 4127–4150. DOI: 10.1007/s12668-025-01980-w

Sidorenko, V., Tobi, A., Sugahara, K. N., & Teesalu, T. (2025). Peptide-targeted nanoparticles for tumor therapy. *Journal of Controlled Release : Official Journal of the Controlled Release Society*, *387*, 114195. DOI: 10.1016/j.jconrel.2025.114195 PMID: 40925419

Soukar, J., & Gaharwar, A. K. (2025). Organelle-targeting nanoparticles. *Advanced Science (Weinheim, Baden-Württemberg, Germany)*, *12*(7), 2411720. DOI: 10.1002/advs.202411720 PMID: 39806939

Sun, L., Zhang, Y., Yang, B., Sun, S., Zhang, J., & Wang, X. (2023). Smart nanoparticles for cancer therapy. *Signal Transduction and Targeted Therapy*, *8*(1), 418. DOI: 10.1038/s41392-023-01642-x PMID: 37919282

Sun, Y., Zhou, Y., Rehman, M., Wang, Y.-F., & Guo, S. (2024). Protein corona of nanoparticles: Isolation and analysis. *Chemical & Biomedical Imaging*, *2*(9), 728–744. DOI: 10.1021/cbe.4c00105 PMID: 39974182

Taghavimandi, F., Kim, M. G., Lee, M., & Shin, K. (2025). Beyond PEGylation: Nanoparticle surface modulation for enhanced cancer therapy. *Health Nanotechnology*, *1*(1), 13. DOI: 10.1186/s44301-025-00014-4

Tan, K., Zhang, H., Yang, J., Wang, H., Li, Y., Ding, G., Gu, P., Yang, S., Li, J., & Fan, X. (2025). Organelle-oriented nanomedicines in tumor therapy: Targeting, escaping, or collaborating? *Bioactive Materials*, *49*, 291–339. DOI: 10.1016/j.bioactmat.2025.02.040 PMID: 40161442

Tang, S., Chen, X., Tong, X., & Zhu, L. (2025). Overcoming the delivery challenges in CRISPR/Cas9 gene editing for effective cancer treatment: A review of delivery systems. *International Journal of Medical Sciences*, *22*(14), 3625–3649. DOI: 10.7150/ijms.112724 PMID: 40959562

Thongpon, P., Tang, M., & Cong, Z. (2025). Peptide-based nanoparticle for tumor therapy. *Biomedicines*, *13*(6), 1415. DOI: 10.3390/biomedicines13061415 PMID: 40564134

Torres, J., Valenzuela Oses, J. K., Rabasco-Álvarez, A. M., González-Rodríguez, M. L., & García, M. C. (2025). Innovations in cancer therapy: Endogenous stimuli-responsive liposomal nanocarriers. *Pharmaceutics*, *17*(2), 245. DOI: 10.3390/pharmaceutics17020245 PMID: 40006612

Venturini, J.. (2025). Developments in nanotechnology approaches for the treatment of cancer. *Cancer Nanotechnology*, *16*, 32. DOI: 10.1186/s12645-025-00315-7

Wang, X., & Zhang, Y. (2022). The Janus of protein corona on nanoparticles for tumor targeting, immunotherapy and diagnosis. *Journal of Controlled Release : Official Journal of the Controlled Release Society*, *345*, 832–850. DOI: 10.1016/j.jconrel.2022.03.056 PMID: 35367478

Wen, F., Wang, L., Li, X., Zhao, J., Xu, T., Zhu, J., Ma, L., & Wang, X. (2025). Precision nanomedicine for cancer: Innovations, strategies, and translational challenges. *OncoTargets and Therapy*, *18*, 1125–1148. DOI: 10.2147/OTT.S550104 PMID: 41098351

Wong, K. Y.. (2024). Aptamer-functionalized liposomes for targeted drug delivery. *Biomaterials Advances*, *156*, 213716. DOI: 10.1016/j.bioadv.2023.213716

Yang, M., Ou, J., Yan, H., He, Y., Gan, L., & Zhang, S.-L. (2025). Triphenylphosphonium-based mitochondrial targeting in cancer therapy: Mechanisms, progress, and perspectives. *Chemical Communications (Cambridge)*, *61*(96), 18937–18950. DOI: 10.1039/D5CC05709D PMID: 41236308

Younas, A., Wang, S., Asad, M., Al Mamun, A., Majeed, S., Sharif, A., Zhou, Q., Liu, Y., Geng, P., Shao, C., & Xiao, J. (2025). Recent advances in cancer nanomedicine: From smart targeting to personalized therapeutics-pioneering a new era in precision oncology. *Materials Today. Bio*, *31*, 101232. DOI: 10.1016/j.mtbio.2025.101232 PMID: 41560793

Zalba, S., & Garrido, M. J. (2022). Stealth nanoparticles in oncology: Facing the PEG dilemma. *Journal of Controlled Release : Official Journal of the Controlled Release Society*, *351*, 22–36. DOI: 10.1016/j.jconrel.2022.09.002 PMID: 36087801

Zeinali, R., Zaeifi, D., Zolfaghari-Moghaddam, S. Y., Paul, M., & Biazar, E. (2025). Current advances in nanocarriers for cancer therapy. *International Journal of Nanomedicine*, *20*, 12217–12262. DOI: 10.2147/IJN.S548088 PMID: 41078404

Zhang, J.. (2025). Targeted cancer treatment using folic acid-functionalized carbohydrate polymer nanocarriers. *Industrial Crops and Products*, *226*, 120653. DOI: 10.1016/j.indcrop.2025.120653

Zhang, S., Zhang, X., Gao, H., Zhang, X., Sun, L., Huang, Y., Zhang, J., & Ding, B. (2024). Cell membrane-coated biomimetic nanoparticles in cancer treatment. *Pharmaceutics*, *16*(4), 531. DOI: 10.3390/pharmaceutics16040531 PMID: 38675192

Zhao, N., Shi, Y., Liu, P., & Lv, C. (2025). pH-responsive carbohydrate polymer-based nanoparticles in cancer therapy. *International Journal of Biological Macromolecules*, *306*, 141236. DOI: 10.1016/j.ijbiomac.2025.141236 PMID: 39978518

Zheng, J., Sun, Y., Shen, Y., & Zhou, Z. (2025). Surface engineering of nanoparticles for precision medicine. *Precision Medicine and Engineering*, *2*(3), 100037. DOI: 10.1016/j.preme.2025.100037

Zhong, Z., Deng, W., Wu, J., Shang, H., Tong, Y., He, Y., Huang, Q., Ba, X., Chen, Z., & Tang, K. (2024). Cell membrane coated nanoparticles as a biomimetic drug delivery platform for enhancing cancer immunotherapy. *Nanoscale*, *16*(18), 8708–8738. DOI: 10.1039/D4NR00284A PMID: 38634521

# Chapter 8
# Nano-Gene Therapy:
## CRISPR/Cas9 and Beyond

**Kasim Sakran Abass**
http://orcid.org/0000-0002-5796-7170
*University of Kirkuk, Iraq*

**Saad Ullah**
*PPG Cell and Structural Biology, Department of General Biology, Federal University of Viçosa, Brazil*

**Muhammad Ali**
http://orcid.org/0009-0002-7593-0395
*Department of Zoology, Faculty of Biological Sciences, Quaid-i-Azam University, Pakistan*

**Sheikh Ahmed Gull**
http://orcid.org/0009-0000-9450-4119
*Department of Zoology, Faculty of Biological Sciences, Quaid-i-Azam University, Pakistan*

**Amjad Khan**
*Department of Zoology, Institute of Biological Sciences, University of Lakki Marwat, Pakistan*

**Bilal Khan**
*Department of Bioscience Technology, Chung Yuan Christian University, Taoyuan, Taiwan*

**Muhammad Tahir Ullah**
http://orcid.org/0009-0007-6025-4044
*Department of Biochemistry, Faculty of Biological Sciences, Quaid-i-Azam University, Pakistan*

## ABSTRACT

*Nano-gene therapy is an emerging approach for precision cancer treatment that involves integration of Advanced nanotechnologies with genome editing. Compared to the other existing platforms, the CRISPR/Cas9 system has proven to be a revolution, which allows selective disruption of oncogenes, correction of pathogenic mutations, bypass of drug resistance, and development of immune cells with stronger anticancer effects. The clinical application of CRISPR/Cas9 has been hindered, however, due to delivery inefficiency, lack of stability of the nucleases, off-target editing,*

DOI: 10.4018/979-8-3373-9848-8.ch008

*and immune-related toxicity. Nanoparticle-based delivery systems can effectively address these issues, including protecting the gene-editing cargo, enhancing the accumulation at tumor sites, achieving intracellular delivery and controlling release in the tumor microenvironment.*

## INTRODUCTION

Cancer is one of the most important causes of morbidity and mortality globally, with almost 20 million new cases and nearly 10 million deaths reported each year. Global Cancer Observatory (GLOBOCAN) and Global Burden of Disease (GBD) estimates project that cancer cases will escalate significantly in the next few decades as a result of population growth and increasing urbanization, ageing and changing lifestyles, that by 2050 could reach more than 30 million new cases annually (Sung et al., 2021; Bray et al., 2012). This is especially true in low and middle income countries, where the lack of access to timely diagnostic and advanced treatment facilities leads to adverse clinical outcomes.

The main treatment strategies for cancer include surgery, chemotherapy, radiotherapy and immunotherapy. Surgical resection is the main treatment for localised tumours but may not be possible if the tumour invades surrounding tissues and/or contains small areas of cancer that can't be removed, which will make it more likely to develop again. Although chemotherapy and radiotherapy have increased the survival rate of patients, there have been various side effects of the treatment, such as non-specific toxicity, damage to healthy tissues, multi-drug resistance and low effectiveness against metastatic and heterogeneous tumors. While this breakthrough in cancer treatment has ushered in new hope for patients, many are not achieving sustainable results due to immune evasion strategies and the profound suppression within the tumor microenvironment (TME) (Sabit et al., 2025; Li et al., 2024). The constraints have led to a pressing demand for more accurate and personalized therapeutic approaches which selectively target the molecular abnormalities driving tumor development while avoiding general toxicity.

With the advance of genome-engineering technologies, nano-gene therapeutics have been rapidly developed. Clustered regularly interspaced short palindromic repeats (CRISPR)/Cas9 system has revolutionized biomedical research and therapeutic development by enabling programmable, efficient and highly specific modification of genomic DNA. The CRISPR/Cas9 system is based on a single-guide RNA (sgRNA) that guides the Cas9 endonuclease to a complementary target sequence, leading to specific DNA cleavage and the activation of endogenous cellular repair mechanisms for gene disruption, correction, and insertion (Janik et al., 2020; Li et al., 2021; Li et al., 2023). Overall, the CRISPR/Cas9 system is simpler, more flexible, more scalable, and more powerful genome engineering system than previous genome editing technologies such as zinc finger nucleases (ZFNs) and transcription activator-like effector nucleases (TALENs), and has become the most popular genome engineering system for current biomedical research (Pickar-Oliver & Gersbach, 2019; Li et al., 2023).

The delivery of the components of the CRISPR/Cas9 system to the target cells and tissues is a critical step for the successful therapeutic uses of the system. To maintain immune system silence and prevent off target disturbances, delivery systems have to pass through multiple physiological barriers for delivering large and sensitive cargoes like Cas proteins, guide RNAs, and donor repair templates (Deng et al., 2019; Kazemian et al., 2022; Farsani et al., 2024). Nanotechnology is one of the potential avenues for tackling these problems. Rational design of particle size, charge, composition, and surface functionalization, such as the introduction of lipid nanoparticles (LNP), can help shield the genome editing cargo, can enhance its intracellular delivery, can promote its endosomal escape, and can enable tissue-specific targeting (Kazemian et al., 2022; Rauf et al., 2025; Jiang et al., 2025). All of these advances have helped make nanotechnology an essential enabling platform for genome engineering in the clinic.

Apart from the classical CRISPR/Cas9 systems, the field of genome editing is constantly evolving with emergence of 'next generation' genome editing technologies. New precision tools brought from new platforms, such as CRISPR/Cas12, CRISPR/Cas13 and base and prime editing and transposase associated systems and RNA-editing technologies, are bringing new editing capabilities (Anzalone et al., 2020; Kantor et al., 2020; Pickar-Oliver & Gersbach, 2019). Mostly critical, both base-editing and prime-editing methods do not produce any double-strand DNA (dsDNA) break(s) so that the risk of unwanted genomic changes is reduced, and more mutations can be treated. (Molla & Yang, 2019; Kantor et al., 2020). The combination of these potent genome-engineering capabilities and ingenious nano-carrier solution is ushering in a new generation of highly personalized and targeted therapeutic strategies in the battle against cancer.

Nano-gene therapy" is a new approach that combines nanotechnology with powerful genome-editing technologies in cancer treatment. Delivery of genetic material via nanocarriers can improve targeted delivery to tumour cells, improving the therapeutic efficiency and reducing toxicity and off-target effects. The basic principles of gene therapy and genome editing, design of nanocarriers, targeted therapeutics, cancer immunotherapies, theranostic cancer therapeutics and implications for biosafety and regulatory issues are addressed in this chapter along with future perspectives for clinical translation.

# NANOTECHNOLOGY IN CANCER THERAPY

## Introduction to Nanotechnology

Nanotechnology is the process of designing, manipulation and use of materials that ranges, generally from 1 to 100 nm in size, or less. Materials at this scale have physical, chemical and biological properties that are distinct from those of bulk materials, which makes them useful for biomedical applications. In oncology, nanotechnology has become a pivotal area of precision medicine, allowing for the targeted delivery of therapeutic and diagnostic agents to tumor tissues with minimal impact on healthy cells (Gavas et al., 2021; Wang et al., 2024).

The principle on which nanomedicine is based is the use of the enhanced permeability and retention (EPR) phenomenon, which allows the preferential accumulation of nanoparticles in the tumor environment due to the leaky tumor vasculature and the defective tumor lymphatics. Furthermore, nanoparticles can be engineered to carry targeting agents like antibodies, peptides, or small molecules that selectively bind to certain cancer markers, thereby enabling active targeting and enhancing the therapeutic benefits and minimizing systemic side effects (Raj et al., 2019; Sun et al., 2023).

There are many different types of nanoparticles that have been explored for cancer treatment including polymeric nanoparticles, lipid-based nanoparticles (liposomes and solid lipid nanoparticles), dendrimers, metallic nanoparticles including gold and iron oxide nanoparticles, mesoporous silica nanoparticles, carbon-based nanoparticles and nature derived extracellular vesicles. Nanotechnology is highly versatile for the development of cancer therapeutics of the next generation due to the fact that each class has unique physicochemical properties which can be used to develop next-generation cancer therapeutic applications such as drug delivery, gene therapy, molecular imaging, photothermal therapy, and theranostics (Sell et al., 2023; Zhu et al., 2025).

## Nanoparticles Used in Cancer Treatment

The ability of nanoparticles to enhance drug delivery, increase accumulation of drugs at tumour sites and lower the toxicity of drugs to normal tissues makes them an important tool in modern cancer therapy. The most common types of nanocarriers are lipid nanoparticles, polymeric nanoparticles, gold nanoparticles, magnetic nanoparticles, and dendrimers (Ajith et al., 2023; Sell et al., 2023). The physicochemical properties that each nanocarrier system has make it suitable for a particular therapeutic and/or diagnostic purposes (Figure 1).

**Lipid nanoparticles (LNPs)**, are highly biocompatible and have the ability to load anticancer drugs to enhance their stability and circulation time (Waheed et al., 2024).

**Polymeric nanoparticles** are synthesized from the biodegradable polymers, such as PLGA and PEG. They release controlled drugs and deliver better targeting to tumor sites, and are appealing drug carriers for chemotherapy and gene delivery (Xiao et al., 2022).

**Gold nanoparticles** The optical properties of gold nanoparticles (AuNPs) allow their application in drug delivery, imaging and photothermal therapy, where the generation of local heat leads to the destruction of cancer cells (Gavas et al., 2021).

**Magnetic nanoparticles** specifically iron oxide nanoparticles, can be employed in the targeted delivery of drugs, magnetic resonance imaging (MRI) and magnetic hyperthermia, have both diagnostic and therapeutic applications (Luiz et al., 2023).

**Dendrimers** are highly branched polymeric nanostructures with numerous surface functional groups that allow efficient drug loading and attachment of targeting molecules, improving the precision of anticancer therapies (Ambekar et al., 2020).

These nanoplatforms offer many advantages, including enhanced targeting, controlled drug release, and multifunctional therapeutic properties, making them potential candidates for future cancer treatment strategies.

*Figure 1. Nanoparticle used in cancer therapy with their primary biomedical applications*

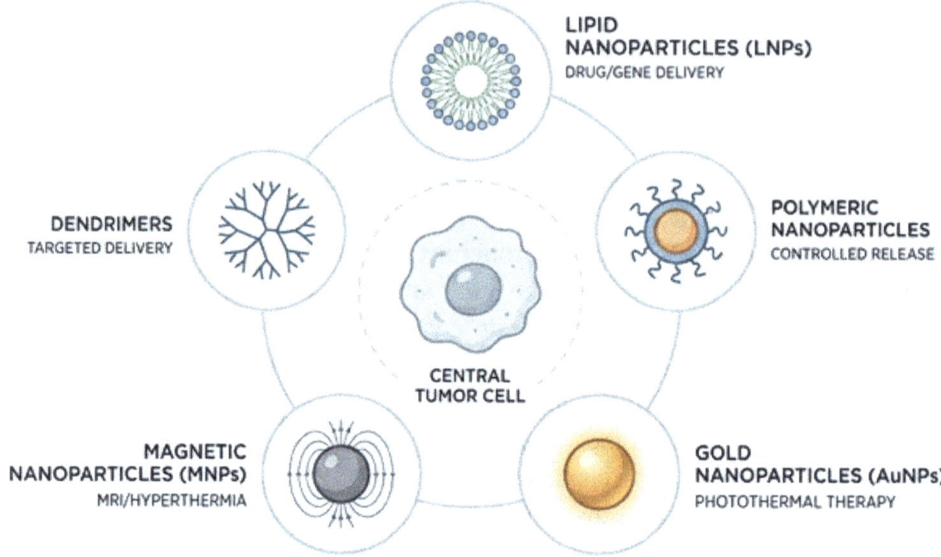


## Advantages of Nanotechnology in Cancer Treatment

Nanotechnology has already had a profound impact on cancer treatment, by improving the delivery of drugs and their efficacy in treating the disease. Nanocarriers are capable of delivering drugs specifically to tumor sites, thus enhancing the therapeutic benefits and reducing the side effects of the treatments compared to conventional chemotherapy (Yao et al., 2020; Gavas et al., 2021).

A key benefit of nanotechnology is the ability to deliver drugs specifically to the area of the body where they are needed. Enhanced permeability and retention (EPR) effect and the ability of modifying nanoparticles with targeting ligands, which selectively target cancer cells, lead to higher concentrations of drugs inside the tumors (Tian et al., 2022).

A second great advantage is less systemic toxicity. Nanoparticle-based formulations reduce the exposure of drugs to healthy tissues, which helps to minimize the common side effects of the conventional chemotherapy, providing added patient safety and tolerability (Zhu et al., 2025).

Another advantage of nanocarriers is their ability to improve the stability of therapeutic agents, shielding them from early degradation and quick clearance from the blood. This helps to extend circulation time and enhance treatment efficiency (Gavas et al., 2021).

Moreover, many nanoparticle systems can also provide controlled drug release, enabling the gradual delivery of a drug or its release in response to certain stimuli of the TME, such as pH or temperature. This allows for a longer period of time that the drug levels are maintained in the therapeutic range at the target site (Manzari-Tavakoli et al., 2024).

Last but not least, nanotechnology improves the bioavailability of poorly water-soluble anticancer drugs when they are made more soluble, absorbed, and taken up by cells, which results in better therapeutic outcomes (Yao et al., 2020).

Overall, the unique properties of nanotechnology-based delivery systems are a valuable asset for creating a safer and more effective cancer treatment strategy.

## Clinically Approved Nanomedicines in Cancer Therapy

Clinical translation of nanotechnology efforts has resulted in approval of multiple formulations of nanomedicines for cancer treatment. The majority of clinically approved nanomedicines consist of chemically altered versions of already approved chemotherapeutic drugs (Table 1). The nanomedicines were primarily developed to enhance drug solubility, increase circulation time, enable better tumor accumulation, and decrease toxicity related to treatment (Tran et al., 2017; Rodríguez et al., 2022).

Some of the most commonly used approved formulations include Doxil® (Caelyx®), a pegylated liposomal formulation of doxorubicin approved for ovarian cancer, metastatic breast cancer, multiple myeloma and Kaposi sarcoma. Another albumin bound nanoparticle formulation Abraxane® is approved for use in metastatic breast cancer, non-small cell lung cancer and pancreatic cancer. Liposomal formulation of irinotecan named Onivyde® is approved for advanced pancreatic cancer treatment as well (Gavas et al., 2021; Nirmala et al., 2023).

Nanoformulations possess improved pharmacokinetics compared to conventional drugs with higher tumor accumulation and reduced systemic adverse effects. While the majority of clinically approved nanoformulations are focused on improving drug delivery rather than developing new mechanism of action specific drugs, they still enable safer and more effective therapeutic options for cancer patients (Wolfram & Ferrari, 2019).

*Table 1. Clinically approved nanomedicines for cancer therapy*

| Nanomedicine | Type | Active Drug | Approved Indications | Clinical Advantage | References |
|---|---|---|---|---|---|
| Doxil® / Caelyx® | PEGylated liposome | Doxorubicin | Ovarian cancer, metastatic breast cancer, multiple myeloma, Kaposi sarcoma | Prolonged circulation, enhanced tumor accumulation, reduced cardiotoxicity | Tran et al., 2017; Gavas et al., 2021; Rodríguez et al., 2022 |
| Abraxane® | Albumin-bound nanoparticle | Paclitaxel | Metastatic breast cancer, non-small cell lung cancer (NSCLC), pancreatic cancer | Improved solubility, solvent-free formulation, enhanced tumor penetration | Gavas et al., 2021; Rodríguez et al., 2022; Nirmala et al., 2023 |
| Onivyde® | Liposomal nanoparticle | Irinotecan | Advanced/metastatic pancreatic cancer | Sustained drug release, improved pharmacokinetics, enhanced efficacy | Gavas et al., 2021; Rodríguez et al., 2022 |
| Fyarro® | Albumin-bound nanoparticle | Sirolimus | Advanced malignant PEComa | Improved bioavailability and targeted drug delivery | Rodríguez et al., 2022 |
| Eligard® | Polymeric (PLGA depot) | Leuprolide acetate | Advanced prostate cancer | Controlled and sustained drug release | Giri et al., 2023; Venturini et al., 2025 |
| NanoTherm® | Iron oxide nanoparticle | Iron oxide | Glioblastoma (hyperthermia adjunct therapy) | Magnetic hyperthermia and image-guided treatment | Soetaert et al., 2020; Rodríguez et al., 2022 |
| Hensify® (NBTXR3) | Hafnium oxide nanoparticle | Hafnium oxide | Soft tissue sarcoma (radiotherapy enhancer) | Enhances local radiotherapy efficacy | Rodríguez et al., 2022; Nirmala et al., 2023 |

**Abbreviations:** PEG, polyethylene glycol; PLGA, poly(lactic-co-glycolic acid); NSCLC, non-small cell lung cancer; PEComa, perivascular epithelioid cell tumor.

## NANO-GENE THERAPY: BRIDGING NANOTECHNOLOGY AND GENETICS

Nano-gene therapy couple's gene therapy technology with nanotechnology to help transport nucleic acids to cells with higher specificity and safety. Nanocarriers loaded with therapeutic nucleic acids such as DNA, mRNA, siRNA, miRNA, or

CRISPR/Cas systems, have shown improved protection of the genetic cargo delivered more efficiently to tumor tissues (Wang et al., 2025; Dhuri et al., 2024).

There are two main types of gene delivery systems, viral vectors and non-viral vectors. Viral vectors have high transfection efficiency but are prone to provoke immune reactions, have limited cargo capacity, and there are concerns regarding safety for clinical translation. Non-viral vectors have low immunogenicity, ease of production, and have no limitations on the size and type of genetic cargo that they can carry. (Sung & Kim, 2019; Dogbey et al., 2023).

Nanoparticles have very important role in modern gene delivery. They can protect nucleic acids from enzyme degradation, make them stay longer in blood circulation, and help cells take up the genetic materials more easily. By modifying the surface with targeting ligands, nanoparticles can accumulate more selectively in tumor tissues. Also, stimuli-responsive nanoparticles can release their cargo when meeting special conditions in TME, such as changes in pH value or temperature (Jiang et al., 2023; Zhang et al., 2023).

Generally, nano-gene therapy combines the high specificity of genetic medicine and the delivery advantages of nanotechnology. This provides a very promising platform for safer and more effective cancer treatment.

## CRISPR/CAS9 IN CANCER THERAPY

CRISPR/Cas9 is a revolutionary genome-editing technology. Originally, it is derived from adaptive immune system of bacteria and archaea. This system consists of two main components: first is the Cas9 endonuclease, which is responsible to cut DNA; second is single-guide RNA—we call it sgRNA—which directs the Cas9 to target specific genomic sequence. After DNA cleavage happens, cell will repair this break by itself. It has two pathways: one is non-homologous end joining (NHEJ), which leading to gene disruption; another is homology-directed repair (HDR), this one can enabling very precise gene correction (Gull et al, 2025; Ding et al., 2023; Zhang et al., 2021).

On the other hand, in the field of cancer treatment, CRISPR/Cas9 has become a very powerful gene-editing tool to target such genes that are associated with the initiation or progression of cancer. An important application is taking out oncogenes, or other harmful genes which is involved in the growth of a tumor and cause cells to develop a tendency to grow and spread. Also, mutations in tumor suppressors can be corrected or replaced by CRISPR technology. This provides great opportunities for clinical personalized treatment strategies (Wang et al., 2022; Salavatipour et al., 2024).

One action point of particular importance that we should mention is the targeting of genes for drug-resistance. In effect, as pathways linked to chemotherapy resistance are disrupted, the CRISPR/Cas9 technology can enhance the tumor cells' sensitivity to traditional therapy. It is very significant in this context, as it enhances therapeutic failure in the clinic that were consistent with some study by Vaghari-Tabari et al. (2022).

Moreover, the CRISPR/Cas9 system also has high significance to cancer immunotherapy. The technology is used in the engineering of immune cells, notably, chimeric antigen receptor T cells (CAR-T cells). This can greatly enhance their ability to recognize and eliminate cancer cells. Furthermore, modifying genes encoding with immune checkpoints, such as PD-1, has demonstrated highly promising results in enhancing antitumor immunity (Chen et al., 2023; Feng et al., 2024).

However, its tremendous promise remains hampered by several difficulties such as ability to deliver into tissues effectively, the propensity for off-target effects, and potential long-term side effects. Nevertheless, we can say CRISPR/Cas9 represents one of most promising approaches for development of precise and personalized cancer therapies.

## Nanoparticle-Mediated Delivery of CRISPR/Cas9

One of the significant hurdles for using the CRISPR/Cas9 system in cancer therapy is the efficient delivery of the genome-editing tools. Nanoparticle-based systems have proven to be interesting non-viral vectors because they can shield the CRISPR components from degradation, enhance delivery to the cells, and minimize the limitation of viral-based vectors.

The most widely explored platforms are lipid nanoparticles (LNPs) that can be used to deliver Cas9 mRNA, guide RNA, or ribonucleoprotein (RNP) complexes without causing significant immunogenicity (Farsani et al., 2024; Kazemian et al., 2022). The PLGA- and PEG-based polymer nanoparticles can be loaded with drugs and functional molecules, including targeting agents, to achieve controlled release and increase the tumor-specific targeting of the systems (Kim et al. 2024; Aghamiri et al., 2020). On the other hand, gold nanoparticles (AuNPs) can be easily surface functionalized and offer high loading capacity, which makes them suitable for CRISPR delivery and theranostic applications (Duan et al., 2021; Lee et al., 2025).

Recent advances have focused on stimuli-responsive nanocarriers that selectively release CRISPR cargo under specific conditions in the TME, which enhances the accuracy of editing and minimizes systemic toxicity (Rauf et al., 2025).

## Challenges and Limitations

Although the promise of nanoparticle mediated CRISPR/Cas9 therapy is great, there are some challenges to overcome. Some potential off-target edits are still possible, but the risk of such edits can be reduced by improving the design of the guide RNAs, and by the use of high fidelity variants of Cas9 (Balasubramanian et al., 2024; Lee et al., 2025).

One of the other problems is Delivery Efficiency, as biological barriers and the low penetration of tumors hinder the delivery of the CRISPR cargo into target cells (Duan et al., 2021; Xu et al., 2021). Moreover, the immunogenicity of Cas 9 proteins and/or the nanoparticle material could impact the therapeutic safety and efficacy (Farsani et al., 2024).

Finally, there are ethical concerns, such as the long-term safety of genomic editing, the unexpected advent of new forms of genomic modification, and the possible exploitation of genome editing technologies before they are used in the clinic (Banda et al., 2024; Youssef et al., 2025).

While the delivery of nanoparticles-based CRISPR has held promise as a route to precision cancer therapy, additional development to make the system safer, more effective and more specific is required before it can be successfully translated into the clinic.

Overall, the use of nanoparticles for the delivery of CRISPR is an interesting strategy for precision therapy of cancer, and future optimization of these systems in terms of targeting and delivery efficiency, along with safety, is needed for successful clinical translation.

## BEYOND CRISPR/CAS9: NEXT-GENERATION GENOME EDITING TECHNOLOGIES

While CRISPR/Cas9 has transformed cancer gene therapy, next-generation genome editing technologies have been developed because the requirement of double-strand DNA breaks and potential for off-target mutations. These next generation platforms are designed to make more precise genetic modifications and less genomic instability. Their combination with nanotechnology-based delivery systems represents the next generation of nano-gene therapy, with greater therapeutic potential, versatility, and safety for precision oncology.

## Base Editing

Base editing is a CRISPR-derived technology that lets researchers directly swap out a single DNA base for another without triggering double-strand breaks or needing a donor template. The tool relies on cytosine base editors (CBEs) to handle C-to-T transitions and adenine base editors (ABEs) to drive A-to-G substitutions (Kantor et al., 2020).

In the cancer field, researchers leverage base editing to fix harmful point mutations, build hyper-accurate disease models, and map out the mechanics behind drug resistance. Because it avoids cutting completely through both DNA strands, the tech slashes the risk of chromosomal rearrangements—making it an incredibly promising candidate for real-world therapies (Wang et al., 2022; Chehelgerdi et al., 2024).

## Prime Editing

Prime editing is an advanced technique for genome editing that involves the use of a Cas9 nickase along with reverse transcriptase enzyme and a pegRNA (prime editing guide RNA). Unlike CRISPR/Cas9 technology, prime editing can introduce any type of point mutation as well as deletions/insertions without cutting both strands of DNA (Kantor et al., 2020; Li et al., 2021).

The prime editing technique provides more scope in terms of genetic correction compared to traditional CRISPR/Cas9 technology due to the high efficiency and less off-target effect.

## CRISPR-Cas12 and Cas13 Systems

CRISPR tools have evolved from Cas9 and now include other types of programmable nucleases with special features. While Cas12 predominantly acts on DNA, it has unique binding sites and therefore is more specific and easier to deliver as it has a relatively small size (Paul & Montoya, 2020).

Unlike Cas12, Cas13 acts on RNA instead of DNA. Therefore, it can regulate gene expression temporarily without making permanent changes to the DNA. Hence, Cas13 is extremely useful in targeting oncocentric RNAs and formulating RNA-based treatments against cancer cells (Villiger et al., 2024; Yang & Patel, 2024).

## Epigenome Editing

In epigenome editing, the Cas endonucleases are rendered catalytically inactive as dCas9 or dCas12, with the addition of transcription or epigenetic activators to influence the expression of genes without modifying their genome sequence (Cai et al., 2023). CRISPRa refers to a gene activation method, whereas CRISPR is used to inhibit gene expression through changes in chromatin or DNA methylation.

The use of epigenome editing can be particularly applicable to cancer research, which is characterized by improper epigenetic regulation associated with cancer progression and development.

### RNA-Based Therapeutics

RNA therapies are also another significant evolution of nano-gene therapy. siRNAs and miRNAs may be used to inhibit oncogenes' functions or influence several pathways involved in cancer development post-transcriptionally by gene regulation (Hussen et al., 2023).

Likewise, mRNA therapies have the potential to produce therapeutic proteins, such as tumor suppressors and immune modulators, in a transitory manner. The recent achievements in mRNA delivery via lipid nanoparticles have fueled the interest in developing RNA therapies with nanomaterials in mind.

## FUTURE PERSPECTIVES

Today the future of cancer treatment increasingly heads toward the combination of nanotechnology, genome editing and artificial intelligence (AI). Nano-gene therapy is expected to revolutionize medicine with precision oncology by its development in these fields (Figure 2).

### AI-Guided Nanoparticle Design

Machine learning and AI are gaining recognition as potential technologies for rationally designing nanocarriers. Machine learning algorithms can help in analyzing biological data sets to achieve optimal size, composition, surface charge, and targeting ligands of nanoparticles, leading to enhanced drug loading, biodistribution, and targeting of tumors. Predictive models may also be developed using AI to facilitate interactions between the nanoparticles and cells, as well as protein corona generation (Das & J, 2023; Wang et al., 2025).

### Personalized Nano-Gene Therapy

The increasing use of genomics and molecular profiling approaches heralds the dawn of personalized nanogene therapy. These nanomaterials can be designed to deliver a personalized cargo that includes siRNA, mRNA, CRISPR/Cas systems, and the latest editing platforms like base editors and prime editors. The customization of such delivery platforms depending on a patient's genome and molecular signature could enhance their effectiveness without side effects (Roma-Rodrigues et al., 2020; Wang et al., 2025).

## Combination Therapies

Nanogene therapies of the future will most probably entail multi-faceted treatment paradigms that include genome editing in combination with conventional and novel anticancer modalities. The versatility of nanoparticles makes them an ideal vehicle for delivering both conventional chemotherapy, immunotherapy, gene-editing technology, and RNA-based medicines using one and the same system (Ashrafizadeh et al., 2023; Mousavi-Kiasary et al., 2025).

## Clinical Translation Challenges

Although a considerable number of hurdles have been overcome, there remain some critical bottlenecks that prevent the clinical use of nano-gene treatments. Among these are mass production, interbatch variability, biosafety concerns in terms of long-term effects, non-specific delivery, regulatory clearance, and costs. Furthermore, the intricate nature of tumor biology and patient heterogeneity often restricts the therapeutic efficiency of target-specific nanomaterials (Gavas et al., 2021; Liu et al., 2024; Venturini et al., 2025).

The future of cancer therapies appears likely to involve the integration of AI technology, nanomedicine, and genome editing. The potential of nanoparticle design using AI, personalized genomics, and combination therapy is immense and can be used to enhance the efficacy of cancer treatments. Nevertheless, tackling the translational, manufacturing, and regulatory challenges associated with advanced nano-gene technology is paramount before such treatments become clinically viable.

*Figure 2. AI-guided workflow for precision cancer treatment*

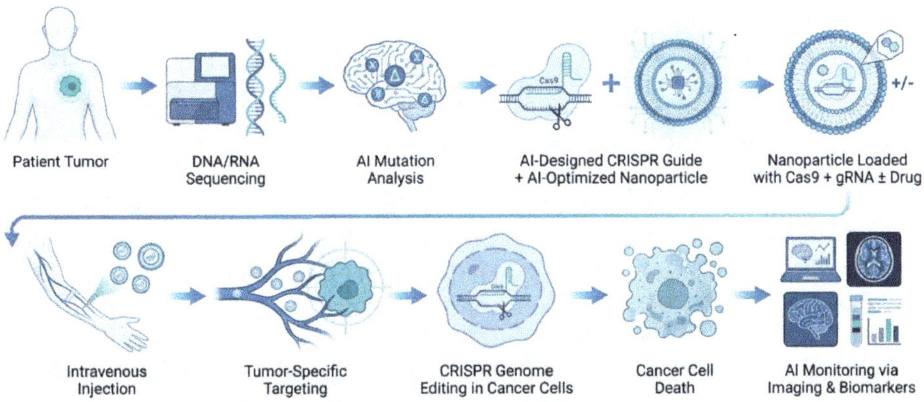


## CONCLUSION

Incorporation of nanotechnology into genome editing techniques has paved the way to new possibilities in precision oncology by addressing limitations of traditional cancer therapies. The advent of nanoparticles in the delivery of drugs and genetic cargo has facilitated better targeting while minimizing side effects.

CRISPR/Cas9 is a revolutionary genome-editing system that is utilized in manipulating cancer-related genes, eliminating drug resistance, and improving immune responses against tumors. Nonetheless, some obstacles including off-target effects, ineffective delivery mechanisms, and immune response issues require more effort. Nanoparticles can serve as a solution in addressing most of the above-mentioned problems via employing non-viral delivery mechanisms.

Furthermore, the implementation of techniques, such as base editing, prime editing, CRISPR Cas12/Cas13 technologies, epigenomic editing, and RNA-mediated approaches, opens the door to greater opportunities in the field of precision medicine. With regard to nanotechnology, these techniques and approaches will revolutionize cancer therapy in the future due to intelligent design of nanoparticles and personalized treatment. Even though some improvements should be done in order to make these innovations more efficient and safer, their combination might be considered an important milestone in the development of the future of cancer therapy.

## REFERENCES


Aghamiri, S., Talaei, S., Ghavidel, A., Zandsalimi, F., Masoumi, S., Hafshejani, N. H., & Jajarmi, V. (2020). Nanoparticles-mediated CRISPR/Cas9 delivery: Recent advances in cancer treatment. *Journal of Drug Delivery Science and Technology, 56*, 101533. DOI: 10.1016/j.jddst.2020.101533

Ajith, S., Almomani, F., Elhissi, A., & Husseini, G. A. (2023). Nanoparticle-based materials in anticancer drug delivery: Current and future prospects. *Heliyon, 9*(11), e21227. Advance online publication. DOI: 10.1016/j.heliyon.2023.e21227 PMID: 37954330

Ambekar, R. S., Choudhary, M., & Kandasubramanian, B. (2020). Recent advances in dendrimer-based nanoplatform for cancer treatment: A review. *European Polymer Journal, 126*, 109546. DOI: 10.1016/j.eurpolymj.2020.109546

Anzalone, A., Koblan, L., & Liu, D. (2020). Genome editing with CRISPR–Cas nucleases, base editors, transposases and prime editors. *Nature Biotechnology, 38*(7), 824–844. DOI: 10.1038/s41587-020-0561-9 PMID: 32572269

Ashrafizadeh, M., Zarrabi, A., Bigham, A., Taheriazam, A., Saghari, Y., Mirzaei, S., Hashemi, M., Hushmandi, K., Karimi-Maleh, H., Zare, E. N., Sharifi, E., Ertas, Y. N., Rabiee, N., Sethi, G., & Shen, M. (2023). (Nano)platforms in breast cancer therapy: Drug/gene delivery, advanced nanocarriers and immunotherapy. *Medicinal Research Reviews, 43*(6), 2115–2176. DOI: 10.1002/med.21971 PMID: 37165896

Balasubramanian, A., Veluswami, K., Rao, S., Aggarwal, S., & Mani, S. (2024). Exploring Clustered Regularly Interspaced Short Palindromic Repeats-CRISPR-Associated Protein 9 (CRISPR-Cas9) as a therapeutic modality for cancer: A scoping review. *Cureus, 16*. Advance online publication. DOI: 10.7759/cureus.64324 PMID: 39130943

Banda, A., Impomeni, O., Singh, A., Baloch, A. R., Hu, W., & Jaijyan, D. (2024). Precision in action: The role of Clustered Regularly Interspaced Short Palindromic Repeats/Cas in gene therapies. *Vaccines, 12*(6), 636. DOI: 10.3390/vaccines12060636 PMID: 38932365

Bray, F., Jemal, A., Grey, N., Ferlay, J., & Forman, D. (2012). Global cancer transitions according to the Human Development Index (2008–2030): A population-based study. *The Lancet. Oncology, 13*(8), 790–801. DOI: 10.1016/S1470-2045(12)70211-5 PMID: 22658655

Cai, R., Lv, R., Shi, X., Yang, G., & Jin, J. (2023). CRISPR/dCas9 tools: Epigenetic mechanism and application in gene transcriptional regulation. *International Journal of Molecular Sciences*, *24*(19), 14865. DOI: 10.3390/ijms241914865 PMID: 37834313

Chehelgerdi, M., Chehelgerdi, M., Khorramian-Ghahfarokhi, M., Shafieizadeh, M., Mahmoudi, E., Eskandari, F., Rashidi, M., Arshi, A., & Mokhtari-Farsani, A. (2024). Comprehensive review of CRISPR-based gene editing: Mechanisms, challenges, and applications in cancer therapy. *Molecular Cancer*, *23*(1), 25. DOI: 10.1186/s12943-023-01925-5 PMID: 38195537

Chen, C., Wang, Z.-K., & Qin, Y. (2023). CRISPR/Cas9 system: Recent applications in immuno-oncology and cancer immunotherapy. *Experimental Hematology & Oncology*, *12*(1), 76. DOI: 10.1186/s40164-023-00457-4 PMID: 37964355

Das, K., & J, C. (2023). Nanoparticles and convergence of artificial intelligence for targeted drug delivery for cancer therapy: Current progress and challenges. *Frontiers in Medical Technology*, *4*, 1067144. Advance online publication. DOI: 10.3389/fmedt.2022.1067144 PMID: 36688144

Deng, H., Huang, W., & Zhang, Z. (2019). Nanotechnology based CRISPR/Cas9 system delivery for genome editing: Progress and prospect. *Nano Research*, *12*(10), 2437–2450. DOI: 10.1007/s12274-019-2465-x

Dhuri, A., Kanp, T., Rode, K., Gupta, U., Guru, S. K., & Singh, P. K. (2024). Current advances in non-viral nanoparticle-based gene therapy for effective management of cancer. *Journal of Drug Delivery Science and Technology*, *100*, 106083. DOI: 10.1016/j.jddst.2024.106083

Ding, S., Liu, J., Han, X., & Tang, M. (2023). CRISPR/Cas9-mediated genome editing in cancer therapy. *International Journal of Molecular Sciences*, *24*(22), 16325. DOI: 10.3390/ijms242216325 PMID: 38003514

Dogbey, D. M., Torres, V. E. S., Fajemisin, E., Mpondo, L., Ngwenya, T., Akinrinmade, O. A., Perriman, A. W., & Barth, S. (2023). Technological advances in the use of viral and non-viral vectors for delivering genetic and non-genetic cargos for cancer therapy. *Drug Delivery and Translational Research*, *13*(11), 2719–2738. DOI: 10.1007/s13346-023-01362-3 PMID: 37301780

Duan, L., Kan, O., Xu, X., Xu, L., Wen, C., Zhou, X., Qin, Z., Xu, Z., Sun, W., & Liang, Y. (2021). Nanoparticle delivery of CRISPR/Cas9 for genome editing. *Frontiers in Genetics*, *12*, 673286. DOI: 10.3389/fgene.2021.673286 PMID: 34054927

Farsani, A. M., Mokhtari, N., Nooraei, S., Bahrulolum, H., Akbari, A., Farsani, Z. M., Khatami, S., Ebadi, M. S., & Ahmadian, G. (2024). Lipid nanoparticles: The game-changer in CRISPR-Cas9 genome editing. *Heliyon, 10*(2), e24606. DOI: 10.1016/j.heliyon.2024.e24606 PMID: 38288017

Feng, X., Li, Z., Liu, Y., Chen, D., & Zhou, Z. (2024). CRISPR/Cas9 technology for advancements in cancer immunotherapy: From uncovering regulatory mechanisms to therapeutic applications. *Experimental Hematology & Oncology, 13*(1), 58. DOI: 10.1186/s40164-024-00570-y PMID: 39427211

Gavas, S., Quazi, S., & Karpiński, T. (2021). Nanoparticles for Cancer Therapy: Current Progress and Challenges. *Nanoscale Research Letters, 16*(1), 173. Advance online publication. DOI: 10.1186/s11671-021-03628-6 PMID: 34866166

Giri, P., Banerjee, A., & Layek, B. (2023). A recent review on cancer nanomedicine. *Cancers (Basel), 15*(8), 2256. DOI: 10.3390/cancers15082256 PMID: 37190185

Gull, S. A., Khan[1], M. A., Hashmi, S., & Noor, T. (2025). A COMPREHENSIVE REVIEW OF TREATMENT APPROACHES FOR CONGENITAL AND HEREDITARY ANOMALIES. *survival, 3*(2). DOI: 10.5281/zenodo.15286622

Hussen, B., Rasul, M. F., Abdullah, S. R., Hidayat, H. J., Faraj, G. S. H., Ali, F. A., Salihi, A., Baniahmad, A., Ghafouri-Fard, S., Rahman, M., Glassy, M. C., Branicki, W., & Taheri, M. (2023). Targeting miRNA by CRISPR/Cas in cancer: Advantages and challenges. *Military Medical Research, 10*(1), 47. DOI: 10.1186/s40779-023-00468-6 PMID: 37460924

Janik, E., Niemcewicz, M., Ceremuga, M., Krzowski, L., Saluk-Bijak, J., & Bijak, M. (2020). Various aspects of a gene editing system—CRISPR–Cas9. *International Journal of Molecular Sciences, 21*(24), 9604. DOI: 10.3390/ijms21249604 PMID: 33339441

Jiang, M.. (2025). Nanomaterials in gene therapy and genome editing: Challenges and emerging directions. *Journal of Nanobiotechnology*, •••, 24. DOI: 10.1186/s12951-025-03124-4 PMID: 41402770

Jiang, T., Gonzalez, K. M., Cordova, L., & Lu, J. (2023). Nanotechnology-enabled gene delivery for cancer and other genetic diseases. *Expert Opinion on Drug Delivery, 20*(4), 523–540. DOI: 10.1080/17425247.2023.2200246 PMID: 37017558

Kantor, A., McClements, M. E., & MacLaren, R. E. (2020). CRISPR-Cas9 DNA base-editing and prime-editing. *International Journal of Molecular Sciences, 21*(17), 6240. DOI: 10.3390/ijms21176240 PMID: 32872311

Kazemian, P., Yu, S.-Y., Thomson, S. B., Birkenshaw, A., Leavitt, B., & Ross, C. J. D. (2022). Lipid-nanoparticle-based delivery of CRISPR/Cas9 genome-editing components. *Molecular Pharmaceutics*, *19*(5), 1669–1686. DOI: 10.1021/acs.molpharmaceut.1c00916 PMID: 35594500

Kim, M., Hwang, Y., Lim, S., Jang, H.-K., & Kim, H.-O. (2024). Advances in nanoparticles as non-viral vectors for efficient delivery of CRISPR/Cas9. *Pharmaceutics*, *16*(9), 1197. DOI: 10.3390/pharmaceutics16091197 PMID: 39339233

Lee, H., Rho, W., Kim, Y.-H., Chang, H., & Jun, B.-H. (2025). CRISPR-Cas9 gene therapy: Non-viral delivery and stimuli-responsive nanoformulations. *Molecules (Basel, Switzerland)*, *30*(3), 542. DOI: 10.3390/molecules30030542 PMID: 39942646

Li, C., Brant, E. J., Budak, H., & Zhang, B. (2021). CRISPR/Cas: A Nobel Prize award-winning precise genome editing technology for gene therapy and crop improvement. *Journal of Zhejiang University. Science. B*, *22*(3), 253–284. DOI: 10.1631/jzus.B2100009 PMID: 33835761

Li, C., Yang, T., Weng, Y., Zhang, M., Zhao, D., Guo, S., Hu, B., Shao, W., Wang, X., Hussain, A., Liang, X., & Huang, Y. (2021). Ionizable lipid-assisted efficient hepatic delivery of gene editing elements for oncotherapy. *Bioactive Materials*, *9*, 590–601. DOI: 10.1016/j.bioactmat.2021.05.051 PMID: 34853819

Li, T., Yang, Y., Qi, H., Cui, W., Zhang, L., Fu, X., He, X., Liu, M., Li, P., & Yu, T. (2023). CRISPR/Cas9 therapeutics: Progress and prospects. *Signal Transduction and Targeted Therapy*, *8*(1), 36. DOI: 10.1038/s41392-023-01309-7 PMID: 36646687

Li, Y. R., Halladay, T., & Yang, L. (2024). Immune evasion in cell-based immunotherapy: Unraveling challenges and novel strategies. *Journal of Biomedical Science*, *31*(1), 5. DOI: 10.1186/s12929-024-00998-8 PMID: 38217016

Luiz, M. T., Dutra, J. A. P., Viegas, J. S. R., de Araújo, J. T. C., Tavares, A. G.Junior, & Chorilli, M. (2023). Hybrid magnetic lipid-based nanoparticles for cancer therapy. *Pharmaceutics*, *15*(3), 751. DOI: 10.3390/pharmaceutics15030751 PMID: 36986612

Manzari-Tavakoli, A., Babajani, A., Tavakoli, M. M., Safaeinejad, F., & Jafari, A. (2024). Integrating natural compounds and nanoparticle-based drug delivery systems: A novel strategy for enhanced efficacy and selectivity in cancer therapy. *Cancer Medicine*, *13*(5), e7010. DOI: 10.1002/cam4.7010 PMID: 38491817

Molla, K. A., & Yang, Y. (2019). CRISPR/Cas-mediated base editing: Technical considerations and practical applications. *Trends in Biotechnology*, *37*(10), 1121–1142. DOI: 10.1016/j.tibtech.2019.03.008 PMID: 30995964

Mousavi-Kiasary, S. S., Senabreh, A., Zandi, A., Pena, R., Cruz, F., Adibi, A., & Hooshmand, N. (2025). Synergistic cancer therapies enhanced by nanoparticles: Advancing nanomedicine through multimodal strategies. *Pharmaceutics*, *17*(6), 682. DOI: 10.3390/pharmaceutics17060682 PMID: 40573996

Nirmala, M., Kizhuveetil, U., Johnson, A., G, B., Nagarajan, R., & Muthuvijayan, V. (2023). Cancer nanomedicine: A review of nano-therapeutics and challenges ahead. *RSC Advances*, *13*(13), 8606–8629. DOI: 10.1039/D2RA07863E PMID: 36926304

Paul, B., & Montoya, G. (2020). CRISPR-Cas12a: Functional overview and applications. *Biomedical Journal*, *43*(1), 8–17. DOI: 10.1016/j.bj.2019.10.005 PMID: 32200959

Pickar-Oliver, A., & Gersbach, C. A. (2019). The next generation of CRISPR–Cas technologies and applications. *Nature Reviews. Molecular Cell Biology*, *20*(8), 490–507. DOI: 10.1038/s41580-019-0131-5 PMID: 31147612

Raj, S., Khurana, S., Choudhari, R., Kesari, K. K., Kamal, M., Garg, N., Ruokolainen, J., Das, B., & Kumar, D. (2019). Specific targeting cancer cells with nanoparticles and drug delivery in cancer therapy. *Seminars in Cancer Biology*. Advance online publication. DOI: 10.1016/j.semcancer.2019.11.002 PMID: 31715247

Rauf, M. A., Rao, A., Sivasoorian, S. S., & Iyer, A. K. (2025). Nanotechnology-based delivery of CRISPR/Cas9 for cancer treatment: A comprehensive review. *Cells*, *14*(15), 1136. DOI: 10.3390/cells14151136 PMID: 40801569

Rodriguez, F., Caruana, P., De la Fuente, N., Espanol, P., Gamez, M., Balart, J., Llurba, E., Rovira, R., Ruiz, R., Martín-Lorente, C., Corchero, J. L., & Céspedes, M. V. (2022). Nano-based approved pharmaceuticals for cancer treatment: Present and future challenges. *Biomolecules*, *12*(6), 784. DOI: 10.3390/biom12060784 PMID: 35740909

Rodríguez, F., Caruana, P., De La Fuente, N., Español, P., Gámez, M., Balart, J., Llurba, E., Rovira, R., Ruiz, R., Martín-Lorente, C., Corchero, J. L., & Céspedes, M. V. (2022). Nano-based approved pharmaceuticals for cancer treatment: Present and future challenges. *Biomolecules*, *12*(6), 784. DOI: 10.3390/biom12060784 PMID: 35740909

Roma-Rodrigues, C., Rivas-García, L., Baptista, P. V., & Fernandes, A. R. (2020). Gene therapy in cancer treatment: Why go nano? *Pharmaceutics*, *12*(3), 233. DOI: 10.3390/pharmaceutics12030233 PMID: 32151052

Sabit, H., Pawlik, T. M., Radwan, F., Abdel-Hakeem, M., Abdel-Ghany, S., Wadan, A. H. S., Elzawahri, M., El-Hashash, A., & Arneth, B. (2025). Precision nanomedicine: Navigating the tumor microenvironment for enhanced cancer immunotherapy and targeted drug delivery. *Molecular Cancer*, *24*(1), 160. DOI: 10.1186/s12943-025-02357-z PMID: 40457437

Salavatipour, M. S., Poursalehi, Z., Rouzbahani, N. H., Mohammadyar, S., & Vasei, M. (2024). CRISPR-Cas9 in basic and translational aspects of cancer therapy. *BioImpacts*, *14*(2), 237–252. DOI: 10.34172/bi.2024.30087 PMID: 39493894

Sell, M., Lopes, A. R., Escudeiro, M., Esteves, B., Monteiro, A. R., Trindade, T., & Cruz-Lopes, L. (2023). Application of Nanoparticles in Cancer Treatment: A Concise Review. *Nanomaterials (Basel, Switzerland)*, *13*(21), 2887. Advance online publication. DOI: 10.3390/nano13212887 PMID: 37947732

Soetaert, F., Korangath, P., Serantes, D., Fiering, S., & Ivkov, R. (2020). Cancer therapy with iron oxide nanoparticles: Agents of thermal and immune therapies. *Advanced Drug Delivery Reviews*, *163–164*, 65–83. DOI: 10.1016/j.addr.2020.06.025 PMID: 32603814

Sun, L., Liu, H., Ye, Y., Lei, Y., Islam, R., Tan, S., Tong, R., Miao, Y., & Cai, L. (2023). Smart nanoparticles for cancer therapy. *Signal Transduction and Targeted Therapy*, *8*(1), 418. Advance online publication. DOI: 10.1038/s41392-023-01642-x PMID: 37919282

Sung, H., Ferlay, J., Siegel, R. L., Laversanne, M., Soerjomataram, I., Jemal, A., & Bray, F. (2021). Global cancer statistics 2020: GLOBOCAN estimates of incidence and mortality worldwide for 36 cancers in 185 countries. *CA: a Cancer Journal for Clinicians*, *71*(3), 209–249. DOI: 10.3322/caac.21660 PMID: 33538338

Sung, Y. K., & Kim, S. W. (2019). Recent advances in the development of gene delivery systems. *Biomaterials Research*, *23*(1), 8. DOI: 10.1186/s40824-019-0156-z PMID: 30915230

Tian, H., Zhang, T., Qin, S., Huang, Z., Zhou, L., Shi, J., Nice, E. C., Xie, N., Huang, C., & Shen, Z. (2022). Enhancing the therapeutic efficacy of nanoparticles for cancer treatment using versatile targeted strategies. *Journal of Hematology & Oncology*, *15*(1), 132. DOI: 10.1186/s13045-022-01320-5 PMID: 36096856

Tran, S., DeGiovanni, P. J., Piel, B., & Rai, P. (2017). Cancer nanomedicine: A review of recent success in drug delivery. *Clinical and Translational Medicine*, *6*(1), 44. DOI: 10.1186/s40169-017-0175-0 PMID: 29230567

Vaghari-Tabari, M., Hassanpour, P., Sadeghsoltani, F., Malakoti, F., Alemi, F., Qujeq, D., Asemi, Z., & Yousefi, B. (2022). CRISPR/Cas9 gene editing: A new approach for overcoming drug resistance in cancer. *Cellular & Molecular Biology Letters, 27*(1), 16. DOI: 10.1186/s11658-022-00348-2 PMID: 35715750

Venturini, J., Chakraborty, A., Baysal, M. A., & Tsimberidou, A. M. (2025). Developments in nanotechnology approaches for the treatment of solid tumors. *Experimental Hematology & Oncology, 14*(1), 56. DOI: 10.1186/s40164-025-00656-1 PMID: 40390104

Villiger, L., Joung, J., Koblan, L., Weissman, J. S., Abudayyeh, O. O., & Gootenberg, J. S. (2024). CRISPR technologies for genome, epigenome and transcriptome editing. *Nature Reviews. Molecular Cell Biology, 25*(6), 464–487. DOI: 10.1038/s41580-023-00697-6 PMID: 38308006

Waheed, I., Ali, A., Tabassum, H., Khatoon, N., Lai, W. F., & Zhou, X. (2024). Lipid-based nanoparticles as drug delivery carriers for cancer therapy. *Frontiers in Oncology, 14*, 1296091. DOI: 10.3389/fonc.2024.1296091 PMID: 38660132

Wang, B., Hu, S., Teng, Y., Chen, J., Wang, H., Xu, Y., Wang, K., Xu, J., Cheng, Y., & Gao, X. (2024). Current advance of nanotechnology in diagnosis and treatment for malignant tumors. *Signal Transduction and Targeted Therapy, 9*(1), 200. Advance online publication. DOI: 10.1038/s41392-024-01889-y PMID: 39128942

Wang, M., Liu, H., Huang, J., Cai, T., Xu, Z. P., & Zhang, L. (2025). Advancing cancer gene therapy: The emerging role of nanoparticle delivery systems. *Journal of Nanobiotechnology, 23*(1), 362. DOI: 10.1186/s12951-025-03433-8 PMID: 40394591

Wang, S.-W., Gao, C., Zheng, Y.-M., Yi, L., Lu, J.-C., Huang, X.-Y., Cai, J.-B., Zhang, P.-F., Cui, Y.-H., & Ke, A.-W. (2022). Current applications and future perspective of CRISPR/Cas9 gene editing in cancer. *Molecular Cancer, 21*(1), 57. DOI: 10.1186/s12943-022-01518-8 PMID: 35189910

Wolfram, J., & Ferrari, M. (2019). Clinical cancer nanomedicine. *Nano Today, 25*, 85–98. DOI: 10.1016/j.nantod.2019.02.005 PMID: 31360214

Xiao, X., Teng, F., Shi, C., Chen, J., Wu, S., Wang, B., Meng, X., Essiet Imeh, A., & Li, W. (2022). Polymeric nanoparticles—Promising carriers for cancer therapy. *Frontiers in Bioengineering and Biotechnology, 10*, 1024143. DOI: 10.3389/fbioe.2022.1024143 PMID: 36277396

Xu, X., Liu, C., Wang, Y., Koivisto, O., Zhou, J.-N., Shu, Y., & Zhang, H. (2021). Nanotechnology-based delivery of CRISPR/Cas9 for cancer treatment. *Advanced Drug Delivery Reviews*, *168*, 113891. DOI: 10.1016/j.addr.2021.113891 PMID: 34324887

Yang, H., & Patel, D. J. (2024). Structures, mechanisms and applications of RNA-centric CRISPR-Cas13. *Nature Chemical Biology*, *20*(6), 673–688. DOI: 10.1038/s41589-024-01593-6 PMID: 38702571

Yao, Y., Zhou, Y., Liu, L., Xu, Y., Chen, Q., Wang, Y., Wu, S., Deng, Y., Zhang, J., & Shao, A. (2020). Nanoparticle-based drug delivery in cancer therapy and its role in overcoming drug resistance. *Frontiers in Molecular Biosciences*, *7*, 193. DOI: 10.3389/fmolb.2020.00193 PMID: 32974385

Youssef, E., Fletcher, B., & Palmer, D. (2025). Enhancing precision in cancer treatment: The role of gene therapy and immune modulation in oncology. *Frontiers in Medicine*, *11*, 1527600. DOI: 10.3389/fmed.2024.1527600 PMID: 39871848

Zhang, H., Qin, C., An, C., Zheng, X., Wen, S., Chen, W., Liu, X., Lv, Z., Yang, P., Xu, W., Gao, W., & Wu, Y. (2021). Application of the CRISPR/Cas9-based gene editing technique in basic research, diagnosis, and therapy of cancer. *Molecular Cancer*, *20*(1), 126. DOI: 10.1186/s12943-021-01431-6 PMID: 34598686

Zhang, Q., Kuang, G., Li, W., Wang, J., Ren, H., & Zhao, Y. (2023). Stimuli-responsive gene delivery nanocarriers for cancer therapy. *Nano-Micro Letters*, *15*(1), 44. DOI: 10.1007/s40820-023-01018-4 PMID: 36752939

Zhu, J., Lee, H., Huang, R., Zhou, J., Zhang, J., Yang, X., Zhou, W., Jiang, W., & Chen, S. (2025). Harnessing nanotechnology for cancer treatment. *Frontiers in Bioengineering and Biotechnology*, *12*, 1514890. Advance online publication. DOI: 10.3389/fbioe.2024.1514890 PMID: 39902172

Chapter 9

# AI-Guided Omics and Nano-CRISPR Platforms for Overcoming Therapeutic Resistance in Cancer

**Maryam Nawaz**
*Department of Pathobiology, Bahuddin Zikarya University, Pakistan*

**Muhammad Saad Naeem**
*University of Agriculture, Faisalabad, Pakistan*

**Muhammad Zulqarnain**
*Department of Zoolgy, The Islamia University, Bahwalpur, Pakistan*

**Amina Mehak Hasnat**
*Superior University, Faisalabad, Pakistan*

**Daneyal Rafique**
http://orcid.org/0009-0001-4426-6578
*NUMS Department: Internal Medicne, Army Medical College, Pakistan*

**Izza Noor**
*University of Agriculture, Faisalabad, Pakistan*

**Manahil Ramzan**
*University of Agriculture, Faisalabad, Pakistan*

**Uswa Muhammad Akram**
*University of Agriculture, Faisalabad, Pakistan*

**Adnan Afzal**
http://orcid.org/0009-0001-4359-5695
*Government College University, Faisalabad, Pakistan*

**Abu Sulman**
http://orcid.org/0009-0008-7194-5219
*University of Agriculture, Faisalabad, Pakistan*

DOI: 10.4018/979-8-3373-9848-8.ch009

**Muhammad Sagheer**
*University of Agriculture, Faisalabad, Pakistan*

**Iqra Rafique**
*Government College University, Faisalabad, Pakistan*

## ABSTRACT

*Therapeutic resistance remains a major challenge in effective cancer management, leading to treatment failure and disease progression. This study highlights the integration of artificial intelligence (AI), multi-omics technologies, and Nano-CRISPR platforms as a novel strategy to overcome resistance mechanisms. AI-driven analytics enable the identification of key molecular drivers, predictive biomarkers, and resistance pathways from complex biological data. Concurrently, Nano-CRISPR systems facilitate precise gene editing and targeted delivery to tumour cells, enhancing therapeutic efficacy. The convergence of these technologies supports personalized and adaptive treatment approaches. Despite challenges such as off-target effects and ethical concerns, this integrated framework offers significant potential for advancing precision oncology.*

## INTRODUCTION

The problem of therapeutic resistance is one of the main obstacles to the contemporary cancer treatment since it compromises the effectiveness of both conventional and targeted cancer treatment methods and leads to the recurrence and mortality rates among the diseases. The ability of cancer cells to survive therapeutic stress is incredibly high, in part because of their genetic instability and adaptive evolution. In general, resistance can be divided into intrinsic and acquired types, which are both motivated by complex and in many cases overlapping biological processes (Russo et al., 2021). Intrinsic resistance Intrinsic resistance is an inherent property of tumour cells that make them intrinsically resistant to treatment. These can be genetic mutations, constitutive activation of survival signals, heterogeneity of the tumour, and the presence of cancer stem-like cells. These cells tend to be dormant, highly versatile, and can circumvent the impacts of cytotoxicity, thus a reservoir of relapse (Kim, 2026).

Acquired resistance, on the other hand, occurs with time as a reaction to prolonged exposure to therapeutic agents. This is a dynamic process that entails selective pressure that promotes survival of resistant clones, resulting in tumour evolution. Multidrug resistance (MDR), where cancer cells become resistant to a broad spectrum of structurally unrelated drugs, is one of the best-documented types of acquired resistance (Alhazza et al., 2025). The overexpression of ATP-binding cassette (ABC) transporters, including P-glycoprotein, which actively extracellularly release drugs, often mediates this effect, decreasing intracellular drug levels and efficacy. Epithelial-mesenchymal transition (EMT) is another important mechanism which facilitates the acquisition of mesenchymal properties by epithelial cancer cells, increasing their migratory ability, invasiveness, and resistance to apoptosis. Tumour progression, metastasis and unfavourable clinical outcomes are closely related with EMT (Manfioletti & Fedele, 2023).

Moreover, changes in the DNA damage response and repair pathways are important in resistance, specifically to therapies based on catalysis of genomic instability, including chemotherapy and radiotherapy. The DNA repair mechanisms such as non-homologous end joining and homologous recombination can be up regulated by cancer cells enabling them to escape the fatal consequences of DNA damage. This bypass of DNA repair allows the tumour cells to sustain genomic integrity amidst sustained therapeutic attack. Moreover, the epigenetic alterations, restructuring of metabolic functions, and tumour microenvironment interactions are further contributors to resistance, building a highly multifaceted and highly adaptive system that is challenging to attack by conventional means. These complications underscore the necessity of new approaches that have the potential to mitigate resistance on more than one biological level at the same time (Jurkovicova et al., 2022; Kang, 2023).

Over the past few years, the interplay between high-tech technology, like Artificial Intelligence (AI), multi-omics, and Nano-CRISPR, has created new opportunities in determining and overcoming therapeutic resistance. Machine learning and deep learning algorithms, which are part of AI, have transformed how it is possible to process large and complex biomedical data. It allows the detection of concealed trends, forecasting drug reactions, and finding new treatment targets. Predictive models can be used to predict the dynamic behaviour of tumours and how its response to treatment varies by integrating various types of data, which AI is capable of producing. This is especially useful in the case of resistance, where there are several interacting pathways (Khalighi et al., 2024; Kharl, 2026).

Multi-omics technologies can be used to complement AI to give a comprehensive understanding of the molecular picture of cancer. Such methods include genomics, transcriptomics, proteomics, metabolomics, and epigenomics, each providing a distinct layer of data. Combined, multi-omics data provide a comprehensive picture of tumour biology, showing the relationship between genetic changes and functional changes on a cellular and systemic scale. This systems-level knowledge is critical in determining the important causes of resistance and discovering weaknesses that may exist. To give an example, combining genomic and proteomic data may be used to identify mutations that do not just exist on the DNA level but are also active, and thus are more feasible as therapeutic targets (Kant et al., 2025).

Nano-CRISPR technology is one of the most remarkable innovations in the field of precision medicine as the technology integrates the precision editing of the CRISPR systems with the delivery benefits of nanotechnology. Off-target effects and limited delivery to particular tissues are some of the challenges of traditional CRISPR strategies. Lipid nanoparticles, polymeric nanoparticles, and exosomes are examples of nanocarriers that can be developed to transport the CRISPR components to the tumour cells with high specificity. This allows the targeted editing of genes in resistance pathways, including those governing drug efflux, apoptosis or DNA repair. Nano-CRISPR with AI-controlled target identification and confirmed by multi-omics can be utilized to selectively disrupt the resistance mechanisms, thus, recovering treatment sensitivity (Sahel et al., 2023).

Combined with these technologies, the framework becomes strong and synergistic in its efforts to fight therapeutic resistance. The AI is the workhorse of analytical capabilities, as it analyses large amounts of data and produces predictive value. Multi-omics gives the biological context; these findings are based on a detailed comprehension of tumour biology. Nano-CRISPR is the (execution) instrument, which allows accurate and targeted interference. Collectively, they facilitate transition between empirical and generalized treatment strategies to rational and individualized ones. The paradigm shift shifts cancer treatment beyond the conventional one-size-fits-all paradigm to precision-engineered treatments specific to the individual molecular profile of the tumour of each patient (Mao et al., 2025; Riaz et al., 2026).

This combined strategy can also be used to create combination therapies that can be applied to different resistance mechanisms at the same time. The following are examples: AI can find synergistic combinations of drugs, multi-omics can support their mechanisms, and Nano-CRISPR can be utilized to improve their activity by changing the key genes. This is not only enhancing treatment results but also minimizes the chances of developing resistance. Moreover, this framework has major implications on the drug discovery and development, allowing the identification of new targets much quicker and conducting the validation process more efficiently (da Silva Correia et al., 2025).

The adoption of this integrated approach has some challenges to be overcome in order to achieve its promise; these issues include the standardization of data, complexity of computation and ethical issues. Nevertheless, with the ongoing improvement in technology and the interdisciplinary collaboration, these limitations will be resolved, and it is likely to become widely used in clinical practice (Harle, 2025).

"Although previous reviews have separately discussed artificial intelligence, multi-omics technologies, CRISPR gene editing, and nanomedicine in cancer research, limited attention has been given to their integration as a unified strategy for overcoming therapeutic resistance. This chapter addresses this gap by presenting a comprehensive framework that combines AI-driven data analytics, multi-omics profiling, and Nano-CRISPR-based therapeutic interventions. The chapter highlights how these technologies can collectively identify resistance mechanisms, predict treatment responses, and enable targeted genome editing for precision oncology."

*Table 1. Overview of therapeutic resistance mechanisms and emerging technologies for precision oncology*

| Technology/Mechanism | Function in Resistance or Therapy | Key Features | Challenges | References |
|---|---|---|---|---|
| Intrinsic Resistance | Pre-existing insensitivity to therapy | Tumour heterogeneity, stem-like cells | Difficult early detection | (Lytle et al., 2018) |
| Acquired Resistance | Develops after treatment exposure | Genetic mutations, adaptive evolution | Leads to relapse | (La Rosa et al., 2022) |
| MDR | Drug efflux and cross-resistance | ABC transporters, reduced drug accumulation | Limits multiple therapies | (Hughes & Andersson, 2015) |
| EMT | Enhances invasion and survival | Cellular plasticity, metastasis | Associated with poor prognosis | (Chaffer et al., 2016) |
| DNA Repair Bypass | Repairs therapy-induced damage | Enhanced repair pathways | Reduces therapy effectiveness | (Jurkovicova et al., 2022) |
| Artificial Intelligence | Predicts targets and responses | Data integration, modelling | Requires large datasets | (Aldoseri et al., 2023) |
| Multi-omics | Comprehensive molecular profiling | Systems-level understanding | Complex analysis | (Jung et al., 2020) |
| Nano-CRISPR | Precise gene editing | Targeted delivery, high specificity | Off-target risks | (Baig & Chien, 2025) |
| Integrated Framework | Combines all approaches | Personalized, multi-target therapy | High cost and complexity | (Makhoba et al., 2020) |

## MAPPING THE RESISTANCE LANDSCAPE: THE ROLE OF MULTI-OMICS

Therapeutic resistance in cancer needs to be explored in a multidimensional and thorough manner, involving the biology of the tumour, and this is now becoming achievable due to the emergence of multi-omics technologies. The presence of complex molecular changes, which occur across the genome, transcriptome, proteome, and metabolome, defines resistant tumours, each of which plays a different role in the formation and maintenance of resistance. Genomics can be used to obtain high-dimensional data that can be used to identify mutations, copy number variations, and structural changes that can confer resistance to targeted therapies (Figure 1). An example of this is that the treatment can become ineffective due to secondary mutation of drug targets or activation of different signalling pathways. Nevertheless, not all genetic changes are functional and, therefore, genomic data cannot be used to fully account for resistance. Transcriptomics helps to overcome this limitation by offering information about the dynamic processes of gene expression and illuminating the mechanisms of transcriptional adaptation of cancer cells to therapeutic pressure. Transcriptional programs can be altered in response to the activation of survival programs, stress programs, and phenotypic plasticity, which can all be considered characteristics of resistant cells (Chakraborty et al., 2018; Nevedomskaya & Haendler, 2022; Pettini et al., 2021).

Proteomics extends this knowledge by determining the functional state of the cell since proteins are the immediate actioners of the biological processes. Proteomic analyses can identify post-translational modifications, protein interactions, and signalling cascades that can indicate active resistance mechanisms that are not apparent at the genomic or transcriptomic level. As an illustration phosphorylation-mediated activation of kinases is capable of maintaining proliferative signalling despite the presence of inhibitors (Monti et al., 2019). In addition to these layers, metabolomics gives a picture of cellular metabolism, which in resistant tumours is frequently reprogrammed to enable survival in stress conditions. Changes in metabolic pathways like accelerated glycolysis or lipid metabolism allow cancer cells to cope with nutrient deprivation and therapy-related insults. The combination of these omics layers provides a system and holistic view of the resistance, enabling researchers to pinpoint important regulatory nodes and interactions that promote tumour adaptation (Chen et al., 2025).

*Figure 1. Mapping the resistance landscape: The role of multi-omics*

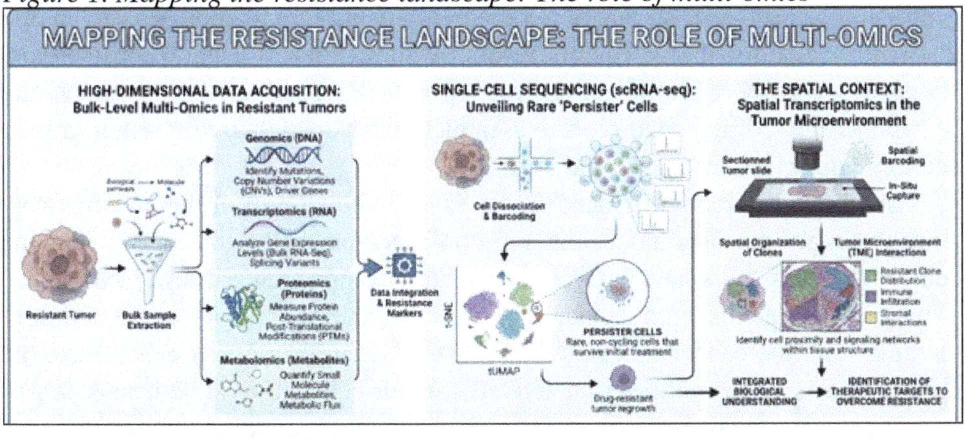


The most significant innovation in this sphere is single-cell RNA sequencing (scRNA-seq), which has transformed the possibility to decompose the heterogeneity of tumours on an unprecedented level. Conventional bulk sequencing methods mean signals of millions of cells in average, which may obscure the existence of rare but clinically significant subpopulations. By contrast, scRNA-seq allows identifying specific subtypes of cells in a tumour, such as rare so-called persister cells that survive initial therapy and eventually develop into resistant clones. Such persister cells are usually in a transient and reversible condition which is low proliferative, metabolic profile changes and higher tolerance to stress. They do not have to be characterized by fixed genetic mutations but are dynamic transcriptional programs that enable them to survive therapeutic pressure (Gilson et al., 2022).

The discovery of these rare populations has important implications on the study of the mechanisms of resistance. scRNA-seq analysis has shown that persister cells may acquire characteristics of epithelial mesenchymal transition, stemness, and immune evasion, thus being especially challenging to destroy. In addition, the cells are able to communicate with the environment around them to form protective niches that further increase their survival. This can also be used to map the evolution of resistance by comparing the transcriptional trajectories of individual cells as a function of time since they have been exposed to the drug. This time aspect offers important clues of the order of events in the sequence of the molecular process that causes resistance, which has the potential of being intervened in advance before the resistant clones become prevalent (Lim, 2024; Sehgal et al., 2021).

Besides cellular heterogeneity, spatial organization of tumours is also instrumental in determining resistance which has gained more and more significance with the evolution of spatial transcriptomics. In contrast to traditional sequencing techniques which dissociate cells, and thus lose localization data, spatial transcriptomics maintains the architecture of the tissue, enabling scientists to map gene expression patterns in their native microenvironment. This method has shown that resistance is not distributed equally across the tumour but is usually confined to a few areas where there are special microenvironmental conditions. These areas can be hypoxic areas, drug penetration areas or niches that are rich in supporting stromal and immune cells (Nagasawa et al., 2021).

The tumour microenvironment (TME) is an essential part of therapeutic resistance that represents a complex chain of non-cancerous cells (fibroblasts, immune cells, endothelial cells, and extracellular matrix components). These factors have a dynamic interaction with cancer cells affecting their behaviour and reaction to therapy. Take the cancer-associated fibroblasts as an example, it can release the growth factors and cytokines that activate tumour cells survival pathways, which in turn can promote or suppress the progression of tumours based on their phenotype. Metabolic reprogramming and increased expression of drug resistance genes can be triggered by hypoxia in the TME, making the treatment process even more challenging (Khalaf et al., 2021).

Spatial transcriptomics has facilitated the identification of certain cellular interactions and gradients of signalling in the TME that drive resistance. As an example, it may show the extent of proximity of blood vessels on drug exposure or the degree of immune cell infiltration within different regions of tumours. This spatially resolved data is vital in the development of specific therapies capable of penetrating resistant niches and interfering with supportive interactions in the TME. Moreover, the combination of spatial data and the other omics layers increases the capacity to build a multifaceted model of tumour behavior, which entails both molecular and environmental factors of resistance (Khalaf et al., 2021; Trédan et al., 2007).

Multi-omics, single-cell sequencing, and spatial transcriptomics are an effective approach to map the resistance landscape in cancer. With a combination of high-dimensional molecular information with cellular and spatial resolutions, researchers can gain a more detailed insight into the mechanisms that lead to resistance emergence and evolution. This combined method not only determines which forces are driving the resistance but also determines the possible therapeutic weak points which can be used to enhance the treatment results. With further advancement of these technologies, they promise to change the face of cancer research and allow the creation of more targeted and efficient ways of overcoming therapeutic resistance (Chen et al., 2024).

## AI AND MACHINE LEARNING: DECIPHERING COMPLEXITY

The recent advances in artificial intelligence (AI) and machine learning (ML) have become transformative resources in contemporary biomedical science, allowing scientists to break the huge complexity of biological processes underlying multidrug resistance (MDR). The AI-based methods can analyse large, high-dimensional datasets and identify non-linear interactions among genetic, epigenetic, proteomic, and environmental factors, unlike the traditional methods of analysis, which usually assume a linear relationship. This is especially important in studying drug resistance, where several pathways can work together in a dynamic fashion to affect the effects of therapy. Using a combination of genomics, transcriptomics, proteomics, and clinical records, AI can offer a broad framework through which it can identify patterns and produce actionable insights that can be used to inform precision medicine (Halder et al., 2025).

Predictive modelling is one of the most effective AI uses in this area, as it allows forecasting the drug response and identifying new biomarkers related to resistance. Random forests, support vectors machines and deep neural networks are machine learning algorithms trained on large datasets of molecular and clinical data. These models acquire intricate correlations between biological aspects and therapy results, and can be utilized to forecast whether a patient will respond to a particular medication (Khan et al., 2026; Patel & Patel, 2024). Notably, predictive modelling is more than mere classification as it establishes important molecular determinants that promote resistance. An example of this is the detection of gene mutations, changes in drug transporters, or activation of compensatory signalling pathways, which are some of the causes of therapeutic failure that AI can identify. Such insights would be precious in creating specific therapies and enhancing diagnoses (Fojo, 2007).

Moreover, there is an increased use of predictive models in clinical decision-making. In cancer treatment and other fields of treatment, AI-based systems are capable of analysing the individual molecular profile of the patient and prescribe individual treatment plan. This methodology removes the use of empirical selection of treatment and the chances of negative effects that could come along with ineffective treatments. The utility of these models can also be increased by continuous learning mechanisms that can also incorporate new information as time passes and improve their predictions. Consequently, predictive modelling is not only enhancing patient outcomes but also hastening the shift towards personalized medicine (Tiwari et al., 2025).

Simultaneously, network pharmacology offers a systems-level view to supplement the understanding obtained through predictive modelling. Biological systems are complex and many genes, proteins and metabolites do not operate independently but interact in complex networks. The re-programming of these networks can often result in drug resistance as cells evade the action of therapeutic agents. AI is important in rebuilding and analysing these networks, combining multi-omics information and pinpointing important regulatory interactions. Network inference methods and graph modelling enable researchers to visualize signalling pathways and identify key nodes that regulate network behaviour (Lopatkin & Collins, 2020).

These are key nodes that are commonly called vulnerability nodes, strategic targets of therapeutic intervention. When it comes to gene editing via CRISPR, it is crucial to define such nodes to create effective and accurate interventions. Researchers can no longer disrupt single genes alone but can instead disrupt whole pathways or network hubs that are vital in resistance maintenance. This method is further extended by AI-based simulations which help forecast the downstream effects of gene editing, thus minimizing the occurrence of unintended consequences and enhancing safety. Also, network pharmacology enables drug repurposing, by identifying already available drugs that can regulate resistance-related pathways, thereby saving a lot of time and cost involved in developing a drug (Selvaraj et al., 2022).

Deep learning, a subfield of machine learning, has also contributed to research on drug resistance with the help of sophisticated image analysis. CNNs are most useful with medical images, where they can extract the complex features that might not be identified with traditional methods. Quantitative features in radiomics are based on imaging data (MRI, CT, PET) and they consist of data on tumour shape, texture, and intensity. AI can be used to predict the response to treatment, disease progression, and resistance probability by analysing these features. Examples of radiomic signatures include identifying tumours that are predisposed to resistant chemotherapy or radiotherapy, which can be useful in deciding the treatment (Forghani et al., 2019).

Digital pathology also extends this power by providing high-resolution images of tissue samples, and which can be processed with deep learning algorithms to provide measurements of cellular morphology and microenvironmental features. The patterns of tumour heterogeneity, immune cell infiltration, and stromal interactions are all patterns that AI models can detect, and each of these factors is important in drug response determination. Combination of radiomics and digital pathology with molecular data leads to a multimodal framework which encompasses both phenotypic and genotypic disease. This holistic method enhances the validity of predictive models and helps to create more efficient and individualized treatment plans (Chen et al., 2016).

In addition, explainable AI is resolving one of the biggest problems, which come with deep learning models, i.e., they are not interpretable. Explainable AI should be used to build trust and help implement these technologies in the clinical setting by offering insights into prediction generation methods. Clinicians have a greater opportunity to comprehend the logic behind AI-made suggestions and implement them into patient care more easily (Ali et al., 2025).

Altogether, AI and machine learning in biomedical research is a paradigm shift in how we comprehend and fight drug resistance. These technologies provide a comprehensive view of understanding the complex biological systems by integrating predictive modelling, network pharmacology, and image analysis by deep learning. They allow the discovery of new biomarkers, mapping of complex signalling pathways, and the creation of individualized treatment plans. With these techniques still in their development and incorporation into the new technologies, they promise the medical field a lot of potential in enhancing therapeutic results and advancing precision medicine (Kalita et al., 2025).

## NANO-CRISPR PLATFORMS: DELIVERY AND PRECISION

Nano-CRISPR platforms are a revolutionary breakthrough in genome editing, as they solve one of the most important issues in therapeutic use: effective and precise delivery. Although the CRISPR-Cas system has unmatched gene-editing potential, its clinical implementation, especially in systemic cancer treatment, has been hampered by delivery system constraints. Traditional approaches, especially viral vectors such as lentiviruses and adeno-associated viruses (AAVs), have been widely used due to their high transfection efficiency. Nonetheless, these systems have significant disadvantages, such as immunogenicity, low cargo capacity, insertional mutagenesis, and lack of control over biodistribution. These limitations are even more intense in the case of systemic cancer therapy, when the therapeutic agent is to be circulated in the blood and selectively act on tumour cells (Figure 2). Viral vectors tend to build up in non-target organs, including the liver and spleen, decreasing efficacy of therapy and increasing the risk of adverse effects. In addition, immune responses render them difficult to administer repeatedly, thus not suitable in chronic or large-scale interventions (Song et al., 2024; Tang et al., 2025).

Nanocarrier-based delivery systems have become a potent alternative, providing a better safety profile, flexibility, and targeting. Of these, the lipid nanoparticles (LNPs) have established themselves as the standard of delivering CRISPR components, especially the mRNA and single-guide RNA (sgRNA). The LNPs consist of ionizable lipids, phospholipids, cholesterol, and polyethylene glycol (PEG)-lipids, which combine to enable effective encapsulation and of nucleic acids against enzyme degradation. The major benefit of LNPs is that they are able to experience charge transitions dependent on pH, which is why they are able to stay stable in circulation and endosomal escape in target cells. This feature guarantees that the CRISPR payload is delivered into the cytoplasm, where it can have its gene-editing effect. The clinical potential of LNPs has also been confirmed by the success of their use in mRNA vaccine delivery, and thus the use of LNPs in CRISPR-based therapies.

Also, LNPs can be designed to maximize size, charge and surface properties, thus improving their pharmacokinetics and tissue distribution (Guzmán-Sastoque et al.).

"By integrating multi-omics datasets with machine learning algorithms, AI enables the identification of complex resistance-associated signatures that may not be detected through conventional analytical approaches. This capability provides a critical link between molecular profiling and targeted therapeutic interventions, thereby supporting the development of precision oncology strategies."

In addition to lipid-based systems, polymeric and inorganic nanoparticles offer further flexibility of CRISPR delivery. CRISPR components can be released in a controlled and sustained manner with the aid of polymeric nanoparticles, which can be made of biodegradable polymers like poly (lactic-co-glycolic acid) (PLGA) or polyethyleneimine (PEI). Researchers can control the degradation rates and release profiles by modifying polymer composition and structure to guarantee long-term therapeutic activity. These systems also have enhanced stability in storage and circulation and hence can be used in a broad spectrum of clinical applications. Inorganic nanoparticles, such as gold nanoparticles, silica-based carriers, and magnetic nanoparticles, offer distinct physicochemical characteristics that can be utilized towards targeted delivery and imaging. As an example, gold nanoparticles can be functionalized with nucleic acids and targeting ligands, and magnetic nanoparticles can be delivered with the help of external magnetic fields. Nonetheless, issues of long-term biocompatibility and tissue accumulation should be also considered to make sure that they are safely used (Visan et al., 2021).

*Figure 2. Nano-CRISPR platforms: Delivery and precision*

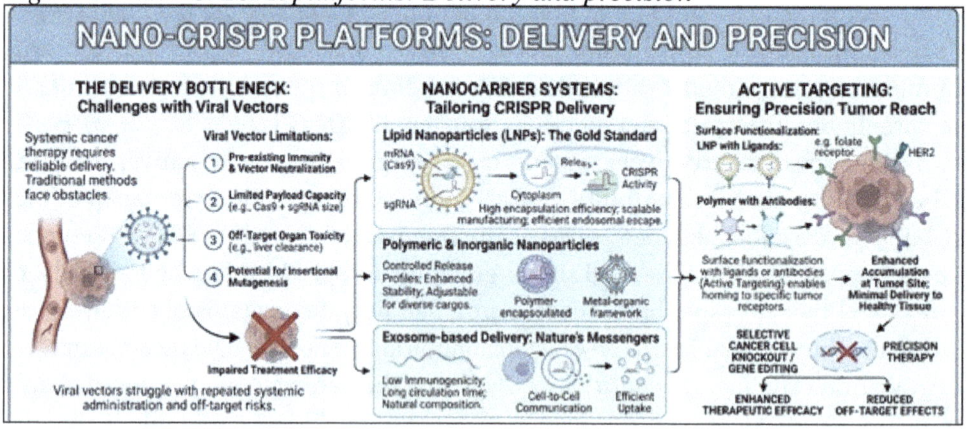


Exosome-based delivery systems are a biologically inspired method, which takes advantage of the natural communication processes in the body. Exosomes are tiny extracellular vesicles released by cells that enable intercellular exchange of proteins, lipids, and nucleic acids. They are very attractive to CRISPR delivery due to their inherent biocompatibility, low immunogenicity, and capability to penetrate biological barriers. Exosomes can avoid immune detection and have natural targeting characteristics depending on the cell of origin, unlike synthetic nanoparticles. Incorporating CRISPR components into exosomes by engineering donor cells will allow researchers to develop highly efficient delivery vehicles that can deliver gene-editing tools to particular tissues. Moreover, exosomes may be engineered to increase specificity in targeting, such as surface proteins that identify tumour-associated markers. Although these benefits exist, issues like mass production, standardization, and loading efficiency are yet to be completely addressed (Rabiee, 2023).

Active targeting is a key feature of Nano-CRISPR platforms, as it makes sure that the therapeutic payload is targeted to tumour cells and off-target effects are reduced in healthy tissues. This is done by surface functionalizing nanocarriers with ligands, antibodies, peptides or small molecules that interact with receptors that are overexpressed on cancer cells. As an illustration, folate receptors, transferrin receptors, and HER2 are usually targeted in different cancers. By attaching these targeting moieties to the surface of nanoparticles, scientists are able to increase cellular uptake via receptor-mediated endocytosis. This does not only enhance the effectiveness of CRISPR delivery but also lowers systemic toxicity and non-target tissue gene editing. Moreover, stimuli-responsive systems are being created to further narrow down targeting precision. Such systems can be triggered to release their payload based on certain tumour microenvironment conditions like acidic pH, hypoxia or enzymatic activity, which adds an extra level of control (Anarjan, 2019).

Altogether, Nano-CRISPR platforms combine nanotechnology, molecular biology, and bioengineering to address the bottleneck of delivery that has traditionally limited the therapeutic potential of CRISPR. These platforms provide a promising avenue to safe and effective genome editing in cancer treatment by integrating effective nanocarriers, biologically inspired delivery systems, and targeted delivery approaches. With the ongoing advancement of these technologies, Nano-CRISPR systems will be at the centre of the future of precision medicine, which will allow treating complex diseases with targeted, lasting, and minimally invasive therapies (Baig & Chien, 2025).

## OVERCOMING RESISTANCE: STRATEGIC APPLICATIONS

Therapeutic resistance is one of the most urgent problems in the contemporary oncology, and CRISPR-based genome editing provides very tactical solutions to re-sensitize tumours and increase the effectiveness of the treatment. Among the most straightforward methods is to attack genes that cause drug efflux and DNA repair, which are two major ways that cancer cells avoid the cytotoxic impact of treatment. Efflux transporters like ABCB1 contribute significantly to the extra-cellular pumping of chemotherapeutic drugs out of the cancer cells, thus lowering the concentration of drugs within the cells. CRISPR-Cas systems can be used to knock out or downregulate such genes, which can dramatically raise drug levels in tumour cells, reinstating chemotherapy sensitivity. Likewise, inhibiting genes that participate in DNA repair mechanisms, including PARP, can prevent cancer cells to repair radiation-induced or cytotoxic drug-induced DNA damage. Interference with these repair processes results in a condition of synthetic lethality, in which cancer cells are extremely susceptible to therapy and normal cells are spared, which still have intact repair systems (Hou et al., 2022; Yadav & Yadav, 2015).

In addition to direct gene knockout, CRISPR technology can also be used to regulate gene expression by epigenetic remodelling with high precision. dCas9 (catalytically inactive Cas9) can be fused with transcriptional activators or repressors to control gene expression without making any changes to the DNA sequence. The method is especially useful in the re-activation of the epigenetic silencing of tumour suppressor genes, which is a frequent characteristic of resistant cancers. dCas9-based systems can reactivate silenced genes by targeting their promoter regions, which can restore gene expression and reactivate pathways regulating cell cycle, apoptosis, and DNA repair. This type of epigenetic reprogramming offers a reversible, highly regulated way to overcome resistance with benefits over irreversible genome editing in terms of safety and flexibility (Cai et al., 2023).

The other potential use is in potentiation of immunotherapy, where CRISPR is applied to overcome immune evasion strategies used by tumours. PD-1 and PD-L1 immune checkpoint pathway is a target in cancer immunotherapy as its activation inhibits the activity of T-cells and permits tumour cells to evade immune surveillance. PD-1 inhibition in T cells or PD-L1 in tumour cells mediated by CRISPR can boost immune responses and increase the effectiveness of checkpoint blockade therapies. Moreover, the combination of Nano-CRISPR platforms with chimeric antigen receptor T-cell therapy has provided new opportunities to optimize CAR-T cell therapy. Through the manipulation of genes that inhibit T-cell survival or induce exhaustion, researchers can develop stronger and longer-lasting CAR-T cells with anti-tumour response (Blank et al., 2005).

Taken together, these strategic uses of CRISPR show that it can be used to directly target resistance mechanisms as well as reprogram tumour biology and improve immune-mediated clearance. Combining gene editing with sophisticated delivery methods like nanoparticles, these methods offer a comprehensive system of overcoming resistance and enhancing the therapeutic effects of cancer treatment (Wang et al., 2025).

*Table 2. Strategic CRISPR applications for overcoming drug resistance*

| Strategy | Target/Example | Mechanism of Action | Therapeutic Outcome | References |
|---|---|---|---|---|
| Tumour Sensitization (Drug Efflux) | ABCB1 | Knockout reduces drug efflux from cancer cells | Increased intracellular drug accumulation and restored chemotherapy sensitivity | (Lei et al., 2021) |
| Tumour Sensitization (DNA Repair) | PARP | Disruption of DNA repair pathways | Enhanced susceptibility to DNA-damaging agents (synthetic lethality) | (Li et al., 2022) |
| Epigenetic Remodelling | dCas9-based systems | Activation of silenced tumour suppressor genes | Restoration of apoptosis and cell cycle control | (Rajanathadurai et al., 2024) |
| Immunotherapy Potentiation | PD-1 / PD-L1 | Disruption of immune checkpoint signalling | Enhanced T-cell activity and improved immune response | (Peng et al., 2020) |
| CAR-T Optimization | CAR-T cell therapy | Editing genes controlling T-cell exhaustion and persistence | Increased durability and anti-tumour efficacy of engineered T cells | (Lei et al., 2024) |

## CHALLENGES AND ETHICAL CONSIDERATIONS

The accelerated development of CRISPR-based technology, nanomedicine, and AI-driven precision medicine have brought revolutionary opportunities to the healthcare sector, but they are also accompanied by major challenges and ethical issues that should be taken into account. The problem of off-target effects of CRISPR genome editing is among the most prominent issues. Although CRISPR-Cas systems are meant to be highly specific in targeting a specific DNA sequence, unexpected edits at adjacent genomic loci may happen, which can cause detrimental mutations, genomic instability, or oncogenes activation. These off-target effects are potentially dangerous, especially when used in clinical practice with human subjects. To address these impacts, scientists are working on high-fidelity Cas variants, better guide RNA design algorithms, and novel sequencing methods to identify unintended edits. Also, the incorporation of AI-based predictive models can be used to increase the specificity of targets by detecting possible off-target sites prior to therapeutic intervention, thus enhancing the safety profile of genome editing interventions (Nidhi et al., 2021; Sander & Joung, 2014).

The other major issue is nanotoxicity, especially with regard to the long-term safety of synthetic nanocarriers in Nano-CRISPR platforms. Although nanoparticles like lipid nanoparticles, polymeric carriers, and inorganic materials have been shown to be effective in delivering CRISPR components, their behaviour in biological systems is complicated and not completely comprehended. Accumulation in vital organs, the creation of oxidative stress, inflammation, and unintended immune responses are some of the issues that are of concern as far as chronic toxicity is concerned. An example is that inorganic nanoparticles can be retained in tissues because they are not readily biodegraded, and some polymeric systems generate toxic degradation products (Rabiee, 2023). Even lipid-based systems, which are typically regarded as safer, may induce immune activation based on their composition and dose rate. To overcome these issues, strict preclinical and clinical testing, such as long-term toxicity tests, biodistribution, and designing biodegradable and biocompatible materials, are needed. A promising avenue to enhance safety is the design of so-called smart nanoparticles that react to certain biological signals and break down into harmless constituents (Lima & Videira, 2018).

Simultaneously, the application of AI and machine learning to patient-specific omics data leads to major data privacy and ethical issues. Precision medicine is based on the gathering and examination of delicate genetic, transcriptomic, and clinical information to personalize treatment to the patient. The storage, sharing and use of such data, however, brings up issues of confidentiality, consent and misuse. Illegal use of genomic data can result in job or insurance discrimination, and insufficient anonymity can threaten patient identity. Moreover, AI algorithms may lead to unequal healthcare outcomes due to biases that may occur because of the unrepresentative training datasets (Ali, 2023). To ensure that AI-driven omics profiles are used ethically, it is necessary to have strong data governance measures, such as secure data storage, open consent procedures, and stringent regulatory compliance. Federated learning and differential privacy are the techniques that are being considered to allow analysing data without revealing sensitive information (Chalamala et al., 2022).

## FUTURE ADVANCEMENTS

Future advancements in AI, single-cell multi-omics, spatial biology, and next-generation CRISPR systems are expected to further improve the precision and effectiveness of cancer therapies. The integration of real-time patient monitoring, digital twins, and adaptive therapeutic platforms may enable dynamic treatment optimization. Continued interdisciplinary collaboration and clinical validation will be essential for translating these technologies into routine oncology practice.

## CONCLUSION

To sum up, AI-guided omics together with Nano-CRISPR platforms will be a paradigm shift in combating therapeutic resistance in cancer. Using the ability of artificial intelligence to process high-dimensional and complex omics data, scientists can discover novel and obscured molecular causes of resistance and pinpoint specific therapeutic targets. When combined with the latest CRISPR-based genome editing tools, these insights can be used to make highly targeted interventions that can re-sensitize tumours, disrupt resistance pathways, and re-establish treatment efficacy. This method is further improved by the introduction of nanotechnology that helps to overcome the long-standing problem of safe and efficient delivery, whereby CRISPR components can be delivered to target cells with high precision with minimum off-target effects. Furthermore, the combination of these technologies promotes a more individualized and adaptive paradigm of cancer treatment, in which the treatment is based on the individual molecular profile of a patient. This integrated framework can be used to tackle the multifactorial aspect of resistance, with predictive modelling and network pharmacology to immunotherapy potentiation and epigenetic remodelling. Nevertheless, the issues of safety, nanotoxicity, and ethical concerns should be properly addressed to provide responsible clinical translation. Finally, AI-controlled Nano-CRISPR systems have a tremendous potential in changing the way cancer is treated, leading to more efficient, long-term and personalized therapeutic approaches.

## REFERENCES


Aldoseri, A., Al-Khalifa, K. N., & Hamouda, A. M. (2023). Re-thinking data strategy and integration for artificial intelligence: Concepts, opportunities, and challenges. *Applied Sciences (Basel, Switzerland), 13*(12), 7082. DOI: 10.3390/app13127082

Alhazza, A., Oyegbesan, A., Bousoik, E., & Montazeri Aliabadi, H. (2025). Multidrug resistance: Are we still afraid of the big bad wolf. *Pharmaceuticals (Basel, Switzerland), 18*(6), 895. DOI: 10.3390/ph18060895 PMID: 40573290

Ali, H. (2023). Artificial intelligence in multi-omics data integration: Advancing precision medicine, biomarker discovery and genomic-driven disease interventions. *Int J Sci Res Arch, 8*(1), 1012–1030. DOI: 10.30574/ijsra.2023.8.1.0189

Ali, M., Benfante, V., Basirinia, G., Alongi, P., Sperandeo, A., Quattrocchi, A., Giannone, A. G., Cabibi, D., Yezzi, A., Di Raimondo, D., Tuttolomondo, A., & Comelli, A. (2025). Applications of artificial intelligence, deep learning, and machine learning to support the analysis of microscopic images of cells and tissues. *Journal of Imaging, 11*(2), 59. DOI: 10.3390/jimaging11020059 PMID: 39997561

Anarjan, F. S. (2019). Active targeting drug delivery nanocarriers: Ligands. *Nano-Structures & Nano-Objects, 19*, 100370. DOI: 10.1016/j.nanoso.2019.100370

Baig, M. M. F. A., Chien, W. T., & Chair, S. Y. (2025). Nanotechnological approaches for the targeted delivery of CRISPR-Cas systems for genomic modifications, biomolecular sensing, and precision medicine. *Biomaterials Science, 13*(17), 4597–4638. DOI: 10.1039/D5BM00711A PMID: 40704512

Blank, C., Gajewski, T. F., & Mackensen, A. (2005). Interaction of PD-L1 on tumor cells with PD-1 on tumor-specific T cells as a mechanism of immune evasion: Implications for tumor immunotherapy. *Cancer Immunology, Immunotherapy : CII, 54*(4), 307–314. DOI: 10.1007/s00262-004-0593-x PMID: 15599732

Cai, R., Lv, R., Shi, X., Yang, G., & Jin, J. (2023). CRISPR/dCas9 tools: Epigenetic mechanism and application in gene transcriptional regulation. *International Journal of Molecular Sciences, 24*(19), 14865. DOI: 10.3390/ijms241914865 PMID: 37834313

Chaffer, C. L., San Juan, B. P., Lim, E., & Weinberg, R. A. (2016). EMT, cell plasticity and metastasis. *Cancer Metastasis Reviews, 35*(4), 645–654. DOI: 10.1007/s10555-016-9648-7 PMID: 27878502

Chakraborty, S., Hosen, M. I., Ahmed, M., & Shekhar, H. U. (2018). Onco-multi-OMICS approach: A new frontier in cancer research. *BioMed Research International, 2018*(1), 9836256. DOI: 10.1155/2018/9836256 PMID: 30402498

Chalamala, S. R., Kummari, N. K., Singh, A. K., Saibewar, A., & Chalavadi, K. M. (2022). Federated learning to comply with data protection regulations. *CSI Transactions on ICT*, *10*(1), 47–60. DOI: 10.1007/s40012-022-00351-0

Chen, D., Sarkar, S., Candia, J., Florczyk, S. J., Bodhak, S., Driscoll, M. K., Simon, C. G.Jr, Dunkers, J. P., & Losert, W. (2016). Machine learning based methodology to identify cell shape phenotypes associated with microenvironmental cues. *Biomaterials*, *104*, 104–118. DOI: 10.1016/j.biomaterials.2016.06.040 PMID: 27449947

Chen, H., Zhang, L., Liu, M., Li, Y., & Rao, P. (2025). Multidimensional perspectives on cancer therapy-induced cardiotoxicity: Characteristics, mechanisms, biomarkers, psychological stress effects, and treatment progress. *Cancer Pathogenesis and Therapy*, *3*, E036–E135. DOI: 10.1016/j.cpt.2025.12.003

Chen, J., Larsson, L., Swarbrick, A., & Lundeberg, J. (2024). Spatial landscapes of cancers: Insights and opportunities. *Nature Reviews. Clinical Oncology*, *21*(9), 660–674. DOI: 10.1038/s41571-024-00926-7 PMID: 39043872

da Silva Correia, A. B. A., de Souza Melo, B. M., da Silva, B. A., Maciel, G. C., Cavalcante, G. M. S. M., Almeida dos Santos, J. P., Silva, L. M. B., Rocha, L. F. P., Brêda, M. B. C., de Oliveira, M. E. T., Lima, K. F. B., de Oliveira Wanderley Nepomuceno, L. B.-H., da Silva, T. H., Rodrigues, G. H. B., dos Santos, M. L., da Silva Fernandes Duarte, A. K., Gomes, E. S. B., de Araújo, S. T., de França-Júnior, R. R., & de Carvalho Fraga, C. A. (2025). Physical Interactions between Cancer Cells and Tumor Microenvironment. *Cell Biochemistry and Biophysics*, *84*(1), 1–16. DOI: 10.1007/s12013-025-01900-5 PMID: 40956489

Fojo, T. (2007). Multiple paths to a drug resistance phenotype: Mutations, translocations, deletions and amplification of coding genes or promoter regions, epigenetic changes and microRNAs. *Drug Resistance Updates*, *10*(1-2), 59–67. DOI: 10.1016/j.drup.2007.02.002 PMID: 17350322

Forghani, R., Savadjiev, P., Chatterjee, A., Muthukrishnan, N., Reinhold, C., & Forghani, B. (2019). Radiomics and artificial intelligence for biomarker and prediction model development in oncology. *Computational and Structural Biotechnology Journal*, *17*, 995–1008. DOI: 10.1016/j.csbj.2019.07.001 PMID: 31388413

Gilson, P., Merlin, J.-L., & Harlé, A. (2022). Deciphering tumour heterogeneity: From tissue to liquid biopsy. *Cancers (Basel)*, *14*(6), 1384. DOI: 10.3390/cancers14061384 PMID: 35326534

Guzmán-Sastoque, P., Monsalve, M. C., Rodríguez, C. F., Castellanos, S., Cruz, J. C., & Reyes, L. H. Delivery of Nucleic Acids Using Nanocarriers: siRNA and miRNA Delivery, mRNA and DNA Delivery, CRISPR-Cas Systems. In *Nanocarriers for Nucleic Acids and Proteins* (pp. 248–291). CRC Press.

Halder, A. K., Biswas, A., Saha, M., & Sadi, S. M. (2025). *Machine learning-based prediction of acquired antimicrobial resistance in multiple bacterial species using K-mer analysis, mutation detection, and AMR gene profiling*. Brac University.

Harle, S. (2025). Integrative approaches to tackle multidisciplinary challenges: A review of multi-science problem analysis. *Current Materials Science*, *18*(6), 703–728. DOI: 10.2174/0126661454279515231225151023

Hou, J., He, Z., Liu, T., Chen, D., Wang, B., Wen, Q., & Zheng, X. (2022). Evolution of molecular targeted cancer therapy: Mechanisms of drug resistance and novel opportunities identified by CRISPR-Cas9 screening. *Frontiers in Oncology*, *12*, 755053. DOI: 10.3389/fonc.2022.755053 PMID: 35372044

Hughes, D., & Andersson, D. I. (2015). Evolutionary consequences of drug resistance: Shared principles across diverse targets and organisms. *Nature Reviews. Genetics*, *16*(8), 459–471. DOI: 10.1038/nrg3922 PMID: 26149714

Jung, G. T., Kim, K.-P., & Kim, K. (2020). How to interpret and integrate multi-omics data at systems level. *Animal Cells and Systems*, *24*(1), 1–7. DOI: 10.1080/19768354.2020.1721321 PMID: 32158610

Jurkovicova, D., Neophytou, C. M., Gašparović, A. Č., & Gonçalves, A. C. (2022). DNA damage response in cancer therapy and resistance: Challenges and opportunities. *International Journal of Molecular Sciences*, *23*(23), 14672. DOI: 10.3390/ijms232314672 PMID: 36499000

Kalita, I., Bhattacharjee, S., & Saharia, M. (2025). Advancements in medicinal plant research: Harnessing artificial intelligence, machine learning, deep learning, and bioinformatics. In *Biotechnology, Multiple Omics, and Precision Breeding in Medicinal Plants* (pp. 135-145). CRC Press.

Kang, T.-H. (2023). *DNA damage, repair, and cancer metabolism* (Vol. 24). MDPI.

Kant, S., Deepika, , & Roy, S. (2025). Integrative Multi-Omics and Artificial Intelligence: A New Paradigm for Systems Biology. *Omics : A Journal of Integrative Biology*, *29*(12), 576–587. DOI: 10.1177/15578100251392371 PMID: 41203247

Khalaf, K., Hana, D., Chou, J. T.-T., Singh, C., Mackiewicz, A., & Kaczmarek, M. (2021). Aspects of the tumor microenvironment involved in immune resistance and drug resistance. *Frontiers in Immunology*, *12*, 656364. DOI: 10.3389/fimmu.2021.656364 PMID: 34122412

Khalighi, S., Reddy, K., Midya, A., Pandav, K. B., Madabhushi, A., & Abedalthagafi, M. (2024). Artificial intelligence in neuro-oncology: Advances and challenges in brain tumor diagnosis, prognosis, and precision treatment. *NPJ Precision Oncology*, *8*(1), 80. DOI: 10.1038/s41698-024-00575-0 PMID: 38553633

Khan, M. Z., Naeem, M. S., Khalid, S., Gill, Z., Ali, H., Iqbal, S., Shahid, S., Sulman, A., Awan, A., & Kharl, H. A. A. (2026). AI-GUIDED QSAR, MOLECULAR DYNAMICS, AND PHARMACOPHORE MODELING. *PharmA AI: Revolutionizing Drug Discovery and Healthcare with Artificial Intelligence*.

Kharl, H. A. A. (2026). *PharmA AI: Revolutionizing Drug Discovery and Healthcare with Artificial Intelligence*. IBH Publisher.

Kim, D. J. (2026). Molecular and Immune Mechanisms Governing Cancer Metastasis, Including Dormancy, Microenvironmental Niches, and Tumor-Specific Programs. *International Journal of Molecular Sciences*, *27*(2), 875. DOI: 10.3390/ijms27020875 PMID: 41596524

La Rosa, R., Johansen, H. K., & Molin, S. (2022). Persistent bacterial infections, antibiotic treatment failure, and microbial adaptive evolution. *Antibiotics (Basel, Switzerland)*, *11*(3), 419. DOI: 10.3390/antibiotics11030419 PMID: 35326882

Lei, T., Wang, Y., Zhang, Y., Yang, Y., Cao, J., Huang, J., Chen, J., Chen, H., Zhang, J., Wang, L., Xu, X., Gale, R. P., & Wang, L. (2024). Leveraging CRISPR gene editing technology to optimize the efficacy, safety and accessibility of CAR T-cell therapy. *Leukemia*, *38*(12), 2517–2543. DOI: 10.1038/s41375-024-02444-y PMID: 39455854

Lei, Z. N., Teng, Q. X., Wu, Z. X., Ping, F. F., Song, P., Wurpel, J. N., & Chen, Z. S. (2021). Overcoming multidrug resistance by knockout of ABCB1 gene using CRISPR/Cas9 system in SW620/Ad300 colorectal cancer cells. *MedComm*, *2*(4), 765–777. DOI: 10.1002/mco2.106 PMID: 34977876

Li, S., Wang, L., Wang, Y., Zhang, C., Hong, Z., & Han, Z. (2022). The synthetic lethality of targeting cell cycle checkpoints and PARPs in cancer treatment. *Journal of Hematology & Oncology*, *15*(1), 147. DOI: 10.1186/s13045-022-01360-x PMID: 36253861

Lim, Z.-F. (2024). *Personalized Molecular Therapies for Advanced Non-Small Cell Lung Cancer: Overcoming Heterogeneity to Optimize Treatment Response and Clinical Outcomes*. West Virginia University.

Lima, B. S., & Videira, M. A. (2018). Toxicology and biodistribution: The clinical value of animal biodistribution studies. *Molecular Therapy. Methods & Clinical Development*, *8*, 183–197. DOI: 10.1016/j.omtm.2018.01.003 PMID: 29541646

Lopatkin, A. J., & Collins, J. J. (2020). Predictive biology: Modelling, understanding and harnessing microbial complexity. *Nature Reviews. Microbiology*, *18*(9), 507–520. DOI: 10.1038/s41579-020-0372-5 PMID: 32472051

Lytle, N. K., Barber, A. G., & Reya, T. (2018). Stem cell fate in cancer growth, progression and therapy resistance. *Nature Reviews. Cancer*, *18*(11), 669–680. DOI: 10.1038/s41568-018-0056-x PMID: 30228301

Makhoba, X. H., Viegas, C.Jr, Mosa, R. A., Viegas, F. P., & Pooe, O. J. (2020). Potential impact of the multi-target drug approach in the treatment of some complex diseases. *Drug Design, Development and Therapy*, *14*, 3235–3249. DOI: 10.2147/DDDT.S257494 PMID: 32884235

Manfioletti, G., & Fedele, M. (2023). *Epithelial–mesenchymal transition (EMT)* (Vol. 24). MDPI.

Mao, Y., Shangguan, D., Huang, Q., Xiao, L., Cao, D., Zhou, H., & Wang, Y.-K. (2025). Emerging artificial intelligence-driven precision therapies in tumor drug resistance: Recent advances, opportunities, and challenges. *Molecular Cancer*, *24*(1), 123. DOI: 10.1186/s12943-025-02321-x PMID: 40269930

Monti, C., Zilocchi, M., Colugnat, I., & Alberio, T. (2019). Proteomics turns functional. *Journal of Proteomics*, *198*, 36–44. DOI: 10.1016/j.jprot.2018.12.012 PMID: 30553948

Nagasawa, S., Kashima, Y., Suzuki, A., & Suzuki, Y. (2021). Single-cell and spatial analyses of cancer cells: Toward elucidating the molecular mechanisms of clonal evolution and drug resistance acquisition. *Inflammation and Regeneration*, *41*(1), 22. DOI: 10.1186/s41232-021-00170-x PMID: 34271973

Nevedomskaya, E., & Haendler, B. (2022). From omics to multi-omics approaches for in-depth analysis of the molecular mechanisms of prostate cancer. *International Journal of Molecular Sciences*, *23*(11), 6281. DOI: 10.3390/ijms23116281 PMID: 35682963

Nidhi, S., Anand, U., Oleksak, P., Tripathi, P., Lal, J. A., Thomas, G., Kuca, K., & Tripathi, V. (2021). Novel CRISPR–Cas systems: An updated review of the current achievements, applications, and future research perspectives. *International Journal of Molecular Sciences*, *22*(7), 3327. DOI: 10.3390/ijms22073327 PMID: 33805113

Patel, R., & Patel, A. (2024). Revolutionizing drug development: AI-driven predictive modeling for accelerated small molecule and biologic therapeutics. *Well Testing Journal*, *33*(S2), 668–691.

Peng, Q., Qiu, X., Zhang, Z., Zhang, S., Zhang, Y., Liang, Y., Guo, J., Peng, H., Chen, M., Fu, Y.-X., & Tang, H. (2020). PD-L1 on dendritic cells attenuates T cell activation and regulates response to immune checkpoint blockade. *Nature Communications*, *11*(1), 4835. DOI: 10.1038/s41467-020-18570-x PMID: 32973173

Pettini, F., Visibelli, A., Cicaloni, V., Iovinelli, D., & Spiga, O. (2021). Multi-omics model applied to cancer genetics. *International Journal of Molecular Sciences*, *22*(11), 5751. DOI: 10.3390/ijms22115751 PMID: 34072237

Rabiee, N. (2023). Natural components as surface engineering agents for CRISPR delivery. *Environmental Research*, *232*, 116333. DOI: 10.1016/j.envres.2023.116333 PMID: 37286127

Rajanathadurai, J., Perumal, E., & Sindya, J. (2024). Advances in targeting cancer epigenetics using CRISPR-dCas9 technology: A comprehensive review and future prospects. *Functional & Integrative Genomics*, *24*(5), 164. DOI: 10.1007/s10142-024-01455-3 PMID: 39292321

Riaz, M. U., Naeem, M. S., Zohaib, M., Ahmad, B., Ijaz, M. T., Tahir, M. U., Sulman, A., Awan, A., & Kharl, H. A. A. (2026). AI IN TARGET IDENTIFICATION AND VALIDATION. *PharmA AI: Revolutionizing Drug Discovery and Healthcare with Artificial Intelligence*.

Russo, M., Sogari, A., & Bardelli, A. (2021). Adaptive evolution: How bacteria and cancer cells survive stressful conditions and drug treatment. *Cancer Discovery*, *11*(8), 1886–1895. DOI: 10.1158/2159-8290.CD-20-1588 PMID: 33952585

Sahel, D. K., Vora, L. K., Saraswat, A., Sharma, S., Monpara, J., D'Souza, A. A., Mishra, D., Tryphena, K. P., Kawakita, S., Khan, S., Azhar, M., Khatri, D. K., Patel, K., & Singh Thakur, R. R. (2023). CRISPR/Cas9 genome editing for tissue-specific in vivo targeting: Nanomaterials and translational perspective. *Advanced Science (Weinheim, Baden-Württemberg, Germany)*, *10*(19), 2207512. DOI: 10.1002/advs.202207512 PMID: 37166046

Sander, J. D., & Joung, J. K. (2014). CRISPR-Cas systems for genome editing, regulation and targeting. *Nature Biotechnology*, *32*(4), 347–355. DOI: 10.1038/nbt.2842 PMID: 24584096

Sehgal, K., Portell, A., Ivanova, E. V., Lizotte, P. H., Mahadevan, N. R., Greene, J. R., Vajdi, A., Gurjao, C., Teceno, T., Taus, L. J., Thai, T. C., Kitajima, S., Liu, D., Tani, T., Noureddine, M., Lau, C. J., Kirschmeier, P. T., Liu, D., Giannakis, M., & Barbie, D. A. (2021). Dynamic single-cell RNA sequencing identifies immunotherapy persister cells following PD-1 blockade. *The Journal of Clinical Investigation*, *131*(2), e135038. DOI: 10.1172/JCI135038 PMID: 33151910

Selvaraj, N., Swaroop, A. K., Nidamanuri, B. S. S., Kumar, R. R., Natarajan, J., & Selvaraj, J. (2022). Network-based drug repurposing: A critical review. *Current Drug Research Reviews*, *14*(2), 116–131. DOI: 10.2174/258997751466622021420403 PMID: 35156575

Song, Z., Tao, Y., Liu, Y., & Li, J. (2024). Advances in delivery systems for CRISPR/Cas-mediated cancer treatment: A focus on viral vectors and extracellular vesicles. *Frontiers in Immunology*, *15*, 1444437. DOI: 10.3389/fimmu.2024.1444437 PMID: 39281673

Tang, S., Chen, X., Tong, X., & Zhu, L. (2025). Overcoming the delivery challenges in CRISPR/Cas9 gene editing for effective cancer treatment: A review of delivery systems. *International Journal of Medical Sciences*, *22*(14), 3625–3649. DOI: 10.7150/ijms.112724 PMID: 40959562

Tiwari, A., Mishra, S., & Kuo, T.-R. (2025). Current AI technologies in cancer diagnostics and treatment. *Molecular Cancer*, *24*(1), 159. DOI: 10.1186/s12943-025-02369-9 PMID: 40457408

Trédan, O., Galmarini, C. M., Patel, K., & Tannock, I. F. (2007). Drug resistance and the solid tumor microenvironment. *Journal of the National Cancer Institute*, *99*(19), 1441–1454. DOI: 10.1093/jnci/djm135 PMID: 17895480

Visan, A. I., Popescu-Pelin, G., & Socol, G. (2021). Degradation behavior of polymers used as coating materials for drug delivery—A basic review. *Polymers*, *13*(8), 1272. DOI: 10.3390/polym13081272 PMID: 33919820

Wang, M., Liu, H., Huang, J., Cai, T., Xu, Z. P., & Zhang, L. (2025). Advancing cancer gene therapy: The emerging role of nanoparticle delivery systems. *Journal of Nanobiotechnology*, *23*(1), 362. DOI: 10.1186/s12951-025-03433-8 PMID: 40394591

Yadav, A., & Yadav, K. (2015). Nanomedicines in the treatment of methicillin-resistant Staphylococcus aureus. Academia Drug Development and Pharmacotherapy 2026; 2. In (Vol. 33, pp. 941-951).

# Chapter 10
# From In-Vitro Systems to Patient-Derived Xenografts:
## Translational Platforms for Sustainable Nano-Oncology

**Sheikh Ahmed Gull**
*Department of Zoology, University of Education, Attock, Pakistan*

**Amjad Khan**
*Institute of Biological Sciences, Department of Zoology, University of Lakki Marwat, Pakistan*

**Muhammad Ali**
http://orcid.org/0009-0002-7593-0395
*Department of Zoology, Quaid-i-Azam University, Pakistan*

**Ayesha Sani**
*School of Engineering and Sciences, Tecnológico de Monterrey, Mexico*

**Yara Cecilia Almanza Arjona**
*School of Engineering and Sciences, Tecnológico de Monterrey, Mexico*

**Chaudhry Muhammad Furqan**
*School of Photovoltaic and Renewable Energy Engineering, University of New South Wales, Australia*

## ABSTRACT

*Traditional preclinical tumor models have limited predictive capabilities, hindering effective translation in nano-oncology. To address this, integration of advanced in vitro systems with patient-derived xenograft (PDX) models provides a robust pathway. This chapter presents an overview of two-dimensional and three-dimensional in vitro models, microfluidic tumor-on-chip systems, and in vivo approaches, with emphasis on sustainability-driven strategies. These platforms assess efficacy, biodistribution, pharmacokinetics, and safety of green-synthesized and biodegradable nanomaterials. PDX models are highlighted for their ability to preserve tumor*

DOI: 10.4018/979-8-3373-9848-8.ch010

*heterogeneity and patient-specific molecular features, enabling more precise therapeutic evaluation. Key challenges, including scalability, ethical considerations, and regulatory requirements, are discussed alongside emerging trends such as humanized PDX models and artificial intelligence-based translational approaches. Overall, this chapter highlights a continuum bridging preclinical research and clinical application in nano-oncology.*

## 1 INTRODUCTION

*Figure 1. Translational platforms for sustainable nano-oncology*

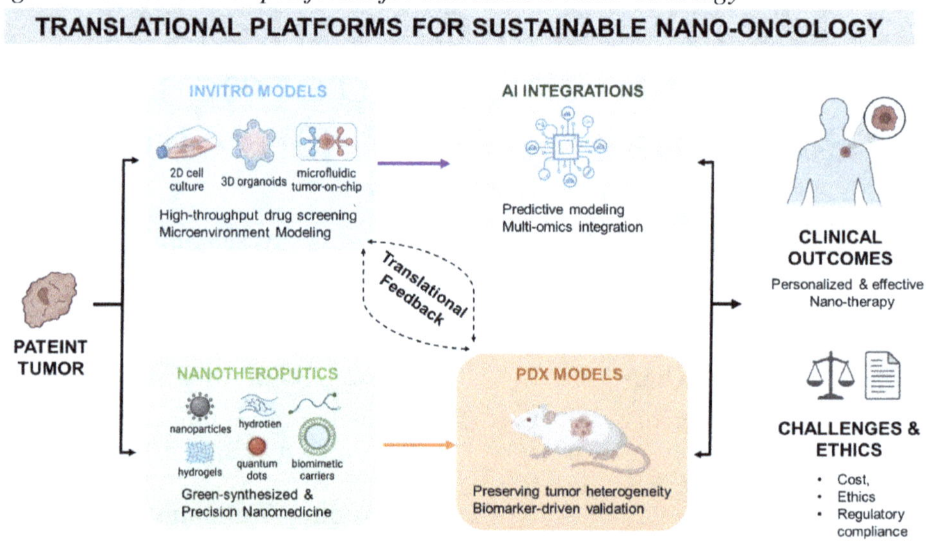


Cancer is a major global health problem and remains one of the leading causes of death worldwide. It is estimated to account for nearly 10 million deaths annually, which is roughly one out of every six deaths (Sung et al., 2021). More recent data suggest that in 2022 alone, there were about 20 million new cancer cases and close to 9.7 million deaths. These statistics indicate the rate of cancer burden rising. Regarding lifetime risk, every person has an estimated risk of one in five to develop cancer, and the risk of death due to the disease death has been reported to be one in nine in men and one in twelve in women (Bray et al., 2024). The main factors that are typically associated with this trend on the upswing are related to the growing life expectancy and population, yet lifestyle and the exposures of the environment are also a significant factor. Tobacco use, obesity, infections, poor diet, and air pollution are among the most commonly reported contributors (Liu et al., 2024). At the same time, there are clear differences between regions. Although higher incidence rates are often reported in more developed countries, mortality tends to be higher in low- and middle-income regions. This is mainly due to late diagnosis and limited access to proper treatment facilities (Sung et al., 2021).

Although instead of improvements in cancer treatment, there are still major shortcomings. The dose-limiting treatment-related toxicities can affect heart, lungs, kidneys, nervous system, and other organs that can result in long-term complications (Kaur et al., 2023). Moreover, heterogeneity of tumors, cellular inactivity and resistance to therapy lead to the recurrence of diseases and incurability of a large number of advanced malignancies (Zafar et al., 2025). The rising cost of treatment and infrastructure is also adding to the global cancer disparity, especially in resource-constrained environments where access to necessary surgery, radiotherapy, and systemic therapies is still insufficient (Bizuayehu et al., 2024).

Integrated nano-oncology has become a multidisciplinary area which involves ultrasensitive biomarkers detection, multimodal nano-imaging, engineered nanodrug delivery systems, nano-immunotherapy, and new nanorobotic platforms (Jiang et al., 2024). Cancer nanomedicine has outgrown its initial dependency on the drug carriers based on the enhanced permeability and retention (EPR) effect to more advanced systems that assist with early detection, immune modulation, cancer vaccination, and image-guided surgical operations (Souri et al., 2022).

To provide translational and environmental sustainability, the contemporary development paradigms focus on simplifying nanocarrier structure, decreasing systemic toxicity, setting up of robust and harmonized safety criteria, and scalable manufacture of good manufacturing practice (GMP)-like production. New approaches are the creation of smart and modular nanocarriers, biomimetic delivery systems, and green or biogenic synthesis methods to increase clinical feasibility and ease of production, for example environmentally friendly production of gold nanoparticles (NPs) (Souri et al., 2022).

To enhance the success of translational research, a strategic 5R top-down development model has been suggested, with specific attention to identifying appropriate patient, appropriate tissue, appropriate safety profile, appropriate molecular target and appropriate commercial potential at an early stage of the innovation process (Lammers, 2024). Stratification of patients based on enhanced imaging and biomarker profiling is becoming a mandate for identifying tumors with favorable transport properties and immune contexture, thereby maximizing nanotherapeutic efficacy (van der Meel et al., 2019). More so, selection of rational drugs and carriers, modular prodrug development, and incorporation of nanomedicines into existing combination regimens, such as immunotherapy and radiotherapy, are essential to clinical impact (Peng et al., 2025).Advanced computational modeling, AI, microfluidic platforms, tumor-on-chip systems, and nanoparticle barcoding technologies are now being leveraged to optimize design parameters and to better understand nano–bio interactions across translational stages (Souri et al., 2022).

Conventional two-dimensional (2D) cell lines of cancer are still widely used in cancer studies. Their reasons are mostly in the fact that they are relatively cheap, readily cultivable, and can be used to investigate biological processes and to test drugs. Nevertheless, as time has gone by, there has been a limitation to the culturing of these cell lines such as the loss of tumor heterogeneity, reduction of the capacity to represent actual patient tumors. Due to this fact, responses of drugs in cell lines do not necessarily give good results in clinical outcomes. This is among the factors that lead to the failure of more than 90 percent of oncology drugs undergoing clinical development (Hidalgo et al., 2014).

To overcome these limitations, we use three-dimensional (3D) models, patient-derived organoids (PDOs) and organoids derived from patient-derived xenografts (PDXOs). These models retain genetic features, structural organization, and diversity of the original tumors (Liu et al., 2023). They also make it possible to perform drug screening at a relatively higher throughput while still maintaining stronger biological relevance. In particular, studies have shown that PDXOs and gastric cancer PDX-derived assembloids can reproduce drug response patterns seen in matched in vivo tumors. This includes similar resistance mechanisms and gene expression profiles. Overall, these models help bridge the gap between in vitro experiments and in vivo systems, making it easier to prioritize potential treatment strategies earlier in the research process (Xu et al., 2024).

Patient-derived xenograft (PDX) models have been used as one of the gold standards in translational research in oncology. PDXs maintain tumor histology, genomic landscapes and extensive intratumoral heterogeneity, and minimally limited parts of the tumor microenvironment (TME). PDX exhibit clinical higher responses (concordance of about 70-80%) compared to conventional cell-lines (Wagner et al., 2025). Prospective resolution of genotype-drug response relation and resistance mechanisms are made possible by large PDX consortia and 1x1x1 study designs, which often replicate results of human clinical trials (Zeng et al., 2023).

Regardless of these benefits, PDX models have their own drawbacks, such as high cost, long experimental times engraftment bias in the case of tumor of aggressive phenotype, replacement of human stroma with that of the mouse, and poor immune system assay. To address these drawbacks, faster systems, including mini-PDX systems and chorioallantoic membrane (CAM)/chick-embryo PDX models, have been created, and drug testing can be completed within a few days even though they still offer high predictive value (Zeng et al., 2023). This chapter provides a comprehensive literature review of the translational platforms in two and three dimensional in vitro models, microfluidic tumor-on-chip, and PDX models approach with particular emphasis on sustainability-driven nano-oncology. These models are considered concerning the determination of the efficacy, biodistribution, pharmacokinetics, and safety of green-synthesized and biodegradable nanomaterials.

## 2 SUSTAINABLE NANO-ONCOLOGY: CONCEPTS AND DESIGN PRINCIPLES

Sustainable nano-oncology also called green nano-oncology, is an emerging research direction that combines cancer nanomedicine with environmental responsibility and long-term safety. This concept not only focuses on improving therapeutic effect and reducing toxicity, but also pays attention to environmental sustainability, ethical application, and efficient use of resources in the development of nanodrugs. In this field, researchers emphasize that nanomaterials should be safe not only for humans, but also for the environment. In this way, biomedical development can be consistent with global goals of sustainability (Petrović et al., 2024; Chehelgerdi et al., 2023; Lammers, 2024).

Furthermore, the scope of sustainable nano-oncology is explained by some frameworks, such as the GREENER concept, which provides more requirements for the design of nanopharmaceuticals. These requirements include good biodegradability, avoidance of persistent, bioaccumulative and toxic characteristics, and the use of active risk control strategies. With these criteria, the research focus is no longer only on patient outcomes, but also considers environmental impact and long-term effects. Therefore, it can improve the practical value and application potential of nanomedicine (Petrović et al., 2024).

Green synthesis plays important role in sustainabile in nano-oncology. Biological methods, including plant-based and microbial-based synthesis of NPs such as gold (Au), silver (Ag), and zinc oxide (ZnO), have received much attention because they are environmentally friendly. These methods usually use water as the reaction medium and natural substances as reducing agents, so harmful chemicals can be avoided. In addition, the NPs produced by these methods often show better biocompatibility and also have strong anticancer effects in both in vitro and in vivo studies (Petrović et al., 2024; Zhuang et al., 2025; Bharadwaj et al., 2021; Sukri et al., 2023).

Besides biogenic NPs, green synthesis methods are also used for silica-based nanomaterials and other advanced systems. These are extensively used in cancer imaging, drug delivery and photothermal therapy. Simultaneously, environmental pollution and threats to the workers can be minimized using environmentally friendly preparation techniques. Moreover, some new strategies, such as Nano-Ayurvedic gold NPs, have shown early clinical potential. This is especially observed in metastatic breast cancer when they are used together with traditional treatment methods (Holghoomi et al., 2024; Khoobchandani et al., 2020). For better understanding,

**Table 1** shows the main green synthesis strategies, their environmental advantages, and their application in oncology.

*Table 1. Green synthesis strategies and their relevance in sustainable nano-oncology*

| Strategy | Features | Relevance | References |
|---|---|---|---|
| Plant/microbial-mediated NPs (Au, Ag, ZnO) | Use of aqueous media, natural reducing agents, low toxicity | Enhanced biocompatibility; anticancer and antimicrobial activity | Petrović et al., 2024; Zhuang et al., 2025; Bharadwaj et al., 2021 |
| Green silica-based NPs | Eco-friendly solvents and synthesis conditions | Imaging, drug delivery, photothermal therapy | Holghoomi et al., 2024 |
| Nano-Ayurvedic gold NPs | Phytochemical capping, oral delivery, minimal chemical processing | Demonstrated benefit in pilot clinical study (breast cancer) | Khoobchandani et al., 2020 |

To design nanomaterials for clinical use, biocompatibility, biodegradability and scalability should be taken into consideration. Biocompatibility is very important because it helps to reduce harmful effects to the human body. Therefore, many researchers focus on using naturally derived or degradable materials, such as poly-glutamic acid and covalent organic frameworks (COFs). These materials can be designed in a way to lower toxicity, while still keeping good therapeutic performance. In this way, their potential for clinical application can be improved (Petrović et al., 2024; Rizawan et al., 2025; Singh et al., 2022).

Biodegradability and effective clearance are also important factors in nanomaterial design. Other high-tech nanocarriers such as the metal-organic frameworks and COFs are characterized by degradation in certain tumor conditions, such as acidic pH or enzyme presence. Such specific degradation can assist in targeted more release of drugs in the tumor site, and also can minimize the accumulation within the body in the long term. Consequently, it is possible to reduce toxicity and make the system of nanomedicine safer and more sustainable (Wang et al., 2024; Chehelgerdi et al., 2023; Rizawan et al., 2025; Singh et al., 2022).

Although these improvements have been made, scalability is a significant issue in the sustainable nano-oncology. Although green synthesis methods have obvious benefits to the environment, their reproducibility and compatibility with GMP remain to be enhanced. Increasing the synthesis of biogenic NPs particularly the gold NPs, usually comes with variability and challenges in process control. Solving these problems is critical to the realization of a large-scale manufacturing process and the possibility of clinical translation (Zhuang et al., 2025; Chehelgerdi et al., 2023; Souri et al., 2022; Mundekkad and Cho, 2022).

Preclinical models are also to be considered in terms of sustainability. The use of animals in experiments can be reduced by the use of plant-derived scaffold models, modern in vitro systems, and systems that are environmentally friendly. Clinically predictive devices, such as organ-on-chip systems and integrated translational systems can be used to minimize clinical trial failure, save resources, and enhance the efficiency of research (Petrovic et al., 2024; Wang et al., 2024; Lammers, 2024; Souri et al., 2022).

Sustained nano-oncology studies have been increasingly becoming common. The production of NPs is the main subject of initial research. Recent studies are more in the focus of the translational design, sustainability evaluation, and clinical application. This indicates that the discipline is increasingly growing and can be used to help advance the field of healthcare as well as the environment.

Finally, sustainable nano-oncology offers a paradigm that brings together green chemistry, design of novel nanomaterials and translational approaches. Using eco-friendly synthesis, biodegradable materials, and clinically relevant models supports development of safe and effective cancer therapy while reducing environmental impact. Future development depends on solving problems in scalability, regulatory approval, and large-scale production.

## 3 IN-VITRO SYSTEMS IN TRANSLATIONAL NANO-ONCOLOGY

In vitro systems are critical platforms in translational nano-oncology to assess behavior of NPs such as cellular uptake, cytotoxicity, and therapeutic efficacy before validation in In-vivo Models. The transition from 2D monocultures to sophisticated 3D and microfluidic systems underscores the need for physiologically relevant platforms that faithfully recapitulate TME and bridge the gap to clinical translation. (Vakhshiteh et al., 2023; Alsheleh et al., 2025). **Table 2** give a concise overview of key in vitro systems in translational nano-oncology with their application and limitations.

*Table 2. Comparison of in vitro models in translational nano-oncology*

| Model Type | Role in Nano-oncology | Limitations | References |
|---|---|---|---|
| 2D cell culture (monolayers) | High-throughput NP screening, cytotoxicity, uptake, mechanistic studies | Lack of TME, no 3D structure, overestimates efficacy | Pinto et al., 2020; Habanjar et al., 2021; Vakhshiteh et al., 2023 |
| 3D (spheroids/ organoids) | NPs penetration, resistance, tumor-like gradients, precision studies | Lower throughput, complexity, standardization issues | Nayak et al., 2023; Esposito et al., 2024; Tosca et al., 2023 |
| Tumor-on-chip / bioprinted systems | Flow-based NP delivery, tumor–stroma–immune interaction | Technical complexity, cost, incomplete TME | De Roode et al., 2024; Gaebler et al., 2024; Moghimi et al., 2023 |

## 3.1 Two-Dimensional Cell Culture Models

The most common cell culture systems used in nano-oncology are 2D models because of their simplicity, low cost, and capability of using for high-throughput screening systems. Such models allow assessing the NPs cytotoxicity, uptake, and mechanisms of action in a short time, which makes them applicable in the early stages of screening (Vakhshiteh et al., 2023; Alsheleh et al., 2025). Nevertheless, 2D systems do not recapitulate tumor architecture in 3D and do not recapitulate important aspects of TME including extracellular matrix interactions, hypoxia and nutrient gradients. As a result, the NPs uptake and therapeutic effect are overrated, and thus cannot be predicted accurately when applied to clinical translation (Pinto et al., 2020; Habanjar et al., 2021; Oliveira et al., 2024; Simelane and Abrahamse, 2024).

## 3.2 Three-Dimensional Tumor Models

Three dimensional tumor models (spheroids and organoids) give a better physiological context of tumors with architecture, hypoxic gradients, and cell-cell interactions. The models can evaluate NPs penetration, distribution and therapeutic effects. In comparison with 2D cultures, 3D cultures can have reduced NPs uptake and drug resistance, which itself is more reminiscent of tumor behavior in vivo (Shah et al., 2025; Jubelin et al., 2022; Nayak et al., 2023; Vakhshiteh et al., 2023). PDOs are also becoming popular to assess personalized treatment responses due to recapitulating tumor heterogeneity (Esposito et al., 2024; Stojanov et al., 2025; Zhang et al., 2026).

## 3.3 Advanced In-Vitro Platforms

Tumor-on-chip and 3D bio-printing models are most advanced models for providing translational insights in nano-oncology. Microfluidics mimic blood flow, vascular barriers, and dynamic TME conditions, are also built into these systems. This system allows a more accurate prediction of nanoparticle transport, extravasation, and distribution (De Roode et al., 2024; Jouybar et al., 2024; Gaebler et al., 2024). Besides, the co-culture systems of tumor, stromal, and immune cells also provide further understanding of the relationship of NPs in a complex biology. Parameters of hypoxia, pH and matrix stiffness can be controlled through these platforms and can also be used to assess advanced and stimuli-responsive nanocarriers and minimize the use of animal models (Abuwatfa et al., 2024; Moghimi et al., 2023).

## 4 TRANSLATIONAL EVALUATION USING IN-VIVO MODELS

Within the context of translational nano-oncology, in vivo evaluation remains a critical step for validating nanocarrier performance beyond in vitro systems. The models are useful to comprehend the biodistribution, pharmacokinetics, tumor uptake and therapeutic efficacy of NPs in an entire biological system. At the same time, they provide essential information regarding safety, toxicity, and long-term effects, which are necessary for successful clinical translation. Traditional tumor models, when used alongside more advanced nano-toxicology approaches, give researchers a well-rounded way to assess both how effective nanomedicines are and how safe they might be (Boix-Montesinos et al., 2021; Shen et al., 2023). To make their roles clearer, **Table 3** breaks down the main in vivo models used in translational nano-oncology, highlighting where they're useful and where their limitations lie.

*Table 3. In vivo models and their role in translational nano-oncology*

| Model | Key Applications | Major Limitations | References |
|---|---|---|---|
| CDX | Tumor growth, NP biodistribution, efficacy studies | Low heterogeneity, exaggerated EPR effect | Boix-Montesinos et al., 2021; Shen et al., 2023 |
| Syngeneic | Immune response studies, immunotherapy evaluation | Species differences, murine tumor biology | Boix-Montesinos et al., 2021 |
| PDX | Patient-specific response, tumor heterogeneity | Immunodeficient system, stromal replacement | Yoshida, 2020; Liu et al., 2023 |

### 4.1 Conventional Xenograft and Syngeneic Models

Xenograft and syngeneic tumor models are traditional in vivo tumor models that are broadly used to assess nanomedicine efficacy. Cell-line-derived xenograft (CDX) models are implantation models that use human cancer cells into immunodeficient mice and are mainly popular because of their simplicity and reproducibility. These models allow bio-distribution, pharmacokinetics, and tumor growth prevention of NPs (Boix-Montesinos et al., 2021; Shen et al., 2023).

Instead, syngeneic models employ tumor cells of the mouse line which are implanted into immune-competent mice and the immune system is maintained. This renders them especially effective to examine immunomodulatory nanotherapies and comprehending the interaction between NPs and components of the immune system. Syngeneic xenograft models also have limitations such as species differences and changes in TME, which might result in overestimating the therapeutic effect (Boix-Montesinos et al., 2021; Cheng et al., 2021).

A more advanced solution for PDX models is implanting tumor tissues from patients to mice. These models do not harm tumor heterogeneity and architecture; they provide improved clinical responses prediction to nanomedicine. Nevertheless, such drawbacks as human immune components and the slow substitution of human stroma are still essential issues (Yoshida, 2020; Liu et al., 2023).

## 4.2 Nano-Toxicological and Safety Assessment

Nano-toxicological analysis plays a critical role for making nanomedicines safe before clinical use. Measurement of acute and chronic toxicity in vivo is normally done using body weight, organ histopathology, hematology and biochemical parameters. Through this we can detect possible side effects and establish the safe dose of nanocarriers (Chen et al., 2020; Liu et al., 2020).

NPs can cause immunogenicity, oxidative stress, and inflammation based on compositions and physicochemical characteristics. Another factor that affects nanoparticle behavior and safety profiles is the protein corona formation and buildup in targeted organs (liver, spleen). Therefore, these interactions are essential to reduce off-target effects and enhance therapy to maximize therapeutic outcomes (Damasco et al., 2020; Xuan et al., 2023).

A safe-by-design approach is becoming part of the nano-oncology to cope with these challenges. This will entail the inclusion of physicochemical characterization, predictive models in vitro and specific in vivo studies into a single framework. These approaches assist in creating safer and more sustainable nanomedicines and comply with the requirements of regulations and translation to clinical applications (Xuan et al., 2023; Shen et al., 2023). In vivo testing involving tumor models with nano-toxicological testing is crucial in the efficacy and safety of nanomedicine in translational nano-oncology.

# 5 PDX AS TRANSLATIONAL PLATFORMS IN NANO-ONCOLOGY

PDX have emerged as powerful translational models capable of retaining the histologic, genomic and phenotypic heterogeneity of human tumors that is superior to the traditional CDX. PDX models provide an immunodeficient or humanized mouse model of interest as one of the biologically relevant systems, evaluating targeted therapies, such as precision nano-formulations, prior to bridging the translational gap between preclinical studies and clinical implementation.

## 5.1 Development and Characterization of PDX Models

PDX models are created by implanting tumor tissue into immunocompromised mice. This can be done subcutaneously or at the original tumor site (orthotopic). After implantation, the tumors are passaged multiple times. Samples are also stored in biobanks. This helps maintain stability and ensures long-term availability of the models (Liu et al., 2023; Jin et al., 2023).

One of the most Standardized frameworks PDX-MI plays an important role. They guide the collection of key clinical, methodological, and molecular data. This improves reproducibility and consistency across different laboratories (Liu & Yang, 2025).

For confirming the accuracy of these models, multi-omics approaches (whole-exome sequencing, RNA sequencing, single-cell profiling, and methylation analysis) are applied. These methods help compare histology, gene expression, and driver mutations with the original patient tumor. They also help monitor any changes during serial passaging (Gu et al., 2025). Overall, these comprehensive analyses ensure that PDX models closely represent patient-specific tumor biology.

## 5.2 Preservation of Tumor Heterogeneity

The possibility to maintain inter- and intra-patient heterogeneity across passages is one of the main benefits of PDX models (Liu et al., 2023; Abdolahi et al., 2022). PDX models of large panels represent diversity of patients with a range of different types of cancers, also provide opportunity to conduct powerful translational studies (Li et al., 2025).

Nevertheless, the formation of PDX may cause a subclonal bottleneck: xenografts of an individual usually possess only a fraction of the subclonal structure of the tumor. Multi-region sampling or parallel generation of PDX per patient are some of the strategies that could be suggested to better represent the intratumoral heterogeneity. Besides, the human stroma and immune cells are replaced step by step by the murine ones, which can change the TME and affect therapeutic results (Jin et al., 2023).

# 6 PDX MODELS IN SUSTAINABLE AND PRECISION NANO-ONCOLOGY

PDX models are useful for testing NPs based therapeutics which can allow co-clinical testing, identify biomarkers, and conduct mechanistic research on resistance (Liu et al., 2023; Li et al., 2025; Gu et al., 2025). Genomically annotated PDX models can be used to prioritize therapeutic candidates such as nanocarriers, by aligning oncogenic changes with treatment plans on the panels. In receptor-mediated endocytosis, humanized PDX also make it possible to test immune-modulating NPs using human tumor-immune interactions in partially recapitulated human tumor-immune systems.

In Precision nano-oncology PDX models are useful because they maintain patient TM and can be used to investigate complex nano-drug delivery systems (Liu et al., 2023; Gu et al., 2025; Liu and Yang, 2025). Novel nanotherapeutics (hydrogels, hyaluronic acid carriers, quantum dots, and enzyme-responsive polymeric prodrug) show significant anti-tumor activity and survival results in PDX Models (Tan et al., 2021; Wu et al., 2023). PDX also make it possible to provide co-clinical testing in the "avatar" of patients, i.e., nanoformulations can be designed and tested before clinical trials to enhance efficacy and safety (Abdolahi et al., 2022; Invrea et al., 2020; Yoshida, 2020).

Combination of PDX with multi-omics, AI and machine learning (ML) enhances even more precision in nano-oncology methods by connecting molecular profiles to treatment response patterns and resistance patterns (Janitri et al., 2024; Perron et al., 2024; Zanella et al., 2022). The PDX-based experiments enable longitudinal monitoring of clonal evolution in response to selective pressure due to the presence of nano-encapsulated chemotherapeutics, photodynamic and photothermal systems, enabling rational design of combination therapy and resistance mitigation strategies. Translational pursuant to these models. Case studies, such as enzyme-responsive polymeric prodrugs achieving >98% tumor growth inhibition in bladder cancer PDXs and biomimetic cancer-cell-membrane-coated nanosystems in HNSCC, highlight the translational potential of these models for validating complex nanomedicines (Tan et al., 2021; Wu et al., 2023).

In vitro Complementary in vitro screening of nanoformulations can be conducted on complementary in vitro platforms, including organoids, tumor spheroids, tumor-on-chip, and zebrafish PDX models, which offer rapid and higher-throughput screening with no need for animals (Ong et al., 2022; Kumari et al., 2022; Ali et al., 2022). With mouse PDX studies, these platforms can be used consecutively to create a translational continuum: initial in vitro screening can be used to narrow candidate nanotherapeutics, and follow-up PDX studies can be used to establish pharmacokinetics, biodistribution, microenvironment interactions, and therapy response. These tiers of approaches contribute to strengthening sustainability, minimizing wasteful use of animals, and improving the chances of clinical success of precision nanomedicines (Liu et al., 2023; Gu et al., 2025; Liu and Yang, 2025).

Altogether, PDX models represent an area of focus of sustainable and precision nano-oncology. Their capability of reproducing human tumor biology combined with multi-omics with AI-based assessment allows rigorous preclinical validation of nanotherapeutics, enhanced predictive ability, and rationalized patient-based therapeutic design. One of the key discovery platforms in nano-oncology is patient-derived xenografts. Maintaining histologic structure, genomic faithfulness, and tumor heterogeneity, PDX enable a predictive platform of preclinical assessment of targeted and precision treatment. Even though it also presents its own problems, such as subclonal bottlenecks, stromal replacement and resource intensity, PDX plays a key role in the discovery of the laboratory success in cancer nanomedicine.

## 7 CHALLENGES, ETHICS, AND REGULATION IN PDX-BASED NANO-ONCOLOGY

PDX organoids and hi-tech in vitro systems provide powerful platforms for sustainable nano-oncology. They allow the precise study of the tumor heterogeneity, drug activation and TME dynamics. Their widespread usage is limited by practical, ethical and regulatory concerns. These issues affect scalability, reproducibility and translation use of nanotherapy. The key issues and future perspectives of PDX applications in nano-oncology are mentioned below.

### 7.1 Key Challenges

PDX models are time-consuming and Expensive. The success rates differ depending on the type of tumor. In the majority of the cases, a stable passage line involves months of passaging to achieve. Hence, these models are slow, and they also postpone the screening of nanoformulations (Liu et al., 2023; Janitri et al., 2024; Yoshida, 2020).

Organoids and 3D culture are faster alternatives. Nevertheless, there are still limitations to these ways. Batch-to-batch Variability is typical and culturing process is complicated. These complicate the reproducibility of the experimental results and affects comparison of the various studies (Heinzelmann et al., 2024; Luce and Duclos-Vallee, 2025; Zhao et al., 2025). In recent years, miniature organoid spheres and microfluidic platforms have gradually developed. These systems can complete drug testing in about 14 days while retaining key matrix and immunomodulatory features. Therefore, they offer a more ideal balance between experimental speed, biological realism, and scalability (Ding et al., 2022).

Ethical and regulatory issues also limit the application of PDX models. Animal experiments must strictly adhere to the "3R principle": Replacement, Reduction, and Refinement. Simultaneously human organoids and induced pluripotent stem cell (iPSC)-derived systems require rigorous ethical management. This includes informed consent, privacy protection, and intellectual property management, especially in the construction of long-term biobanks (Heinzelmann et al., 2024; Zhao et al.,

2025). From a regulatory perspective, relevant institutions are gradually increasing requirements for human relevance and data standardization. Therefore, organoids and organ-on-a-chip systems are being seen as important supplements to animal experiments, and even as alternatives in drug development in some cases (Ingber, 2022; Luce & Duclos-Vallée, 2025; Tosca et al., 2023). However, there is still a lack of unified experimental procedures, testing standards, and reporting norms, which to some extent limits the reproducibility of results and the comparability between research institutions (Yoshida, 2020; Singh et al., 2025; Wang et al., 2025; Thorel et al., 2024).

Altogether high cost, limited scalability, long development cycles, and stringent ethical and regulatory requirements remain the main bottlenecks for the application of PDX in nano-oncology. To address these challenges, a tiered research strategy is necessary.

## 7.2 Emerging Trends and Future Directions

Several emerging strategies are reshaping PDX-based nano-oncology research and overcome existing limitations. Humanized PDX models, by partially reconstructing human TME, enable the evaluation of nanoimmunotherapy. Simultaneously, these models can improve the predictive accuracy of immunomodulatory nanomedicines (Chuprin et al., 2023; Liu et al., 2023). Furthermore, the integration of AI with computational modeling is accelerating the field's development. These methods can be used to predict nanomedicine responses, optimize PDX and organoid model design, and integrate multi-omics data to support biomarker-guided precision medicine (Janitri et al., 2024; Luce & Duclos-Vallée, 2025; Zhao et al., 2025; Wang et al., 2025). Meanwhile, hybrid platforms are emerging. For example, organoid-PDX combined systems (such as PDO-from-PDX and AVI-PDO) combine rapid in vitro screening with in vivo validation. This model not only speeds up the research process but also makes the findings more relevant and applicable in real-world settings (Kumari et al., 2022; Costechareyre et al., 2025; Singh et al., 2025; Ding et al., 2022).

Generally according to developmental point of view, sustainable nano-oncology studies demand a multi-level approach. First, initial screening with high throughput must be done in standardized systems in three 3D organoids. Second, intermediate level assessments are to be conducted on organ-on-a-chip systems to recreate more complicated physiological conditions. The reproducibility, efficiency, and ethical soundness of study can be further enhanced by creating a standardized system of biobanks, developing co-clinical trials, and implementing AI-based data analysis (Liu et al., 2023; Yoshida, 2020; Singh et al., 2025; Tosca et al., 2023; Thorel et al., 2024). In general, humanized models, hybrid platforms, and AI integration technologies are developing the outline of the future development. This model enables predictable (highly), sustainable, and ethical PDX nano-oncology studies.

## 8 CONCLUSION

Integration of PDX models in in-vitro systems is a more practical and effective platform for sustainable nano-oncology. Advanced 3D models better represent real tumor conditions, while PDX models keep tumor complexity and patient-specific responses, helping connect lab research to clinical results. Although there are challenges like cost and scalability, new approaches such as hybrid models and AI tools are making the process more accurate and sustainable from research to treatment.

## REFERENCES


Abdolahi, S., Ghazvinian, Z., Muhammadnejad, S., Saleh, M., Aghdaei, A., & Baghaei, K. (2022). Patient-derived xenograft (PDX) models, applications and challenges in cancer research. *Journal of Translational Medicine*, *20*(1), 206. Advance online publication. DOI: 10.1186/s12967-022-03405-8 PMID: 35538576

Abuwatfa, W., Pitt, W., & Husseini, G. (2024). Scaffold-based 3D cell culture models in cancer research. *Journal of Biomedical Science*, *31*(1), 7. Advance online publication. DOI: 10.1186/s12929-024-00994-y PMID: 38221607

Ali, Z., Vildevall, M., Rodriguez, G., Tandiono, D., Vamvakaris, I., Evangelou, G., Lolas, G., Syrigos, K., Villanueva, A., Wick, M., Omar, S., Erkstam, A., Schueler, J., Fahlgren, A., & Jensen, L. (2022). Zebrafish patient-derived xenograft models predict lymph node involvement and treatment outcome in non-small cell lung cancer. *Journal of Experimental & Clinical Cancer Research : CR*, *41*(1), 58. Advance online publication. DOI: 10.1186/s13046-022-02280-x PMID: 35139880

Alsheleh, T., Zraikat, M., Daoud, F., Alqudah, D., Abdelghany, S., Siniyeh, A., & Alshaer, W. (2025). In vitro cellular interaction of drug-loaded liposomes with 2D and 3D cell culture. *PLoS One*, *20*(3), e0320374. Advance online publication. DOI: 10.1371/journal.pone.0320374 PMID: 40131912

Bharadwaj, K., Rabha, B., Pati, S., Sarkar, T., Choudhury, B., Barman, A., Bhattacharjya, D., Srivastava, A., Baishya, D., Edinur, H. A., Abdul Kari, Z., & Mohd Noor, N. H. (2021). Green synthesis of gold nanoparticles using plant extracts as beneficial prospect for cancer theranostics. *Molecules (Basel, Switzerland)*, *26*(21), 6389. Advance online publication. DOI: 10.3390/molecules26216389 PMID: 34770796

Bizuayehu, H., Ahmed, K., Kibret, G., Dadi, A., Belachew, S., Bagade, T., Tegegne, T., Venchiarutti, R., Kibret, K., Hailegebireal, A., Assefa, Y., Khan, M., Abajobir, A., Alene, K., Mengesha, Z., Erku, D., Enquobahrie, D., Minas, T., Misgan, E., & Ross, A. (2024). Global disparities of cancer and its projected burden in 2050. *JAMA Network Open*, *7*(4), e2443198. DOI: 10.1001/jamanetworkopen.2024.43198 PMID: 39499513

Boix-Montesinos, P., Soriano-Teruel, P., Armiñán, A., Orzáez, M., & Vicent, M. (2021). The past, present, and future of breast cancer models for nanomedicine development. *Advanced Drug Delivery Reviews*, *173*, 306–330. DOI: 10.1016/j.addr.2021.03.018 PMID: 33798642

Bray, F., Laversanne, M., Sung, H., Ferlay, J., Siegel, R. L., Soerjomataram, I., & Jemal, A. (2024). Global cancer statistics 2022: GLOBOCAN estimates of incidence and mortality worldwide for 36 cancers in 185 countries. *CA: a Cancer Journal for Clinicians, 74*(3), 229–263. DOI: 10.3322/caac.21834 PMID: 38572751

Chehelgerdi, M., Chehelgerdi, M., Allela, O., Pecho, R., Jayasankar, N., Rao, D., Thamaraikani, T., Vasanthan, M., Viktor, P., Lakshmaiya, N., Saadh, M. J., Amajd, A., Abo-Zaid, M. A., Castillo-Acobo, R. Y., Ismail, A. H., Amin, A. H., & Akhavan-Sigari, R. (2023). Progressing nanotechnology to improve targeted cancer treatment: Overcoming hurdles in its clinical implementation. *Molecular Cancer, 22*(1), 169. Advance online publication. DOI: 10.1186/s12943-023-01865-0 PMID: 37814270

Chen, R., Chen, Y., Liao, M., Lee, Y., Chen, Z., Yan, S., Yeh, Y., Yang, L., Lee, Y., Wu, Y., & Wang, Y. (2020). Autophagy in nanomaterial toxicity. *International Journal of Molecular Sciences, 21*(7), 2387. Advance online publication. DOI: 10.3390/ijms21072387 PMID: 32235610

Cheng, Z., Li, M., Dey, R., & Chen, Y. (2021). Nanomaterials for cancer therapy. *Journal of Hematology & Oncology, 14*(1), 85. Advance online publication. DOI: 10.1186/s13045-021-01096-0 PMID: 34059100

Chuprin, J., Buettner, H., Seedhom, M., Greiner, D., Keck, J., Ishikawa, F., Shultz, L., & Brehm, M. (2023). Humanized mouse models for immuno-oncology research. *Nature Reviews. Clinical Oncology, 20*(3), 192–206. DOI: 10.1038/s41571-022-00721-2 PMID: 36635480

Costechareyre, C., Teinturier, R., Loskutov, J., Mongellaz, M., Nadolny, M., Lacourrège, M., Jarrosson, L., Delloye-Bourgeois, C., Wedeken, L., & Castellani, V. (2025). Integrating AVI-PDX and patient-derived organoid technologies for enhanced preclinical modeling in immuno-oncology. *Cancer Research, 85*(8_Supplement_1), 6491. Advance online publication. DOI: 10.1158/1538-7445.AM2025-6491

Damasco, J., Ravi, S., Perez, J., Hagaman, D., & Melancon, M. (2020). Safe-by-design nanomedicine. *Nanomaterials (Basel, Switzerland), 10*. Advance online publication. DOI: 10.3390/nano10112186 PMID: 33147800

De Roode, K., Hashemi, K., Verdurmen, W., & Brock, R. (2024). Tumor-on-a-chip models for predicting nanoparticle behavior. *Small, 20*(35), 2402311. Advance online publication. DOI: 10.1002/smll.202402311 PMID: 38700060

Ding, S., Hsu, C., Wang, Z., Natesh, N., Millen, R., Negrete, M., Giroux, N., Rivera, G., Dohlman, A., Bose, S., Rotstein, T., Spiller, K., Yeung, A., Sun, Z., Jiang, C., Xi, R., Wilkin, B., Randon, P., Williamson, I., & Shen, X. (2022). Patient-derived micro-organospheres enable clinical precision oncology. *Cell Stem Cell*, *29*(6), 905–917. e6. Advance online publication. DOI: 10.1016/j.stem.2022.04.006 PMID: 35508177

Esposito, A., Ferraresi, A., & Vallino, L.. (2024). 3D cell cultures as alternatives in cancer research. *International Journal of Biological Sciences*, *20*, 5293–5311. DOI: 10.7150/ijbs.96469 PMID: 39430243

Gaebler, D., Hachey, S., & Hughes, C. (2024). Tumor microenvironment in chip systems. *Frontiers in Bioengineering and Biotechnology*, •••, 12. PMID: 39386043

Gu, A., Li, J., Li, M., & Liu, Y. (2025). Patient-derived xenograft model in cancer: Establishment and applications. *MedComm*, *6*(2), e70059. Advance online publication. DOI: 10.1002/mco2.70059 PMID: 39830019

Habanjar, O., Diab-Assaf, M., Caldefie-Chézet, F., & Delort, L. (2021). 3D cell culture systems in cancer. *International Journal of Molecular Sciences*, •••, 22.

Heinzelmann, E., Piraino, F., Costa, M., Roch, A., Norkin, M., Garnier, V., Homicsko, K., & Brandenberg, N. (2024). iPSC-derived and Patient-Derived Organoids: Applications and challenges in scalability and reproducibility as pre-clinical models. *Current Research in Toxicology*, *7*, 100197. Advance online publication. DOI: 10.1016/j.crtox.2024.100197 PMID: 40276485

Hidalgo, M., Amant, F., Biankin, A. V., Budinská, E., Byrne, A. T., Caldas, C., Clarke, R. B., de Jong, S., Jonkers, J., Mælandsmo, G. M., Roman-Roman, S., Seoane, J., Trusolino, L., & Villanueva, A. (2014). Patient-derived xenograft models: An emerging platform for translational cancer research. *Cancer Discovery*, *4*(9), 998–1013. DOI: 10.1158/2159-8290.CD-14-0001 PMID: 25185190

Holghoomi, R., Kharab, Z., Rahdar, A., Pandey, S., & Ferreira, L. (2024). Harnessing the power of green synthesis of nanomaterials for anticancer applications. *Coordination Chemistry Reviews*, *513*, 215903. Advance online publication. DOI: 10.1016/j.ccr.2024.215903

Ingber, D. (2022). Human organs-on-chips for disease modelling, drug development and personalized medicine. *Nature Reviews. Genetics*, *23*(8), 467–491. DOI: 10.1038/s41576-022-00466-9 PMID: 35338360

Invrea, F., Rovito, R., Torchiaro, E., Petti, C., Isella, C., & Medico, E. (2020). Patient-derived xenografts (PDXs) as model systems for human cancer. *Current Opinion in Biotechnology*, *63*, 151–156. DOI: 10.1016/j.copbio.2020.01.003 PMID: 32070860

Janitri, V., Aruljothi, K., Mythili, V., Singh, S., Prasher, P., Gupta, G., Dua, K., Hanumanthappa, R., Karthikeyan, K., & Anand, K. (2024). The roles of patient-derived xenograft models and artificial intelligence toward precision medicine. *MedComm*, *5*(10), e745. Advance online publication. DOI: 10.1002/mco2.745 PMID: 39329017

Jiang, J., Cui, X., Huang, Y., Yan, D., Wang, B., Yang, Z., Chen, M., Wang, J., Zhang, Y., Liu, G., Zhou, C., Cui, S., Ni, J., Yang, F., & Cui, D. (2024). Advances and prospects in integrated nano-oncology. *Nano Biomedicine and Engineering*, *16*(2), 152–187. Advance online publication. DOI: 10.26599/NBE.2024.9290060

Jin, J., Yoshimura, K., Sewastjanow-Silva, M., Song, S., & Ajani, J. (2023). Challenges and Prospects of Patient-Derived Xenografts for Cancer Research. *Cancers (Basel)*, *15*(17), 4352. Advance online publication. DOI: 10.3390/cancers15174352 PMID: 37686627

Jouybar, M., de Winde, C. M., Wolf, K., Friedl, P., Mebius, R. E., & den Toonder, J. M. (2024). Cancer-on-chip models for metastasis: Importance of the tumor microenvironment. *Trends in Biotechnology*, *42*(4), 431–448. DOI: 10.1016/j.tibtech.2023.10.001 PMID: 37914546

Jubelin, C., Muñoz-García, J., & Griscom, L.. (2022). 3D in vitro models in oncology. *Cell & Bioscience*, •••, 12.

Kaur, R., Bhardwaj, A., & Gupta, S. (2023). Cancer treatment therapies: Traditional to modern approaches to combat cancers. *Molecular Biology Reports*, *50*(11), 9663–9676. DOI: 10.1007/s11033-023-08809-3 PMID: 37828275

Khoobchandani, M., Katti, K., Karikachery, A., Thipe, V., Srisrimal, D., & Mohandoss, D.. (2020). New approaches in breast cancer therapy through green nanotechnology and nano-Ayurvedic medicine. *International Journal of Nanomedicine*, *15*, 181–197. DOI: 10.2147/IJN.S219042 PMID: 32021173

Kumari, R., Xu, X., & Li, H. (2022). Translational and Clinical Relevance of PDX-Derived Organoid Models in Oncology Drug Discovery and Development. *Current Protocols*, *2*(7), e431. Advance online publication. DOI: 10.1002/cpz1.431 PMID: 35789132

Lammers, T. (2024). Nanomedicine tumor targeting. *Advanced Materials*, *36*(26), e2312169. DOI: 10.1002/adma.202312169 PMID: 38361435

Li, J., Gu, A., Tang, N., Zengin, G., Li, M., & Liu, Y. (2025). Patient-derived xenograft models in pan-cancer: From bench to clinic. *Interdisciplinary Medicine*, *3*(5), e20250016. Advance online publication. DOI: 10.1002/INMD.20250016

Liu, B., Zhou, H., Tan, L., Siu, K., & Guan, X. (2024). Exploring treatment options in cancer: Tumor treatment strategies. *Signal Transduction and Targeted Therapy, 9*(1), 1856. Advance online publication. DOI: 10.1038/s41392-024-01856-7 PMID: 39013849

Liu, M., & Yang, X. (2025). Patient-derived xenograft models: Current status, challenges, and innovations in cancer research. *Genes & Diseases, 12*(5), 101520. Advance online publication. DOI: 10.1016/j.gendis.2025.101520 PMID: 40548062

Liu, X., Tang, I., Wainberg, Z., & Meng, H. (2020). Safety considerations in cancer nanomedicine. *Small, 16*(36), 2000673. Advance online publication. DOI: 10.1002/smll.202000673 PMID: 32406992

Liu, Y., Wu, W., Cai, C., Zhang, H., Shen, H., & Han, Y. (2023). Patient-derived xenograft models in cancer therapy: Technologies and applications. *Signal Transduction and Targeted Therapy, 8*(1), 160. Advance online publication. DOI: 10.1038/s41392-023-01419-2 PMID: 37045827

Luce, E., & Duclos-Vallée, J. (2025). Stem Cells and Organoids: A Paradigm Shift in Preclinical Models Toward Personalized Medicine. *Pharmaceuticals (Basel, Switzerland), 18*(7), 992. Advance online publication. DOI: 10.3390/ph18070992 PMID: 40732281

Moghimi, N., Hosseini, S., Dalan, A., Mohammadizadeh, D., Goldman, A., & Kohandel, M. (2023). Controlled tumor heterogeneity in a co-culture system by 3D bio-printed tumor-on-chip model. *Scientific Reports, 13*(1), 13648. Advance online publication. DOI: 10.1038/s41598-023-40680-x PMID: 37607994

Mundekkad, D., & Cho, W. (2022). Nanoparticles in clinical translation for cancer therapy. *International Journal of Molecular Sciences, 23*(3), 1685. Advance online publication. DOI: 10.3390/ijms23031685 PMID: 35163607

Nayak, P., Bentivoglio, V., Varani, M., & Signore, A. (2023). Tumor spheroid models for anticancer evaluation. *Cancers (Basel)*, •••, 15. PMID: 37835541

Oliveira, B., Fernandes, A., & Baptista, P. (2024). Gene silencing in 2D vs 3D cultures. *Frontiers in Bioengineering and Biotechnology*, •••, 12.

Ong, L., Chia, S., Wong, S., Zhang, X., Chua, H., Loo, J., Chua, W., Chua, C., Tan, E., Hentze, H., Tan, I., Dasgupta, R., & Toh, Y. (2022). A comparative study of tumour-on-chip models with patient-derived xenografts for predicting chemotherapy efficacy in colorectal cancer patients. *Frontiers in Bioengineering and Biotechnology, 10*, 952726. Advance online publication. DOI: 10.3389/fbioe.2022.952726 PMID: 36147524

Peng, Y., Yu, M., Li, B., Zhang, S., Cheng, J., Wu, F., Du, S., Miao, J., Hu, B., Olkhovsky, I., & Li, S. (2025). Advances in nanocarrier-mediated cancer therapy: Progress in immunotherapy, chemotherapy, and radiotherapy. *Chinese Medical Journal, 138*(16), 1927–1944. DOI: 10.1097/CM9.0000000000003703 PMID: 40545573

Perron, U., Grassi, E., Chatzipli, A., Viviani, M., Karakoc, E., Trastulla, L., Brochier, L., Isella, C., Zanella, E., Klett, H., Molineris, I., Schueler, J., Esteller, M., Médico, E., Conte, N., McDermott, U., Trusolino, L., Bertotti, A., & Iorio, F. (2024). Integrative ensemble modelling of cetuximab sensitivity in colorectal cancer patient-derived xenografts. *Nature Communications, 15*(1), 9139. Advance online publication. DOI: 10.1038/s41467-024-53163-y PMID: 39528460

Petrović, S., Bita, B., & Barbinta-Patrascu, M. (2024). Nanoformulations in Pharmaceutical and Biomedical Applications: Green Perspectives. *International Journal of Molecular Sciences, 25*(11), 5842. Advance online publication. DOI: 10.3390/ijms25115842 PMID: 38892030

Pinto, B., Henriques, A., Silva, P., & Bousbaa, H. (2020). 3D spheroids as preclinical models. *Pharmaceutics*, •••, 12.

Rizawan, A., Rehman, U., Gupta, G., Alsayari, A., Wahab, S., & Kesharwani, P. (2025). Polyglutamic acid in cancer nanomedicine. *International Journal of Pharmaceutics, 676*, 125623. Advance online publication. DOI: 10.1016/j.ijpharm.2025.125623 PMID: 40254191

Shah, S., & D'Souza, G. (2025). Tumor microenvironment modeling in vitro. *Cells*, •••, 14. PMID: 40422235

Shen, X., Pan, D., Gong, Q., Gu, Z., & Luo, K. (2023). Nanomedicine in solid tumors. *Bioactive Materials, 32*, 445–472. DOI: 10.1016/j.bioactmat.2023.10.017 PMID: 37965242

Simelane, N., & Abrahamse, H. (2024). Photodynamic therapy in 2D and 3D systems. *Frontiers in Molecular Biosciences*.

Singh, H., Mijaković, I., & Singh, P. (2025). Advances in precision oncology using patient-derived organoids and functional biomaterials. *Frontiers in Cell and Developmental Biology, 13*, 1670328. Advance online publication. DOI: 10.3389/fcell.2025.1670328 PMID: 41098309

Singh, N., Kim, J., Kim, J., Lee, K., Zunbul, Z., Lee, I., Kim, E., Chi, S.-G., & Kim, J. S. (2022). Covalent organic framework nanomedicines. *Bioactive Materials, 21*, 358–380. DOI: 10.1016/j.bioactmat.2022.08.016 PMID: 36185736

Souri, M., Soltani, M., Kashkooli, F., Shahvandi, M., Chiani, M., Shariati, F., Mehrabi, M., & Munn, L. L. (2022). Towards principled design of cancer nanomedicine to accelerate clinical translation. *Materials Today. Bio*, *13*, 100208. DOI: 10.1016/j.mtbio.2022.100208 PMID: 35198957

Stojanov, J., Grozdanić, M., & Ljujić, M.. (2025). Cancer 3D models and drug resistance. *Oncology Research*, •••, 33. PMID: 41050078

Sukri, S., Shameli, K., Teow, S., Chew, J., Ooi, L., & Soon, M.. (2023). Enhanced antibacterial and anticancer activities of green synthesized nanoparticles. *Frontiers in Microbiology*, *14*. Advance online publication. DOI: 10.3389/fmicb.2023.1194292 PMID: 37577438

Sung, H., Ferlay, J., Siegel, R. L., Laversanne, M., Soerjomataram, I., Jemal, A., & Bray, F. (2021). Global cancer statistics 2020: GLOBOCAN estimates of incidence and mortality worldwide for 36 cancers in 185 countries. *CA: a Cancer Journal for Clinicians*, *71*(3), 209–249. DOI: 10.3322/caac.21660 PMID: 33538338

Tan, P., Cai, H., Wei, Q., Tang, X., Zhang, Q., Kopytynski, M., Yang, J., Yi, Y., Zhang, H., Gong, Q., Gu, Z., Chen, R., & Luo, K. (2021). Enhanced chemo-photodynamic therapy of an enzyme-responsive prodrug in bladder cancer patient-derived xenograft models. *Biomaterials*, *277*, 121061. DOI: 10.1016/j.biomaterials.2021.121061 PMID: 34508957

Thorel, L., Perréard, M., Florent, R., Divoux, J., Coffy, S., Vincent, A., Gaggioli, C., Guasch, G., Gidrol, X., Weiswald, L., & Poulain, L. (2024). Patient-derived tumor organoids: A new avenue for preclinical research and precision medicine in oncology. *Experimental & Molecular Medicine*, *56*(7), 1531–1551. DOI: 10.1038/s12276-024-01272-5 PMID: 38945959

Tosca, E., Ronchi, D., Facciolo, D., & Magni, P. (2023). 3D models in drug development. *Biomedicines*, •••, 11. PMID: 37189676

Vakhshiteh, F., Bagheri, Z., & Soleimani, M.. (2023). Heterotypic tumor spheroids for nanomedicine. *Journal of Nanobiotechnology*, •••, 21. PMID: 37533100

van der Meel, R., Sulheim, E., Shi, Y., Kiessling, F., Mulder, W. J. M., & Lammers, T. (2019). Smart cancer nanomedicine. *Nature Nanotechnology*, *14*(11), 1007–1017. DOI: 10.1038/s41565-019-0567-y PMID: 31695150

Wagner, B., Ettner-Sitter, A., Ihlo, N., Behr, M., Koelbl, S., Brunner, S., Weber, F., Rau, B., Schlitt, H., Brochhausen, C., Schoenmehl, R., Artinger, A., Schott, D., Pizon, M., Pachmann, K., Aung, T., Haerteis, S., & Hackl, C. (2025). Patient-derived xenografts from circulating cancer stem cells as a preclinical model for personalized pancreatic cancer research. *Scientific Reports*, *15*(1), 87054. Advance online publication. DOI: 10.1038/s41598-025-87054-z PMID: 39843495

Wang, B., Hu, S., Teng, Y., Chen, J., Wang, H., Xu, Y., Wang, K., Xu, J., Cheng, Y., & Gao, X. (2024). Current advance of nanotechnology in diagnosis and treatment for malignant tumors. *Signal Transduction and Targeted Therapy*, *9*(1), 200. Advance online publication. DOI: 10.1038/s41392-024-01889-y PMID: 39128942

Wu, Q., Chen, L., Huang, X., Lin, J., Gao, J., Yang, G., Wu, Y., Wang, C., Kang, X., Yao, Y., Wang, Y., Xue, M., Luan, X., Chen, X., Zhang, Z., & Sun, S. (2023). A biomimetic nanoplatform for customized photothermal therapy of HNSCC evaluated on patient-derived xenograft models. *International Journal of Oral Science*, *15*(1), 9. Advance online publication. DOI: 10.1038/s41368-022-00211-2 PMID: 36765028

Xu, X., Gao, Y., Dai, J., Wang, Q., Wang, Z., Liang, W., Zhang, Q., Liu, Z., Luo, H., Qiao, Z., Li, L., Wang, Z., Chen, L., Zhang, Y., & Xiong, Z. (2024). Gastric cancer assembloids derived from patient-derived xenografts: A preclinical model for therapeutic drug screening. *Small Methods*, *8*(9), e2400204. DOI: 10.1002/smtd.202400204 PMID: 38948952

Xuan, L., Ju, Z., Skonieczna, M., Zhou, P., & Huang, R. (2023). Nanoparticle toxicity and evaluation models. *MedComm*, *4*. Advance online publication. DOI: 10.1002/mco2.327 PMID: 37457660

Yoshida, G. (2020). Applications of patient-derived tumor xenograft models and tumor organoids. *Journal of Hematology & Oncology*, *13*(1), 4. Advance online publication. DOI: 10.1186/s13045-019-0829-z PMID: 31910904

Zafar, A., Khatoon, S., Khan, M., Abu, J., & Naeem, A. (2025). Advancements and limitations in traditional anti-cancer therapies: A comprehensive review of surgery, chemotherapy, radiation therapy, and hormonal therapy. *Discover. Oncology*, *16*(1), 2198. Advance online publication. DOI: 10.1007/s12672-025-02198-8 PMID: 40272602

Zanella, E., Grassi, E., & Trusolino, L. (2022). Towards precision oncology with patient-derived xenografts. *Nature Reviews. Clinical Oncology*, *19*(11), 719–732. DOI: 10.1038/s41571-022-00682-6 PMID: 36151307

Zeng, M., Ruan, Z., Tang, J., Liu, M., Hu, C., Fan, P., & Dai, X. (2023). Generation, evolution, interfering factors, applications, and challenges of patient-derived xenograft models in immunodeficient mice. *Cancer Cell International*, *23*(1), 2953. Advance online publication. DOI: 10.1186/s12935-023-02953-3 PMID: 37344821

Zhang, Y., Payab, N., Weigelin, B., & Schürch, C. (2026). Evolution of cancer models from 2D to 3D systems. *Theranostics*, *16*, 4042–4057. DOI: 10.7150/thno.127053 PMID: 41695491

Zhao, R., Feng, Q., Xia, Y., Liao, L., & Xie, S. (2025). Iteration of Tumor Organoids in Drug Development: Simplification and Integration. *Pharmaceuticals (Basel, Switzerland)*, *18*(10), 1540. Advance online publication. DOI: 10.3390/ph18101540 PMID: 41155654

Zhuang, L., Lian, Y., & Zhu, T. (2025). Multifunctional gold nanoparticles in cancer. *Molecular Cancer*, *24*. Advance online publication. DOI: 10.1186/s12943-025-02431-6 PMID: 40993729

# Chapter 11
# The Role of Hypothermia and Cryo-Nanomedicine in Precision Oncology

**Bushra Ashiq**
http://orcid.org/0009-0002-2975-3121
*Taiyuan University of Technology, China*

**Amjid Khan**
*University of South Africa, South Africa*

**Huang Di**
*Taiyuan University of Technology, China*

**Muzammil Sattar**
*National Institute for Biotechnology and Genetic Engineering (NIBGE), Pakistan*

**Niu Xiaolian**
*Taiyuan University of Technology, China*

## ABSTRACT

*This chapter per the authors explores the emerging field of cryo-nanomedicine as an advanced strategy in precision oncology for improving cancer treatment outcomes. The chapter discusses how the integration of therapeutic hypothermia and nanotechnology enables targeted drug delivery, enhanced tumor ablation, and reduced systemic toxicity. By applying the Q10 principle, controlled cooling suppresses metabolic activity and oxidative stress within the tumor microenvironment. The chapter further examines the use of advanced nanomaterials, including phase-change nanoparticles and thermally conductive systems, for temperature-responsive*

DOI: 10.4018/979-8-3373-9848-8.ch011

*therapeutic action. In addition, Cryo-Electron Microscopy (Cryo-EM) is presented as an essential tool for validating the structural stability and functionality of nano-formulations. Overall, the chapter highlights cryo-nanomedicine as a sustainable and precise approach for next-generation cancer interventions.*

## INTRODUCTION

### 1.1 Thermal Biology and Nanotechnology: A Perfect Match

The convergence of thermal biology and nanotechnology has significantly advanced oncology, shifting therapeutic practices from systemic physiological regulation toward controlled molecular-scale thermal engineering (Gupta et al., 2026; Khan, Anas, et al., 2025; Khan, Ashiq, et al., 2025). Thermal biology primarily concerns about the response of biological systems (such as enzymes and membranes) to variations in temperature. In precision oncology, this interaction is closely associated with the Q10 principle, which states that metabolic processes typically slow down by two- to three-fold for every 10°C drop in temperature (Yaramiri et al., 2025). Nanotechnology is the key enabling platform in this context by providing engineered "smart" materials that can respond, amplify or sense thermal fluctuations. Contemporary modern nanocarriers possess tunable physiochemical properties that enable them to respond to the tumor microenvironment (TME), in temperature-sensitive interactions (Ali Mohammad et al., 2023). For example, cold-sensitive nanocarriers can exploit reduced metabolic activity in hypothermic tissues to improve hypoxic drugs stability or trigger phase transitions in lipid or polymer systems, enabling localized and controlled drug release (Wang et al., 2018). This combination enables improved spatial control of therapeutic action, thereby reducing systemic toxicity while enhancing the precision of conventional chemotherapy approaches (Rulten et al., 2023).

### 1.2 From Hypothermia to Cryo-Nanomedicine

The use of cold for therapeutic purposes has come a long way, since its early non-specific systemic applications to its precision-based targeted approaches in oncology. The first clinical application of systemic hypothermia was for systemic neuroprotection and metabolic arrest during cardiac surgery and in critical care. Hypothermia reduces cellular metabolism, disrupting ischemic pathways, regulating inflammatory processes, and protects cellular integrity at multiple levels (Gonzalez-Ibarra et al., 2011). However, despite its protective effects, its use in oncology remains limited due to poor spatial control and systemic side effects including coagulopathy, arrhythmias and pharmacokinetic variations (Ma et al., 2024). Clinical studies also indicated that hypothermia can significantly alter drug absorption, distribution, metabolism, and excretion, making it difficult to achieve therapeutic accuracy, in critically ill patients (Gijsen et al., 2021). These drawbacks highlighted the need for more targeted and responsive cold-based therapeutic approaches.

This led to the evolution of conventional cryosurgery, which introduced localised tumor destruction through controlled freezing. Cryoablation, involving cryogens like argon or liquid nitrogen, allows for minimally invasive destruction of solid tumors in organs like liver, prostate, lung and breast. This approach has clinical advantages including minimal invasiveness, quick recovery and high tolerability, and is particularly beneficial for treating unresectable tumors and early-stage cancers (Ma et al., 2024). Nevertheless, its effectiveness is constrained by thermal heterogeneity within the ice ball, leading to an "incomplete kill zone" at the tumor edges. This occurs when temperatures are not sufficiently low to cause irreversible damage to the cells around the periphery of the tumor, leading to an increased risk of recurrence (Kwak et al., 2022). Additionally, although cryoablation can trigger local tumor eradication, it is still unreliable in producing systemic anti-tumor immune responses, such as the abscopal effect, remains limited (Ali Mohammad et al., 2023).

The emergence of precision cryo-nanomedicine is the latest and most advanced stage in this evolution, which involves the use of nanotechnology to improve the spatial and functional precision of cryotherapy. Nanoparticles, such as liposomes, polymeric systems, inorganic and hybrid nanoparticles, are engineered to improve tumor targeting, drug delivery and treatment response, forming a key component of precision oncology (Gupta et al., 2026). In cryo-nanomedicine, nanomaterials function as "cryo-sensitizers" by improving thermal efficiency, facilitating ice formation within cells, and allowing for the modulation of freezing conditions (Gupta et al., 2026; Kwak et al., 2022). Moreover, recent developments in smart nanocarriers show that thermo-responsive nanoparticles can selectively release drugs in cold environments, allowing precise spatiotemporal control over drug release and reducing systemic toxicity (Yaramiri et al., 2025). Cold-sensitive nanoparticles that undergo disassembly at low temperatures further refine this control and improve tumor cell killing while sparing normal tissues by releasing drugs in response to cryoablation-induced stress (Wang et al., 2018). More recently, novel approaches like cryo-immune engineering harness nanoparticle-mediated freezing to transform immunologically "cold" tumors into "hot" tumors, thus triggering systemic anti-tumor immunity and enhancing therapeutic efficacy (Ali Mohammad et al., 2023; Ou et al., 2023). Collectively, this progression in paradigm represents the evolution from a non-specofic physical cooling to a targeted molecular therapy that harnesses the unique vulnerabilities of cancer cells under cold stress, in line with the objectives of precision oncology.

The evolution of systemic hypothermia to conventional cryoablation and ultimately to precision cryo-nanomedicine represents a transition between the cooling of the whole body to the localized, molecularly directed, cancer therapy. **Table 1** provides an overview of this evolution by comparing the level of control and biological basis as well as immunological effects of each approach on the targeting scale and the primary mechanism. As shown in **Table 1** precision cryo-nanomedicine overcomes some of the limitations of the previous cold-based therapies by integrating local thermal control and drug delivery mediated by nanoparticles, cryo-sensitization and immune modulation.

*Table 1. Evolution and mechanistic progression of cold-based therapeutics in oncology*

| Feature | Conventional Hypothermia | Conventional Cryosurgery (Cryoablation) | Precision Cryo-Nanomedicine | References |
|---|---|---|---|---|
| **Targeting & Spatial Scale** | Systemic / Whole-body: Used for systemic neuroprotection and metabolic arrest. | Localized (Organ-level): Destruction of solid tumors in specific organs (e.g., liver, prostate, lung). | Molecular / Spatiotemporal: Targeted molecular therapy exploiting unique cellular vulnerabilities. | (Gonzalez-Ibarra et al., 2011; Gupta et al., 2026; Kwak et al., 2022) |
| **Primary Mechanism** | Physiological Regulation: Metabolic suppression via the Q10 principle; regulation of inflammatory processes. | Physical Freezing: Direct cellular destruction via "ice ball" formation using cryogens (argon/nitrogen). | "Cryo-sensitization": Use of smart materials to amplify thermal fluctuations and trigger drug release. | (Baust et al., 2023; Gonzalez-Ibarra et al., 2011; Wang et al., 2018) |
| **Level of Control & Limitations** | Poor Spatial Control: Risk of systemic "Lethal Triad" (acidosis, coagulopathy, arrhythmia). | Thermal Heterogeneity: "Incomplete kill zone" at tumor edges; risk of recurrence. | Precise Thermal Compartmentalization: Adaptive cooling and real-time monitoring via MRI/AI. | (Gonzalez-Ibarra et al., 2011; Kim et al., 2024; Kwak et al., 2022) |
| **Biological Basis** | Slows metabolic processes by two- to three-fold for every 10°C drop in temperature. | Intracellular Ice Formation (IIF): Physical membrane disruption through freezing. | Synergistic Engineering: Remodeling of the TME; cold-triggered disassembly of nanocarriers. | (Baust et al., 2023; Yaramiri et al., 2025) |
| **Immunological Impact** | Non-specific Protection: Aimed at protecting cellular integrity during ischemic events. | Unreliable Immunity: Limited ability to produce systemic responses such as the abscopal effect. | In-situ Cryo-Immune Engineering: Transforming "cold" tumors into "hot" (immuno-active) ones. | (Ali Mohammad et al., 2023; Ou et al., 2023; Waitz et al., 2012) |

## 1.3 Sustainable Scope in Cancer Synthesis and Therapeutics

The growing emphasis on sustainability in healthcare has positioned cryo-nanomedicine as a promising approach within "green nano-oncology" integrating environmentally responsible material design with precision therapeutic strategies. This idea encompasses both sustainable synthesis and sustainable therapeutic application.

### 1.3.1. Sustainable Synthesis:

Cryo-nanocarrier development increasingly embraces "safe-by-design" principles and green chemistry approaches aimed at minimizing environmental impact and improving biocompatibility. These strategies involve bio-based polymers and supercritical fluid technologies that reduce reliance on toxic organic solvents (Gupta et al., 2026; Mao et al., 2024). In addition, advances in ecofriendly nanofabrication, such as plant derived nanomaterials and biologically synthesized nanoparticles, have demonstrated potential for scalable and sustainable production while maintaining functional performance (Wen et al., 2025). Such approaches not only reduce environmental burden but also enhance regulatory acceptance and long-term safety profiles of nanomedicines.

### 1.3.2. Sustainable Therapeutics:

From a therapeutic perspective, cryo-nanomedicine increases the therapeutic index (balancing efficacy and systemic toxicity) and lowers the long-term costs of health care by preventing side effects from chemotherapy. Temperature-responsive nanocarriers allow localized and controlled drug release, thereby minimizing off-target effects and reducing cumulative drug exposure (Yaramiri et al., 2025). Moreover, stimuli-responsive features enable optimal use of drugs, thus reducing pharmaceutical waste by releasing therapeutic agents only under defined conditions, such as temperature or tumor-specific triggers (S. Imani et al., 2025). With precision oncology shifting towards patient-calibrated models, the deployment of sustainable, temperature-sensitive nanoplatforms represents a critical step toward developing cost-effective, environmentally responsible, and clinically efficient cancer therapies (Gupta et al., 2026). This chapter presents cryo-nanomedicine as a next-generation precision oncology platforms that integrates thermal biology, nanotechnology and artificial intelligence to enable spatially controlled, immunologically active and environmentally sustainable cancer therapy.

## 2. BIOLOGICAL BASIS OF THERMOREGULATION THERAPY

This section examines fundamental physiological mechanisms that drive the efficacy of temperature-controlled therapy for cancer. It begins with metabolic control through the Q10 principle, which addresses the effect of temperature on cellular metabolic activities and the energy requirements of cancer cells. It then discusses the role of mitochondrial integrity and regulation of oxidative stress in determining cancer cell survival under hypothermic conditions. Finally, it evaluates the influence of controlled cooling on the tumor microenvironment (TME), with its vascular, biochemical, oxidative status, and immunological characteristics, thereby enhancing the efficacy of therapies using cryo-nanomedicine.

### 2.1 Metabolic Control and the Q10 Principle in Cancer

Cancer cells exhibit a hyper-glycolytic phenotype (Warburg effect), a metabolic adaptation that fuels the growth of cancer cells while also offering a unique physiological target for thermal therapy. The fundamental principle underpinning temperature-controlled therapy is the Q10 law, which describes the rate of biological and chemical reactions, typically decrease by two to three-fold for every 10°C reduction in temperature. In precision oncology, the induction of localised hypothermia leverages this principle to slow down the hyper-metabolic cells of a tumor (Yaramiri et al., 2025). The induced local hypothermia allows clinicians to "turn off" the high metabolic rate of the tumor by decreasing its adenosine triphosphate (ATP) demand and slowing down enzyme activity involved in DNA replication and protein synthesis (Amin et al., 2020; Yaramiri et al., 2025). This metabolic suppression enhances the sensitivity of tumor cells adjunctive therapies, including chemotherapy and cryoablation, while relatively preserving normal tissue function due to reduced metabolic demand. Consequently, localized cooling may expand the therapeutic window by limiting off-target toxicity and improving treatment selectively (Gupta et al., 2026; Mao et al., 2024).

### 2.2 Mitochondrial Integrity and Redox State

Mitochondria play a key role in determining cell fate under thermal stress both as an energy source and as the initiators of apoptotic cell death. Normothermic cancer cells are frequently associated with mitochondrial dysfunction, including excessive generation of reactive oxygen species (ROS) production, contributing to genomic instability and tumor progression. The use of precision hypothermia profoundly alters this prooxidant environment, by preserving mitochondrial membranes and electron leakage from the respiratory chain, thereby modulating oxidative stress (Ali

Mohammad et al., 2023; Gonzalez-Ibarra et al., 2011). In combination with cryo-nanomedicine, a biphasic thermal strategy is often employed. Firstly, mild cooling is employed to protect healthy cell mitochondria and avoid necrosis; secondly, targeted, nanoparticle-based freezing is applied to irreversible rupture of mitochondria and release of cytochrome c in the tumor cells. This includes mitochondrial membrane disruption, cytochrome c release and activation of apoptotic pathway (Kwak et al., 2022). Importantly, this temperature-dependent modulation of oxidative stress enables selective "toxic dose" of heat remains confined to the tumor, taking advantage of the mitochondrial vulnerability of cancer cells to initiate apoptosis, and avoiding the bystander effects of oxidative stress caused by the ablation therapies (Dai et al., 2024; Wang et al., 2018).

## 2.3 Hypothermia's Effects on the Tumor Microenvironment (TME)

The tumor microenvironment (TME) is characterised by hypoxia, acidosis, abnormal vasculature, and elevated interstitial pressure, all of which contribute therapeutic resistance. Modulated hypothermia has a profound effect on this ecosystem by modifying the physical and immunological features of the stroma. Cooling the TME results in a temporary drop in blood flow and vessel permeability, which, while seemingly paradoxical, can be harnessed to "lock" nanocarriers in the tumor interstitium after they have left the bloodstream (Gupta et al., 2026). Additionally, exposure to cold stress can affect the infiltration of tumor-associated macrophages (TAMs) and other immune cells. Recent studies have shown that controlled cooling, in combination with nano-sensitizers, can transform a "cold" (non-immunogenic) TME into a "hot" (immunogenic) one by triggering the release of damage-associated molecular patterns (DAMPs) during the rewarming phase (Ali Mohammad et al., 2023; Ou et al., 2023). This process contributes to the conversion of immunologically cold tumors into hot tumors, thereby improving responsiveness to immunotherapy. The biological basis of in situ cryo-immune engineering is the immunological consequences of controlled cooling, which is the biological basis of cryo-immune engineering in in vivo studies. This pathway connects remodels in tumor microenvironment induced by hypothermia to the therapeutic approach depicted in **Figure 2** below. Overall, the combination of metabolic suppression, structural disruption and immune activation through controlled thermal cycling remodels the TME, reducing biological barriers and enhancing the penetration and efficacy of therapeutic agent (Rulten et al., 2023).

# 3. ADVANCED MATERIALS FOR CRYO-NANOMEDICINE

This section examines the design and engineering of advanced nanomaterials for cryo-nanomedicine. These materials are classified based on their roles in precision oncology, such as cold-responsive nanocarriers for temperature-controlled drug release and phase-change materials (PCMs) for precise temperature control and energy storage. Furthermore, the section covers the role of thermally conductive nanoparticles to enhance heat conduction and increase thermal efficiency during cryoablation. Lastly, we discuss temperature-sensitive polymers as regulatory materials, that allow spatiotemporal control of therapeutic payloads in the tumor microenvironment (TME) (Mao et al., 2024; Wen et al., 2025).

## 3.1 Cold-Responsive Nanomaterials: Mechanisms of Action

Cold-responsive nanomaterials are a class of stimuli-responsive materials that undergo structural or chemical changes upon cooling to a temperature below room temperature. These nanomaterials differ from conventional hyperthermic systems, which use lower critical solution temperature (LCST) or phase-transition triggers to trigger therapeutic actions (Ali Mohammad et al., 2023). The primary mechanism involves temperature-induced phase transitions or structural destabilization of lipid bilayers or polymeric shells, leading to increased membrane permeability or carrier disassembly (Wang et al., 2018). Cooling can induce lipid phase transitions (gel-to-liquid crystalline) or polymer dehydration, resulting in controlled drug release. Recent advances have demonstrated that cold-triggered nanocarriers can undergo irreversible disassembly below -10°C, enabling site-specific payload release during cryoablation (Ou et al., 2023; Wang et al., 2018). In addition, polymer-based nanostructures exploit cold-induced conformational changes and chain contraction to expel encapsulated drugs selectively within the cooled tumor region, while maintaining stability at physiological temperature (S. Imani et al., 2025). This selective responsiveness significantly reduces systemic toxicity and improves therapeutic precision.

## 3.2 Phase-Change Materials (PCMs): Thermal Energy Storage for Precise Temperature Control

Phase-change materials (PCMs) are incorporated into nanoplatforms to regulate thermal distribution during cryotherapy through their high latent heat capacity. These materials absorb or release large amounts of energy during solid-liquid phase transitions, thereby stabilizing temperature fluctuations during cryoablation. In precision cryo-nanomedicine, organic PCMs like paraffin wax or fatty acid derivatives are embedded in biocompatible shells to form nanocapsules that can store or release considerable amounts of heat during phase transitions (solid-liquid and vice versa) (Sharma et al., 2009; Zhang et al., 2018). These materials play a key role in protecting healthy tissues by providing a thermal barrier that prevents freezing outside the tumor margin, while also maintaining a controlled temperature gradient within the "killing zone" during the cryoablation process. Numerical simulations in recent studies have shown that PCM-integrated nanospheres can significantly improve thermal uniformity and reduce collateral tissue damage during cryoablation procedures (Xie et al., 2025). This thermal regulation is particularly important for tumors located near critical structures such as nerves and large blood vessels.

*Figure 1. Dual mechanism of cryo-nanomedicine involving cold-triggered drug release and PCM-mediated thermal buffering. Cold-responsive nanocarriers can be stable at physiological temperatures but destabilize in response to cold temperature through destabilization of the membrane or disassembly of the polymer shell. In the meantime, the temperature of a phase-change material in the locality of the tumor margin, buffer temperature varies by absorbing or releasing latent heat, which helps to control the cryoablation boundary and protect adjacent healthy tissue*

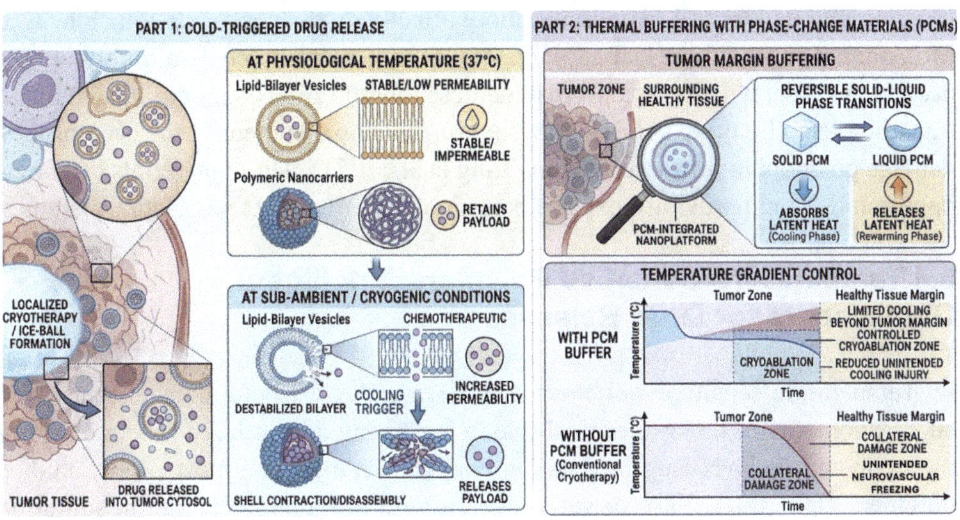


Cold-responsive nanocarriers can significantly enhance the spatiotemporal regulation of chemotherapy, as illustrated in Figure 1, by enabling the local release of contents at sub-ambient or cryogenic temperatures (Wang et al., 2018; Xie et al., 2025). Conversely, phase-change materials primarily facilitate the thermal regulation of cryotherapy by absorbing and giving up latent heat during solid-liquid transitions, thereby moderating the cryoablation boundary and reducing damage to surrounding healthy tissue (Xie et al., 2025; Zhang et al., 2018). Combined, these platforms provide the basis of sophisticated cryo-nanomedicine systems which integrates localized drug delivery with localized tumor freezing. Based on these mechanisms, the thermally conductive nanoparticles can further increase the efficacy of cryoablation by enhancing the heat transfer and promoting a more uniform formation of ice in the tumor mass (Kwak et al., 2022).

## 3.3 Thermally Conductive Nanoparticles: Enhancing Cryoablation Efficiency

Cryoablation efficiency is often compromised by the uneven and slow growth of the ice ball in the non-uniform tumor tissue. To address this issue, thermally conductive nanoparticles, such as gold nanospheres, superparamagnetic iron oxide nanoparticles (SPIONs) and carbon nanotubes, are used as "cryo-sensitizers" (Kwak et al., 2022). Infusing the tumor with these highly conductive materials increases the speed of the heat transfer process, facilitating rapid formation of ice crystals and more even freezing of the tumor. For instance, nanoparticles such as $Fe_3O_4$ have been demonstrated to modify intracellular ice formation (IIF) resulting in physical disruption of cancer cell membranes more effectively than cryoablation alone (Ali Mohammad et al., 2023; Dai et al., 2024). Emerging research also highlights the use of liquid metal nanoparticles that increase their thermal conductivity and phase transition from liquid to solid, facilitating further enhancement of the cryo-probe's damage potential in the target region (Tang et al., 2021). These materials offer new opportunities for improving the spatial precision and efficacy of cryoablation therapies.

## 3.4 Temperature-Sensitive Polymers: Intelligent Gatekeepers for Drug Release

Temperature-sensitive polymers (TSPs) are the "structural gatekeepers" in cryo-nanomedicines, play a critical role in regulating drug release. The most common polymer, Poly(N-isopropylacrylamide) (PNIPAm)is the most widely studied polymers, undergo a sharp phase transition with a dramatic change in solubility and permeability properties (Amin et al., 2020). Once attached to the surface of nanocarriers such as mesoporous silica nanoparticles or liposomes, they form a "brush" or "gate" that is closed at body temperature but opens the pores during a cold trigger. This spatiotemporal selectivity is crucial for "precision dosing" in which drug release follows first-order kinetics only in the thermal zone (S. Imani et al., 2025; Valentino et al., 2024). Moreover, "dual-responsive" copolymers, both in temperature and the acidic TME, so even if the nanocarrier extravasates outside the treatment zone, the drug is still locked until the specific triggers are met. The functional diversity of cryo-nanomedicine depends on the integration of nanomaterials with distinct thermal, structural and drug release properties. **Table 2** summarizes the major categories used in cryo-nanomedicine, their representatives' examples, key mechanisms and functional roles in precision oncology. As summarized in **Table 2**, each nanomaterial class contributes a different function to cryo-nanomedicine. Phase-change materials regulate thermal gradients, thermally conductive nanoparticles improve freezing efficiency, temperature-sensitive polymers control drug release, cold responsive nanocarriers enable on demand chemotherapy and liquid metal nanoparticles provide additional opportunities for thermal modulation and multimodal therapy.

*Table 2. Advanced nanomaterials for cryo-nanomedicine and their functional roles*

| Material Category | Representative Examples | Key Mechanism | Functional Role in Precision Oncology | References |
|---|---|---|---|---|
| **Phase-Change Materials (PCMs)** | Organic fatty acids, paraffin wax, 1-tetradecanol | Absorption and release of latent heat during solid–liquid phase transitions, enabling thermal buffering | Acts as a thermal regulator to prevent unintended cooling of surrounding healthy tissue; stabilizes the cryoablation zone | (Kwak et al., 2022; Sharma et al., 2009; Zhang et al., 2018) |
| **Thermally Conductive Nanoparticles** | Gold nanospheres, SPIONs, carbon nanotubes, graphene oxide | High thermal conductivity enhances heat transfer and promotes intracellular ice formation (IIF) | Acts as a cryo-sensitizer to improve uniform ice formation and enhance tumor cell destruction | (Ali Mohammad et al., 2023; Dai et al., 2024; Kwak et al., 2022) |
| **Temperature-Sensitive Polymers** | Poly(N-isopropylacrylamide) (PNIPAm), Pluronic F-127 | Reversible phase transition at the lower critical solution temperature (LCST), leading to polymer collapse or swelling | Functions as a smart gatekeeper, enabling controlled drug release under temperature-specific conditions | (Amin et al., 2020; Torchilin, 2014) |

continued on following page

*Table 2. Continued*

| Material Category | Representative Examples | Key Mechanism | Functional Role in Precision Oncology | References |
|---|---|---|---|---|
| **Cold-Responsive Nanocarriers** | pNIPAAm-BA nanogels, phospholipid vesicles | Temperature-induced phase separation or membrane destabilization under hypothermic conditions | Enables on-demand drug release specifically within the cooled tumor region, reducing systemic toxicity | (Gupta et al., 2026; Ou et al., 2023; Wang et al., 2018) |
| **Liquid Metal Nanoparticles** | Gallium-based alloys, EGaIn (eutectic gallium–indium) | Phase transition and high thermal conductivity with dynamic structural adaptability | Enhances cryoablation efficiency and enables multimodal imaging and thermal modulation | (Li et al., 2020; Tang et al., 2021; Wen et al., 2025) |

## 4. NANO-ONCOLOGY APPLICATIONS

This section highlights the translational and clinical applications of cryo-nanomedicine to show how nanoscale thermal engineering overcomes several limitations in cancer treatment. We explore how nanoparticles sharpen the physics of the cryoablation zone and cold trigger drug delivery. Collectively, these innovations enhance therapeutic efficacy while minimizing systemic toxicity, also explore the use of antioxidant nano-formulations to prevent unintentional tissue damage from oxidative stress and the new paradigm of thermal cycling to induce immunogenic cell death (Gong et al., 2024; S. Imani et al., 2025; Khan et al., 2026; Mao et al., 2024; Ullah et al., 2026; Wen et al., 2025).

### 4.1 Nano-Enhanced Cryoablation: Minimizing Collateral Tissue Damage

This section discusses the use of high-thermal-conductivity materials to enhance the "freeze-thaw" cycle during nano-enhanced cryoablation, and circumvent problems associated with conventional cryosurgery. Conventional cryoablation techniques often limited by non-uniform ice-ball formation, resulting in an incomplete tumor destruction on the outer edge of the tumor and unintentional freezing of surrounding neurovascular tissues.

This can be achieved by loading a tumor with metallic supra-structured cryo-nanocatalysts (MSCNs) or gold nanostructures, which increase the freezing point of the tissue and the creation of the ice-nuclei (Dai et al., 2024; Kwak et al., 2022). These materials facilitate intracellular ice formation (IIF), which is the primary mechanism of irreversible cell damage during cryoablation (Baust et al., 2023). These nano-catalysts serve as heat transfer agents that concentrate the energy within the tumor mass. The mechanism of cell death is shifted from simple cryo-induced cell damage to a more enhanced biological effect; that is, the nano-catalysts enhance cryo-induced physical cell damage and induce necroptosis rather than accidental necrosis. This selective cell death is accomplished by the RIPK3-MLKL pathway, which assures the destruction of the primary tumour while preserving the healthy stroma by the sharp thermal gradient created by the nanoparticles (Galluzzi et al., 2017; Ou et al., 2023).

## 4.2 Cold-Triggered Drug Release: Chemotherapy on Demand

Sub-ambient temperatures are a prime example of an external trigger in "on-demand" drug delivery. In contrast to systemic chemotherapy, which is based on constant circulation, cold-triggered systems hold potent drugs (such as Doxorubicin or Irinotecan) in nanocarriers that are tightly encapsulated at physiological temperature (37°C). Cold-responsive nanocarriers, particularly polymeric and lipid-based systems, exhibit temperature dependent permeability changes or disassembly, allowing rapid and localized drug release during cryoablation (Ou et al., 2023; Wang et al., 2018). This "burst release" allows the maximum concentration of the drug to be released in the cytosol of cryo-stressed cells. Recent advances in nanogels and stimuli-responsive systems demonstrate that cold-triggered release can provide faster and more direct drug activation compared to pH or enzyme-responsive systems, particularly in overcoming tumor microenvironment barriers (Amin et al., 2020; Torchilin, 2014). These systems improve therapeutic precision and reduce off-target toxicity.

## 4.3 Antioxidant Nano-formulations: Preserving Healthy Cells during Thermal Insult

A major challenge in cryo-nanomedicine is the formation of reactive oxygen species (ROS) that may cause secondary injury to healthy cells during freeze-thaw cycle. Antioxidant nano-formulations are currently under investigation to provide localised protection, including FDA-approved antioxidant agents such as Tempol,

Edaravone or Coenzyme Q10. Co-administered with therapeutic hypothermia, these nano-antioxidants function as quenchers of ROS that help maintain the structural integrity of mitochondria in non-malignant cells (Gijsen et al., 2021; Gonzalez-Ibarra et al., 2011; Peng et al., 2024). These systems modulate oxidative stress (Dai et al., 2024). By co-administering these protective compounds to the healthy stroma adjacent to a tumor, clinicians can adopt aggressive cryoablation protocols without the risk of systemic oxidative stress and neurotoxicity, thus extending the therapeutic window treatment (Liu et al., 2020; Valentino et al., 2024).

## 4.4 Hyperthermia/Hypothermia Cycles: A Novel Paradigm in Tumor Eradication

The new approach of "In-situ Cryo-Immune Engineering" (ICIE) uses synergetic thermal cycling to convert tumor microenvironment (TME) from immunologically "cold" (immunosuppressive) to "hot" (immuno-active). This approach consists of a first cycle of local cryoablation to trigger immunogenic cell death (ICD) and a second warming phase (or mild hyperthermia) to promote immune cell infiltration (Ou et al., 2023).Cryoablation induces the release of damage-associated molecular patterns (DAMPs) and pro-inflammatory cytokines during the phase-change from ice to liquid that drive dendritic cell maturation and activation of $CD8^+$ cytotoxic T cells (Ali Mohammad et al., 2023; Ou et al., 2023; Waitz et al., 2012).

Experimental evidence demonstrates that thermal cycling can boost the ratio of T-cells to regulatory T cells by over 100-fold. This process not only eliminates the primary tumor but also induces an abscopal effect, where the systemic immune response targets and destroys distant metastatic sites that were not frozen, thereby enhancing thermal cycling as a valuable complement to current immunotherapy regimes (Ali Mohammad et al., 2023; Ou et al., 2023; Waitz et al., 2012). This integration of thermal therapy with immunomodulation positions cryo-nanomedicine as a promising adjunct to modern immunotherapies, including immune checkpoint blockade strategies(Gong et al., 2024; Wen et al., 2025).

Nano-enhanced cryoablation offers a mechanistic intermediation of localized tumor destruction and systemic immune activation. Thermally conducting nanoparticles enhance the freezing step by increasing the transfer of heat, enhancing the formation of ice-balls of uniform shape, improving intracellular ice formation in tumor cells. This local freeze-thaw injury disrupts tumor-cell membranes and intracellular organelles, with the subsequent rewarming phase promoting the release of damage-associated molecular patterns, such as HMGB1, ATP and calreticulin. These immunogenic signals stimulate the maturation of dendritic cells, antigen presentation, and $CD8^+$ cytotoxic T-cell activation mechanisms, therefore, supporting the conversion of an immunosuppressive/cold tumor microenvironment to an immunologically active/

hot tumor microenvironment. The general process of nanoparticle-assisted freezing, immunogenic cell death and ICIE-mediated tumor remodeling are outlined in **Figure 2.** As illustrated in **Figure 2**, ICIE enhances the effect of cryoablation on direct tumor destruction by physical means, by tying together the effect of thermal injury with the effect of immune priming. It is especially relevant to precision oncology since it fuses localized ablation with tumor microenvironment remodeling and it may be more responsive to immunotherapeutic treatments.

*Figure 2. Therapeutic hypothermia and in situ cryo-immune engineering in cryo-nanomedicine. Nanoparticles that are thermally conductive can be used to enhance localized cryoablation by improving the transfer of heat, promoting uniform ice-ball formation, and inducing intracellular ice formation. This leads to freeze-thaw damage to tumor-cells, causing immunogenic cell death, and releasing DAMP during rewarming. These signals trigger the dendritic cells and CD8$^+$ cytotoxic T-cells, which help convert a cold immunosuppressive tumor microenvironment to a hot immunologically active one with possible systemic anti-tumor effects*

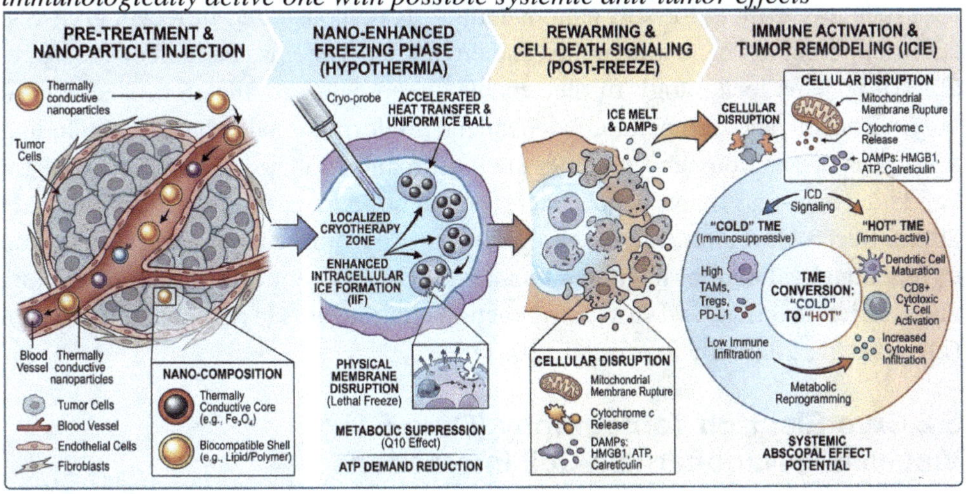


## 5. STRUCTURAL VALIDATION AND IMAGING

Here, we discuss the advanced techniques for characterization and monitoring needed to advance cryo-nanomedicines from lab research to clinical practice. We examine the role of high-resolution Cryo-Electron Microscopy (Cryo-EM) as the primary tool to confirm the native structure of thermo-sensitive nanocarriers in solution. In addition, we discuss the use of Cryo-Electron Tomography (Cryo-ET) in tracking the 3D intracellular pathway of nanoparticles and the clinical need for real-time thermal imaging to provide spatial targeting and therapeutic control of nano-mediated cooling.

## 5.1 Cryo-Electron Microscopy (Cryo-EM): Nanostructures in Their Native State

Confirmation of the structure of temperature-sensitive nanocarriers is intrinsically difficult because conventional electron microscopy includes dehydration and staining, which can result in collapse or aggregation of soft materials, such as liposomes and polymeric micelles. Cryo-Electron Microscopy (Cryo-EM) has addressed this challenge by allowing the imaging of macromolecular assemblies and nanomaterials in their "vitrified" (amorphous frozen) native state (Takechi-Haraya, 2025). Through quick freezing of the sample, usually by plunge-freezing in liquid ethane, the water is vitrified without forming crystalline, thus preserving the exact 3D structure of the sample. This preservation of hydrated morphology is particularly critical for soft nanomaterials, as demonstrated in recent cryo-EM studies of complex nano assemblies and emulsified systems (Fernandez Ainaga et al., 2025). This technology further enables nanoscale visualization of structural polymorphism and temperature dependent conformational, changes, including lipid phase transitions and polymer shell reorganization, which are essential for validating cold-responsive nanocarriers (Fernandez Ainaga et al., 2025; Gupta et al., 2026). Recent developments have extended its use from structural characterization to functional validation, including quantitative assessment of drug encapsulation, nanocarriers stability, and structural function relationships in nanomedicine design (Lovatt et al., 2026; Sheibani et al., 2021).

## 5.2 Cryo-Electron Tomography (Cryo-ET): Visualising Nanoparticle-Cell Interactions

Although single-particle Cryo-EM is ideal for obtaining the average structure of isolated particles, Cryo-Electron Tomography (Cryo-ET) can deliver a label-free and non-disturbing view of these particles in their sophisticated, three-dimensional cellular environment. Cryo-ET allows label free imaging of intact cells in a near-native hydrated state, providing nanometre-scale resolution (~2-4 nm) of cellular architecture (Sheibani et al., 2021). This technique enables detailed mapping of nanoparticle trafficking and interactions with sub-cellular organelles such as mitochondria, endosomes or the nucleus. Notably, cryo-ET has been used to directly visualize the biomolecular corona surrounding nanoparticles, which plays a critical role in determining cellular uptake, biodistribution and therapeutic efficacy

(Sheibani et al., 2021). In cryo-nanomedicine, scientists produce thin sections or lamellae of cells using focused ion beam (FIB) milling and then image at several tilt angles to generate a 3D tomogram. This precise mapping is essential to confirm that cryo-sensitizing nanoparticles have docked at the desired target sites, such as mitochondrial membranes, before thermal trigger is applied (Z. Imani et al., 2025; Sheibani et al., 2021). Cryo-EM and Cryo-ET offer complementary structural validation concepts of cryo-nanomedicine. Cryo-EM maintains soft nanocarriers in a near-native hydrated, vitrified state, allowing visualization of morphology, structural integrity, and polymorphism without dehydrating or staining artifacts (Takechi-Haraya, 2025). Cryo-ET further examines the above analysis into intact cellular contexts by mapping nanoparticle localization, biomolecular corona formation, intracellular trafficking, and interactions with subcellular targets such as mitochondrial membranes (Sheibani et al., 2021). These complementing workflows can be summarized as in **Figure 3**. By confirming both nanocarrier structure and intracellular localization, Cryo-EM and Cryo-ET provide the structural basis for image-guided cryo-nanomedicine. The present validation justifies further application of real-time thermal imaging to track the distribution of nanoparticles and therapeutic response using nano-mediated therapy.

*Figure 3. Structural validation and visualization of nanocarrier-cell interactions using Cryo-EM and Cryo-ET. Cryo-EM involves rapid vitrification to maintain soft nanocarriers in a near-native hydrated form, permitting the study of morphology, structural integrity and polymorphism without dehydrating or staining artifacts. Cryo-ET facilitates visualization of nanoparticles in cellular environments in three dimensions, including biomolecular corona formation, intracellular trafficking, and docking at subcellular targets such as mitochondrial membranes. Combined, these strategies aid structural validation and functional localization of cryo-nanomedicine platforms*

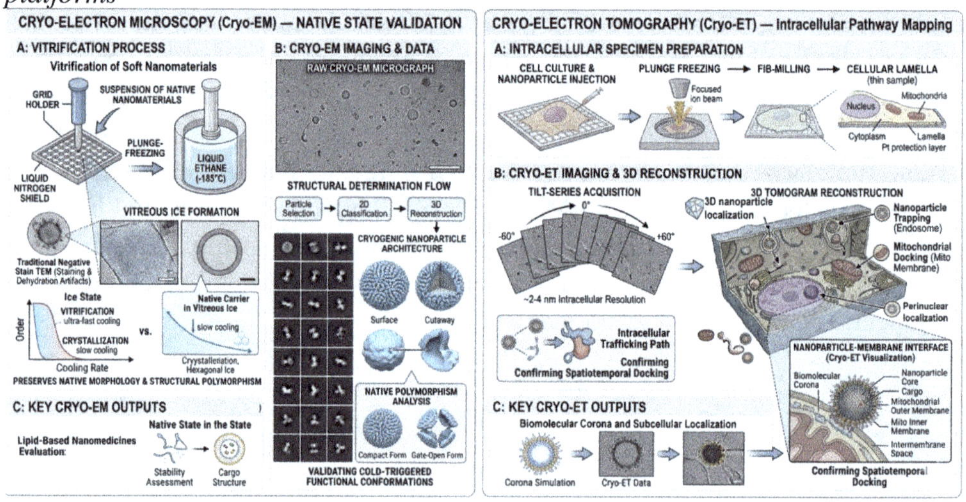


## 5.3 Real-Time Thermal Imaging in Nano-Mediated Therapy

Real-time temperature monitoring is critical for the success of cryo-nanomedicine therapy, as it ensures that the temperature gradient is sufficient to target tumor tissue without affecting surrounding tissues. Real-time thermal imaging and MRI-based thermometry are essential to provide feedback for the containment of the nanoparticle-mediated ice ball within the desired therapeutic margins. MRI thermometry, based on proton resonance frequency (PRF) shifts, enables non-invasive ang high-resolution mapping of temperature changes in deep tissues, making it particularly suitable for guiding cryoablation and nano-particle-mediated thermal therapies (Blackwell et al., 2022; Kim et al., 2024; Rieke & Butts Pauly, 2008). These imaging techniques are especially crucial when employing in-situ cryo-Immune engineering" (ICIE) strategies, which need to establish specific thermal boundaries to induce immunogenic cell death (Ou et al., 2023). Recent advances have introduced nanotheranostics constructs, which combine therapeutic agents with imaging agents, enabling the real-time monitoring of nanoparticle distribution and local temperature changes (Ashique et al., 2023; Habeeb et al., 2024). Furthermore, the integration of these real-time data with artificial intelligence (AI) enables closed-loop cryo-therapy systems, whereby the cooling rate is dynamically adjusted based on the structural and thermal responses of the nanostructures themselves, significantly improving treatment precision and reproductivity (Gupta et al., 2026; Wen et al., 2025).

## 6. CHALLENGES AND SUSTAINABLE IMPLEMENTATION

This section critically assesses the translation barriers restricting the clinical adoption of cryo-nanomedicine. Key challenges include (i) the long-term biodistribution and potential immunogenicity of cold-responsive materials, (ii) the physiological handling of local thermal insults and (iii) the need for sustainable and scalable manufacturing strategies aligned with green nanotechnology principles.

### 6.1 Long-term Safety of Cryo-nanomaterials

The clinical translation of cryo-nanomedicine is strictly controlled by the long-term biocompatibility of the nanomaterials used as sensitizers or drug carriers. One key issue is the protein corona, that forms on nanoparticles once they are injected into the bloodstream. Recent studies suggest that this corona is highly temperature sensitive. Non-ambient temperatures during the cryo-therapy treatment can modify the protein density at the surface of the nanoparticles, potentially resulting in undesired activation of the macrophages or changes in the biodistribution of the nanomaterials (S. Imani et al., 2025; Sheibani et al., 2021). In addition, metallic nanoparticles (e.g., gold and iron oxide) offer excellent heat transfer properties, their retention in the mononuclear phagocyte system (MPS) may lead to chronic inflammation or oxidative stress (Gupta et al., 2026; Liu et al., 2020; Mitchell et al., 2021). To overcome these limitations, contemporary safe-by-design (SbD) strategies emphasize the development of biodegradable and hybrid nanomaterials with controlled degradation profiles and efficient clearance via renal or hepatobiliary pathways (Fadeel et al., 2018; Mao et al., 2024). These approaches aim to balance therapeutic performance with long-term biosafety, which is essential for regulatory approval and clinical translation.

### 6.2 "Lethal Triad" vs. Localized Cooling

A major physiological concern with cryo-based therapies is avoiding systemic complications typically associated with hypothermia, referred to as the "Lethal Triad" (acidosis, coagulopathy and arrhythmia). In conventional trauma management, systemic cooling interferes with enzymes involved in the blood-clotting process and metabolic regulation (Gonzalez-Ibarra et al., 2011). Precision cryo-nanomedicine addresses this limitation through thermal compartmentalization, where cooling is confined to the tumor region. By selectively seeding the tumor volume with phase-change materials (PCMs) or thermoconductive nanostructures, the cooling effect can be delivered to achieve the super-cooled temperatures necessary for ablation, while maintaining the surrounding tissue at a normothermic temperature (Kwak et al., 2022). Advance image guided systems, including MRI-based thermometry, further enable precise monitoring and control of temperature gradients during treatments, minimizing systemic thermal drift and reducing the risk of whole-body hypothermic complications (Kim et al., 2024; Rieke & Butts Pauly, 2008). This targeted effect eliminates the systemic drift that occurs via shock, resulting in the Lethal Triad. In addition, integration of real-time physiological sensors in the nanoplatform also enables adaptive cooling to maintain the patient's core temperature, even during intense local cryoablation (Wen et al., 2025).

### 6.3 Sustainable and Scalable Synthesis of Smart Nanomaterials

Scaling-up the synthesis of complex, thermo-responsive nanomedicines from lab to plant scale is a challenging task. The Quality-by-Design (QbD) concept is now being extended to modular production, to ensure that cold-responsive nanocarriers have uniform size, drug loading and transition temperatures over large-scale production (Mao et al., 2024). The quality by design framework is increasingly applied to nanomedicine manufacturing to ensure consistency and scalability (ICH Q8(R2) guidelines). Sustainability is a key driver of this transformation. Recent advances emphasize solvent free or low toxicity synthesis methods, including supercritical fluid processing, microfluidic fabrication and biologically derived nanomaterials, such as exosome-like vesicles and plant-based nanoparticles (Ashique et al., 2023; Wen et al., 2025). These environmentally responsible approaches for nanomedicine synthesis not only mitigate the environmental footprint of pharmaceutical production but also improve the biocompatibility of the final product, making regulatory approval easier (Fadeel et al., 2018; Mitchell et al., 2021).

## 7. FUTURE PERSPECTIVES AND TRANSLATIONAL MEDICINE

This chapter examines the future promise of cryo-nanomedicine as it moves from concept to reality. It highlights the integration of Artificial Intelligence (AI) in the design of thermal treatment protocols and nanocarriers, the transition to eco-friendly nutraceutical-based nanoformulations to improve biocompatibility, and a pragmatic approach to clinical trials with regard to the regulatory and physiological challenges associated with temperature-sensitive therapies.

## 7.1 AI in Precision Cryo-Nanomedicine

The complexity of the tumor microenvironment (TME) and its adaptive response to temperature changes approaches for a digital approach to treatment design. Artificial Intelligence (AI) and machine learning (ML) are now being used to develop "Digital Twins"- realistic computer models of a patient's biological system in the context of cryo-nanomedicine (Corral-Acero et al., 2020; Xie et al., 2025). These AI-based models are used to integrate real-time multi-omics data and spatial transcriptomics to model how particular nanocarriers will respond in the TME under thermal stress (Mitchell et al., 2021; Wen et al., 2025). By using generative protein design software (e.g. RFdiffusion), researchers can now design nanocarriers with Chameleon Coronas, surface motifs that are AI-predicted to capture specific host proteins at sub-ambient temperatures, thus facilitating tumor targeting and evasion of immune recognition (Watson et al., 2023). In addition, AI plays a critical role in closed-loop therapy, where feedback from thermal imaging is used to regulate the flow of cryoprobe to maintain the ice ball at lethal threshold and minimize damage to surrounding normal tissue (Kim et al., 2024; Rieke & Butts Pauly, 2008).

## 7.2 Next-Generation Nutraceutical-Based Nanotherapeutics

In line with sustainable nano-medicine strategies, there is an emerging trend in the use of nutraceuticals (nutrients and other health-promoting compounds found in foods) as the structural or therapeutic component of cryo-nanoplatforms. Examples include the encapsulation of flavonoids (e.g., Quercetin), Curcumin, and Coenzyme Q10 in lipid or polymer-based nanocarriers as dual-functioning cryoprotectants and sensitizers. These green nanomedicine formulations harness the natural biocompatibility and antioxidant capacity of nutraceuticals to counteract the oxidative damage of rewarming during (Liu et al., 2020; Zhang et al., 2024). Nano-encapsulation enhances their bioavailability and stability, enabling targeted delivery and controlled release within the tumor microenvironment (Ashique et al., 2023; S. Imani et al., 2025). Importantly, plant-based nanomedicines not only make pharmaceutical production more eco-friendly but also overcomes the long-term toxicity issues associated with purely synthetic or metallic nanostructures, supporting their use in sustainable nano-oncology (Fadeel et al., 2018; Mitchell et al., 2021).

## 7.3 Roadmap to Clinical Trials

The roadmap for translating cryo-nanomedicines must be robust, multimodal and comply with regulatory requirements. Bench-to-bedside transition is currently driven by Quality-by-Design (QbD) principles, which provide the batch-to-batch reproducibility needed for nanocarrier manufacturing in Phase I clinical trials (Mao et al., 2024). An important landmark in this journey is the adoption of N-of-1 clinical trial designs, whereby the cyclical development of AI-designed nanomedicines are tailored to the individual neo-epitope fingerprint of a patient in a four-week rapid-feedback cycle (Bansal, 2025; Corral-Acero et al., 2020).

Translational Pathways:

- **Pre-Clinical:**
  Emphasis on "Safe-by-Design" (SbD) confirmation in patient-derived organoids, platforms, and 3D-bioprinted TMEs (Mao et al., 2024; Shah et al., 2024).
- **Phase I/II trails:**
  Assessment of safety, dose escalation and therapeutic testing, using real-time thermal imaging and biomarker monitoring (Kim et al., 2024; Rieke & Butts Pauly, 2008).
- **Phase III and beyond:**
  Large cohorts of patients treated with nano-enhanced cryoablation vs. current standard of care, focusing on overall survival and the induction of an abscopal effect (Mitchell et al., 2021; Waitz et al., 2012).

*Table 3. Roadmap to clinical translation of cryo-nanomedicines*

| Translational Stage | Key Objective | Core Approach | Main Challenge | Strategy | Key References |
|---|---|---|---|---|---|
| **Preclinical development** | Validate safety and efficacy | SbD, organ-on-chip/TME models, corona analysis | Protein corona, biodistribution, toxicity | Biodegradable design and early corona profiling | (Bhatia & Ingber, 2014; Fadeel et al., 2018; Sheibani et al., 2021) |
| **Phase I/II translation** | Define safety and thermal dose | MRI thermometry, dose escalation, thermal-window mapping | Ice-ball heterogeneity and thermal drift | PRF-based MRI monitoring and localized cooling | (Kim et al., 2024; Kwak et al., 2022; Rieke & Butts Pauly, 2008) |
| **Phase III evaluation** | Test clinical benefit | Nano-cryoablation ± immunotherapy | Variable immune response and tumor heterogeneity | Stratified trials with survival and immune endpoints | (Mitchell et al., 2021; Waitz et al., 2012)) |
| **AI-enabled personalization** | Optimize patient-specific therapy | Digital twins, RFdiffusion, closed-loop control | Multi-omics complexity and model validation | Predictive modeling with real-time imaging feedback | (Björnsson et al., 2019; Corral-Acero et al., 2020; Watson et al., 2023) |
| **Manufacturing and regulation** | Ensure reproducibility and scale-up | QbD, microfluidics, green synthesis | Batch variability and solvent burden | ICH-guided QbD and scalable low-solvent processing | (Mao et al., 2024; Use, 2009; Wen et al., 2025) |

**Table 3** summarizes the key translational requirements for cry nanomedicine, including preclinical safety testing, imaging-guided dose optimization, AI-based personalization, scalable manufacturing, and regulatory validation needed for clinical application.

## 8. CONCLUSION

The integration of therapeutic hypothermia and nanotechnology, known as cryo-nanomedicine, represents a transformative advancement in precision oncology. The development of this field, which transcends the constraints of whole-body cooling and traditional cryosurgery, takes advantage of the biochemical specificity of the Q10 effect and the physical adaptability of stimuli-responsive nanomaterials to achieve targeted therapy. This chapter, has discussed how next-generation nanoplatforms, including phase-change nanocapsules and thermally conductive cryo-sensitizers, enable selective eradication of tumour cells while preserving the physiological function of the stromal compartment. Additionally, the integration of tools for structural validation, such as Cryo-EM, and real-time thermal feedback has established the foundation for validation of stability and spatial precision of these methods. With the advent of AI-driven Digital Twins and the future of sustainable, nutraceutical-based production, cryo-nanomedicine will soon break the translational barriers of today. These innovations will enable patient-specific optimization of treatment protocols while addressing long-standing challenges related to toxicity, scalability and environmental impact. In summary, the creation of patient-specific, thermo-sensitive, drug-delivery engines provides a strong and eco-friendly approach to enhance cancer destruction while minimising systemic side effects and improving overall therapeutic outcomes.

## REFERENCES


Ali Mohammad, S., Hak, A., Pogu, S. V., & Rengan, A. K. (2023). Radiotherapy, photodynamic therapy, and cryoablation-induced abscopal effect: Challenges and future prospects. *Cancer Innovation*, *2*(5), 323–345. 10.1002/cai2.53. DOI: 10.1002/cai2.53 PMID: 38090387

Amin, M., Huang, W., Seynhaeve, A. L. B., & ten Hagen, T. L. M. (2020). Hyperthermia and Temperature-Sensitive Nanomaterials for Spatiotemporal Drug Delivery to Solid Tumors. *Pharmaceutics*, *12*(11), 1007. DOI: 10.3390/pharmaceutics12111007 PMID: 33105816

Ashique, S., Garg, A., Mishra, N., Raina, N., Ming, L. C., Tulli, H. S., Behl, T., Rani, R., & Gupta, M. (2023). Nano-mediated strategy for targeting and treatment of non-small cell lung cancer (NSCLC). *Naunyn-Schmiedeberg's Archives of Pharmacology*, *396*(11), 2769–2792. DOI: 10.1007/s00210-023-02522-5 PMID: 37219615

Bansal, D. (2025). Personalized medicine in oncology: advancements, difficulties, and prospects. *International journal of therapeutic innovation, 3*(4).

Baust, J. G., Santucci, K. L., Snyder, K. K., Robilotto, A., & Baust, J. M. (2023). Mechanisms of Tissue Injury in Cryosurgery. In *The Application of Heat in Oncology* (pp. 45-71). 10.1002/9781119799627.ch4

Bhatia, S. N., & Ingber, D. E. (2014). Microfluidic organs-on-chips. *Nature Biotechnology*, *32*(8), 760–772. DOI: 10.1038/nbt.2989 PMID: 25093883

Björnsson, B., Borrebaeck, C., Elander, N., Gasslander, T., Gawel, D. R., Gustafsson, M., Jörnsten, R., Lee, E. J., Li, X., Lilja, S., Martínez-Enguita, D., Matussek, A., Sandström, P., Schäfer, S., Stenmarker, M., Sun, X. F., Sysoev, O., Zhang, H., & Benson, M.Swedish Digital Twin. (2019). Digital twins to personalize medicine. *Genome Medicine*, *12*(1), 4. DOI: 10.1186/s13073-019-0701-3 PMID: 31892363

Blackwell, J., Kraśny, M. J., O'Brien, A., Ashkan, K., Galligan, J., Destrade, M., & Colgan, N. (2022). Proton Resonance Frequency Shift Thermometry: A Review of Modern Clinical Practices. *Journal of Magnetic Resonance Imaging*, *55*(2), 389–403. DOI: 10.1002/jmri.27446 PMID: 33217099

Corral-Acero, J., Margara, F., Marciniak, M., Rodero, C., Loncaric, F., Feng, Y., Gilbert, A., Fernandes, J. F., Bukhari, H. A., Wajdan, A., Martinez, M. V., Santos, M. S., Shamohammdi, M., Luo, H., Westphal, P., Leeson, P., DiAchille, P., Gurev, V., Mayr, M., & Lamata, P. (2020). The 'Digital Twin' to enable the vision of precision cardiology. *European Heart Journal*, *41*(48), 4556–4564. DOI: 10.1093/eurheartj/ehaa159 PMID: 32128588

Dai, X., Dai, Y., Zheng, Y., & Lv, Y. (2024). Magnetic nanoparticles and possible synergies with cold atmospheric plasma for cancer treatment. *RSC Advances, 14*(40), 29039–29051. DOI: 10.1039/D4RA03837A PMID: 39282063

Fadeel, B., Bussy, C., Merino, S., Vázquez, E., Flahaut, E., Mouchet, F., Evariste, L., Gauthier, L., Koivisto, A. J., Vogel, U., Martín, C., Delogu, L. G., Buerki-Thurnherr, T., Wick, P., Beloin-Saint-Pierre, D., Hischier, R., Pelin, M., Candotto Carniel, F., Tretiach, M., & Bianco, A. (2018). Safety Assessment of Graphene-Based Materials: Focus on Human Health and the Environment. *ACS Nano, 12*(11), 10582–10620. DOI: 10.1021/acsnano.8b04758 PMID: 30387986

Fernandez Ainaga, D. L., Roncal-Herrero, T., Ilett, M., Aslam, Z., Cheng, C., Hitchcock, J. P., Cayre, O. J., & Hondow, N. (2025). Native state structural and chemical characterisation of Pickering emulsions: A cryo-electron microscopy study. *Journal of Microscopy, 298*(1), 92–105. DOI: 10.1111/jmi.13391 PMID: 39887717

Galluzzi, L., Buqué, A., Kepp, O., Zitvogel, L., & Kroemer, G. (2017). Immunogenic cell death in cancer and infectious disease. *Nature Reviews. Immunology, 17*(2), 97–111. DOI: 10.1038/nri.2016.107 PMID: 27748397

Gijsen, M., Vlasselaers, D., Spriet, I., & Allegaert, K. (2021). Pharmacokinetics of Antibiotics in Pediatric Intensive Care: Fostering Variability to Attain Precision Medicine. *Antibiotics (Basel, Switzerland), 10*(10), 1182. https://www.mdpi.com/2079-6382/10/10/1182. DOI: 10.3390/antibiotics10101182 PMID: 34680763

Gong, N., Zhong, W., Alameh, M.-G., Han, X., Xue, L., El-Mayta, R., Zhao, G., Vaughan, A. E., Qin, Z., Xu, F., Hamilton, A. G., Kim, D., Xu, J., Kim, J., Teng, X., Li, J., Liang, X.-J., Weissman, D., Guo, W., & Mitchell, M. J. (2024). Tumour-derived small extracellular vesicles act as a barrier to therapeutic nanoparticle delivery. *Nature Materials, 23*(12), 1736–1747. DOI: 10.1038/s41563-024-01961-6 PMID: 39223270

Gonzalez-Ibarra, F. P., Varon, J., & Lopez-Meza, E. G. (2011). Therapeutic Hypothermia: Critical Review of the Molecular Mechanisms of Action [Review]. *Frontiers in Neurology, volume 2 - 2011*. DOI: 10.3389/fneur.2011.00004

Gupta, V., Kumar, D., Gupta, S., Tanwar, R., Jaiswal, N. K., Islam, M. M., Singh, S., Choudhary, N., Gowri, S., Webster, T. J., & Faiyazuddin, M. (2026). Nanotechnology-Driven Cancer Therapies for Precision Oncology: Advances and Clinical Outlook. *International Journal of Nanomedicine, 21*(null), 568254. DOI: 10.2147/IJN.S568254

Habeeb, M., Deepthi, K. L., Vengateswaran, H. T., Arsey, S., You, H. W., & Aher, K. B. (2024). Innovations in Metal-Based Nanomedicines to Interfere with the Signaling Communications of Prostate Cancer. *BioNanoScience*, *15*(1), 123. DOI: 10.1007/s12668-024-01723-3

Imani, S., Moradi, S., Faraj, T. A., Hassanpoor, P., Musapour, N., Najmaldin, S. K., Abdulhamd, A. H., Mohammadi, A. T., Taha, C. H., & Aminnezhad, S. (2025). Nanoparticle technologies in precision oncology and personalized vaccine development: Challenges and advances. *International Journal of Pharmaceutics: X*, *10*, 100353. DOI: 10.1016/j.ijpx.2025.100353 PMID: 40688030

Imani, Z., Amini, M., Ghahremani, M. H., Khonsari, F., Shahbaz, S., Atyabi, F., & Dinarvand, R. (2025). Docetaxel-dimer prodrug as a glutathione-responsive nano drug delivery system for improved chemotherapy. *Journal of Pharmaceutical Sciences*, *114*(9), 103874. DOI: 10.1016/j.xphs.2025.103874 PMID: 40541678

Khan, A., Anas, M., Ullah, N., Khattak, W. A., Ali, M., Maaza, M., & Shinwari, Z. K. (2025). Harnessing Green Nanotechnology: Exploring the Potential of Silver, Copper, and their Nanocomposites for Biomedical and Agricultural Applications. *Pakistan Journal of Botany*, *57*(6), 6. DOI: 10.30848/PJB2025-6(5)

Khan, A., Ashiq, B., Qadri, T. A., Khattak, W. A., Anas, M., Khan, R. A., Abbas, S., Shinwari, Z. K., & Maaza, M. (2025). Nanotechnology Solutions For Advancing Sustainable Development Goals (SDGs): Integrating Nanomaterials For Global Sustainability. *BioNanoScience*, *15*(4), 590. DOI: 10.1007/s12668-025-02214-9

Khan, A., Khan, R. A., Anas, M., Qadri, T. A., Ashiq, B., Younis, T., Ullah, N., Khan, A. M., Nasreen, Z., & Shinwari, Z. K. (2026). Biogenic Nanoparticles and Green Nanocomposites as Sustainable Antimicrobial Strategies Against Bacteria, Fungi and Viruses. *Nano-Horizons: Journal of Nanosciences and Nanotechnologies*, *5*, 31 pages. DOI: 10.25159/3005-2602/20247

Kim, K., Narsinh, K., & Ozhinsky, E. (2024). Technical advances in motion-robust MR thermometry. *Magnetic Resonance in Medicine*, *92*(1), 15–27. DOI: 10.1002/mrm.30057 PMID: 38501903

Kwak, K., Yu, B., Lewandowski, R. J., & Kim, D. H. (2022). Recent progress in cryoablation cancer therapy and nanoparticles mediated cryoablation. *Theranostics*, *12*(5), 2175–2204. DOI: 10.7150/thno.67530 PMID: 35265206

Li, H., Qiao, R., Davis, T. P., & Tang, S.-Y. (2020). Biomedical Applications of Liquid Metal Nanoparticles: A Critical Review. *Biosensors (Basel)*, *10*(12), 196. DOI: 10.3390/bios10120196 PMID: 33266097

Liu, Y., Liu, Y., Zang, J., Abdullah, A. A. I., Li, Y., & Dong, H. (2020). Design Strategies and Applications of ROS-Responsive Phenylborate Ester-Based Nanomedicine. *ACS Biomaterials Science & Engineering*, *6*(12), 6510–6527. DOI: 10.1021/acsbiomaterials.0c01190 PMID: 33320631

Lovatt, C., O'Sullivan, T., Ortega-de San Luis, C., Ryan, T. J., & Frank, R. A. W. (2026). Memory engram synapse 3D macromolecular architecture visualized by cryoCLEM-guided cryoET. *Structure (London, England)*, *34*(1), 100–112.e103. DOI: 10.1016/j.str.2025.10.014 PMID: 41240908

Ma, B., Teixeira, W., & Jiang, L. (2024). Cryoablation techniques in bladder cancer [Frigid Zone Medicine]. *RE:view*, *4*(2), 72–77. doi:10.1515/fzm-2024-0008

Mao, Y., Xie, J., Yang, F., Luo, Y., Du, J., & Xiang, H. (2024). Advances and prospects of precision nanomedicine in personalized tumor theranostics [Review]. *Frontiers in Cell and Developmental Biology, Volume 12 - 2024*. DOI: 10.3389/fcell.2024.1514399

Mitchell, M. J., Billingsley, M. M., Haley, R. M., Wechsler, M. E., Peppas, N. A., & Langer, R. (2021). Engineering precision nanoparticles for drug delivery. *Nature Reviews. Drug Discovery*, *20*(2), 101–124. DOI: 10.1038/s41573-020-0090-8 PMID: 33277608

Ou, W., Stewart, S., White, A., Kwizera, E. A., Xu, J., Fang, Y., Shamul, J. G., Xie, C., Nurudeen, S., Tirada, N. P., Lu, X., Tkaczuk, K. H. R., & He, X. (2023). In-situ cryo-immune engineering of tumor microenvironment with cold-responsive nanotechnology for cancer immunotherapy. *Nature Communications*, *14*(1), 392. DOI: 10.1038/s41467-023-36045-7 PMID: 36693842

Peng, Y., Yang, Z., Li, J., & Liu, S. (2024). Research progress on nanotechnology of traditional Chinese medicine to enhance the therapeutic effect of osteoarthritis. *Drug Delivery and Translational Research*, *14*(6), 1517–1534. DOI: 10.1007/s13346-024-01517-w PMID: 38225521

Rieke, V., & Butts Pauly, K. (2008). MR thermometry. *Journal of Magnetic Resonance Imaging*, *27*(2), 376–390. DOI: 10.1002/jmri.21265 PMID: 18219673

Rulten, S. L., Grose, R. P., Gatz, S. A., Jones, J. L., & Cameron, A. J. M. (2023). The Future of Precision Oncology. *International Journal of Molecular Sciences*, *24*(16), 12613. https://www.mdpi.com/1422-0067/24/16/12613. DOI: 10.3390/ijms241612613 PMID: 37628794

Shah, M., Prakashan, D., & Gandhi, S. (2024). Microfluidics, organs-on-a-chip, and 3D printing. In Mohanan, P. V. (Ed.), *Human Organs-on-a-Chip Technology* (pp. 91–112). Academic Press., DOI: 10.1016/B978-0-443-13782-2.00008-5

Sharma, A., Tyagi, V. V., Chen, C. R., & Buddhi, D. (2009). Review on thermal energy storage with phase change materials and applications. *Renewable & Sustainable Energy Reviews, 13*(2), 318–345. DOI: 10.1016/j.rser.2007.10.005

Sheibani, S., Basu, K., Farnudi, A., Ashkarran, A., Ichikawa, M., Presley, J. F., Bui, K. H., Ejtehadi, M. R., Vali, H., & Mahmoudi, M. (2021). Nanoscale characterization of the biomolecular corona by cryo-electron microscopy, cryo-electron tomography, and image simulation. *Nature Communications, 12*(1), 573. DOI: 10.1038/s41467-020-20884-9 PMID: 33495475

Takechi-Haraya, Y. (2025). Cryogenic transmission electron microscopy for the evaluation of lipid-based nanomedicines: Principles, applications, and challenges. *Biological & Pharmaceutical Bulletin, 48*(11), 1641–1651. DOI: 10.1248/bpb.b25-00158 PMID: 41183888

Tang, S.-Y., Tabor, C., Kalantar-Zadeh, K., & Dickey, M. D. (2021). Gallium liquid metal: The devil's elixir. *Annual Review of Materials Research, 51*(1), 381–408. DOI: 10.1146/annurev-matsci-080819-125403

Torchilin, V. P. (2014). Multifunctional, stimuli-sensitive nanoparticulate systems for drug delivery. *Nature Reviews. Drug Discovery, 13*(11), 813–827. DOI: 10.1038/nrd4333 PMID: 25287120

Ullah, N., Khan, A., Maqbool, H., Ali, M., Munawar, K. S., Fallatah, A. M., Mohamed, H. E. A., Hkiri, K., Maaza, M., & Shinwari, Z. K. (2026). Green synthesis and biomedical evaluation of graphene oxide nanoparticles using Monotheca buxifolia extract. *3 Biotech, 16*(4), 144. DOI: 10.1007/s13205-026-04758-3

Use, I. C. H. T. R. R. P. H. (2009). ICH harmonised tripartite guideline: Pharmaceutical development Q8 (R2). In: International Council for Harmonisation of Technical Requirements for….

Valentino, A., Yazdanpanah, S., Conte, R., Calarco, A., & Peluso, G. (2024). Smart Nanocomposite Hydrogels as Next-Generation Therapeutic and Diagnostic Solutions. *Gels (Basel, Switzerland), 10*(11), 689. DOI: 10.3390/gels10110689 PMID: 39590045

Waitz, R., Solomon, S. B., Petre, E. N., Trumble, A. E., Fassò, M., Norton, L., & Allison, J. P. (2012). Potent induction of tumor immunity by combining tumor cryoablation with anti–CTLA-4 therapy. *Cancer Research, 72*(2), 430–439. DOI: 10.1158/0008-5472.CAN-11-1782 PMID: 22108823

Wang, H., Agarwal, P., Liang, Y., Xu, J., Zhao, G., Tkaczuk, K. H. R., Lu, X., & He, X. (2018). Enhanced cancer therapy with cold-controlled drug release and photothermal warming enabled by one nanoplatform. *Biomaterials*, *180*, 265–278. DOI: 10.1016/j.biomaterials.2018.07.021 PMID: 30055400

Watson, J. L., Juergens, D., Bennett, N. R., Trippe, B. L., Yim, J., Eisenach, H. E., Ahern, W., Borst, A. J., Ragotte, R. J., Milles, L. F., Wicky, B. I. M., Hanikel, N., Pellock, S. J., Courbet, A., Sheffler, W., Wang, J., Venkatesh, P., Sappington, I., Torres, S. V., & Baker, D. (2023). De novo design of protein structure and function with RFdiffusion. *Nature*, *620*(7976), 1089–1100. DOI: 10.1038/s41586-023-06415-8 PMID: 37433327

Wen, F., Wang, L., Li, X., Zhao, J., Xu, T., Zhu, J., Ma, L., & Wang, X. (2025). Precision Nanomedicine for Cancer: Innovations, Strategies, and Translational Challenges. *OncoTargets and Therapy*, *18*(null), 1125-1148. DOI: 10.2147/OTT.S550104

Xie, Y., Zou, Y., Lyu, L., Yang, M., Liu, Y.-N., Zeng, K., & Chen, W. (2025). Nanomaterials for Cryomedicine: Classification, Mechanisms, and Applications. *Advanced Functional Materials*, *35*(41), 2506488. DOI: 10.1002/adfm.202506488

Yaramiri, A., Asalh, R. A., Asalh, M. A., AlSawaftah, N., Abuwatfa, W. H., & Husseini, G. A. (2025). A Comprehensive Review of Smart Thermosensitive Nanocarriers for Precision Cancer Therapy. *International Journal of Molecular Sciences*, *26*(15), 7322. DOI: 10.3390/ijms26157322 PMID: 40806454

Zhang, D., Liu, X., Li, X., Cai, X., Diao, Z., Qiu, L., Chen, X., Liu, Y., Sun, J., Cui, D., Ye, Q., & Yin, T. (2024). A Multifunctional Low-Temperature Photothermal Nanomedicine for Melanoma Treatment via the Oxidative Stress Pathway Therapy. *International Journal of Nanomedicine*, *19*(0), 11671–11688. DOI: 10.2147/IJN.S487683 PMID: 39553457

Zhang, N., Yuan, Y., Cao, X., Du, Y., Zhang, Z., & Gui, Y. (2018). Latent Heat Thermal Energy Storage Systems with Solid–Liquid Phase Change Materials: A Review. *Advanced Engineering Materials*, *20*(6), 1700753. DOI: 10.1002/adem.201700753

Chapter 12
# Hyperthermia in Nano-Oncology


**Tauqeer Ahmed Qadri**
http://orcid.org/0009-0000-9617-4512
*University of South Africa, South Africa*

**Nabeel ur Rehman**
http://orcid.org/0009-0009-6632-6218
*COMSATS University Islamabad, Pakistan*

**Ayesha Fazal**
http://orcid.org/0009-0007-5328-267X
*COMSATS University Islamabad, Pakistan*

**Sakandar Khan**
*KBCMA College of Veterinary and Animal Sciences, Narowal, Pakistan*

**Shab Niaz Khan**
*COMSATS University Islamabad, Pakistan*

**Shahid Hussain**
http://orcid.org/0000-0002-9820-7477
*COMSATS University Islamabad, Pakistan*

**Asad Ali Chohan**
*University of Wah, Wah Cantt, Pakistan*

**Shabbar Abbas**
http://orcid.org/0000-0003-3435-1266
*COMSATS University Islamabad, Pakistan*



## ABSTRACT

*Hyperthermia in nano-oncology integrates controlled heat therapy with nanotechnology for precise cancer treatment. This chapter reviews nano-enabled hyperthermia systems, including magnetic, photothermal, radiofrequency, microwave, and ultrasound approaches, along with metallic, magnetic, carbon-based, semiconductor, and polymeric nanoplatforms. It highlights green synthesis strategies that improve biocompatibility and sustainability. The synergistic use of hyperthermia with chemotherapy, radiotherapy, immunotherapy, and gene therapy is discussed, as are theranostic applications enabling simultaneous imaging and treatment. Challenges*


DOI: 10.4018/979-8-3373-9848-8.ch012

*related to toxicity, biodistribution, and clinical translation remain, while AI-driven designs and smart nanoplatforms offer promising directions for future precision oncology*

## 1. INTRODUCTION

Cancer is among the major causes of morbidity and mortality across the globe, and a major burden to civic health and socioeconomic status (Sung et al., 2024). With significant diagnosis and treatment improvement efforts, the world cancer prevalence trends are increasing because of the expanding population, ageing, environmental and lifestyle-based risks (Deivayanai et al., 2025). Traditional therapeutic measures such as surgery, chemotherapy, and radiotherapy have played a great role in increasing patient survival, but they are usually limited in their selection and radiation, lack systemic tolerance, and are toxic and destructive to the normal body tissues (Sayed et al., 2024). Tumor location and metastatic spread often limit the options of surgical intervention, whereas chemotherapy and radiotherapy may have severe adverse effects undermining the quality of life of a patient (Fan et al., 2023). These constraints have intensified efforts into seeking novel therapeutic modalities that can carefully target tumors with a low toxicity to non-target organs (Deivayanai et al., 2025).

Here, the concept of nano-oncology has proven to be a revolutionary branch which combines nanotechnology in cancer diagnosis, imaging and therapy (Ciftci et al., 2025). Nanoscale materials and delivery systems have also made possible the enhancement of drug solubility, increased circulation, enhanced tumor remittance and controlled drug release (Fan et al., 2023). Nanotechnology has provided new prospects of treating cancer precisely through passive and active targeting of the tumor using the enhanced permeability and retention (EPR) effect and by the use of tumor specific ligand respectively (Pallares et al., 2025). Moreover, multifunctional nanoplatforms have the capability of both therapeutic and diagnostic tasks and are already the basis of the emerging theranostics of cancer (Akpe et al., 2025).

Hyperthermia is one of the alternative therapies that have attracted significant interest in recent years as a potential measure in managing cancer (Bravo et al., 2024). Hyperthermia is the premeditated increase in temperature of tissues, normally between 39 and 45°C, to create cellular stress, augmenting tumor vulnerability, and increasing treatment results (Bala et al., 2023). Unreversible cell damage and tumor ablation could take place at elevated temperatures. The application of heat as a treatment method can be traced back to the centuries, but the recent developments in nanotechnology contributed to the precision and effectiveness of hyperthermia-based therapy to a significant extent (Szwed and Marczak, 2024). Nanoparticles, including gold nanoparticles, iron oxide nanoparticles, carbon nanostructures and hybrid nanocomposites in nanomaterials are capable of producing localized heat in response to external stimuli that include magnetic fields, near-infrared light, radiofrequency waves, and ultrasound (Shakeri-Zadeh et al., 2025). Hypothermia and nanomedicine approaches can be completely synergized to produce a highly local tumor destruction with minimum damage to adjacent normal tissues (Appidi et al., 2025).

The chapter offers an extensive review of hyperthermia within the context of nano-oncology including sustainable production of nanomaterial, nano-mediated heat generation mechanisms, advanced hyperthermia platforms as well as their utilization in targeted cancer therapy. Also, the existing challenges, clinical translation opportunities, and future perspectives of next-generation nano-hyperthermia systems are discussed.

## 2. FUNDAMENTALS OF HYPERTHERMIA IN CANCER THERAPY

### 2.1. Principles of Therapeutic Hyperthermia

Hyperthermia is a treatment technique that entails the controlled increase of the temperature of the tissues to intensify the killing of cancerous cells whilst maintaining the healthy tissues around (Szwed and Marczak, 2024). In contrast to thermal ablation which involves such high temperatures that they cause an immediate necrosis to the tissues, therapeutic hyperthermia normally involves temperatures between 39°C to 45 o C without causing any immediate tissue necrosis but rather act to sensitize tumor cells and enhance the effectiveness of other treatment modalities (Bala et al., 2023). The foreign condition of cancer cells makes them more vulnerable to the harmful effects of heat than normal cells because of their abnormal vasculature, high metabolic rate, hypoxia and defective heat dissipation systems (Bravo et al., 2024). As a result, hyperthermia has the potential to be selective towards tumors and can improve therapy outcomes alone or in combination with chemotherapy, radiotherapy, immunotherapy, and nanomedicine-based interventions (Molaei et al., 2024).

### 2.2. Classification of Hyperthermia

Hyperthermia may be categorized based on body exposure and area being targeted (Szwed and Marczak, 2024).

### 2.2.1. Local Hyperthermia

Local hyperthermia concentrates the heat on a lesion or tumor. Local heating is produced using energy sources which includes lasers, radiofrequency waves, microwaves, ultrasound, and magnetic fields (Bala et al., 2023). This is especially applicable when dealing with easily reachable solid tumors and eliminates a lot of thermal damage on other healthy tissues (Gavilán et al., 2025).

### 2.2.2. Regional Hyperthermia

Regional hyperthermia is applied to large parts of the body or organs, or body cavities that have malignancy. The procedure can usually be applied to cancers in the extremities, abdomen, liver and the pelvis (Vlastou et al., 2025). The regional hyperthermia is effective in increasing the penetration of drugs and therapeutic responses in the treatment area through increasing the blood flow and the permeability of tissues in the area (Appidi et al., 2025).

### 2.2.3. Whole-Body Hyperthermia

Whole-body hyperthermia refers to raising body temperature in its entirety and is mostly applied in the treatment of metastatic/disseminated cancers. Manual systemic heating induces immunity, circulation, and could increase the performance of systemic anticancer treatments (Bravo et al., 2024). Though, there is a need to be closely monitored in order to prevent negative physiological impacts (Coene and Leliaert, 2022).

## 2.3. Temperature-Dependent Cellular Effects

Hyperthermia has astronomical biological effects based on temperature and duration of contact (Sayed et al., 2024). Mild hyperthermia (39-43 C) augments the blood perfusion, oxygenation, cellular sensitivity to chemotherapy, and radiotherapy (Vlastou et al., 2025). Intermediate temperatures (43 45C) cause an irreversible cellular stress, inhibit the repair response, and stimulate the programmed cell death. Warmer temperatures cause rapid coagulation of proteins, disruption of the membrane and extensive necrosis of tumors, which causes thermal ablation (Sung et al., 2024). The effects of this nature of temperature can offer a degree of flexibility in the development of treatment schemes that are specific to various types of cancer and targeted at specific clinical goals (Graham et al., 2025).

## 2.4. Molecular Mechanisms of Hyperthermia-Induced Tumor Destruction

Hyperthermia affects the antitumor effect in a variety of interlocking molecular mechanisms. Increase in temperatures will lead to protein degradation and unfolding (Figure 1) and interfere with cellular functions and metabolism (Sung et al., 2024). There is also an effect of heat exposure that elevates the permeability of membrane resulting to loss of cellular integrity and alteration of signaling pathway. Moreover, hyperthermia results in the production of reactive oxygen species (ROS), which leads to oxidative stress, lipid peroxidation, and mitochondrial dysfunction (Chauhan et al., 2026). Damage to DNA and a block of DNA (Figure 1) repair further undermine survival of cancer cells (Chauhan et al., 2025). All of these effects trigger apoptosis, necrosis, and autophagy ultimately decreasing tumor viability and growth (Figure 1). The expression of heat shock proteins (also dual promoting cellular protection and immune-activation) can also be induced by hyperthermia (Garg et al., 2025).

## 2.5. Hyperthermia and Tumor Microenvironment Modulation

In addition to the direct cytotoxicity, hyperthermia also has a major impact on the tumor microenvironment. Regulated heating enhances blood flow and vascular permeability of the tumors, which in effect raises oxygen supply, decreases hypoxia, a significant factor in treatment resistance (Sayed et al., 2024). Perfusion is also improved enabling the localization of treatment agents in tumors and increasing their penetration of the drug (Hwang et al., 2026). Hyperthermia also regulates immune, as it stimulates antigen presentation, activates immune effecter cells and promotes the release of danger-associated molecular patterns of damaged tumor cells (Li et al., 2024). These alterations have the potential to modify an immunosuppressive tumor microenvironment into a more immunologically active one and thus enhances the efficacy of immunotherapeutic and nanoparticle-based treatment approach (Figure 1). As such hyperthermia is not only a direct anticancer modality but also an efficacious potentiator of precision cancer therapeutics (Xu et al., 2025).

*Figure 1. Cellular and molecular mechanisms underlying hyperthermia-mediated tumor destruction and therapeutic sensitization*

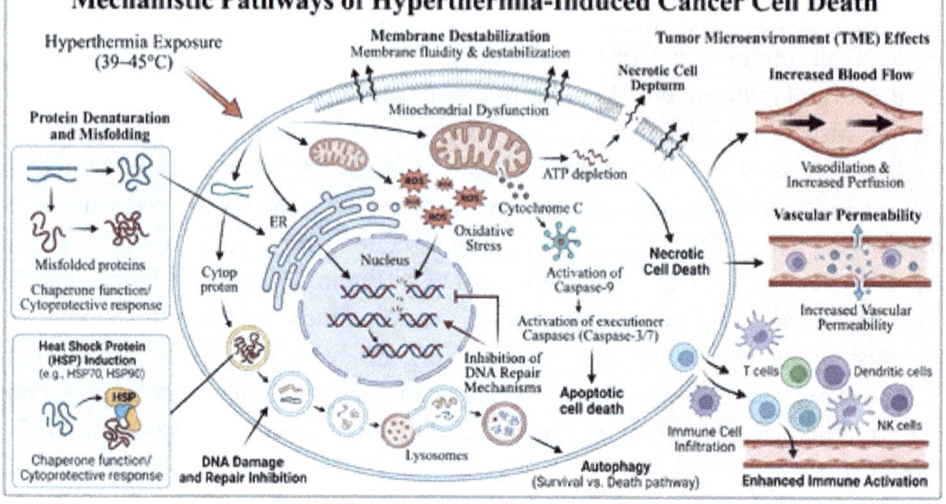


## 3. NANOTECHNOLOGY-ENABLED HYPERTHERMIA PLATFORMS

### 3.1. Role of Nanomaterials in Hyperthermia

The combination of nanotechnology and hyperthermia has transformed the process of cancer treatment to allow the localized, controlled and specific application of heat in tumor tissues (Molaei et al., 2024). The traditional forms of hyperthermia are usually characterized by lack of sufficient heat distribution, low tumor selectivity and unwanted destruction of other soft tissues (Chen et al., 2025). Nanomaterials elute these limitations by means of its physicochemical characteristics, such as variable size, high surface area-volume ratio, high biocompatibility as well as responsive to external stimuli. Engineered nanoparticles have the capacity to convert external energy to heat when subjected to magnetic fields, near-infrared (NIR) light, radiofrequency waves, or ultrasound energy thereby specifically destroying tumors (Koç et al., 2025). Additionally, nanomaterials may also be targeted with targeting ligands and accumulate selectively with the support of enhanced permeability and retention (EPR) effect in tumors (Patnaik et al., 2025). All of this is achieved with nanoparticle-mediated hyperthermia, which becomes a promising tool in precision oncology with improved therapeutic potencies with reduced systemic toxicity and is expected to have a major role (Aggarwal et al., 2025).

## 3.2. Types of Hyperthermia-Generating Nanomaterials

Extensive types of nanomaterials are produced to enable using hyperthermia to treat cancer (Rajan et al., 2024). They are chosen based on the heating efficiency, biocompatibility, ability to target, and sensitivity to particular outside stimuli (Graham et al., 2025).

### 3.2.1. Metallic Nanoparticles

Metallic nanoparticles form one of the best-researched nanomaterials to be used in hyperthermia (Obalola et al., 2025). The interest in gold nanoparticles (AuNPs) is predetermined by their high-biocompatibility, the ability to chemically modify surfaces, and plentiful surface plasmon resonance (Koç et al., 2025). When exposed to NIR light, the gold nanostructures such as nanorods, nanoshells, and nanoshells fluoresce and convert the light into heat, thus they are extremely apt to use in photothermal therapy (Patnaik et al., 2025). The nanoparticles also provide photothermal functionality with silver and platinum being developed but they are still to achieve the level of clinical development as their gold-based counterparts have (Figure 2) (Obalola et al., 2025).

### 3.2.2. Magnetic Nanoparticles

One of the most clinically relevant hyperthermia platforms is magnetic nanoparticles, particularly, superparamagnetic iron oxide nanoparticles (SPIONs) (Chen et al., 2025). With Nefel and Brownian relaxation, these nanoparticles produce heat when subjected to varying and alternative magnetic fields (Baldea et al., 2025). The magnetic hyperthermia has good spatial localization and deep tissue penetration and is specifically used in the treatment of solid tumors that are within internal organs (Shakeri-Zadeh et al., 2025). Besides therapeutic heating, magnetic nanoparticles may be used as a diagnostic and treatment agent by a process called magnetic resonance imaging (MRI), which allows diagnosis and treatment of different diseases at the same time (Graham et al., 2025).

### 3.2.3. Carbon-Based Nanomaterials

Nanomaterials based on carbon have outstanding optical, thermal, and mechanical characteristics that can be attracted to applications in photothermal (Han et al., 2025). Graphene, graphene oxide, carbon nanotubes and carbon dots are good materials that have a high NIR absorption and are effective in photothermal conversion. Their elevated surface area also makes them easy in loading drugs and Mult functionalization to provide combined approaches of hyperthermia and drug delivery (Figure 2) (Koç et al., 2025). Nevertheless, the biocompatibility and biodegradation arguments of the long term are still of great concern to clinical translation (Overchuk et al., 2023).

### 3.2.4. Semiconductor Nanomaterials

The electronic and optical properties of semiconductor nanomaterials have made them promising photothermal agents, which they can be tuned to respond to (Han et al., 2025). Examples are copper sulfide, molybdenum disulfide, tungsten oxide and quantum dots. They are highly NIR absorbing materials that can be effectively used to generate heat when irradiated by laser (Hwang et al., 2026). Moreover, semiconductor nanomaterials can have imaging and sensing properties and therefore, they are good candidates of theranostic applications that combine both diagnosis and therapy (Sathuvan et al., 2026).

### 3.2.5. Polymeric and Hybrid Nanoparticles

Polymeric nanoparticles have been demonstrated to be very biocompatible, biodegradable, and adaptable in drug encapsulation (Shanahan et al., 2025). Thermoresponsive polymers can be designed to deactivate therapeutic agents onto high temperatures, improving therapeutic accuracy (Yaramiri et al., 2025). Hybrid nanoparticles are a combination of a combination of various materials, including metals, polymers, lipids, and magnetic, in one platform (Kumar et al., 2024). These multi-functional systems combine drugs delivery, imaging, and hyperthermia thus maximizing the effect of therapy and minimizing the side effects of treatment (Figure 2) (Kesharwani et al., 2025).

*Figure 2. Major nanomaterials and energy-conversion mechanisms used in nano-hyperthermia for targeted cancer treatment*

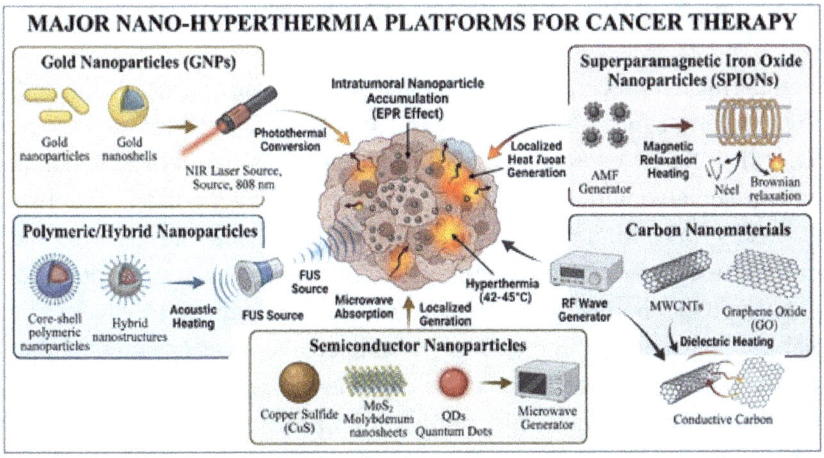


## 3.3. Tumor Targeting Strategies

This can be mostly guaranteed by the effective deposition of the nanoparticles in tumor tissues in order to ensure the success of nano-enabled hyperthermia (Ciftci et al., 2025). Passive targeting takes advantage of the enhanced permeability and retention (EPR) effect, which is the preferential uptake of nanoparticles in tumors because of these two factors explained as leaky vasculature and disrupted lymphatic drainage (Pallares et al., 2025). Specificity can be further boosted by active targeting, where the peptide, aptamer, or small-molecule ligand that recognizes tumor-associated receptors is conjugated with antibodies. Moreover, the stimuli-responsive nanoparticles may be engineered to sense pH, redox potential, enzymes, temperature or any external energy, and thereby be activated only in the tumor microenvironment (Tang et al., 2025). The use of these targeting strategies enhances the retention of nanoparticles, increases localized heat production and reduces off target toxicity (Kuwahata et al., 2025). As a result, the integration of high-end nanomaterials and cutting-edge targeting modalities have enhanced the precision of therapeutic and clinical potential of hyperthermia-based cancer therapy significantly (Tang et al., 2025).

## 4. SUSTAINABLE SYNTHESIS OF HYPERTHERMIA NANOPLATFORMS

### 4.1. Need for Sustainable Nanotechnology in Oncology

The emerging nanotechnology has greatly changed the routine of cancer diagnostics and treatment, especially in the establishment of hyperthermia nanoplatforms (Soomro et al., 2025). Nevertheless, traditional synthesis procedures of nanoparticles are usually associated with toxic substances, energy usage, poisonous solvents, and various wastes that pollute the environment. These issues have led to the growing focus on sustainable nanotechnology directions that can be supported by the principles of green chemistry and environmental stewardship (Nisha et al., 2024). Sustainability, particularly in oncology, is of special concern in that nanomaterials used in biomedical purposes have to be capable of high therapeutic potential and should also have an acceptable biocompatibility profile, with biodegradability and safety (Bhattacharya et al., 2023). Sustainable synthesis methods are designed to minimize ecological footprint and synthesize functional nanomaterials to generate localized heating to treat cancer. Moreover, the problem of residual toxicity can also be minimized, clinical acceptance can be enhanced, and mass manufacturing of hyperthermia nanoplatforms can be achieved due to the eco-friendly fabrication methods (Moacă et al., 2022). In turn, the appreciation of sustainability as a facet of nano-oncology is, therefore, a key move towards crafting safer and more eco-friendly precision cancer therapeutics (El-Khawaga et al., 2026).

### 4.2. Green Synthesis Approaches

Green synthesis has become an appealing alternative to the conventional fabrication techniques of nanoparticles through the use of biological systems and renewable resources as reducing agents, stabilizing agents and capping agents (Soomro et al., 2025). Such methods do not utilize or vastly minimize the utilization of toxic reagents and provide inexpensive and scalable routes of production (Lee et al., 2025).

#### 4.2.1. Plant-Mediated Synthesis

One of the most common methods of green fabricating nanoparticles through plants is the so-called plant-mediated synthesis (Nisha et al., 2024). Plant extracts are highly heterogeneous with regard to bioactive compounds which may include polyphenols, flavonoids, terpenoids, alkaloids, protein encasements and carbohydrates which have the potential to serve as natural reducing factors and stabilizing factors (Lee et al., 2025). These phytochemicals enable changing of metal ions into nanoparticles with mild reaction in the absence of hazardous chemicals. Nanoparticles made of gold, silver, iron oxide and copper have been produced via plant extracts and have proven to have good photothermal and magnetic properties applicable in hyperthermia (Soomro et al., 2025). In addition, surface-localized phytochemicals could be beneficial in improving nanoparticle stability, biocompatibility and biological activity, thereby plant-mediated synthesis could be used to design sustainable hyperthermia nanoplatforms (Figure 3) (Amirinezhadfard et al., 2025).

### 4.2.2. Microbial Synthesis

Bacteria, fungi, yeast, and microalgae are microorganisms widely used as an excellent choice of biological systems to make nanoparticles (Figure 3). These microorganisms can produce nanoparticles by enzyme reduction and metabolism either intra or extracellularly (El-Khawaga et al., 2026). Microbial synthesis has a number of benefits, such as the ability to fully control particle morphology, a distribution of particle sizes, and production of complex nanostructures (Moacă et al., 2022). Nanoparticles of iron oxide produced by means of microbiology have gained much interest regarding magnetic hyperthermia, whereas biologically prepared metal nanoparticles have performed exceptionally well in photothermal application. Also, microfabrication techniques tend to be energy-neutral, and environmentally non-toxic, promotes sustainable manufacturing efforts within nano-oncology (El-Khawaga et al., 2026).

### 4.2.3. Biopolymer-Assisted Fabrication

Biopolymers have been considered as a material with great importance to form sustainable nanoplatform, and this pertains especially to natural sources. Chitosan, alginate, cellulose, gelatin, dextran, and silk fibroin are all biodegradable, biocompatible and low-toxic materials (Bhattacharya et al., 2023). These biopolymers may serve as nanoparticle templates, nanoparticle stabilizers or nanotherapeutic drug delivery matrices and enhance the overall safety profile of nanotherapeutics (Zheng et al., 2026). Biopolymer-assisted nanoparticles can be used to improve the colloidal stability, increase the circulation time, and release the drug in a controlled manner in hyperthermia applications (Figure 3). Moreover, multifunctional systems that are able to generate heat, deliver drugs and be imaged in the target site can be engineered with the incorporation of metallic or magnetic nanomaterials into biopolymeric matrices (Zheng et al., 2026).

*Figure 3. Green and sustainable synthesis approaches for hyperthermia-generating nanomaterials in nano-oncology*

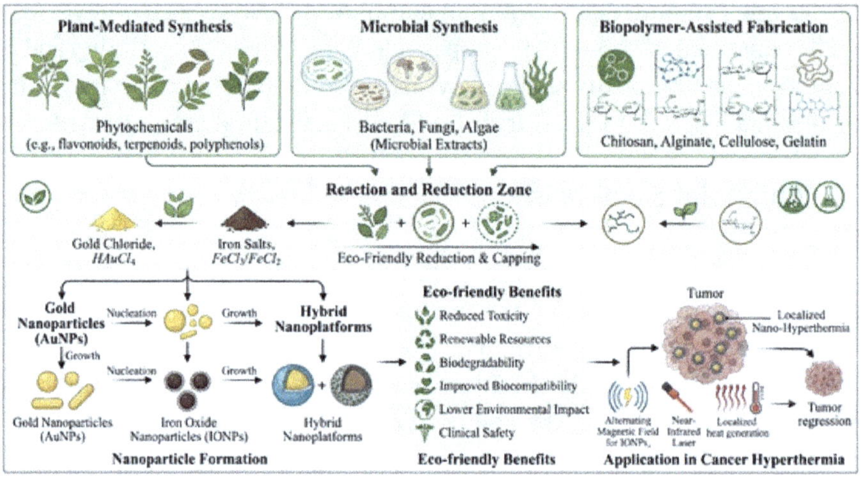


## 4.3. Advantages of Sustainable Nanoparticle Synthesis

The benefits of sustainable nanoparticle synthesis as compared to traditional methods of fabrication are many (Amirinezhadfard et al., 2025). By removing toxic solvents, dangerous reducing agents, the impact on the environment is minimized and the possibility of negative biological consequences is decreased. Green-synthesized nanoparticles are typically associated with better biocompatibility and less cytotoxicity, which seems to be crucial features of nanoparticles in clinical oncology (Nisha et al., 2024). Renewable biological resources are also used through sustainable methods and hence it is economically viable and environmentally acceptable (Amirinezhadfard et al., 2025). Also, biological molecules that are present in the process of synthesis are often of natural origin, and these serve as effective functionalizers of the surface, enhancing nanoparticle stability and therapeutic action. Translational, what concerns sustainable fabrication strategies contribute to compliance with regulations, social acceptance, and producible manufacturing (Bhattacharya et al., 2023). Given that the need to develop safe and more effective cancer therapies is on the rise, green synthesis methods will have an even greater role in development of the next-generation hyperthermia nanoplatforms of next-generation gauging of oncology (Moacă et al., 2022).

## 5. NANO-HYPERTHERMIA MODALITIES AND MECHANISMS

One of the most promising solutions in the field of precision oncology has been identified through nano-hyperthermia that allows targeted and local tissue heating in the tumor (Gavilán et al., 2025). In comparison to the traditional methods of hyperthermia, nanoparticle-based systems are more specific, better heat delivery, and less harm to the normal tissues (Chen et al., 2025). Different nano-hyperthermia modalities have been built depending on the external energy source which includes magnetic fields, near-infrared light, radiofrequency waves, microwaves and ultrasound (Hwang et al., 2026). Such technologies take advantage of the special physicochemical characteristics of nanomaterials by transforming external energy into therapeutic heat that selectively kills tumors leading to an increased treatment efficacy (Kuwahata et al., 2025).

### 5.1. Magnetic Hyperthermia

One of the most researched nano-hyperthermia approaches is magnetic hyperthermia that uses magnetic nanoparticles that are subjected to an alternating magnetic field (AMF) (Zhang et al., 2025). Nanoparticles are able to produce localized heat under influence of the magnetic field, which is able to destroy cancer cells and leave the surrounding healthy tissues intact (Xiong et al., 2025).

#### 5.1.1. Mechanism of Heat Generation

Magnetic hyperthermia is mainly relaxed by two processes: Néel relaxation and Brownian relaxation (Baldea et al., 2025). In Néel relaxation, nanoparticles constantly reorient their magnetic moments with the changing magnetic field, which loses energy as heat (Zhang et al., 2024). The process of relaxation is called Brownian relaxation, which entails the physical rotation of nanoparticles in the medium surrounding it, producing thermal energy, as a result of frictional contact (Molaei et al., 2024). The effectiveness with which heat is produced is determined by the particle sizes, magnetic characteristics, strength of the field, and frequency (Zhang et al., 2024).

#### 5.1.2. Magnetic Nanoparticles in Cancer Therapy

The most extensively used magnetic agents are superparamagnetic iron oxide nanoparticles (SPIONs) that are highly biocompatible agents, respond to magnetic fields, and can be imaged (Overchuk et al., 2023). The accumulation and therapeutic efficacy of the tumor further increases with the surface modification by polymers, antibodies, or targeting ligands (Garg et al., 2025). In addition to causing hyperthermia, the magnetic nanoparticles can also serve as a drug carrier and also as a contrast agent in magnetic resonance imaging (MRI) and hence they can be used as a multifunctional cancer treatment regimen (Aggarwal et al., 2025).

### 5.1.3. Clinical Applications

Magnetic hyperthermia has been shown to possess a lot of potential in the treatment of glioblastoma, prostate cancer, breast cancer, and recurred solid tumors (Shakeri-Zadeh et al., 2025). In multiple clinical studies enhanced tumor control has been reported with a combination of magnetic hyperthermia and chemotherapy or radiotherapy (Kraus et al., 2026). This modality has shown a high potential to be clinically translated to deep tissue heating without invasive methods (Xiong et al., 2025).

## 5.2. Photothermal Therapy (PTT)

Photothermal therapy involves the use of photothermal agents, which are able to transform the absorbed light energy into heat, leading to local destructions of tumors (Zhang et al., 2025). PTT has become a topic of significant interest, because it has high spatial resolution, low invasiveness, and is compatible with nanotechnology-based delivery systems (Gavilán et al., 2025).

### 5.2.1. Principles of Photothermal Conversion

Photothermal conversion is when nanomaterials absorb light, unlikely in the visible light spectrum, and release the absorbed energy in forms of heat by non-radiative processes (Han et al., 2025). This is a local rise in temperature leading to the denaturation of proteins, membrane perturbation, oxidative stress, and tumor cell death (Shang et al., 2025).

### 5.2.2. Near-Infrared Responsive Nanomaterials

Photothermal agents that have been developed have numerous types of nanomaterials, such as gold nanorods, gold nanoshells, graphene oxide, carbon nanotubes, copper sulfide nanoparticles, and transition metal dichalcogenides (Zhang et al., 2025). NIR-responsive materials are specifically beneficial in that the NIR light can better pass through biological tissues as compared to visible wavelengths, thus allowing deeper penetration of tumors without causing much damage to the surrounding tissues (Chauhan et al., 2026; Obalola et al., 2025).

### 5.2.3. Tumor-Specific Heat Generation

Accumulation of nanoparticles selectively in the tumors is the key to the effective PTT (Shang et al., 2025). The photothermal agents have specific localization in cancerous tissues due to passive and active targeting through which they preferentially accumulate in the target tissue as compared to normal tissues (Ciftci et al., 2025). When NIR irradiation is used, heat production is only localized to the nanoparticle rich areas which allows ablation of tumors selectively as a result, sparing healthy cells (Amaral et al., 2025).

## 5.3. Radiofrequency-Induced Hyperthermia

Hyperthermia using radiofrequency (RF) electromagnetic waves uses conductive or dielectric nanomaterials to heat them (Rajan et al., 2024). RF responsiveness has been exhibited to be efficient by gold nanoparticles, carbon nanotubes, and metallic nanocomposites. The RF waves have a deeper tissue penetration as compared to optical meaning that they can be used in the treatment of internal tumors (Li et al., 2023). Also, radiofrequency hyperthermia may be used with drug-loaded nanoparticles to form thermal therapy and the release of drugs simultaneously (Kazemi Siyanaki et al., 2025).

## 5.4. Microwave Hyperthermia

Microwave hyperthermia involves use of electromagnetic radiation within the microwave frequency to heat body tissues (Fakhradini et al., 2024). Nanomaterials like the iron oxide nanoparticles, carbon nanostructures and metal based composite aid in maximizing microwave absorption as well as enhancing heating efficiency. Microwave induced hyperthermia provides fast and even heating and has been found to treat liver, lung and breast cancers (Schreiner et al., 2026). The use of nanomaterials also enhances the selectivity of tumors and specificity in therapy (Fakhradini et al., 2024).

## 5.5. Ultrasound-Mediated Hyperthermia

Ultrasound-mediated hyperthermia makes use of focused ultrasound waves to produce focal thermal heating effects on tumors (Kashkooli et al., 2023). Nanoparticles have a great potential at amplifying the absorption of ultrasound energy leading to better heat generation and treatment effects (Al Refaai et al., 2024). It offers great spatial accuracy and penetration in the deep tissues without any ionizing radiations (Vicente et al., 2024). Furthermore, synergies of hyperthermia and chemotherapy can be achieved with the help of ultrasound-responsive nanocarriers along with controlled release of drugs (Yang et al., 2025).

*Table 1. Comparative overview of nano-hyperthermia platforms for precision cancer therapy*

| Hyperthermia Modality | Nanomaterials Used | Mechanism of Heat Generation | Advantages | Limitations | Clinical/Research Status | References |
|---|---|---|---|---|---|---|
| **Magnetic Hyperthermia** | Iron oxide nanoparticles (SPIONs), ferrites | Néel & Brownian relaxation under alternating magnetic field | Deep tissue penetration, MRI compatibility, high precision | Potential long-term retention, field optimization required | Advanced preclinical + ongoing clinical trials | Pawar and Prabhu, 2025; Kuwahata et al., 2025 |
| **Photothermal Therapy (PTT)** | Gold nanorods, graphene, carbon nanotubes, CuS nanoparticles | NIR light → non-radiative heat conversion | High spatial control, minimal invasiveness | Limited tissue penetration of NIR light | Strong preclinical, early clinical studies | Overchuk et al., 2023; Amaral et al., 2025 |
| **Radiofrequency Hyperthermia** | Metal nanoparticles, carbon-based nanostructures | RF-induced dielectric and conductive heating | Deep tissue heating, non-invasive | Heating control complexity | Preclinical to early clinical stage | Lodi et al., 2025; Zhang et al., 2024 |
| **Microwave Hyperthermia** | Metal oxide nanoparticles, hybrid nanocomposites | Microwave energy absorption → heat generation | Rapid and uniform heating | Risk of non-specific heating | Experimental/early research | Fakhradini et al., 2024; Schreiner et al., 2026 |
| **Ultrasound-Mediated Hyperthermia** | Lipid nanoparticles, polymeric carriers, nanobubbles | Acoustic cavitation and absorption heating | Deep penetration, imaging compatibility | Cavitation-related tissue variability | Preclinical development | Kashkooli et al., 2023; Yang et al., 2025 |
| **Hybrid/Theranostic Systems** | Multifunctional nanoparticles (metal-polymer hybrids) | Multi-stimuli (magnetic + NIR + RF) response | Imaging + therapy integration, precision control | Complex synthesis, regulatory hurdles | Emerging advanced research | Coene and Leliaert, 2022; Akpe et al., 2025 |

## 6. HYPERTHERMIA-ENHANCED PRECISION CANCER THERAPEUTICS

Combination of hyperthermia and sophisticated treatment tools has brought about a wide range of accuracy oncology (Wang et al., 2025). In addition to direct cytotoxic action, hyperthermia increases the delivery rate of drugs, maximizes the responsiveness of the treatment, activates antitumor immunities, and enables the transfer of genes (Wang et al., 2025). These advantages are further complemented by the introduction of nanotechnology in that it allows the attributes of heat generation, targeted delivery and real-time therapeutic monitoring (Wang et al., 2025). As a result, hyperthermia has transformed into a supportive therapy to be used as a multipurpose platform that can enhance the efficacy and specificity of contemporary cancer therapeutics (Damavandi et al., 2026).

## 6.1. Hyperthermia-Assisted Drug Delivery

The ability to enhance drug delivery and intratumoral drug accumulation is one of the greatest benefits of hyperthermia (Yaramiri et al., 2025). High temperatures enhance blood flow, vascular permeability and fluidity of cellular membrane in tumors, allowing therapeutic agents to penetrate and accumulate in the malignant tissues (Yaramiri et al., 2025). In combination with nanoparticles-based drug delivery systems, hyperthermia allows a more efficient delivery of anticancer agents to their target sites with reducing systemic toxicity (Torres et al., 2025).

### 6.1.1. Thermosensitive Nanocarriers

Nanosensors Nanocarriers are designed to trigger their therapeutic cargoes in response to certain temperatures (Li et al., 2023). Nanocarriers that can be developed include liposomes, polymeric nanoparticle, hydrogels, and lipid-based nanocarriers that can stay stable in the physiological condition but can quickly release drugs when subjected to hyperthermic conditions (Torres et al., 2025). These systems enable the delivery of drugs to a local site at the tumor location being maximized which would increase therapy response and decrease side effects on normal tissues (Al Refaai et al., 2024).

### 6.1.2. Stimuli-Responsive Drug Release

Besides temperature sensitivity, more sophisticated nanocarriers may have the capability to respond to several stimuli, such as pH, enzymes, redox, magnetic fields, and near-infrared irradiation (Torres et al., 2025). The release mechanisms in hyperthermia give a tight spatiotemporal response to drug delivery, which avoids the release of therapeutic agents outside of the tumor microenvironment (Al Refaai et al., 2024). This guided administration is a major step towards enhancing the effectiveness of therapy and promoting individual therapy approaches (Kumar et al., 2024).

## 6.2. Hyperthermia and Chemotherapy Synergy

There have been impressive synergistic effects of hyperthermia and chemotherapy (Damavandi et al., 2026). High temperatures improve cellular uptake of chemotherapeutic agents, membrane permeability and drug diffusion in tumors (Einolghasi et al., 2025). Heat can also prevent cellular repair processes and subvert mechanisms of multidrug resistance and therefore sensitize cancer cells to the effect of anticancer drugs. These effects are further amplified by nanoparticle-mediated hyperthermia allowing the generation of heat and the delivery of a drug at a single platform (Damavandi et al., 2026). This synergistic interaction enables low dosages of drugs to produce better treatment with minimal systemic toxicity and complications of the treatment (Einolghasi et al., 2025).

## 6.3. Hyperthermia and Radiotherapy Enhancement

Radiation therapy has been a mainstay in the treatment of cancer, but usually it is the hypoxia in the tumor that limits its efficacy (Vlastou et al., 2025). Hyperthermia enhances radiosensitivity by enhancing blood circulation and decreasing hypoxia in the tumors, promoting oxygen levels in tumors (Appidi et al., 2025). Moreover, heat disrupts pathways of DNA damage repair, thus cancerous cells are more prone to the adverse effects of radiations. It has been repeatedly shown by clinical studies that hyperthermia combined with radiotherapy is associated with a better tumor control, long survive time and better therapeutic responses regarding the situation than radiotherapy (Mousavi-Kiasary et al., 2025). Hyperthermia can also be aided by nanotechnology, which provides more accuracy by heating up only the tumors and saving lives of other tissues (Freitas et al., 2023).

## 6.4. Hyperthermia-Mediated Immunotherapy

The recent focus on immunotherapy to treat cancer has identified the possibility of hyperthermia as a mode of immune-modulation (Xu et al., 2025). By causing tumor damage with heat, both innate and adaptive immune responses may be raised and turn an immunologically cold tumor into a more favorable therapeutic target (Li et al., 2024).

### 6.4.1. Immunogenic Cell Death

Hyperthermia stimulates immunogenic cell death (ICD), which is an event where tumor-associated antigens and danger-associated molecular patterns (DAMPs) are released (Xu et al., 2025). These are the signals which point the immune system to the existence of malignant cells and help in the activation of antigen presenting cells (Chauhan et al., 2025). Consequently, hyperthermia may improve antitumor immunity and has long-term immune monitoring (Figure 4) (Jiao et al., 2024).

### 6.4.2. Immune Activation Pathways

The stress effect results of heat cause several immune pathways that include dendritic cells, macrophages, natural killer cells, and cytotoxic T lymphocytes (Li et al., 2024). Hyperthermia may also enhance the expression of heat shock proteins, which are molecular chaperones of tumor antigens and are immune recognition enhancers (Chauhan et al., 2026). Hyperthermia due to combination with immune checkpoint blockers or cancer vaccines can have a strong therapeutic effect and eliminate cancer immune avoidance (Jiao et al., 2024).

*Figure 4. Hyperthermia-induced immunogenic cell death promotes antigen presentation, T-cell activation, and systemic antitumor immune responses*

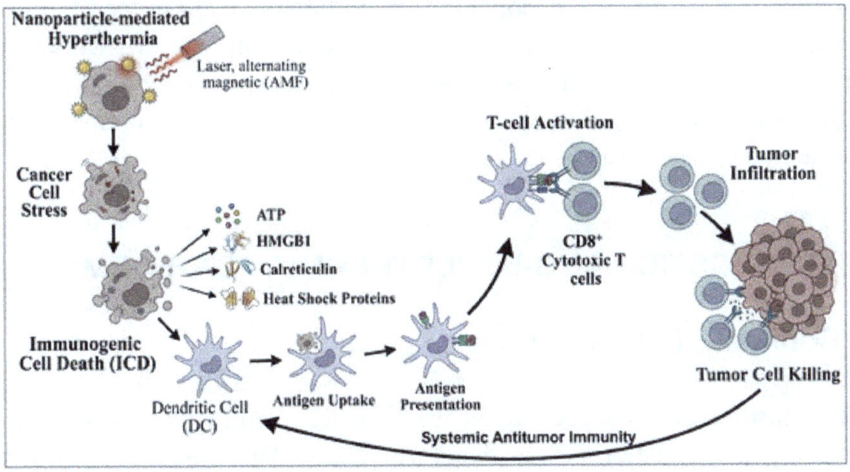


## 6.5. Hyperthermia and Gene Therapy

One potential future direction in the treatment of cancers on a molecular scale is through gene therapy, a potential approach that is yet to be fully realized due to the inefficient delivery of genes into the patients (Mousavi-Kiasary et al., 2025). Hyperthermia can enhance efficiency of transfection by enhancing cell permeability and promoting uptake of nucleic acids into cells (Tang et al., 2025). Nanocarriers, which become active in the presence of heat, have been designed to deliver therapeutic genes, small interfering RNA (siRNA), microRNA, and CRISPR-based gene-editing elements in a controlled way (Rauf et al., 2025). Hyperthermia and gene therapy together have the potential to provide greater regulation of gene expression, as well as to increase therapeutic responses (Rauf et al., 2025).

### 6.6. Personalized and Precision Nano-Oncology Approaches

The merging of hyperthermia, nanotechnology, molecular diagnostics, and personalized medicine is recasting the future of cancer treatment (Dhanapal et al., 2026). Nano-hyperthermia platforms in the present day can be designed based on tumor-specific biomarkers, genetic profiles, as well as microenvironmental features (Gupta et al., 2024). Multidimensional nanostructures that can be imaged, deliver drugs, generate heat, and monitor drugs, allow personalized treatment regimens to be provided to each patient with their disease profile (Li et al., 2023). The development of new technologies in the area of artificial intelligence, machine learning, and the integration of a biosensor in nanoplatforms additionally leads to the improvement of treatment planning and real-time response evaluation (Mousavi-Kiasary et al., 2025). Such developments are propelling the shift to precise nano-oncology with hyperthermia being a core element of personalized, effective, and minimally invasive cancer therapies (Sathuvan et al., 2026).

## 7. THERANOSTIC NANO-HYPERTHERMIA PLATFORMS

### 7.1. Concept of Theranostics

Theranostics is a new concept of precision medicine uniting the roles of diagnosis and therapy in one platform (Shanahan et al., 2025). Theranostic systems in nano-oncology are imaging agents with therapeutic payloads plus hyperthermia enabling nanomaterials to detect cancer, treat it, and in real-time tracking the treatment (Gupta et al., 2024). Nanotechnology and hyperthermia have similarly come together to hasten the creation of versatile nanoplatforms through which real-time information about nanoparticle biodistribution, therapeutic response and therapy efficacy can be obtained (Salgueiro and Zubillaga, 2025). The effect of such systems of treating cancer is accuracy of treatment, reducing unnecessary treatment and personalized treatment to cancer (Menon et al., 2025).

## 7.2. Imaging-Guided Hyperthermia

Use of imaging-guided hyperthermia is one of the important developments in cancer treatment as it allows visualization of tumors carefully, during and after the treatment (Zhu et al., 2025). Imaging functionalities of nanoplatforms of hyperthermia enable clinician to track nanoparticle uptake and optimize treatment modalities and to assess therapeutic effects more accurately (Zhu et al., 2025).

### 7.2.1. MRI-Guided Systems

One of the popular modalities of imaging in theranostic hyperthermia is magnetic resonance imaging (MRI) (Zhu et al., 2025). Superparamagnetic iron oxide nanoparticles (SPIONs) can be used both as MRI contrast agent and magnetic hyperthermia mediator (Vicente et al., 2024). MRI-based systems can achieve high spatial resolution and have great soft-tissue contrast, which allows the accurate localization of tumors and real-time tracking of the distribution of nanoparticles during therapy (Vicente et al., 2024).

### 7.2.2. CT Imaging

Computed tomography (CT) imaging provides high resolution anatomical images and can be combined with metallic nanoparticles like gold and bismuth-based nanomaterials (Parthasarathy et al., 2025). Such nanoparticles have a high level of X-ray attenuation, which makes it possible to image and provide hyperthermia-based therapy at the same time (Najdian et al., 2024). Hyperthermia with CT guidance is especially valuable to determine the morphology of tumors and structural alterations caused by treatment (Sathuvan et al., 2026).

### 7.2.3. Photoacoustic Imaging

Photoacoustic imaging is a technique that integrates optical stimulation with ultrasonic signal detection, and it involves visualizations that are both high-contrast and deep-tissue (Salgueiro and Zubillaga, 2025). Photothermal nanomaterials like gold nanorods, derivatives of graphene, and semiconductor nanoparticles with great photothermal properties can be used as photoacoustic contrast agents and heat-generating agents (Najdian et al., 2024). This two-mode capability is useful to ensure that the photothermal treatment of tumors is accurately localized and that the treatment is monitored in real-time (Menon et al., 2025).

### 7.2.4. Fluorescence Imaging

Fluorescence imaging employs fluorescent dyes, quantum dots, or fluorescent nanomaterials to visualize tumor tissues, and nanoparticle accumulation (Najdian et al., 2024). Despite the weaknesses caused by the depth of tissue penetration, fluorescence imaging is very sensitive and finds immense use in image-guided surgery and detection of superficial tumors (Salgueiro and Zubillaga, 2025). Combination with hyperthermia systems also improves the treatment accuracy and the treatment evaluation (Amaral et al., 2025).

## 7.3. Real-Time Monitoring of Treatment Response

One of the key benefits of theranostic nano-hyperthermia systems can be the real-time feedback that can be received during treatment (Parthasarathy et al., 2025). Monitoring of localization of nanoparticles in real-time, temperature distribution, and tumor response enables clinicians to adjust therapeutic parameters optimally and prevent overheating of normal tissue (Parthasarathy et al., 2025). The real-time imaging also allows prompt evaluation of treatment optimization, and adaptive treatment provision and enhances clinical outcomes (Shanahan et al., 2025). These abilities are crucial towards realizing the aims of precision medicine in cancer treatment (Gawne et al., 2023).

## 7.4. Artificial Intelligence and Smart Nano-Hyperthermia

The future of theranostic hyperthermia is being changed by the recent development of artificial intelligence (AI) and smart nanotechnology (Shirzad et al., 2025). The AI-based algorithms can be used to analyze images, predict the reactions of treatment procedures, optimize heat distribution, and suggest personalized planning of treatment (Dhanapal et al., 2026). The smart nano-hyperthermia systems with built-in biosensors and stimulas-responsive units can change dynamic therapeutic activity with respect to alterations in the tumor microenvironment (Shang et al., 2025). Such intelligent systems can make automated drug delivery, regulated heat production, and real-time therapeutic corrections, thus improving the safety and effectiveness of treatments (Coene and Leliaert, 2022). The combination of AI, next-generation imaging, and multipurpose nanoplatforms are likely to drive innovation in next-generation precision oncology systems, which can provide highly personalized and adaptive cancer treatment (Gawne et al., 2023).

## 8. CLINICAL TRANSLATION, SAFETY CHALLENGES, AND FUTURE PERSPECTIVES

This combined area considers both the bench-to-bedside translation of nano-hyperthermia systems and how they relate to clinical, regulatory, and safety issues, and also identifies innovations that are poised to define the next generation of more targeted therapeutic approaches of cancer (Rajan et al., 2024). Although nano-hyperthermia has achieved considerable success in preclinical studies, there has yet to be a metaclinical expression of nano-hyperthermia, since there are present gaps between laboratory-scale cellular efficacy and actual biomedical uses (Pallares et al., 2025). Early and preclinical trials have proven to have promising tumor regression, drug sensitivity, and survival benefit with using magnetic, photothermal, and hybrid nanoplatforms (Aggarwal et al., 2025). Nevertheless, few of these formulations have ever advanced to have clinical trials, mainly because of the issues on long-term safety, reproducibility and scalability in production (Kraus et al., 2026).

Nanotoxicity and biosafety uncertainty is a significant hindrance to clinical implementation (EMA, 2025). Despite their superb therapeutic properties, interactions of most nanomaterials with biological systems may result in unintended cytotoxicity, oxidative stress and inflammatory responses based on the dosage, surface chemistry, and accumulation profiles (Baldea et al., 2025). On the same note, is the question of biodistribution and clearance where nanoparticles can build up in the liver, spleen or kidney, which can be concerning on long term retention and organ toxicity (Patnaik et al., 2025). Attainable predictable pharmacokinetics and safe biodegradation is a crucial feature demanded to be clinically approved (Gawne et al., 2023).

Translational-wise, regulatory acceptance and scalable production are also seen as a challenge (EMA, 2025). Nanoparticle production, quality, batch-to-batch reproducibility and regulatory frameworks concern Standardization Standardization of production and quality of nanoparticles, along with regulatory compliance, remain challenging because of the physicochemical diversity of nanomaterials (EMA, 2025). Moreover, the barriers to cost-efficiency and commercialization decreases its uptake especially in healthcare systems with lower-to-middle-income statuses where state-of-the-art nanomedicine facilities might be scarce (Xiong et al., 2025). Ethical and environmental issues should also be favorable, especially the long-term ecological effects of disposal of nanoparticles and safety of patients in general in precision nanomedicine (Pawar and Prabhu, 2025).

Under the microscope of the future, nano-hyperthermia is highly pegged on smart and adaptable and AI-oriented systems (Dhanapal et al., 2026). Growing self-regulating nanoplatforms have the capability to control the production of heat depending on tumor microenvironment feedbacks to enhance therapeutic safety and effectiveness (Shirzad et al., 2025). Treatment planning is becoming more thoroughly based on artificial intelligence, which allows one to predictively model the distribution of heat, optimize it based on specifics, and modify therapy in real-time (Rauf et al., 2025). Nanoplatforms that are multifunctional and sustainable are also under development to combine imaging, drug delivery and hyperthermia in the same platform to enhance the ideal of really theranostic oncology (Gupta et al., 2024).

Moreover, convergence of nano-hyperthermia with immuno-oncology is creating new therapeutic prospects, especially, by hyperthermia-induced immunogenic cell death and immune system activation (Kraus et al., 2026). This synergy increases recognition of tumors and systemic immunity against cancer (Shirzad et al., 2025). Finally, a broader strategy of next-generation nano-hyperthermia platforms is anticipated to be more customized, smarter, and clinically adaptable and can represent a paradigm shift in precise treatment of cancer with greater safety, more efficacy, and sustainably (Chauhan et al., 2025).

## 9. CONCLUSION

Hyperthermia in nano-oncology is a rapidly developing and extremely promising approach to precise cancer treatment that combines controlled heat inactivation with the most sophisticated nanotechnology to attain local tumour killing. The hyperthermia takes advantage of the intrinsic weakness of cancer cells to high temperatures, thus increasing cellular loading, impacting molecular restorative pathways, and altering the tumor micro-environment, which introduces enhanced therapy sensitivity. The integration of nanomaterials has greatly optimized this method, facilitating localized, efficient and externally controllable heat production via magnetic, photothermal, radiofrequency, microwave, and ultrasound induced processes.

The future of nanoplatforms in the context of clinical relevance of nano-hyperthermia lies in sustainable synthesis of nanoparticles, which reinforces ecofriendly, biocompatible, and a cost-effective approach to fabrication. In doing so, green chemistry strategies relying on plant extracts, microbial systems, and biopolymers offer superior biomedical applicability and scalability as well as toxicity concerns are minimized. Moreover, nanoplatforms that are able to perform multiple functions at the same time provide a powerful basis of multifunctional delivery, imaging and hyperthermia, making them an effective theranostic platform.

Nevertheless, notwithstanding impressive advances, concerns like nanotoxicity, lack of biodistribution information, regulation, and insufficient clinical translation still impede mainstream clinical implementation. Nonetheless, these shortcomings will be addressed with continuous innovations in artificial intelligence, intelligent nanomaterials, and treatments that incorporate the immune system.

In sum, nano-hyperthermia is a remarkably promising future of materials science, oncology, and biomedical engineering providing an excellent platform to a personalized and minimally invasive approach toward treating cancer. Further interdisciplinary studies with clinical validation will be needed in order to transform these technologies into safe, efficient and universally available therapeutic means to cure cancer in the future.

## REFERENCES


Aggarwal, R., Magar, A., Dongsar, T. T., Dongsar, T. S., Almoyad, M. A. A., Wahab, S., & Kesharwani, P. (2025). Magnetic nanoparticles in cancer nanomedicine: Advances in targeting, hyperthermia, and theranostics. *International Journal of Pharmaceutics*, •••, 126483. PMID: 41397473

Akpe, V., Cock, I. E., Telukutla, S. R., & Shah, N. (2025). Advances in cancer treatment through nanotheranostics and emerging therapies. *Drugs and Drug Candidates*, *6*(4), 29. DOI: 10.3390/ddc6040029

Al Refaai, K. A., AlSawaftah, N. A., Abuwatfa, W., & Husseini, G. A. (2024). Drug release via ultrasound-activated nanocarriers for cancer treatment: A review. *Pharmaceutics*, *16*(11), 1383. DOI: 10.3390/pharmaceutics16111383 PMID: 39598507

Amaral, M. N., Kumar, P., Faísca, P., Ferreira, H. A., Coelho, J. M. P., Gaspar, M. M., & Reis, C. P. (2025). Gold nanoparticle-mediated photothermal therapy: Expanding the frontiers of cancer treatment and theragnostics. *Biomedicine & Pharmacotherapy = Biomédecine & Pharmacothérapie*, *190*, 118399. DOI: 10.1016/j.biopha.2025.118399 PMID: 40784123

Amirinezhadfard, E., Niazi Tabar, A., Xia, L. J., & Yang, W. C. (2025). Plant-derived nanomaterials in cancer therapy: Advances from green synthesis to application of PANoptosis-mediated tumor suppression. *Industrial Crops and Products*, *236*, 121850. DOI: 10.1016/j.indcrop.2025.121850

Appidi, T.. (2025). Nanoparticle-based triple therapy: A three-in-one approach for cancer treatment. *Molecular Cancer*, *24*, 119. DOI: 10.1186/s12943-025-02266-1 PMID: 40050802

Bala, V. M., Lampropoulou, D. I., Grammatikaki, S., Spanoudaki, M., Zambouli, D., Magkouta, S., Lagopati, N., Gazouli, M., & Efstathopoulos, E. P. (2023). Nanoparticle-mediated hyperthermia and cytotoxicity mechanisms in cancer. *International Journal of Molecular Sciences*, *25*(1), 296. DOI: 10.3390/ijms25010296 PMID: 38203467

Baldea, I., Iacovi ă, C., Gurgu, R. A., Vizitiu, A. S., Râzniceanu, V., & Mitrea, D. R. (2025). Magnetic hyperthermia with iron oxide nanoparticles: From toxicity challenges to cancer applications. *Nanomaterials (Basel, Switzerland)*, *15*(19), 1519. DOI: 10.3390/nano15191519 PMID: 41090863

Bhattacharya, T., Preetam, S., Ghosh, B., Chakrabarti, T., Chakrabarti, P., Samal, S. K., & Thorat, N. (2023). Advancement in biopolymer-assisted cancer theranostics. *ACS Applied Bio Materials*, *6*(9), 3488–3510. DOI: 10.1021/acsabm.3c00458 PMID: 37699558

Bravo, M., Fortuni, B., Mulvaney, P., Hofkens, J., Uji-i, H., Rocha, S., & Hutchison, J. A. (2024). Nanoparticle-mediated thermal cancer therapies: Strategies to improve clinical translatability. *Journal of Controlled Release : Official Journal of the Controlled Release Society*, *372*, 751–777. DOI: 10.1016/j.jconrel.2024.06.055 PMID: 38909701

Chauhan, A., Arya, A., & Kumar, V.. (2025). The evolution of integrated magnetic hyperthermia and chemodynamic therapy: Multifunctional nanomaterials as precision tools in cancer treatment. *Nanoscale Advances*, *7*, 2139–2157. DOI: 10.1039/D4NA01004C PMID: 40708892

Chauhan, D. S., Prasad, R., Dhanka, M., Kaur, N., Vithalani, H., Liyanapathirana, K., Hebbandi Nanjundappa, R., Gao, H., & Umeshappa, C. S. (2026). Nanoparticles-based phototherapy systems: Molecular mechanisms and clinical applications. *Signal Transduction and Targeted Therapy*, *11*(1), 143. DOI: 10.1038/s41392-025-02536-w PMID: 41839827

Chen, Y., Sun, H., Li, Y., Han, X., Yang, Y., Chen, Z., Zhao, X., Qian, Y., Liu, X., Zhou, F., Bai, J., & Qiao, Y. (2025). Magnetic nanomaterials for hyperthermia-based therapy and controlled drug delivery. *Bioactive Materials*, *53*, 591–629. DOI: 10.1016/j.bioactmat.2025.07.033 PMID: 40761549

Ciftci, F., Özarslan, A. C., Kantarci, İ. C., Yelkenci, A., Tavukcuoglu, O., & Ghorbanpour, M. (2025). Advances in drug targeting, drug delivery, and nanotechnology applications: Therapeutic significance in cancer treatment. *Pharmaceutics*, *17*(1), 121. DOI: 10.3390/pharmaceutics17010121 PMID: 39861768

Coene, A., & Leliaert, J. (2022). Magnetic nanoparticles in theranostic applications. *Journal of Applied Physics*, *131*(16), 160902. DOI: 10.1063/5.0085202

Damavandi, S. E., Banihashemi Jozdani, S. M., Elyasigorji, Z., & Vosough, M. (2026). Bridging cancer therapies: The role of magnetic nanoparticles in combination cancer therapy. *Discover Nano*, *21*(1), 49. DOI: 10.1186/s11671-026-04439-3 PMID: 41712065

Deivayanai, V. C., Thamarai, P., Karishma, S., Saravanan, A., Yaashikaa, P. R., Vickram, A. S., Hemavathy, R. V., Rohith Kumar, R., Rishikesavan, S., & Shruthi, S. (2025). Advances in nanoparticle-mediated cancer therapeutics: Current research and future perspectives. *Cancer Pathogenesis and Therapy*, *3*(4), 293–308. DOI: 10.1016/j.cpt.2024.11.002 PMID: 40697517

Dhanapal, Y., Sridhar, D., Khute, S., Subash, P., & AnandaThangadurai, S. (2026). Artificial intelligence-assisted design of nanomedicines for colorectal cancer diagnosis and therapy: Advances, challenges, and future directions. Intelligent Pharmacy. DOI: 10.1016/j.ipha.2026.01.001

Einolghasi, F. B., Zeinali, B., Vafai, K., & Mojra, A. (2025). Development of a prediction model for hyperthermia-enhanced drug delivery using thermosensitive nanoparticles. *Journal of Thermal Biology*, *129*, 104124. DOI: 10.1016/j.jtherbio.2025.104124 PMID: 40300402

El-Khawaga, A. M.. (2026). Advances in green-synthesized magnetic nanoparticles for cancer diagnosis and therapy. *Frontiers in Bioengineering and Biotechnology*, *14*, 1655936. DOI: 10.3389/fbioe.2026.1655936

EMA. (2025). Nanotechnology-based medicinal products for human use: EU-IN Horizon Scanning Report. European Medicines Agency, EMA/20989/2025/Rev.1. https://www.ema.europa.eu/

Fakhradini, S. S., Mosharaf-Dehkordi, M., & Ahmadikia, H. (2024). Enhancing liver cancer treatment: Exploring the frequency effects of magnetic nanoparticles for heat-based tumor therapy with microwaves. *International Journal of Thermal Sciences*, *204*, 109154. DOI: 10.1016/j.ijthermalsci.2024.109154

Fan, D., Cao, Y., Cao, M., Wang, Y., Cao, Y., & Gong, T. (2023). Nanomedicine in cancer therapy. *Signal Transduction and Targeted Therapy*, *8*(1), 293. DOI: 10.1038/s41392-023-01536-y PMID: 37544972

Freitas, S. C., Sanderson, D., Caspani, S., Magalhães, R., Cortés-Llanos, B., Granja, A., Reis, S., Belo, J. H., Azevedo, J., Gómez-Gaviro, M. V., & Sousa, C. T. (2023). New frontiers in colorectal cancer treatment combining nanotechnology with photo- and radiotherapy. *Cancers (Basel)*, *15*(2), 383. DOI: 10.3390/cancers15020383 PMID: 36672333

Garg, P., Krishna, M., Horne, D., Salgia, R., & Singhal, S. S. (2025). Hyperthermia-induced apoptosis: The role of magnetic nanoparticles in targeted delivery for breast cancer treatment. Biochimica et Biophysica Acta (BBA) -. *Revisiones en Cáncer*, *1880*(6), 189490. DOI: 10.1016/j.bbcan.2025.189490 PMID: 41173154

Gavilán, H., Gallo-Cordova, A., Rábade-Chediak, M., Páez-Rodríguez, A., Morales, M. P., & Gutiérrez, L. (2025). Magnetic hyperthermia in focus: Emerging non-cancer applications of magnetic nanoparticles. *Nanoscale*, *17*(48), 27734–27758. DOI: 10.1039/D5NR03329B PMID: 41292379

Gawne, P. J., Ferreira, M., Papaluca, M., Grimm, J., & Decuzzi, P. (2023). New opportunities and old challenges in the clinical translation of nanotheranostics. *Nature Reviews. Materials*, *8*(12), 783–798. DOI: 10.1038/s41578-023-00581-x PMID: 39022623

Graham, W., Torbett, B. E., & Dougherty, J. P. (2025). Magnetic nanoparticles and drug delivery systems for anti-cancer therapy. *Nanomaterials (Basel, Switzerland)*, *15*(4), 285. DOI: 10.3390/nano15040285 PMID: 39997849

Gupta, D., Gupta, T. K., Roy, P., Sharma, R., Kasana, R., & Rathore, P. (2024). Recent nanotheranostic approaches in cancer research. *Clinical and Experimental Medicine*, *24*(1), 8. DOI: 10.1007/s10238-023-01262-3 PMID: 38240834

Han, X., Feng, X., & Zhou, H.. (2025). Functional nanomaterials for enhanced tumor photothermal therapy-The mechanisms and applications. *Frontiers in Pharmacology*, *16*, 1604965. DOI: 10.3389/fphar.2025.1604965 PMID: 40727107

Hwang, J., & Oh, J.. (2026). Nanoparticle-enabled photothermal therapy integrated with gene delivery, immunotherapy, and chemotherapy: A review of recent advances and future perspectives. *International Journal of Nanomedicine*, *21*, 101–132. DOI: 10.2147/IJN.S573093 PMID: 41869393

Jiao, W., Dai, L., Yan, B., Lyu, Y., Fan, H., & Liu, X. (2024). Heating up the immune battle: Magnetic hyperthermia against cancer. *Fundamental Research (Beijing)*. Advance online publication. DOI: 10.1016/j.fmre.2024.08.006 PMID: 41466973

Kashkooli, F. M., Soltani, M., & Momeni, M. M. (2023). Integrating therapeutic ultrasound with nanosized drug delivery systems in the battle against cancer. *Technology in Cancer Research & Treatment*, *22*, 15330338231211472. DOI: 10.1177/15330338231211472 PMID: 37946517

Kazemi Siyanaki, H. S., Rezaei, F., Kabiri, S., & Ashrafkhani, B. (2025). Computational modeling of combined laser-ultrasound hyperthermia therapy for breast cancer using gold nanoparticles: A simulation study. *Lasers in Medical Science*, *40*(1), 430. DOI: 10.1007/s10103-025-04657-8 PMID: 41081940

Kesharwani, P., Kumar, V., Goh, K. W., Gupta, G., Alsayari, A., Wahab, S., & Sahebkar, A. (2025). PEGylated PLGA nanoparticles: Unlocking advanced strategies for cancer therapy. *Molecular Cancer*, *24*(1), 205. DOI: 10.1186/s12943-025-02410-x PMID: 40702523

Koç, M. M.. (2025). Nanoparticles in photothermal therapy-based medical and biological applications. *The European Physical Journal Plus*, *140*, 496. DOI: 10.1140/epjp/s13360-025-06452-4

Kraus, S., Arbib, S., Ventura-Bixenshpaner, H., Rukenstein, P., Hirsh-Kovacs, M., Shalev, B., Eltanani, M., Ron, U., Gumina, O., Shalev, O., & Geschwind, J. (2026). Case report: Treatment with magnetic nanoparticle-based hyperthermia stabilizes metastatic disease in stage IV primary liver cancer patients. *Frontiers in Oncology*, *15*, 1742249. DOI: 10.3389/fonc.2025.1742249 PMID: 41626176

Kumar, N., Singh, S., Sharma, P., Kumar, B., & Kumar, A. (2024). Single-, dual-, and multi-stimuli-responsive nanogels for biomedical applications. *Gels (Basel, Switzerland)*, *10*(1), 61. DOI: 10.3390/gels10010061 PMID: 38247784

Kuwahata, A.. (2025). Magnetic hyperthermia using iron oxide nanoparticles via lymphatic drug delivery for metastatic lymph nodes. *Scientific Reports*, *15*, 40658. DOI: 10.1038/s41598-025-25808-5 PMID: 41290791

Lee, B. J., Kanaan, M. H. G., Abdullah, S. S., & Ghasemian, A. (2025). Plant-derived nanoparticles in cancer therapy: A comprehensive review of recent advances and future prospects. *OBM Genetics*, *9*(3), 308. DOI: 10.21926/obm.genet.2503308

Li, J., Wang, S., Fontana, F., Tapeinos, C., Shahbazi, M.-A., Han, H., & Santos, H. A. (2023). Nanoparticles-based phototherapy systems for cancer treatment: Current status and clinical potential. *Bioactive Materials*, *23*, 471–507. DOI: 10.1016/j.bioactmat.2022.11.013 PMID: 36514388

Li, Y.. (2024). Nanomaterial-mediated photothermal therapy modulates tumor-associated macrophages for cancer treatment. *Journal of Materials Chemistry. B*, *12*, 11732–11749. DOI: 10.1039/D4TB01928H PMID: 39501854

Lodi, M. B., Possidente, R., Melis, A., Di Meglio, A., Fanti, A., & Baccoli, R. (2025). Thermal behavior of magnetic scaffolds for RF-induced hyperthermia. *Applied Sciences (Basel, Switzerland)*, *15*(17), 9782. DOI: 10.3390/app15179782

Menon, G., Anandan, T., Kolage, S. M., Shenoy, S. C., Lohakare, O. K., Ahire, A. S., & Ravindran, D. (2025). Theranostics in the management of colorectal cancer. *Discover Nano*, *20*(1), 214. DOI: 10.1186/s11671-025-04390-9 PMID: 41276733

Moacă, E. A.. (2022). Biologic impact of green synthesized magnetic iron oxide nanoparticles. *Pharmaceutics*, *15*(1), 95. DOI: 10.3390/pharmaceutics15010002 PMID: 36678723

Molaei, M. J.. (2024). Magnetic hyperthermia in cancer therapy, mechanisms, challenges, and perspectives. *Journal of Biomaterials Applications*, *39*, 786–808. DOI: 10.1177/08853282241244707 PMID: 38606627

Mousavi-Kiasary, S. M. S., Senabreh, A., Zandi, A., Pena, R., Cruz, F., Adibi, A., & Hooshmand, N. (2025). Synergistic cancer therapies enhanced by nanoparticles: Advancing nanomedicine through multimodal strategies. *Pharmaceutics*, *17*(6), 682. DOI: 10.3390/pharmaceutics17060682 PMID: 40573996

Najdian, A., Beiki, D., Abbasi, M., Gholamrezanezhad, A., Ahmadzadehfar, H., Amani, A. M., Shafiee Ardestani, M., & Assadi, M. (2024). Exploring innovative strides in radiolabeled nanoparticle progress for multimodality cancer imaging and theranostic applications. *Cancer Imaging; the Official Publication of the International Cancer Imaging Society*, *24*(1), 127. DOI: 10.1186/s40644-024-00762-z PMID: 39304961

Nisha, N.. (2024). Plant-mediated gold nanoparticles in cancer therapy: Green synthesis, mechanisms and applications. *Frontiers in Nanotechnology*, *6*, 1490980. DOI: 10.3389/fnano.2024.1490980

Obalola, A. A., Abrahamse, H., & Dhilip Kumar, S. S. (2025). Recent advances in gold nanoparticle-based targeted photodynamic and photothermal therapies. *Nanoscale Advances*, *7*(23), 5853–5875. DOI: 10.1039/D5NA00767D PMID: 41159015

Overchuk, M., Weersink, R. A., Wilson, B. C., & Zheng, G. (2023). Photodynamic and photothermal therapies: Synergy opportunities for nanomedicine. *ACS Nano*, *17*(9), 7979–8003. DOI: 10.1021/acsnano.3c00891 PMID: 37129253

Pallares, R. M., Barmin, R. A., Wang, A., Kiessling, F., & Lammers, T. (2025). Clinical cancer nanomedicines. *Journal of Controlled Release : Official Journal of the Controlled Release Society*, *385*, 113991. DOI: 10.1016/j.jconrel.2025.113991 PMID: 40582644

Parthasarathy, N.. (2025). Recent progress in multifunctional theranostic inorganic nanoparticles for cancer therapy. *Smart Materials in Medicine*, *6*, 801–825. DOI: 10.1016/j.smaim.2025.07.006

Patnaik, C., Rath, S., & De, A. (2025). Avenues for integrating photothermal therapy in cancer clinic. *Communications Chemistry*, *8*(1), 225. DOI: 10.1038/s42004-025-01705-w PMID: 41188487

Pawar, P., & Prabhu, A. (2025). Smart SPIONs for multimodal cancer theranostics: A review. *Molecular Pharmaceutics*, *22*(5), 2372–2391. DOI: 10.1021/acs.molpharmaceut.5c00411 PMID: 40223773

Rajan, A., Laha, S. S., Sahu, N. K., Thorat, N. D., & Shankar, B. (2024). Recent advancements and clinical aspects of engineered iron oxide nanoplatforms for magnetic hyperthermia-induced cancer therapy. *Materials Today. Bio*, *29*, 101348. DOI: 10.1016/j.mtbio.2024.101348 PMID: 39669801

Rauf, M. A., Rao, A., Sivasoorian, S. S., & Iyer, A. K. (2025). Nanotechnology-based delivery of CRISPR/Cas9 for cancer treatment: A comprehensive review. *Cells*, *14*(15), 1136. DOI: 10.3390/cells14151136 PMID: 40801569

Salgueiro, M. J., & Zubillaga, M. (2025). Theranostic nanoplatforms in nuclear medicine: Current advances, emerging trends, and perspectives for personalized oncology. *Journal of Nanotheranostics*, *6*(4), 27. DOI: 10.3390/jnt6040027

Sathuvan, M., Narayanan, K., Cheong, K.-L., & Thangam, R. (2026). Nanoplatforms for multimodal imaging and targeted cancer therapy: Recent advances and future perspectives. *Bioengineering (Basel, Switzerland)*, *13*(2), 174. DOI: 10.3390/bioengineering13020174 PMID: 41749714

Sayed, Z. S., Ghiaty, A., & Elbaz, N. M. (2024). Cancer treatment approaches within the frame of nanotechnology. *RSC Advances*, *14*, 33645–33669. DOI: 10.1039/D4RA06992G PMID: 39670162

Schreiner, O. D., Schreiner, T. G., Miron, L., & Ciobanu, R. C. (2026). Two-stage microwave hyperthermia using magnetic nanoparticles for optimal chemotherapy activation in liver cancer: Concept and preliminary tests on Wistar rat model. *Cancers (Basel)*, *18*(2), 330. DOI: 10.3390/cancers18020330 PMID: 41595249

Shakeri-Zadeh, A., & Bulte, J. W. M. (2025). Imaging-guided precision hyperthermia with magnetic nanoparticles. *Nature Reviews Bioengineering*, *3*(3), 245–260. DOI: 10.1038/s44222-024-00257-3 PMID: 40260131

Shanahan, K., Coen, D., & Nafo, W. (2025). Polymer-based nanoparticles for cancer theranostics: Advances, challenges, and future perspectives. *Exploration of Biomat-X*, *2*, 101342. DOI: 10.37349/ebmx.2025.101342

Shang, Y., Yi, X., Xiang, D., & Zhou, L. (2025). Nanoagent-mediated photothermal therapy: From delivery system design to synergistic theranostic applications. *International Journal of Nanomedicine*, *20*, 6891–6927. DOI: 10.2147/IJN.S522736 PMID: 40458746

Shirzad, M.. (2025). Artificial intelligence-assisted design of nanomedicines for cancer diagnosis and therapy. *Drug Delivery and Translational Research*, *15*, 4127–4150. DOI: 10.1007/s12668-025-01980-w

Soomro, R.. (2025). A novel plant-based approach for synthesis of iron oxide nanoparticles in cancer theranostics. *Nano Biomedicine and Engineering*, *17*, 241–259. DOI: 10.26599/NBE.2024.9290097

Sung, H., Ferlay, J., Siegel, R. L., Laversanne, M., Soerjomataram, I., Jemal, A., & Bray, F. (2024). Global cancer statistics 2022: GLOBOCAN estimates of incidence and mortality worldwide for 36 cancers in 185 countries. *CA: a Cancer Journal for Clinicians*, *74*(3), 229–263. DOI: 10.3322/caac.21834 PMID: 38572751

Szwed, M., & Marczak, A. (2024). Application of nanoparticles for magnetic hyperthermia for cancer treatment-The current state of knowledge. *Cancers (Basel)*, *16*(6), 1156. DOI: 10.3390/cancers16061156 PMID: 38539491

Tang, S., Chen, X., Tong, X., & Zhu, L. (2025). Overcoming the delivery challenges in CRISPR/Cas9 gene editing for effective cancer treatment: A review of delivery systems. *International Journal of Medical Sciences*, *22*(14), 3625–3649. DOI: 10.7150/ijms.112724 PMID: 40959562

Torres, J., Valenzuela Oses, J. K., Rabasco-Álvarez, A. M., González-Rodríguez, M. L., & García, M. C. (2025). Innovations in cancer therapy: Endogenous stimuli-responsive liposomes as advanced nanocarriers. *Pharmaceutics*, *17*(2), 245. DOI: 10.3390/pharmaceutics17020245 PMID: 40006612

Vicente, T. T., Arsalani, S., Quiel, M. S., Fernandes, G. S., da Silva, K. R., Fukada, S. Y., Gualdi, A. J., Guidelli, É. J., Baffa, O., Carneiro, A. A. O., Ramos, A. P., & Pavan, T. Z. (2024). Improving the theranostic potential of magnetic nanoparticles by coating with natural rubber latex for ultrasound, photoacoustic imaging, and magnetic hyperthermia: An in vitro study. *Pharmaceutics*, *16*(11), 1474. DOI: 10.3390/pharmaceutics16111474 PMID: 39598597

Vlastou, E., Zolota, V., Koukourakis, M. I., & Koukourakis, G. V. (2025). The emerging role of nanoparticles combined with either radiotherapy or immunotherapy in head and neck cancer. *Cancers (Basel)*, *17*(5), 899. DOI: 10.3390/cancers17050899 PMID: 40075746

Wang, Z., Hasan, I., Zhang, Y., Peng, T., & Guo, B. (2025). Photothermal combination therapy for metastatic breast cancer: A new strategy and future perspectives. *Biomedicines*, *13*(10), 2558. DOI: 10.3390/biomedicines13102558 PMID: 41153838

Xiong, L., Liang, B., & Yu, K. (2025). Magnetic hyperthermia in oncology: Nanomaterials-driven combinatorial strategies for synergistic therapeutic gains. *Materials Today. Bio*, *33*, 102070. DOI: 10.1016/j.mtbio.2025.102070 PMID: 40688660

Xu, R., Wang, S., Guo, Q., Zhong, R., Chen, X., & Xia, X. (2025). Anti-tumor strategies of photothermal therapy combined with other therapies using nanoplatforms. *Pharmaceutics, 17*(3), 306. DOI: 10.3390/pharmaceutics17030306 PMID: 40142970

Yang, T., Lou, Y., & Ying, Z. (2025). Nanoparticle-ultrasound synergy: An emerging theranostic paradigm for breast and gynecologic cancers. *Frontiers in Oncology, 15*, 1617939. DOI: 10.3389/fonc.2025.1617939 PMID: 41426324

Yaramiri, A., Asalh, R. A., Asalh, M. A., AlSawaftah, N., Abuwatfa, W. H., & Husseini, G. A. (2025). A Comprehensive Review of Smart Thermosensitive Nanocarriers for Precision Cancer Therapy. *International Journal of Molecular Sciences, 26*(15), 7322. DOI: 10.3390/ijms26157322 PMID: 40806454

Zhang, Y., Paraskeva, C., Chen, Q., Maisuradze, A., Ansari, S. R., Sarkar, T., Koliaraki, V., & Teleki, A. (2025). Flame-made nanoparticles for magnetic hyperthermia and MRI in colorectal cancer theranostics. *Nanoscale Advances, 7*(17), 5284–5299. DOI: 10.1039/D5NA00603A PMID: 40704284

Zhang, Y. F., Wang, Y., Zhi, D., Du, Z., & Zhou, B. (2024). Advances in magnetic induction hyperthermia. *Frontiers in Bioengineering and Biotechnology, 12*, 1432189. DOI: 10.3389/fbioe.2024.1432189 PMID: 39161353

Zheng, L.. (2026). Engineering biopolymer nanoparticles for targeted cancer therapy. *Frontiers in Nutrition, 13*, 1778133. DOI: 10.3389/fnut.2026.1778133 PMID: 42146093

Zhu, L., Jiang, Y., Tian, H., Yu, Y., Gan, Y., Li, H., Yuan, M., Huang, X., & Liu, X. (2025). Recent advancement in MRI-based nanotheranostic agents for tumor diagnosis and treatment. *International Journal of Nanomedicine, 20*, 7251–7284. DOI: 10.2147/IJN.S529003 PMID: 40904647

Chapter 13

# Lab to Clinic Scale-Up and Global Regulatory Frameworks

**Farhad Badshah**
http://orcid.org/0000-0001-6971-0957
*Chinese Academy of Agricultural Sciences, China*

**Hira Maryam**
*COMSATS University Islamabad, Pakistan*

**Muhammad Zakria Iqbal Khan**
*COMSATS University Islamabad, Pakistan*

**Mujtaba ul Hassan**
*COMSATS University Islamabad, Pakistan*

**Nabeel ur Rehman**
http://orcid.org/0009-0009-6632-6218
*COMSATS University Islamabad, Pakistan*

## ABSTRACT

*Nano-oncology has revolutionized the precision of cancer therapy by enhancing drug targeting and pharmacokinetics, and by integrating diagnostics and therapeutics. Progress in nano-carriers (lipid-based, polymeric, inorganic, biomimetic), but obstacles to clinical use remain: scalability, reproducibility, and regulatory compliance. This chapter summarizes the lab-to-clinic transition in the field of Nano-Oncology with a particular focus on scale-up engineering, manufacturing, and international regulations. Critical Quality Attributes (CQAs) and Good Manufacturing Practice (GMP) are key to a successful translation. Consistency of nanoparticles is pro-*

DOI: 10.4018/979-8-3373-9848-8.ch013

*vided by techniques such as microfluidics and high-pressure homogenization. It also highlights the preclinical to clinical transitions, the regulatory aspects of the FDA, EMA, NMPA, PMDA, and sustainability aspects, suggesting a model that combines nanoscience, pharmaceutical engineering, and regulatory science for efficient clinical translation.*

## 1. INTRODUCTION

Nano-oncology has become a paradigm shift in cancer studies, combining nanotechnology with oncology to allow highly specific, effective and multifunctional therapeutic and diagnostic approaches. Nanomedicine-based drug delivery systems have shown extraordinary promise in enhancing solubility and stability of pharmacokinetics and location to tumor of anticancer agents in the last 20 years(Ammar et al., 2025). Liposomes, polymeric nanoparticles, lipid nanoparticles, inorganic nanocarriers, and biomimetic systems have all introduced a new facet to the realm of targeted cancer therapeutic platforms, by allowing drug delivery to be fine-tuned, decreasing systemic toxicity, and improving therapeutic index. Irrespective of these developments, there has been a limited and inconsistent transfer of the nano-oncology systems innovatively developed in labs to clinical ready products(Cheng et al., 2025).

The notable obstacle in this translational route is the seemingly positive disconnect between solitary, viable, preclinical results and effective clinical practice. Nanocarriers used in labs are often optimized under highly in vivo conditions that are not fully similar to the complexity of human physiology, heterogeneity of tumors, interactions between carriers and immune systems, and corral protein in biological fluids. Because of this, the formulations which show good efficacy in vitro and on animal models often do not show the same efficacy when used in human clinical work(M. Chehelgerdi et al., 2023). The result of this disconnect has created what has commonly been known as the valley of death in nanomedicine development: the tough phase of many innovative nanotherapeutics failing to get beyond proof-of-concept because of scalability, reproducibility and regulatory issues(Metselaar & Lammers, 2020).

The second major constraint is the absence of standardized protocols to synthesize nanoparticle, characterize and control quality which leads to large batch-to-batch variability when scaled-up. Moreover, the question of long-term toxicity, biodistribution and clearance mechanisms also pose uncertainties which make the process of regulatory approval even more complicated(El-Shabasy et al., 2026). The lack of globally unified recommendations specifically designed to suit nanomedicines is yet another complexity increment that contributes to the inability of scientists and industry players to navigate the clinical translation pathway smoothly(Zhang, Li, & Ma, 2024).

It is in this regard that the following chapter will be used to offer an overall perspective of the lab-to-clinic translational framework in nano-oncology, specifically, the scale-up engineering perspective, industrial manufacturing view, and regulatory environments across the globe(Đorđević et al., 2022). It specifically discusses the issues surrounding taking nanomedicine out of bench-scale synthesis into Good Manufacturing Practice (GMP)-compliant manufacture without compromising such critical quality considerations as particle size, surface properties, drug loading efficiency, and stability. Moreover, it analyzes the changing regulatory frameworks, put in place by key international organizations, such as the FDA, EMA, NMPA and PMDA, and harmonization efforts, by online organizations like ICH, ISO and OECD(Zhang et al., 2025).

## 2. FUNDAMENTALS OF TRANSLATIONAL NANO-ONCOLOGY

The physics/chemical performance of nanocarriers, in combination with biological performance, basically determines nanocitotherapeutics translation in the field of nano-oncology. Material platforms and how nanomedicines can be optimally transferred into and through clinical use necessitate an in-depth insight into quality criteria defining safety, efficacy, and reproducibility in complex biological systems(Kurul et al., 2025).

### 2.1. Nanocarrier Systems in Cancer Therapeutics

Systems based upon nanocarriers constitute the technological heart of nano-oncology and are designed in such a way that they enhance the efficiency of drug delivery, selectivity towards tumors, and therapeutic outcome (Figure 1). Lipid nanoparticle, such as liposomes and lipid nanoparticles (LNPs), have a high usage rate because the nanoparticles are biocompatible, successfully encapsulate both hydrophilic and hydrophobic drugs and clinical testing has been approved in a number of FDA-approved products(Waheed et al., 2024). PLGA and PEGylated systems, which are polymeric nanoparticles, provide controlled drug release schemes, and a greater ability to stabilize and thus can be used to deliver chemotherapy and targeted therapy over time. Examples of inorganic nanoparticles such as gold, silica, iron oxide-based formulations have special optical, magnetic, and photothermal characteristics, which are incredibly useful in imaging-guided therapy and theranostic applications(Firouzpour et al., 2026). Biomimetic nanoparticles, including exosome-based bioprojectiles and cell membrane-coated system, also increases immune evasion and biological compatibility through the resemblance to natural cellular structures (Figure 1) (Zhang et al., 2026).

*Figure 1. Major nano-carrier platforms in cancer therapy for targeted drug delivery, imaging, and the agnostic applications*

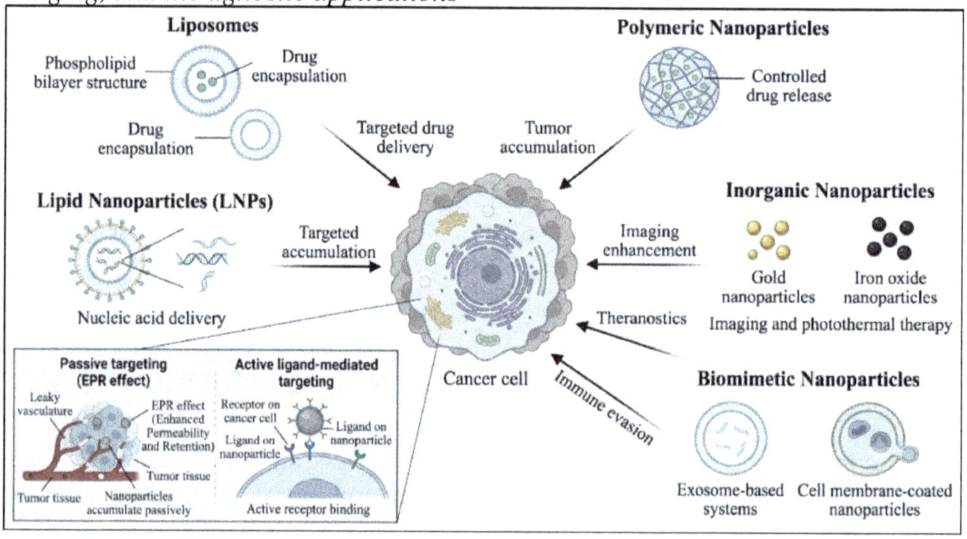


These nanocarriers have two roles in cancer management: (1) as drug delivery constructs, which enhance pharmacokinetics, increase half-life, and allow tumor accumulation due to passive (EPR effect) or active targeting(Jayaraj et al., 2026); and (2) as imaging and diagnostics, which enhance tumor visualization by enhancing contrast, utilizing oral fluorescence, or integrating multimodal biosensing. This fact of multifunctionality makes nanocarriers a part of precision oncology platforms that allow to combination diagnose, treat, and monitor(Shi et al., 2023).

## 2.2. Critical Quality Attributes for Translation

Optimization of Critical Quality Attributes (CQAs), that determine the physico-chemical and biological functionality of nanocarriers is the key to clinical success of nano-oncology systems (Figure 2) . Some of the most significant CQAs are particle size and size distribution that directly affect biodistribution, cellular uptake, and the effectiveness of tumor penetration(Hocaoğlu, Zıkşahna, & Ihlamur, 2026). The range of nanoscale (10200 nm) is thought to be ideal utilizing the maximum possible accumulation of the tumor and eliminating the rapid clearance rate by the renal status and the mononuclear phagocytosis system(Peng, Huang, & Zheng, 2020).

Surface charge which is commonly referred to as zeta potential, is important in nanoparticle stability and interaction with biological membranes. Cationic particles tend to increase cellular uptake, though also tend to be more cytotoxic and serum protein binding, though neutral or slightly negative surfaces tend to elevate systemic stability and circulation times(Hwang et al., 2025). The efficacy of the drug in therapies further depends on the drug loading efficiency and stability of the encapsulation; low loading minimizes their effect, whereas early release of the drug results in systemic toxicity and a lack of ability to target the tumor(Lee & Kim, 2025).

Another aspect that is of paramount importance with regard to clinical performance is stability under physiological conditions, such as aggregation resistance, corona formation, and enzymatic degradation. Notably, these design parameters do not do it in isolation but are instead interconnected, which, when taken collectively affects vivo behavior(Cisneros et al., 2024). As examples, changing surface chemistry to enhance stability can potentially change the degree of cellular uptake, whereas loading more drug can change the size of nanoparticles and its release dynamics(Herdiana et al., 2022).

*Figure 2. Impact of protein corona formation on nanoparticle identity, biodistribution, immune recognition, and clinical efficacy*

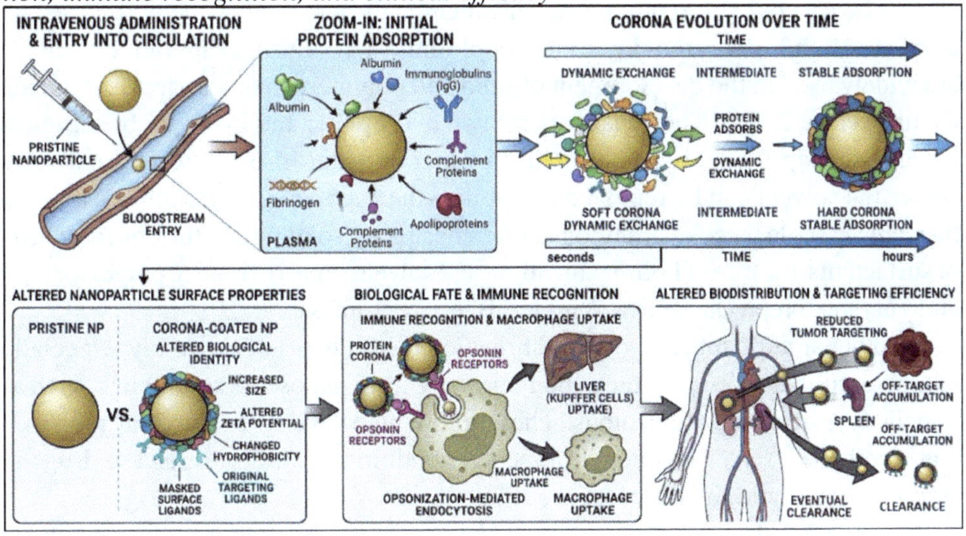


Thus, interaction between the design of nanocarriers and clinical efficacy is highly dynamic and must be optimized on a systems level. The interdependence of these parameters is a key to successful translation, as the balance ensures reproducible, scalable, and regulatory compliant nanomedicine platforms, which can be used in clinical environments and provide consistent therapeutic outcomes(Sebastián, 2026).

## 3. LAB-SCALE DEVELOPMENT AND PRECLINICAL VALIDATION

Laboratory stage is the base stage in the nano-oncology development, wherein nanocarrier systems are conceptualized, synthesized, optimized and the initial spots on therapeutic efficacy are tried. The main goal of this level is to engineer nano-medicine systems with desirable physicochemical characteristics and biological functionality that can be subsequently scaled to manufacturable levels(Xu et al., 2025). Nevertheless, although in laboratory-scale systems they frequently exhibit encouraging anticancer activity, the ability to transfer them to clinically achievable products rests squarely on the strength of the synthesis strategies, and the extrapolative competence of the preclinical models(Bousbaa, 2021).

### 3.2. Laboratory Synthesis Approache:

This method is especially beneficial in preparation of polymeric nanoparticles with small size-distributions and high encapsulation efficacy under gentle circumstances hence could be applied on fragile therapeutic agents. The methods based on emulsification, such as the single and double emulsion solvent evaporation methods are widely used in the development of lipid and polymeric nanocarrier(de Almeida Campos et al., 2024). They enable these two hydrophilic and hydrophobic drugs to be easily encapsulated by creating oil-in-water or water-in-oil-in-water emulsions before the solvent can be removed to stabilize the nanoparticle structure. Although they are versatile, special attention is often necessary to optimize the concentration of surfactants, the rate of homogenization, and solvent mixes to reduce polydispersity and guarantee reproducibility(Winkler, Barai, & Tomassone, 2019).

Another significant category of fabrication methods is self-assembly, especially in amphiphilic molecules which spontaneously form nanoscale assemblies, such as micelles, vesicles and liposomes. These allow high levels of structural precision and also highly sensitive to environmental conditions in order to release drugs in response to stimuli(Kashapov et al., 2020).

Formulation design optimization at the laboratory level is usually done using systematic methods like Design of Experiment (DoE) and response surface methodology (RSM)(Khorsheed et al., 2025). It is possible to assess a variety of variables of formulations and processes at the same time using these statistical tools such as polymer concentration, drug-to-carrier ratio, mixing speed, and solvent composition.

The aim of this optimization is to present optimal nanoparticle properties, such as homogenous size distribution, high drug loading, release kinetics, and increased colloidal stability(Khalbas et al., 2024).

## 3.1. Preclinical Testing and Limitations

After the successful synthesis and optimization, nanocarrier systems do pass thorough preclinical investigation evaluating its biological capabilities and therapeutic advantages. The most common type of in vitro study is cytotoxicity which can be MTT or cell viability assays to find out anticancer efficacy against a particular tumor cell line(Eder et al., 2022). The analysis of cellular uptake by using fluorescence microscopy or flow cytometry can yield information on which pathways the protein is internalized, and how it is trafficked within the cell. Other mechanistic tests can be used to test apoptosis induction, cell cycle arrest, and the reactive oxygen species (ROS) production(Papa et al., 2023).

Preclinical studies in vivo are typically performed using in vivo murine xenograft or orthotopic tumor models to determine biodistribution, tumor accumulation, pharmacokinetics and therapeutic efficacy. Fluorescence imaging, MRI or PET scanning are common imaging techniques that are commonly used to trace the localization of nanoparticles and measure real-time drug-delivery processes in a living organism(Skotland et al., 2022).

Although these have been achieved, there are still severe constraints to preclinical level nano-oncology studies. A significant problem is the inability to reproduce batches and systems between experiments (which frequently happens due to minor differences in synthesis conditions, purification procedures), and storage things(Sharaf et al., 2023). Also, biological translation is extremely unpredictable, as animal models and human tumor physiology are different in terms of immune response, vascular structure, and complexity of the tumor microenvironment(Xu et al., 2024).

*Figure 3. Mechanistic overview of biological barriers affecting nanoparticle transport from systemic circulation to intracellular drug delivery in tumors*

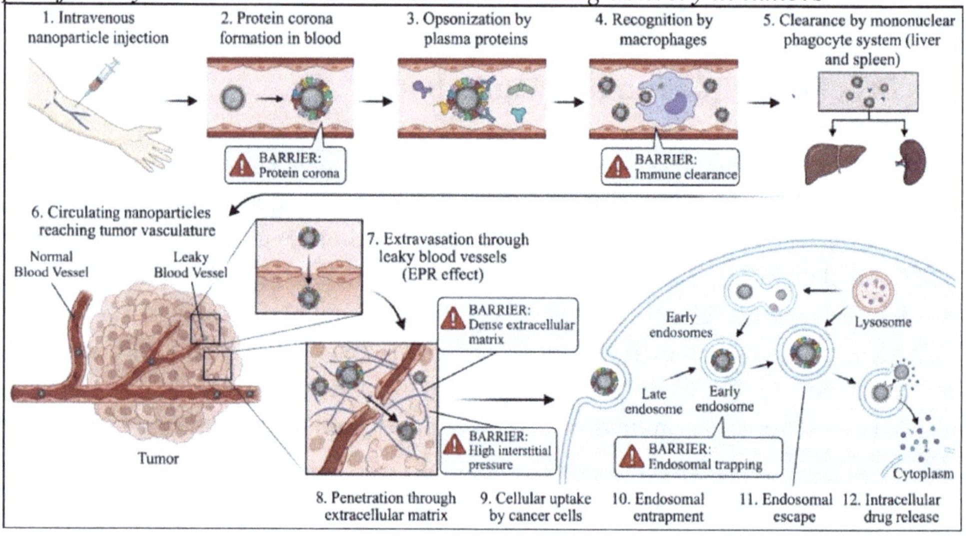


Moreover, the traditional preclinical models do not always represent the clinical conditions sufficiently well, which results in an exaggeration of therapeutic effectiveness at an initial stage of development. Nanoparticle behavior in vivo can also be greatly influenced by protein corona formation in the biological fluids, making it subsequently more difficult to predict clinical outcomes(Ramos et al., 2022). Consequently, numerous nanomedicine systems that work well in controlled experimental environments do not show similar efficacy on clinical translation.

## 4. SCALE-UP ENGINEERING IN NANO-ONCOLOGY

When nano-oncology goes beyond small-scale synthesis in the lab to large scale production in the industry, this is one of the most important and challenging steps in nanomedicine translation. Although laboratory preparations usually favor the best performance in controlled experimental settings, large-scale production must be characterized by rigid requirements on reproducibility, physicochemical uniformity, compliance with regulation and economy(Ji et al., 2025). Scale-up engineering can thus help fill in the gap between experimental nanoscience with clinically deployable nanotherapeutics, so that nanosystems can maintain their functional integrity as they are produced in large quantities in the marketplace(Crecente-Campo & Alonso, 2019).

## 4.1. Industrial Production Technologies

Nanomedicines are industrially fabricated through the application of advanced manufacturing technologies that are aimed to achieve scalability, reproducibility, and uniformity of the product(Kara et al., 2025). Microfluidic systems, in particular, have been a subject of intensive interest in these applications since microfluidic systems can exactly control the mixing at the microscale; this allows the generation of nanoparticles with fine size distributions and good lot-to-lot variability. Microfluidics especially finds application in lipid nanoparticles and polymeric systems, whereby accurate nucleation and fast solvent exchange is critical in a homogeneous production of particles(Bezelya, Kucukturkmen, & bozkır, 2023).

The other popular method of nanomedicine production in the industry is high-pressure homogenization (Figure 4), particularly in lipid-based formulations, e.g., liposomes and solid lipid nanoparticles. It uses high mechanical forces to decrease the particle size to form stable nanosuspensions, thus is suited to scale production of pharmaceutical larger-scale manufacturing(Viegas et al., 2023). On the same note, spray drying is also used in the conversion of liquid nanosuspensions into dry powders, to increase long-term stability and storage capacity as well as transport friendliness but this does not compromise the encapsulation efficiency of the drug(Tsiaxerli et al., 2024).

Supercritical fluid technology is a newer development in green nanomanufacturing which provides solvent-free or low-solvent operating environments. Using this method, a processing medium of the supercritical CO 2 is used to generate nanoparticles with a controlled morphology, which reduces environmental impact, thereby complying with the principle of manufacturing pharmaceuticals sustainably(Kalpage & Isuranga, 2025).

One of the fundamentals of scale-up engineering is the shift of laboratory synthesis (where products are produced in batches) to continuous manufacturing systems. In contrast to traditional batch processes, continuous flow production can provide real-time control over the reaction parameters, enhanced reproducibility, minimized variability in production and increased throughpu(Kashinath et al., 2025)t. This change is especially valuable to clinical translation, where regulatory agencies require uniform quality of products in large batches of production.

*Figure 4. Mechanistic representation of the transition from laboratory nanoparticle synthesis to GMP-compliant industrial manufacturing*

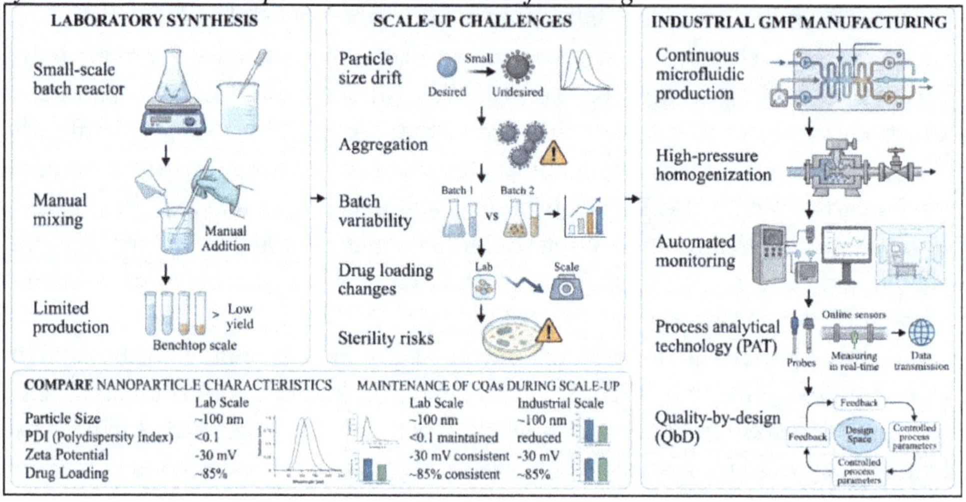


## 4.2. Process Control and Quality Assurance

Control of processes based on advanced strategies like Quality-by-Design (QbD) is necessary to make sure product quality and consistency in scales-up. QbD focuses on a planned methodology of developing pharmaceuticals, where quality is engineered using design and is not determined during the last phase of development(Yang et al., 2025). In nano-oncology, this entails defining Critical Quality Attributes (CQAs) including particle size, surface charge and drug loading and relating them to Critical Process Para-meters (CPPs) including mixing speed, temperature and solvent composition. QbD has enabled developing an effective design space that allows nanomedicines to stay within acceptable quality margins even with variability in the processes used(Soni et al., 2020).

To supplement this style is the Process Analytical technology (PAT) that facilitates real time monitoring and control processes of manufacturing. Continuous evaluation of the formation of nanoparticles, size distribution, and stability is possible with PAT tools like dynamic light scattering, spectroscopy, laser diffraction and online imaging systems(Gerzon, Sheng, & Kirkitadze, 2021). This real time feedback loop fosters better understanding of the process, lowering the failure rate of batches and increasing the compliance with the regulatory requirements by giving traceable and data based quality assurance.

*Table 1. Comparative overview of scale-up challenges, regulatory requirements, and quality control parameters in nano-oncology translation*

| Dimension | Key Aspects | Scale-Up / Clinical Relevance | Regulatory Considerations | Major Challenges | References |
|---|---|---|---|---|---|
| **Nanocarrier Systems** | Lipid, polymeric, inorganic, biomimetic nanoparticles | Determine drug delivery efficiency, targeting, and biodistribution | Must meet safety and biocompatibility standards | Structural instability during scaling | (Khalid-Salako et al., 2025) |
| **Critical Quality Attributes (CQAs)** | Size, PDI, zeta potential, drug loading | Directly influence pharmacokinetics and tumor targeting | Required for regulatory approval (FDA/EMA CMC data) | Batch variability and poor reproducibility | (Németh et al., 2022) |
| **Manufacturing Technologies** | Microfluidics, homogenization, spray drying, supercritical fluids | Enable transition from lab-scale to industrial production | Must comply with GMP and validation protocols | Scale-dependent physicochemical drift | (Wahab et al., 2025) |
| **Process Control** | QbD, PAT systems, real-time monitoring | Ensures consistent nanoparticle quality | Regulatory expectation for manufacturing control | High cost and technical complexity | (Simões, Veiga, & Vitorino, 2024) |
| **Preclinical Evaluation** | In vitro assays, animal models, biodistribution studies | Predicts clinical performance and safety | Required for IND submission | Poor correlation with human outcomes | (Nair & Weiskirchen, 2024) |
| **Regulatory Pathways** | FDA, EMA, NMPA, PMDA frameworks | Governs clinical approval process | Classification as drug/device/ biologic | Lack of global harmonization | (Tan et al., 2025 |
| **Quality & Safety** | GMP, sterility, endotoxin control, stability testing | Ensures safe clinical-grade production | Mandatory for clinical trials | Sensitivity of nanomaterials to processing conditions | (Souto et al., 2020) |

## 5. CLINICAL TRANSLATION PATHWAYS

Clinical translation of nano-oncology systems is a synergistic step in the development cycle, with promising nanotherapeutic systems moving forward out of engineered preparations and preclinical testing, to human trials. It is at this point that is extensively controlled, scientifically intense, and risky because it is directly involved on the onward journey of a nanomedicine to clinical acceptance, and ultimate commercialization(Gameel et al., 2025). In contrast to traditional small-molecule drug candidates, nanomedicines have other factors to consider based around biodistribution, interactions with the immune system, long-term toxicity, and complicated pharmacokinetic abilities, which cannot be lightly considered without a vigorous assessment prior to clinical use(Tripathi et al., 2025).

## 5.1. Preclinical Safety and IND Requirements

Nano-oncology preparations before being put in human trials have to undergo rigorous preclinical safety trials to determine their toxicology, biological compatibility, and therapeutic window(Damasco et al., 2020). Toxicity testing normally involves acute, sub acute and chronic toxicity assays done in the appropriate animal models to test dose related systemic effects. Specific consideration is given to immuno-taxicity, hepatotoxicity, nephrotoxicity, and possible changes in the hematological profile since nanocarriers tend to interact with the mononuclear phagocyte system, and might accumulate in such organs like the liver, spleen, and kidney(Erhirhie, Ihekwereme, & Ilodigwe, 2018).

The pharmacokinetic studies and biodistribution studies are also essential in the in vivo fate of nanomedicines. Such studies assess circulation half-life, tissue localization trends, tumor targeting capacity and clearance processes and typically track the behavior of nanoparticles over time using radiolabeling or high-tech imaging platforms. Notably, nanocarriers are often subject to non-linear pharmacokinetic characteristics caused by size-dependent filtration, formation of protein coronas, and immune cell uptake and hence their study is more difficult than a traditional drug(Skotland et al., 2022).

In order to file clinical trials, developers are required to file an Investigational New Drug (IND) application to the regulatory authorities like the FDA. This dossier contains in-depth Chemistry, Manufacturing and Controls (CMC) report, preclinical safety reports, formulation stability reports, and manufacturing reports(Nugent, Duncan, & Colagiovanni, 2017). In the case of nanomedicines, regulatory oversight is especially keen on reproducibility, particle characterization, sterilization procedures, and batch consistency since the change in nanoscale characteristics by only a little can cause biological behavior to change significantly(Kazi et al., 2025).

## 5.2. Clinical Trial Development

After the approval of IND, nano oncology candidates undergo regulated clinical trial programs aimed at testing safety, effectiveness, and therapeutic excellence. Phase I trials are mainly aimed at safety evaluation, dose escalation and maximum tolerated dose in few patients. The same studies even examine the patterns of biodistribution, and initial pharmacokinetics (nanomedicine) because the behavior of the nanoparticles in humans can be very different to what has been shown to occur in preclinical models(Younas et al., 2026).

Phase II trials increase the number of patients and they are aimed to assess therapeutic efficacy, best dosing levels, and to access initial signs of clinical benefit in particular cancers. This is a critical phase to determine tumor response rates, progression-free survival, and possible biomarkers of treatment response(Torres-Saavedra & Winter, 2022). Phase III trials include large, multicenter studies that compare the nanoformulation to standard-of-care therapies, to determine conclusive clinical benefits, safety margins, and regulatory approval potential(Venturini et al., 2025).

Another characteristic of the nano-oncology clinical development is the reliance on imaging modalities and molecular biomarkers, which is gradually growing to facilitate real-time assessment of therapeutic efficacy(Venturini et al., 2025). The distribution, tumor accumulation and response of nanoparticle are tracked using techniques like magnetic resonance imaging (MRI), positron emission tomography (PET), and fluorescence-based imaging. Also, circulating tumor biomarkers, genetic signatures, and protein expression profiles are also used to determine individualized patient responses and aid precision medicine strategies(Lee et al., 2026).

## 6. GLOBAL REGULATORY FRAMEWORKS

The international regulatory framework on nano-oncology is multifaceted, unequal, and dynamic as a response to a speedy innovation process in nanomedicine, as well as the lack of harmonized international regulatory frameworks(X. Ma et al., 2024). In contrast to traditional pharmaceuticals, nanomedicines can be characterized as hybrid in nature and they often include drug, biologic and device-like capabilities, provide tremendous difficulties in classifying, evaluating, and approving these novel and unique therapeutic agents. Consequently, various regional regulatory bodies have come up with different, albeit overlapping frameworks to evaluate the safety, efficacy and quality of nanotherapeutic products(Kurul et al., 2025).

### 6.1. Regulatory Agencies and Approval Pathways

The United States Food and Drug administration (FDA), European Medicines Agency (EMA), the National Medical Products Administration (NMPA) of China, and the Pharmaceuticals and Medical Devices Agency (PMDA) of Japan are some of the key regulatory bodies that control the approval of nano-oncology therapeutics(Liu et al., 2025). Individually, each pharmaceutical regulatory agency has defined regulatory submission and approval procedures, such as Investigational New Drug

(IND), New Drug Application (NDA), and marketing authorization procedures, and is progressively adding nanotechnology-specific factors to their assessment procedures(Rodríguez-Gómez et al., 2025).

FDA again continues to be among the most very powerful regulators in this field, offering guidance materials under the Nanotechnology Task Force, and focusing Chemistry, Manufacturing, and Controls (CMC) provisions of PFCF products on a nanoscale. Equally, the EMA also uses risk-based approach, especially in the context of its guidelines on Advanced Therapy Medicinal Products (ATMPs) and reflection papers that deal with nanomedicines(Dri et al., 2023). The rapid development of the regulatory framework in China by the NMPA and in Japan by the PMDA is still underdeveloped compared to the western models; however, do resemble in that their nanomedicine-specific regulations are at their initial developmental phase(Song et al., 2019).

One of the key issues in all types of regulatory cultures is the confusion of the classification of nanomedicines. Because they can be used with a variety of purposes, nano-oncology products can be both a small-molecule drug and biologic agent, diagnostic or medical device(Xiaohan Ma et al., 2024). This mixed identity makes regulation classification more difficult, as its approval channels and evaluation methods and data requirements vary in various jurisdictions. Therefore, developers usually enjoy great regulatory ambiguity in their globalization efforts of nanomedicine products(De Jong, Geertsma, & Borchard, 2022).

## 6.2. Standardization and Harmonization Efforts

To overcome such issues, various international agencies have embarked on trying to standardize and harmonize the approaches to evaluating nanomedicine. Guidelines like Q8 (Pharmaceutical Development), Q9 (Quality Risk Management), and Q10 (Pharmaceutical Quality System) offered by the International Council for Harmonisation of Technical Requirements about Pharmaceuticals to Human Use are widely adopted, which indirectly facilitate the development of nanomedicine to happen with the use of structured quality and risk assessment methods(VanDuyse, Fulford, & Bartlett, 2021).

Simultaneously, the International Organization of Standardization (ISO), more specifically Technical Committee 229 (Nanotechnologies), has come up with the set of standards concerning the characterization of nanoparticles, their terminology and protocols of their measurements(Sharifi et al., 2022). The objectives of these standards are to enhance uniformity of reporting and analytical procedures both in research and industrial environments. Organisation for Economic Co-operation and Development (OECD) has also played a crucial role in that they outline the testing guidelines on the safety and environmental impact and toxicological testing of nanomaterial(Malinowska & Whelan, 2025).

In spite of all these activities, there is yet no one fully unified international regulatory system concerning nanomedicine. The varying terminology, standard of characterization, level of safety, and approval continue to impose fragmentation between regulatory jurisdictions(Daniels et al., 2025). Such incompatibility is a major impediment to the commercialization of nanotechnology around the globe, escalates the expenses of developing nano-oncology products, and also delays the translation of promising nano-oncology therapies into clinical practice(Gawne et al., 2023).

Hence, formulating a globally harmonized regulatory system is a vital agenda to warrant the safe, effective, and fair translationation of nanomedicines across the world by taking the laboratory discovery to the clinical practice(Gujjar et al., 2025).

## 7. GOOD MANUFACTURING PRACTICE (GMP) AND QUALITY CONTROL

Effective clinical transfer of nano-oncology systems is not only possible through innovative design of formulations and preclinical validation but also through the application of Good Manufacturing Practice (GMP) and effective quality control measures(Gupta et al., 2026). Owing to the extreme sensitivity of nanomedicines to changes in subtle parameters of processes, consistency, sterility, and reproducibility of products may be crucial in protecting safety, efficacy, and compliance regarding product regulatory standards on a large scale. GMP systems offer the format of organising the development of laboratory-sized nanocarriers into pharmaceutical products that could be used in the clinic(Hertig et al., 2021).

### 7.1. GMP Requirements in Nanomedicine Production

The GMP regulations in the manufacture of nanomedicines are set in such a way that units of each lot of any nanoproduct defines the required quality standards that can be used on human admission (Souto et al., 2020). One of the most important aspects of GMP compliance is the provision of controlled cleanroom conditions, usually categorized under ISO regulations, where the air quality, presence of particulate matter, temperatures and humidity are strictly controller. These sterile manufacturing conditions are necessary to avoid microbial contamination which is especially required with injectable nanomedicines to be delivered intravenously(-Yarahmadi et al., 2026).

Validated aseptic practices, the use of sterile filtration, where necessary, and strict environmental monitoring must also be used in nano-oncology food production to prevent the occurrence of any safety risks to products(Gavrila , 2025). Nanomedicines on average, unlike traditional pharmaceuticals, do not resist terminal sterilization, like a high heat, or radiation because this can change the nanoparticle size, shape, or drug release profiles. Hence, aseptic production turns into a much-needed factor in structural and functional integrity maintenance(Halwani, 2022).

Another significant GMP requirement when manufacturing nanomedicine is batch consistency and scalability. Considering the sensitivity of nanoparticle systems to small alterations in formulation parameters, any slight change in the mixing rate, solvent volumes and temperature ratios may cause considerable changes in the sizes of the particles used as well as the efficiency of loading drugs into them(Abdel-monem et al., 2025). This makes high process control and reproducibility across all production batches of an industry an essential part of the production process to meet regulatory demands in clinical deployment(Afzal et al., 2026).

## 7.2. Quality Assurance and Risk Management

Controlling nano-oncology manufacturing Quality assurance entails an elaborate yet broadly adopted testing strategies aimed at maintaining stability, safety and performance of the product until it finds its place in the shelf(Afridi, 2025). Stability testing is conducted to test physicochemical stability of considered nanoparticles in different environmental conditions such as temperatures, pH, and storage times. These experiments are critical in establishing the shelf life of formulations and also to guarantee therapeutic effects are sustained(Pujitha et al., 2025).

Another important quality assurance factors that are vital to endotoxin control especially of parenteral nanomedicines. Minimal amounts of endotoxins can result in strong immune reactions, and therefore, highly rigorous testing and elimination processes are necessary during the formulation and purification of products during the production and formulation phases. Endotoxin contamination is usually tested and measured by techniques like Limulus Amebocyte Lysate (LAL) assays(Rajpali et al., 2025).

Other forms of risk management frameworks including the Failure Mode and Effects Analysis (FMEA) are being incorporated into the nanomedicine production system to preemptively factor in possible sources of the process failure(El-Awady, 2023). FMEA allows the methodical assessment of every production phase, starting with the original choice of raw material up to the packaging of the end product by assessing the criticality, probability, and traceability of possible failures. This is a systematic method and one that enables manufacturers to take preventive steps, minimize variability and increase the general reliability of products(Vyawhare, 2025).

Taken together, both the GMP compliance and state-of-the-art quality control systems are the key to the safe and effective production of nano-oncology. They make sure that the process of laboratory innovation to clinical application is supported by reproducible, scalable, and regulation-compliant production systems with the ability to deliver the high demands of current precision oncology(Padrela, Mouras, & Killackey, 2025).

## 8. ETHICAL, ENVIRONMENTAL, AND SUSTAINABILITY ASPECTS

The fast growth of nano-oncology has not only presented radical prospects of precision cancer treatment, but it has also presented critical ethical, environmental, and sustainability issues(Badgugar et al., 2026). With an advancing evolution of nanomedicine systems as research into systems that are industrially scaled and employed in clinics, the environmental consequence of nanomedicine systems on ecosystems, healthcare equity, and accessibility across the world becomes more significant. It is also vital to incorporate the concepts of sustainability and ethical governance into nano-oncology development to achieve responsible innovation and benefit the society in the long run(Mohammad Chehelgerdi et al., 2023).

### 8.1. Green Nanotechnology and Environmental Safety

Green nanotechnology is a new trend in paradigm focusing on engineering and manufacturing nanomaterials in a manner that is energy-saving, eco-friendly and sustainable(Dhahri, Shall, & Thabet Mliki, 2024). Sustainable synthesis, in the context nano-oncology, has focused on biological systems, plant extracts, microbial platforms, and water-based solvents as alternatives to more traditional toxic chemical synthesis. These bi-inspired approaches do not only minimise environmental load but also improve biocompatibility and minimise residual toxicity of the end biases nanoproducts(Thipe et al., 2022).

One significant impact of conventional nanoparticles synthesis methods is the widespread application of potentially toxic organic solvents, surfactants, and limiting agents that lead to the generation of chemical waste, as well as contamination of the ecosystem(Ahmed et al., 2026. Sustainable nanomanufacturing aims at reducing or avoiding these toxic reagents, through adopting solvent-free, or aqueous-phase nanomanufacturing methods, using biodegradable polymers and natural lipids, available as biomaterials. There are also other process intensification measures like microfluidics, continuous flow synthesis among others which also assist in reduction of waste through enhanced efficiency of the reaction and reduced consumption of resources(Bendre et al., 2022).

Other areas of environmental safety go into nanomaterials life cycle such as their degradation, persistence and possible bioaccumulation in the ecosystems. Since there is a possibility in the long run that nanomedicines can be found in the wastewater streams or biological waste systems, it is important to determine the ecological fate of the nanomedicines in order to determine their ecological risks in the long term. Thus, all-around environmental impact measurements are becoming part of the initial phases of nanomedicine developmen (Nthunya et al., 2025).

### 8.2. Ethical and Socioeconomic Considerations

In addition to environmental issues, nano-oncology poses significant ethical and socioeconomic questions, such as access to healthcare, cost, and equity throughout the world. State-of-the-art nanomedicines have been linked with substantial development and production expenses, so the avenue may restrict their accessibility to affluent geographic locations and cause an uneven spread in cancer treatment results among various populations(Firoozi, Firoozi, & Maghami, 2025).

One of the leading ethical issues is accessibility; a number of low- and middle-income nations have financial and infrastructural limitations to the implementation of state-of-the-art nanotherapeutic solutions. This brings up the issue of access variability to life-saving cancer therapies and adds a call for cost-effective manufacturing measures and international health policy interventions making sure of equal distribution(Al-Omari et al., 2025).

Global access to cancer treatment must also be embraced equitably by taking into consideration intellectual property rights, the transfer of technologies and joint international systems, which facilitate in sharing innovation(Rid et al., 2024). To make nano-oncology development ethical, inclusive clinical trial representation, clear regulatory procedures, and equitable pricing frameworks should be prioritized to enable nano-oncology development to impact on a broad population, as opposed to a small group of patients(Pourmadadi et al., 2025).

## 9. FUTURE OUTLOOKS IN THE TRANSLATION OF NANO-ONCOLOGY

It is anticipated that the combination of improved computational capabilities, intelligent production, and the next level of precision therapeutics will determine the future of nano-oncology translation, which will collectively transform the approach to the formation and optimization of nanomedicines and their clinical delivering. Nanomedicine development is also becoming increasingly involved with artificial intelligence (AI) and machine learning to speed up formulation optimization such that large datasets based on experiment parameters, physicochemical properties, and biological results are used to predict optimal nanoparticle design associated with better efficacy and safety profiles. The resulting data-based methods can be used to quickly screen formulation variables, minimize experimental redundancy, and reveal high-performing nanocarrier systems with increased translational capabilities. Parallel to this, novel predictive manufacturing systems with digital twin engineering and real-time process monitoring and analytics are becoming potent means of simulating and optimizing large-scale manufacturing, enabling better control over vital process variables and reducing the batch-to-batch variation in manufacturing at industrial scale.

Along with the developments in design and fabrication, the next-generation clinical nanotherapeutics will bring a highly enhanced smart nanoparticles that are able to identify certain biological stimuli including pH, temperature, enzymes or redox conditions of the tumor microenvironment. These stimuli-responsive systems provide control of drug release at specific sites and provide greater precision with regard to therapy, reducing systemic toxicity. Moreover, development of personalized and precision nanomedicine solutions is expected to revolutionize cancer treatment by modulating nanocarrier design using individual patient profiles such as genetic, proteomic and tumor microenvironmental components. This patient-based approach is in line with the overarching objectives of precision oncology, where nanomedicines can be tailored towards an optimal therapeutic response and alleviation of inter-patient differences. Together, these new trends show a future where nano-oncology can be more intelligent, adaptive and personalized, intersecting between experimental nanoscience and universally accepted clinical care of cancer.

## 10. CONCLUSION

Nano-oncology has also developed into a strong tool in the development of precision cancer therapeutics with opportunities that have never been seen before regarding therapy targets and treatment and the ability to deliver drugs directly into its targets, increased therapeutic efficacy and reduced systemic toxicity. Nevertheless, the development of nanomedicines to a clinically approved product on a large scale is still limited by a number of scientific, engineering, regulatory and ethical issues. This chapter points out that the bench to bedside translation does not revolve only around novel nanoparticle design but also a solid scale-up engineering, manufacturing that has to be reproducible, and standards that are globally harmonized are foundational.

One of the most crucial limitations in the translation of nano-oncology is ensuring physicochemical consistency and biological functionality at industrial scale production. The combination of new manufacturing technology, including microfluidics, high-pressure homogenization and continuous flow systems with Quality-by-Design and Process Analytical Technology solutions is key to reproducibility and regulatory compliance. Moreover, preclinical to clinical transition should involve the use of more predictive biological paradigms and uniformed measures of evaluation to curb the rate of translational failure.

Regulatory heterogeneity remains a slowdown on the global clinical adoption, a factor that has necessitated the implementation of harmonized guidelines between different agencies like FDA, EMA, NMPA and PMDA. Other crucial factors are those involving ethics and sustainability, such as the fairness of access to nanomedicine-based treatments and a socially responsible approach to manufacturing.

Generally, the future of nano-oncology hinges on the harmonious collaboration of nanoscience, pharmaceutical and clinical medicine as well as regulatory science. As interdisciplinary cooperation and technological advances continue, nano-oncology will grow its potentials that remain experimental now into a fully realised component of contemporary precision cancer therapy.

## REFERENCES


Abdelmonem, R., Abdelwahab, M., Lofty, S., Shalabi, Y., Nabeah, J., Abdelsalam, H., Hassan, M., Abdelzaher, A., Aboalfa, A., Abdelkhalek, M.-H., Ali, R., Abdelrahman, S., Fathy, M., Ali, A., Abdellatif, M., & Eltahan, M. (2025).. . *Advancing Nanoparticle Production: Scaling Up Techniques, Challenges, and Future Perspectives in Pharmaceutical Applications.*, *2*, 26–39. DOI: 10.21608/jpsdm.2025.374728.1031

Afridi, S. (2025). Quality Control of Nano-pharmaceuticals: Current Scenario and Challenges.

Afzal, S., Fatima, M., Mishra, M., Shukla, N., Singh, N. K., Sarraf, M., & Rastogi, A. (2026). Insights into progress in biogenic synthesis of nanomaterials: A review. *Plant Nano Biology*, *16*, 100272. DOI: 10.1016/j.plana.2026.100272

Al-Omari, A., Altalla, B., Alayyan, O., Abdel-Razeq, H., & Mansour, A. (2025). *Barriers to Cancer Research in Low- and Middle-Income Countries: Findings from the Arab Region and Beyond.* DOI: 10.21203/rs.3.rs-7184575/v1

Ammar, M. M., Ali, R., Abd Elaziz, N. A., Habib, H., Abbas, F. M., Yassin, M. T., Maniah, K., & Abdelaziz, R. (2025). Nanotechnology in oncology: Advances in biosynthesis, drug delivery, and theranostics. *Discover. Oncology*, *16*(1), 1172. DOI: 10.1007/s12672-025-02664-3 PMID: 40542991

Badgugar, L., Maru, S., Sisodiya, N., Maru, R., & Sugunan, S. (2026). Chapter Thirteen - Safety, ethical, and regulatory considerations in nanotechnology for skin cancer. In Kumar Singh, A., Kaur, R., Pareek, A., Maru, S., & Prakash, A. (Eds.), *Advances in Cancer Research* (Vol. 169, pp. 381–415). Academic Press., ,,10.1016/bs.acr.2026.02.008

Bendre, A., Bhat, M. P., Lee, K.-H., Altalhi, T., Alruqi, M. A., & Kurkuri, M. (2022). Recent developments in microfluidic technology for synthesis and toxicity-efficiency studies of biomedical nanomaterials. *Materials Today. Advances*, *13*, 100205. DOI: 10.1016/j.mtadv.2022.100205

Bezelya, A., Kucukturkmen, B., & bozkır, A. (2023). *The Use of Microfluidic Systems for Nanoparticle Production.* ,DOI: 10.20944/preprints202309.1452.v1

Bousbaa, H. (2021). Novel Anticancer Strategies. *Pharmaceutics*, *13*(2), 275. Advance online publication. DOI: 10.3390/pharmaceutics13020275 PMID: 33670469

Chehelgerdi, M., Chehelgerdi, M., Allela, O. Q. B., Pecho, R. D. C., Jayasankar, N., Rao, D. P., Thamaraikani, T., Vasanthan, M., Viktor, P., Lakshmaiya, N., Saadh, M. J., Amajd, A., Abo-Zaid, M. A., Castillo-Acobo, R. Y., Ismail, A. H., Amin, A. H., & Akhavan-Sigari, R. (2023). Progressing nanotechnology to improve targeted cancer treatment: Overcoming hurdles in its clinical implementation. *Molecular Cancer*, *22*(1), 169. DOI: 10.1186/s12943-023-01865-0 PMID: 37814270

Chehelgerdi, M., Chehelgerdi, M., Allela, O. Q. B., Pecho, R. D. C., Jayasankar, N., Rao, D. P., Thamaraikani, T., Vasanthan, M., Viktor, P., Lakshmaiya, N., Saadh, M. J., Amajd, A., Abo-Zaid, M. A., Castillo-Acobo, R. Y., Ismail, A. H., Amin, A. H., & Akhavan-Sigari, R. (2023). Progressing nanotechnology to improve targeted cancer treatment: Overcoming hurdles in its clinical implementation. *Molecular Cancer*, *22*(1), 169. DOI: 10.1186/s12943-023-01865-0 PMID: 37814270

Cheng, Z., Huang, H., Yin, M., & Liu, H. (2025). Applications of liposomes and lipid nanoparticles in cancer therapy: Current advances and prospects. *Experimental Hematology & Oncology*, *14*(1), 11. DOI: 10.1186/s40164-025-00602-1 PMID: 39891180

Cisneros, E. P., Morse, B. A., Savk, A., Malik, K., Peppas, N. A., & Lanier, O. L. (2024). The role of patient-specific variables in protein corona formation and therapeutic efficacy in nanomedicine. *Journal of Nanobiotechnology*, *22*(1), 714. DOI: 10.1186/s12951-024-02954-y PMID: 39548452

Crecente-Campo, J., & Alonso, M. J. (2019). Engineering, on-demand manufacturing, and scaling-up of polymeric nanocapsules. *Bioengineering & Translational Medicine*, *4*(1), 38–50. DOI: 10.1002/btm2.10118 PMID: 30680317

Damasco, J. A., Ravi, S., Perez, J. D., Hagaman, D. E., & Melancon, M. P. (2020). Understanding Nanoparticle Toxicity to Direct a Safe-by-Design Approach in Cancer Nanomedicine. *Nanomaterials (Basel, Switzerland)*, *10*(11), 2186. Advance online publication. DOI: 10.3390/nano10112186 PMID: 33147800

Daniels, B., Gräff, T., Kotschik, P., & Walter-Rohde, S. (2025). High time to Update Statistical Guidance in Ecotoxicology—A Workshop Synthesis on the Revision of OECD Document No.54. *Integrated Environmental Assessment and Management*. DOI: 10.1093/inteam/vjaf113

de Almeida Campos, L. A., de Souza, J. B., de Queiroz Macêdo, H. L. R., Borges, J. C., de Oliveira, D. N., & Cavalcanti, I. M. F. (2024). Synthesis of polymeric nanoparticles by double emulsion and pH-driven: Encapsulation of antibiotics and natural products for combating Escherichia coli infections. *Applied Microbiology and Biotechnology*, *108*(1), 351. DOI: 10.1007/s00253-024-13114-5 PMID: 38819646

De Jong, W. H., Geertsma, R. E., & Borchard, G. (2022). Regulatory safety evaluation of nanomedical products: Key issues to refine. *Drug Delivery and Translational Research, 12*(9), 2042–2047. DOI: 10.1007/s13346-022-01208-4 PMID: 35908133

Dhahri, S., Shall, H., & Thabet Mliki, N. (2024). Towards sustainable Nanomaterials: Exploring green synthesis methods and their impact on electrical properties. *Inorganic Chemistry Communications, 168*, 112872. DOI: 10.1016/j.inoche.2024.112872

Đorđević, S., Gonzalez, M. M., Conejos-Sánchez, I., Carreira, B., Pozzi, S., Acúrcio, R. C., Satchi-Fainaro, R., Florindo, H. F., & Vicent, M. J. (2022). Current hurdles to the translation of nanomedicines from bench to the clinic. *Drug Delivery and Translational Research, 12*(3), 500–525. DOI: 10.1007/s13346-021-01024-2 PMID: 34302274

Dri, D. A., Rinaldi, F., Carafa, M., & Marianecci, C. (2023). Nanomedicines and nanocarriers in clinical trials: Surfing through regulatory requirements and physico-chemical critical quality attributes. *Drug Delivery and Translational Research, 13*(3), 757–769. DOI: 10.1007/s13346-022-01262-y PMID: 36450964

Eder, K. M., Marzi, A., Wågbø, A. M., Vermeulen, J. P., de la Fonteyne-Blankestijn, L. J. J., Rösslein, M., Ossig, R., Klinkenberg, G., Vandebriel, R. J., & Schnekenburger, J. (2022). Standardization of an in vitro assay matrix to assess cytotoxicity of organic nanocarriers: A pilot interlaboratory comparison. *Drug Delivery and Translational Research, 12*(9), 2187–2206. DOI: 10.1007/s13346-022-01203-9 PMID: 35794354

El-Awady, S. M. M. (2023). Overview of Failure Mode and Effects Analysis (FMEA): A Patient Safety Tool. *Global Journal on Quality and Safety in Healthcare, 6*(1), 24–26. DOI: 10.36401/JQSH-23-X2 PMID: 37260856

El-Shabasy, R. M., Zayed, A., Farag, M. A., & Shoueir, K. R. (2026). Green synthesis of relevant and sustainable bio-applications of few-layer graphene: A multi-faceted review and future perspectives. *Materials Today Sustainability, 33*, 101259. 10.1016/j.mtsust.2025.101259

Erhirhie, E. O., Ihekwereme, C. P., & Ilodigwe, E. E. (2018). Advances in acute toxicity testing: Strengths, weaknesses and regulatory acceptance. *Interdisciplinary Toxicology, 11*(1), 5–12. DOI: 10.2478/intox-2018-0001 PMID: 30181707

Firoozi, A. A., Firoozi, A. A., & Maghami, M. R. (2025). Transformative Impacts of Nanotechnology on Sustainable Construction: A Comprehensive Review. *Results in Engineering, 26*, 104973. DOI: 10.1016/j.rineng.2025.104973

Firouzpour, H., Najafi, F., Mohammadgholi, A., Alizadegan, D., Mahmoudi Beram, F., Akhtari, N., Zoghi, M., Ansar, N., & Tavakkoli Yaraki, M. (2026). Nanoparticle-based drug delivery systems for effective cancer treatment: Mechanisms and applications. *Next Nanotechnology, 9*, 100372. 10.1016/j.nxnano.2026.100372

Gameel, M., Elsayed, D. A., Shehab, W. S., Magdy, N., & Al-Erian, M. A., Emwas, A.-H. M., Jaremko, M., & Elmansy, M. (2025). Enzymes reloaded: Nanotechnology's role in the next generation of cancer therapies. *Results in Chemistry, 18*, 102889. 10.1016/j.rechem.2025.102889

Gavrila , S. (2025). Nanomembranes as Eco-Friendly Instruments for Modern Food Processing, from Filtration to Packaging. *Membranes (Basel)*, *15*(6), 167. Advance online publication. DOI: 10.3390/membranes15060167 PMID: 40559345

Gawne, P., Ferreira, M., Papaluca, M., Grimm, J., & Decuzzi, P. (2023). New Opportunities and Old Challenges in the Clinical translation of Nanotheranostics. *Nature Reviews. Materials*, *8*(12), 783–798. Advance online publication. DOI: 10.1038/s41578-023-00581-x PMID: 39022623

Gerzon, G., Sheng, Y., & Kirkitadze, M. (2021). Process Analytical Technologies-Advances in Bioprocess Integration and Future Perspectives. *Journal of Pharmaceutical and Biomedical Analysis*, *207*, 114379. DOI: 10.1016/j.jpba.2021.114379 PMID: 34607168

Gujjar, S., Kukal, S., Jayabal, P., Balaji, N., Sainger, S., Roy, S., Rallapalli, S., Mahadevappa, R., Minocha, S., Kumar, S., & Mathapati, S. (2025). Nanomaterials for biomedical applications: Addressing regulatory hurdles and strategic solutions. *Nano Trends, 11*, 100127. 10.1016/j.nwnano.2025.100127

. Gupta, V., Kumar, D., Gupta, S., Tanwar, R., Jaiswal, N. K., Islam, M. M., Singh, S., Choudhary, N., Gowri, S., Webster, T. J., & Faiyazuddin, M. (2026). Nanotechnology-Driven Cancer Therapies for Precision Oncology: Advances and Clinical Outlook. *International Journal of Nanomedicine, 21*(null), 568254. DOI: 10.2147/IJN.S568254

Halwani, A. A. (2022). Development of Pharmaceutical Nanomedicines: From the Bench to the Market. *Pharmaceutics*, *14*(1), 106. Advance online publication. DOI: 10.3390/pharmaceutics14010106 PMID: 35057002

Herdiana, Y., Wathoni, N., Shamsuddin, S., & Muchtaridi, M. (2022). Scale-up polymeric-based nanoparticles drug delivery systems: Development and challenges. *OpenNano*, *7*, 100048. DOI: 10.1016/j.onano.2022.100048

Hertig, J. B., Shah, V. P., Flühmann, B., Mühlebach, S., Stemer, G., Surugue, J., Moss, R., & Di Francesco, T. (2021). Tackling the challenges of nanomedicines: Are we ready? *American Journal of Health-System Pharmacy*, *78*(12), 1047–1056. DOI: 10.1093/ajhp/zxab048 PMID: 33599767

Hocaoğlu, R., Zıkşahna, K., & Ihlamur, M. (2026). Selecting nanocarrier platforms for drug delivery: A criteria-based, translational review integrating CQA/CMC and immune interactions. *Health Nanotechnology*, *2*(1), 5. DOI: 10.1186/s44301-026-00026-8

Hwang, J., Kang, J., Oh, N., Li, C., Yoo, C.-Y., & Kwak, M. (2025). Effect of Surface Charge of Nanoparticles on Cellular Uptake in HeLa Cells. *Langmuir : The ACS Journal of Surfaces and Colloids*, *41*(21), 13192–13198. DOI: 10.1021/acs.langmuir.5c00832 PMID: 40384298

Jayaraj, S., Jayaprakash, N., Devaraji, M., Murugavel, S., Robinson, A. X., & Murugesan, L. (2026). Nanocarrier-based strategies for advanced cancer drug delivery and therapeutics: Design principles, biological interactions, and clinical potential. *Nano Trends, 14*, 100203. 10.1016/j.nwnano.2026.100203

Ji, J., Li, L., Guo, W., Zhang, J., Yao, Y., Chen, H., Liao, F., Jin, Z., Liu, L., Ouyang, J., & Liang, X.-J. (2025). Nanoscale synthetic biology with innovative medicinal applications. *Fundamental Research (Beijing)*, *5*(5), 1889–1910. DOI: 10.1016/j.fmre.2024.11.006 PMID: 41613444

Kalpage, C., & Isuranga, T. (2025). Supercritical Fluid Technologies for Nanoparticle Production. In (pp. 215-246). ,DOI: 10.1007/978-981-96-9176-0_11

Kara, A., Ongoren, B., Anaya, B. J., Lalatsa, A., & Serrano, D. R. (2025). Continuous manufacturing of nanomedicines using 3D-printed microfluidic devices. *Applied Materials Today*, *43*, 102672. DOI: 10.1016/j.apmt.2025.102672

Kashapov, R., Gaynanova, G., Gabdrakhmanov, D., Kuznetsov, D., Pavlov, R., Petrov, K., Zakharova, L., & Sinyashin, O. (2020). Self-Assembly of Amphiphilic Compounds as a Versatile Tool for Construction of Nanoscale Drug Carriers. *International Journal of Molecular Sciences*, *21*(18), 6961. Advance online publication. DOI: 10.3390/ijms21186961 PMID: 32971917

Kashinath, K. P., Sanjay, L. R., Ashokbhai, M. K., Roy, S., Sardar, M. S., & Kaity, S. (2025). Continuous manufacturing based paradigm shift in pharmaceuticals production and current regulatory framework. *Chemical Engineering Research & Design*, *215*, 1–22. DOI: 10.1016/j.cherd.2025.01.003

Kazi, R. N. A., Hasani, I. W., Khafaga, D. S. R., Kabba, S., Farhan, M., Aatif, M., Muteeb, G., & Fahim, Y. A. (2025). Nanomedicine: The Effective Role of Nanomaterials in Healthcare from Diagnosis to Therapy. *Pharmaceutics*, *17*(8), 987. Advance online publication. DOI: 10.3390/pharmaceutics17080987 PMID: 40871010

Khalbas, A. H., Albayati, T. M., Ali, N. S., & Salih, I. K. (2024). Drug loading methods and kinetic release models using of mesoporous silica nanoparticles as a drug delivery system: A review. *South African Journal of Chemical Engineering*, *50*, 261–280. DOI: 10.1016/j.sajce.2024.08.013

Khalid-Salako, F., Salimi Khaligh, S., Fathi, F., Demirci, O. C., Öncer, N., Kurt, H., & Yüce, M. (2025). The Nanocarrier Landscape—Evaluating Key Drug Delivery Vehicles and Their Capabilities: A Translational Perspective. *ACS Applied Materials & Interfaces*, *17*(26), 37383–37403. DOI: 10.1021/acsami.5c07366 PMID: 40526827

Khorsheed, K., Asadi, F., Asadi, R., Aydoğmuş, E., Erzen, B., Karataş, M., Deniz, Ş., & Orhan, R. (2025). Process Optimization with Response Surface Methodology: A Comprehensive Approach by Experimental Design Types. *International Journal of Advanced Natural Sciences and Engineering Researches*, *9*, 546–558.

Kurul, F., Turkmen, H., Cetin, A. E., & Topkaya, S. N. (2025). Nanomedicine: How nanomaterials are transforming drug delivery, bio-imaging, and diagnosis. *Next Nanotechnology, 7*, 100129. 10.1016/j.nxnano.2024.100129

Lee, J., Kim, J., Cho, E., Jeong, J., Song, J., & Kang, M. (2026). Nanotechnology-Enhanced Approaches for Early Cancer Diagnosis. *Nano Select*, *7*(1), e70105. DOI: 10.1002/nano.70105

Lee, Y. J., & Kim, M. S. (2025). Advances in drug-loaded microspheres for targeted, controlled, and sustained drug delivery: Potential, applications, and future directions. *Biomedicine & Pharmacotherapy = Biomédecine & Pharmacothérapie*, *189*, 118244. DOI: 10.1016/j.biopha.2025.118244 PMID: 40516334

Liu, Y., Zhang, Y., Li, H., & Hu, T. Y. (2025). Recent advances in the bench-to-bedside translation of cancer nanomedicines. *Acta Pharmaceutica Sinica. B*, *15*(1), 97–122. DOI: 10.1016/j.apsb.2024.12.007 PMID: 40041906

Ma, X., Tian, Y., Yang, R., Wang, H., Allahou, L. W., Chang, J., Williams, G., Knowles, J. C., & Poma, A. (2024). Nanotechnology in healthcare, and its safety and environmental risks. *Journal of Nanobiotechnology*, *22*(1), 715. DOI: 10.1186/s12951-024-02901-x PMID: 39548502

Ma, X., Tian, Y., Yang, R., Wang, H., Allahou, L. W., Chang, J., Williams, G., Knowles, J. C., & Poma, A. (2024). Nanotechnology in healthcare, and its safety and environmental risks. *Journal of Nanobiotechnology*, *22*(1), 715. DOI: 10.1186/s12951-024-02901-x PMID: 39548502

Malinowska, J. M., & Whelan, M. (2025). The important role of standards for the uptake of transcriptomics and metabolomics based in vitro methods in regulatory toxicology. *Archives of Toxicology*, *99*(10), 3865–3875. DOI: 10.1007/s00204-025-04119-8 PMID: 40739409

Metselaar, J. M., & Lammers, T. (2020). Challenges in nanomedicine clinical translation. *Drug Delivery and Translational Research*, *10*(3), 721–725. DOI: 10.1007/s13346-020-00740-5 PMID: 32166632

Nair, D. G., & Weiskirchen, R. (2024). Advanced In Vitro Models for Preclinical Drug Safety: Recent Progress and Prospects. *Current Issues in Molecular Biology*, *47*(1), 7. Advance online publication. DOI: 10.3390/cimb47010007 PMID: 39852122

Németh, Z., Csóka, I., Semnani Jazani, R., Sipos, B., Haspel, H., Kozma, G., Kónya, Z., & Dobó, D. G. (2022). Quality by Design-Driven Zeta Potential Optimisation Study of Liposomes with Charge Imparting Membrane Additives. *Pharmaceutics*, *14*(9), 1798. Advance online publication. DOI: 10.3390/pharmaceutics14091798 PMID: 36145546

Nugent, P., Duncan, J. N., & Colagiovanni, D. (2017). Preparation of a Preclinical Dossier to Support an Investigational New Drug (IND) Application and First-In-Human Clinical Trial. In (pp. 189-213). DOI: 10.1016/B978-0-12-803620-4.00008-6

Padrela, L., Mouras, R., & Killackey, D. (2025). Innovation testbeds as enabling ecosystems driving nanopharmaceuticals to market. *Drug Discovery Today*, *30*(8), 104433. DOI: 10.1016/j.drudis.2025.104433 PMID: 40675439

Papa, S., Ortolani, C., Fernández, P., & O'Connor, J. E. (2023). Flow Cytometry and Its Applications to Molecular Biology and Diagnosis 2.0. *International Journal of Molecular Sciences*, *24*(22), 16215. Advance online publication. DOI: 10.3390/ijms242216215 PMID: 38003405

Peng, C., Huang, Y., & Zheng, J. (2020). Renal clearable nanocarriers: Overcoming the physiological barriers for precise drug delivery and clearance. *Journal of Controlled Release : Official Journal of the Controlled Release Society*, *322*, 64–80. DOI: 10.1016/j.jconrel.2020.03.020 PMID: 32194171

Pourmadadi, M., Shabestari, S. M., Abdouss, H., Rahdar, A., Fathi-Karkan, S., & Pandey, S. (2025). Artificial intelligence-driven intelligent nanocarriers for cancer theranostics: A paradigm shift with focus on brain tumors. *Seminars in Oncology*, *52*(6), 152429. DOI: 10.1016/j.seminoncol.2025.152429 PMID: 41218468

Pujitha, K., Harichandhana, N., Suchitra, M., & Chandra, Y. (2025). The Role of Stability Studies in Ensuring Pharmaceutical Product Quality. *International Journal of Allied Medical Sciences and Clinical Research*, *13*(4), 660–675. DOI: 10.61096/ijamscr.v13.iss4.2025.660-675

Rajpali, R., Teotia, A., Thota, P., Khanna, D., Samaniya, V., Kumar, P., Pandey, M., Lagashetti, A., Kalaiselvan, V., & Raghuvanshi, R. (2025). A comparative analysis of pharmacopeial quality standards for antibiotics with respect to bacterial endotoxin limits. *Journal of Biological Methods*, *13*(1), e99010088. DOI: 10.14440/jbm.0127

Ramos, T. I., Villacis-Aguirre, C. A., López-Aguilar, K. V., Santiago Padilla, L., Altamirano, C., Toledo, J. R., & Santiago Vispo, N. (2022). The Hitchhiker's Guide to Human Therapeutic Nanoparticle Development. *Pharmaceutics*, *14*(2), 247. Advance online publication. DOI: 10.3390/pharmaceutics14020247 PMID: 35213980

Rid, A., Aguilera, B., Banda, C., Divi, R., Harris, M., Kim, A., Ossandon, M., Zervos, J., & Rowthorn, V. (2024). Global health reciprocal innovation: Ethical, legal and regulatory considerations. *BMJ Global Health*, *8*(Suppl 7), e014693. Advance online publication. DOI: 10.1136/bmjgh-2023-014693 PMID: 38821558

Rodríguez-Gómez, F. D., Monferrer, D., Penon, O., & Rivera-Gil, P. (2025). Regulatory pathways and guidelines for nanotechnology-enabled health products: A comparative review of EU and US frameworks. *Frontiers in Medicine*, *12*, 1544393. DOI: 10.3389/fmed.2025.1544393 PMID: 40109724

Sebastián, V. (2026). Microfluidic reactors for the synthesis of inorganic and hybrid nanoparticles for drug delivery. *Advanced Drug Delivery Reviews*, *234*, 115872. DOI: 10.1016/j.addr.2026.115872 PMID: 41951054

Sharaf, M., Alhamad, A. A., Ltaief, O. O., & Amor, I. B. (2023). Challenges of nanomaterials-based cancer therapy: A future destination. *International Journal of Surgery*, *109*(6), 1819–1820. DOI: 10.1097/JS9.0000000000000412 PMID: 37072147

Sharifi, S., Mahmoud, N. N., Voke, E., Landry, M. P., & Mahmoudi, M. (2022). Importance of Standardizing Analytical Characterization Methodology for Improved Reliability of the Nanomedicine Literature. *Nano-Micro Letters*, *14*(1), 172. DOI: 10.1007/s40820-022-00922-5 PMID: 35987931

Shi, P., Cheng, Z., Zhao, K., Chen, Y., Zhang, A., Gan, W., & Zhang, Y. (2023). Active targeting schemes for nano-drug delivery systems in osteosarcoma therapeutics. *Journal of Nanobiotechnology*, *21*(1), 103. DOI: 10.1186/s12951-023-01826-1 PMID: 36944946

Simões, A., Veiga, F., & Vitorino, C. (2024). Question-based review for pharmaceutical development: An enhanced quality approach. *European Journal of Pharmaceutics and Biopharmaceutics : Official Journal of Arbeitsgemeinschaft für Pharmazeutische Verfahrenstechnik e.V*, *195*, 114174. DOI: 10.1016/j.ejpb.2023.114174 PMID: 38160986

Skotland, T., Iversen, T. G., Llorente, A., & Sandvig, K. (2022). Biodistribution, pharmacokinetics and excretion studies of intravenously injected nanoparticles and extracellular vesicles: Possibilities and challenges. *Advanced Drug Delivery Reviews*, *186*, 114326. DOI: 10.1016/j.addr.2022.114326 PMID: 35588953

Song, R., Sang, G., Geng, M., & Jiang, H. (2019). China's reform of the regulatory system for medical products and its impact. *National Science Review*, *6*(1), 1. DOI: 10.1093/nsr/nwz001 PMID: 34691816

Soni, G., Kale, K., Shetty, S., Gupta, M. K., & Yadav, K. S. (2020). Quality by design (QbD) approach in processing polymeric nanoparticles loading anticancer drugs by high pressure homogenizer. *Heliyon*, *6*(4), e03846. DOI: 10.1016/j.heliyon.2020.e03846 PMID: 32373744

Souto, E. B., Silva, G. F., Dias-Ferreira, J., Zielinska, A., Ventura, F., Durazzo, A., Lucarini, M., Novellino, E., & Santini, A. (2020). Nanopharmaceutics: Part I-Clinical Trials Legislation and Good Manufacturing Practices (GMP) of Nanotherapeutics in the EU. *Pharmaceutics*, *12*(2), 146. Advance online publication. DOI: 10.3390/pharmaceutics12020146 PMID: 32053962

Thipe, V. C., Karikachery, A. R., Çakılkaya, P., Farooq, U., Genedy, H. H., Kaeokhamloed, N., Phan, D.-H., Rezwan, R., Tezcan, G., Roger, E., & Katti, K. V. (2022). Green nanotechnology—An innovative pathway towards biocompatible and medically relevant gold nanoparticles. *Journal of Drug Delivery Science and Technology*, *70*, 103256. DOI: 10.1016/j.jddst.2022.103256

Torres-Saavedra, P. A., & Winter, K. A. (2022). An Overview of Phase 2 Clinical Trial Designs. *International Journal of Radiation Oncology, Biology, Physics*, *112*(1), 22–29. DOI: 10.1016/j.ijrobp.2021.07.1700 PMID: 34363901

Tripathi, D., Pandey, P., Sharma, S., Rai, A. K., & Prabhu, B. H. M. (2025). Advances in nanomaterials for precision drug delivery: Insights into pharmacokinetics and toxicity. *BioImpacts*, *15*, 30573. DOI: 10.34172/bi.30573 PMID: 40256227

Tsiaxerli, A., Vardaka, E., Moutroupidis, C., Taylor, K. M. G., Kachrimanis, K., & Malamatari, M. (2024). Updates on the conversion of nanosuspensions to solid oral dosage forms. *Journal of Food and Drug Analysis*, *32*(4), 398–411. DOI: 10.38212/2224-6614.3525 PMID: 39752860

VanDuyse, S. A., Fulford, M. J., & Bartlett, M. G. (2021). ICH Q10 Pharmaceutical Quality System Guidance: Understanding Its Impact on Pharmaceutical Quality. *The AAPS Journal*, *23*(6), 117. DOI: 10.1208/s12248-021-00657-y PMID: 34773177

Venturini, J., Chakraborty, A., Baysal, M. A., & Tsimberidou, A. M. (2025). Developments in nanotechnology approaches for the treatment of solid tumors. *Experimental Hematology & Oncology*, *14*(1), 76. DOI: 10.1186/s40164-025-00656-1 PMID: 40390104

Viegas, C., Patrício, A. B., Prata, J. M., Nadhman, A., Chintamaneni, P. K., & Fonte, P. (2023). Solid Lipid Nanoparticles vs. Nanostructured Lipid Carriers: A Comparative Review. *Pharmaceutics*, *15*(6), 1593. Advance online publication. DOI: 10.3390/pharmaceutics15061593 PMID: 37376042

Vyawhare, N. (2025). An FMEA-driven holistic approach to quality risk management for new product introduction in pharmaceutical commercial plant: Navigating global regulatory landscapes. *International Journal of Research in Pharmaceutical Sciences*, *16*(3), 24–45. DOI: 10.26452/ijrps.v16i3.4788

Wahab, W., Alshamsi, R., Alharsousi, B., Alnuaimi, M., Alhammadi, Z., & Al-Zaitone, B. (2025). Recent Developments in Pharmaceutical Spray Drying: Modeling, Process Optimization, and Emerging Trends with Machine Learning. *Pharmaceutics*, *17*(12), 1605. Advance online publication. DOI: 10.3390/pharmaceutics17121605 PMID: 41471119

Waheed, I., Ali, A., Tabassum, H., Khatoon, N., Lai, W. F., & Zhou, X. (2024). Lipid-based nanoparticles as drug delivery carriers for cancer therapy. *Frontiers in Oncology*, *14*, 1296091. DOI: 10.3389/fonc.2024.1296091 PMID: 38660132

Winkler, J. S., Barai, M., & Tomassone, M. S. (2019). Dual drug-loaded biodegradable Janus particles for simultaneous co-delivery of hydrophobic and hydrophilic compounds. *Experimental Biology and Medicine (Maywood, N.J.)*, *244*(14), 1162–1177. DOI: 10.1177/1535370219876554 PMID: 31617755

Xu, H., Wen, J., Yang, J., Zhou, S., Li, Y., Xu, K., Li, W., & Li, S. (2024). Tumor-microenvironment-on-a-chip: The construction and application. *Cell Communication and Signaling : CCS*, *22*(1), 515. DOI: 10.1186/s12964-024-01884-4 PMID: 39438954

. Xu, Z., Xie, Y., Chen, W., & Deng, W. (2025). Nanocarrier-Based Systems for Targeted Delivery: Current Challenges and Future Directions. *MedComm (2020), 6*(9), e70337. DOI: 10.1002/mco2.70337

Yang, S., Hu, X., Zhu, J., Zheng, B., Bi, W., Wang, X., Wu, J., Mi, Z., & Wu, Y. (2025). Aspects and Implementation of Pharmaceutical Quality by Design from Conceptual Frameworks to Industrial Applications. *Pharmaceutics, 17*(5), 623. Advance online publication. DOI: 10.3390/pharmaceutics17050623 PMID: 40430914

Yarahmadi, R., Damiri, Z., Darvishi, M.-M., Alipoor, A., Pourbabaki, R., Moridi, P., Sharifi, J., & Jebeli, M. (2026). Evaluating Clean Room Performance in the Pharmaceutical Industry: A Numerical and Experimental Investigation Based on ISO 14644–3 Standards. *Aerosol and Air Quality Research, 26*(3), 25. Advance online publication. DOI: 10.1007/s44408-026-00101-w

Younas, A., Wang, S., Asad, M., Al Mamun, A., Majeed, S., Sharif, A., Zhou, Q., Liu, Y., Geng, P., Shao, C., & Xiao, J. (2026). Recent advances in cancer nanomedicine: From smart targeting to personalized therapeutics - pioneering a new era in precision oncology. *Materials Today. Bio, 36*, 102660. DOI: 10.1016/j.mtbio.2025.102660 PMID: 41560793

Zhang, H., Li, S., & Ma, X. (2024). Transforming Healthcare with Nanomedicine: A SWOT Analysis of Drug Delivery Innovation. *Drug Design, Development and Therapy, 18*, 3499–3521. DOI: 10.2147/DDDT.S470210 PMID: 39132625

Zhang, X., Chan, H. W., Shao, Z., Wang, Q., Chow, S., & Chow, S. F. (2025). Navigating translational research in nanomedicine: A strategic guide to formulation and manufacturing. *International Journal of Pharmaceutics, 671*, 125202. DOI: 10.1016/j.ijpharm.2025.125202 PMID: 39799998

Zhang, X., Jing, H., Wang, Y., Mu, S., Wang, X., Zheng, H., Chen, Y., Wang, K., Sun, J., Bi, J., & Zhang, G. (2026). Membrane-derived biomimetic nanovesicles in anti-tumor immunotherapy: Advances and outlook. *Acta Pharmaceutica Sinica. B, 16*(3), 1388–1420. DOI: 10.1016/j.apsb.2026.01.019 PMID: 41909733

# Compilation of References


Abdelhamid, A., Zayed, D., Heikal, L., Khattab, S., Mady, O., El-Gizawy, S., & Elzoghby, A. (2021). Recent Advances in Polymer Shell Oily-Core Nanocapsules for Drug-Delivery Applications. *Nanomedicine (London)*, *16*(18), 1613–1625. Advance online publication. DOI: 10.2217/nnm-2021-0037 PMID: 34189946

Abdelmonem, R., Abdelwahab, M., Lofty, S., Shalabi, Y., Nabeah, J., Abdelsalam, H., Hassan, M., Abdelzaher, A., Aboalfa, A., Abdelkhalek, M.-H., Ali, R., Abdelrahman, S., Fathy, M., Ali, A., Abdellatif, M., & Eltahan, M. (2025)... *Advancing Nanoparticle Production: Scaling Up Techniques, Challenges, and Future Perspectives in Pharmaceutical Applications.*, *2*, 26–39. DOI: 10.21608/jpsdm.2025.374728.1031

Abdolahi, S., Ghazvinian, Z., Muhammadnejad, S., Saleh, M., Aghdaei, A., & Baghaei, K. (2022). Patient-derived xenograft (PDX) models, applications and challenges in cancer research. *Journal of Translational Medicine*, *20*(1), 206. Advance online publication. DOI: 10.1186/s12967-022-03405-8 PMID: 35538576

Abdo, T., Alhalabi, A., Yaghi, S., Aloran, M., Li, Y., Herrán, M., Tfayli, R., Samuel, T. A., & Nahleh, Z. (2026). Minimal residual disease in solid tumors: Clinical applications and future directions. *Cancer*, *132*(3), e70286. 10.1002/cncr.70286. DOI: 10.1002/cncr.70286 PMID: 41604278

Abdullah, K. M., Sharma, G., Singh, A. P., & Siddiqui, J. A. (2025). Nanomedicine in Cancer Therapeutics: Current Perspectives from Bench to Bedside. *Molecular Cancer*, *24*(1), 169. Advance online publication. DOI: 10.1186/s12943-025-02368-w PMID: 40490771

Abed, H. F., Abuwatfa, W. H., & Husseini, G. A. (2022). Redox-Responsive Drug Delivery Systems: A Chemical Perspective. *Nanomaterials (Basel, Switzerland)*, *12*(18), 3183. Advance online publication. DOI: 10.3390/nano12183183 PMID: 36144971

Abuwatfa, W., Pitt, W., & Husseini, G. (2024). Scaffold-based 3D cell culture models in cancer research. *Journal of Biomedical Science*, *31*(1), 7. Advance online publication. DOI: 10.1186/s12929-024-00994-y PMID: 38221607

Abuzeid, H. M., Julien, C. M., Zhu, L., & Hashem, A. M. (2023). Green Synthesis of Nanoparticles and Their Energy Storage, Environmental, and Biomedical Applications. *Crystals*, *13*(11), 1576. DOI: 10.3390/cryst13111576

Adepu, S., & Ramakrishna, S. (2021). Controlled Drug Delivery Systems: Current Status and Future Directions. *Molecules (Basel, Switzerland)*, *26*(19), 5905. Advance online publication. DOI: 10.3390/molecules26195905 PMID: 34641447

Adeyemi, J. O., Oriola, A. O., Onwudiwe, D. C., & Oyedeji, A. O. (2022). Plant Extracts Mediated Metal-Based Nanoparticles: Synthesis and Biological Applications. *Biomolecules*, *12*(5), 627. DOI: 10.3390/biom12050627 PMID: 35625555

Afridi, S. (2025). Quality Control of Nano-pharmaceuticals: Current Scenario and Challenges.

Afzal, O., Altamimi, A. S., Nadeem, M., Alzarea, S. I., Almalki, W., Tariq, A., Mubeen, B., Murtaza, B. N., Iftikhar, S., Riaz, N., & Kazmi, I. (2022). Nanoparticles in Drug Delivery: From History to Therapeutic Applications. *Nanomaterials (Basel, Switzerland)*, *12*(24), 4494. Advance online publication. DOI: 10.3390/nano12244494 PMID: 36558344

Afzal, S., Fatima, M., Mishra, M., Shukla, N., Singh, N. K., Sarraf, M., & Rastogi, A. (2026). Insights into progress in biogenic synthesis of nanomaterials: A review. *Plant Nano Biology*, *16*, 100272. DOI: 10.1016/j.plana.2026.100272

Aggarwal, R., Magar, A., Dongsar, T. T., Dongsar, T. S., Almoyad, M. A. A., Wahab, S., & Kesharwani, P. (2025). Magnetic nanoparticles in cancer nanomedicine: Advances in targeting, hyperthermia, and theranostics. *International Journal of Pharmaceutics*, •••, 126483. PMID: 41397473

Aghamiri, S., Talaei, S., Ghavidel, A., Zandsalimi, F., Masoumi, S., Hafshejani, N. H., & Jajarmi, V. (2020). Nanoparticles-mediated CRISPR/Cas9 delivery: Recent advances in cancer treatment. *Journal of Drug Delivery Science and Technology*, *56*, 101533. DOI: 10.1016/j.jddst.2020.101533

Agiba, A. M., Elsayyad, N., ElShagea, H. N., Metwalli, M., Mahmoudsalehi, A. O., Beigi-Boroujeni, S., Lozano, O., Aguirre-Soto, A., Arreola-Ramírez, J. L., Segura-Medina, P., & Hamed, R. R. (2024). Advances in Light-Responsive Smart Multifunctional Nanofibers: Implications for Targeted Drug Delivery and Cancer Therapy. *Pharmaceutics*, *16*(8), 1017. Advance online publication. DOI: 10.3390/pharmaceutics16081017 PMID: 39204362

Agrawal, A. (2022). *Top-down synthesis of 2D nanomaterials. In 2D Nanomaterials*. CRC Press.

Ahmad, A., Mukherjee, P., Senapati, S., Mandal, D., Khan, M. I., Kumar, R., & Sastry, M. (2003). Extracellular biosynthesis of silver nanoparticles using the fungus Fusarium oxysporum. *Colloids and Surfaces. B, Biointerfaces*, *28*(4), 313–318. DOI: 10.1016/S0927-7765(02)00174-1

Ahmad, I., Mobeen, M. F., Sheraz, M. A., Ahmed, S., Anwar, Z., Shaikh, R. S., Hussain, I., & Ali, S. M. (2018). Photochemical interaction of ascorbic acid and nicotinamide in aqueous solution: A kinetic study. *Journal of Photochemistry and Photobiology. B, Biology*, *182*, 115–121. DOI: 10.1016/j.jphotobiol.2018.04.011 PMID: 29660587

Ahmad, N., Ansari, M. A., Al-Mahmeed, A., Joji, R. M., Saeed, N. K., & Shahid, M. (2024). Biogenic silver nanomaterials synthesized from Ocimum sanctum leaf extract exhibiting robust antimicrobial and anticancer activities: Exploring the therapeutic potential. *Heliyon*, *10*(15), e35486. DOI: 10.1016/j.heliyon.2024.e35486 PMID: 39170333

Ahmed, M. J., Perveen, S., Hussain, S. G., Khan, A. A., Ejaz, S. M. W., & Rizvi, S. M. A. (2023). Design of a facile, green and efficient graphene oxide-based electrochemical sensor for analysis of acetaminophen drug. *Chemical Papers*, *77*(4), 2275–2294. DOI: 10.1007/s11696-022-02628-9 PMID: 36589858

Ahmed, N., Abusalah, M. A. H. A., Absar, M., Noor, M. S., Bukhari, B., Anjum, S. A., Singh, K. K. B., & Yean, C. Y. (2025). Carbon-based Nanomaterials as Multifunctional Particles for Cancer Diagnosis and Treatment. *Nano LIFE*, *15*(04), 2430005. DOI: 10.1142/S179398442430005X

Ahmed, S., Ahmad, M., Swami, B. L., & Ikram, S. (2016). A review on plants extract mediated synthesis of silver nanoparticles for antimicrobial applications: A green expertise. *Journal of Advanced Research*, *7*(1), 17–28. DOI: 10.1016/j.jare.2015.02.007 PMID: 26843966

Ajith, S., Almomani, F., Elhissi, A., & Husseini, G. A. (2023). Nanoparticle-based materials in anticancer drug delivery: Current and future prospects. *Heliyon*, *9*(11), e21227. Advance online publication. DOI: 10.1016/j.heliyon.2023.e21227 PMID: 37954330

Ajorlou, E., & Khosroushahi, A. Y. (2017). Trends on polymer-and lipid-based nanostructures for parenteral drug delivery to tumors. *Cancer Chemotherapy and Pharmacology*, *79*(2), 251–265. DOI: 10.1007/s00280-016-3168-6 PMID: 27744564

Akpe, V., Cock, I. E., Telukutla, S. R., & Shah, N. (2025). Advances in cancer treatment through nanotheranostics and emerging therapies. *Drugs and Drug Candidates*, *6*(4), 29. DOI: 10.3390/ddc6040029

Al Jayoush, A. R., Haider, M., Khan, S. A., & Hussain, Z. (2025). Hyaluronic acid-functionalized nanomedicines for CD44-receptors-mediated targeted cancer therapy: A review of selective targetability and biodistribution to the tumor microenvironment. *International Journal of Biological Macromolecules*, *308*(Pt 2), 142486. DOI: 10.1016/j.ijbiomac.2025.142486 PMID: 40139601

Al Refaai, K. A., AlSawaftah, N. A., Abuwatfa, W., & Husseini, G. A. (2024). Drug release via ultrasound-activated nanocarriers for cancer treatment: A review. *Pharmaceutics*, *16*(11), 1383. DOI: 10.3390/pharmaceutics16111383 PMID: 39598507

Al-asbahi, M. G. S. S., Al-Ofiry, B. A., Saad, F. A. A., Alnehia, A., & Al-Gunaid, M. Q. A. (2024). Silver nanoparticles biosynthesis using mixture of Lactobacillus sp. And Bacillus sp. Growth and their antibacterial activity. *Scientific Reports*, *14*(1), 10224. DOI: 10.1038/s41598-024-59936-1 PMID: 38702368

Albertoni, R., Colantonio, S., & Stefanowski, J. (2023). Reproducibility of machine learning: Terminology, recommendations and open issues. *arXiv preprint arXiv:2302.12691*.

Aldoseri, A., Al-Khalifa, K. N., & Hamouda, A. M. (2023). Re-thinking data strategy and integration for artificial intelligence: Concepts, opportunities, and challenges. *Applied Sciences (Basel, Switzerland)*, *13*(12), 7082. DOI: 10.3390/app13127082

Alfryyan, N., Kordy, M. G. M., Abdel-Gabbar, M., Soliman, H. A., & Shaban, M. (2022). Characterization of the biosynthesized intracellular and extracellular plasmonic silver nanoparticles using Bacillus cereus and their catalytic reduction of methylene blue. *Scientific Reports*, *12*(1), 12495. DOI: 10.1038/s41598-022-16029-1 PMID: 35864132

Alhazza, A., Oyegbesan, A., Bousoik, E., & Montazeri Aliabadi, H. (2025). Multidrug resistance: Are we still afraid of the big bad wolf. *Pharmaceuticals (Basel, Switzerland), 18*(6), 895. DOI: 10.3390/ph18060895 PMID: 40573290

Ali Dheyab, M., Tang, J. H., Abdul Aziz, A., Hussein Nowfal, S., Jameel, M. S., Alrosan, M., Oladzadabbasabadi, N., & Ghasemlou, M. (2025). Green synthesis of gold nanoparticles and their emerging applications in cancer imaging and therapy: A review. *Reviews in Inorganic Chemistry*, *45*(4), 663–685. DOI: 10.1515/revic-2024-0048

Ali Mohammad, S., Hak, A., Pogu, S. V., & Rengan, A. K. (2023). Radiotherapy, photodynamic therapy, and cryoablation-induced abscopal effect: Challenges and future prospects. *Cancer Innovation*, *2*(5), 323–345. 10.1002/cai2.53. DOI: 10.1002/cai2.53 PMID: 38090387

Ali, B. A., Allam, R. M., Hasanin, M. S., & Hassabo, A. A. (2024). Biosynthesis of selenium nanoparticles as a potential therapeutic agent in breast cancer: G2/M arrest and apoptosis induction. *Toxicology Reports*, *13*, 101792. DOI: 10.1016/j.toxrep.2024.101792 PMID: 39554610

Ali, H. (2023). Artificial intelligence in multi-omics data integration: Advancing precision medicine, biomarker discovery and genomic-driven disease interventions. *Int J Sci Res Arch*, *8*(1), 1012–1030. DOI: 10.30574/ijsra.2023.8.1.0189

Ali, K., Ahmed, B., Dwivedi, S., Saquib, Q., Al-Khedhairy, A. A., & Musarrat, J. (2015). Microwave Accelerated Green Synthesis of Stable Silver Nanoparticles with Eucalyptus globulus Leaf Extract and Their Antibacterial and Antibiofilm Activity on Clinical Isolates. *PLoS One*, *10*(7), e0131178. DOI: 10.1371/journal.pone.0131178 PMID: 26132199

Ali, M., Azad, M. A. K., Khan, K. A., Rahman, M. O., Chakma, U., & Kumer, A. (2023). Analysis of Crystallographic Structures and Properties of Silver Nanoparticles Synthesized Using PKL Extract and Nanoscale Characterization Techniques. *ACS Omega*, *8*(31), 28133–28142. DOI: 10.1021/acsomega.3c01261 PMID: 37576647

Ali, M., Benfante, V., Basirinia, G., Alongi, P., Sperandeo, A., Quattrocchi, A., Giannone, A. G., Cabibi, D., Yezzi, A., Di Raimondo, D., Tuttolomondo, A., & Comelli, A. (2025). Applications of artificial intelligence, deep learning, and machine learning to support the analysis of microscopic images of cells and tissues. *Journal of Imaging*, *11*(2), 59. DOI: 10.3390/jimaging11020059 PMID: 39997561

Ali, Z., Vildevall, M., Rodriguez, G., Tandiono, D., Vamvakaris, I., Evangelou, G., Lolas, G., Syrigos, K., Villanueva, A., Wick, M., Omar, S., Erkstam, A., Schueler, J., Fahlgren, A., & Jensen, L. (2022). Zebrafish patient-derived xenograft models predict lymph node involvement and treatment outcome in non-small cell lung cancer. *Journal of Experimental & Clinical Cancer Research : CR*, *41*(1), 58. Advance online publication. DOI: 10.1186/s13046-022-02280-x PMID: 35139880

Al-Jumaili, A., Alancherry, S., Bazaka, K., & Jacob, M. V. (2017). Review on the antimicrobial properties of carbon nanostructures. *Materials (Basel)*, *10*(9), 1066. DOI: 10.3390/ma10091066 PMID: 28892011

Al-Mahmud, M., Hossain Shishir, M. K., Ahmed, S., Tabassum, S., Islam Sadia, S., Haque Sachchu, M. M., Tanbin Tama, R., Rahim Miah, A., & Ashraful Alam, M. (2024). Stoichiometry crystallographic phase analysis and crystallinity integration of silver nanoparticles: A Rietveld refinement study. *Journal of Crystal Growth*, *643*, 127815. DOI: 10.1016/j.jcrysgro.2024.127815

Almeida, A., Fernandes, E., Sarmento, B., & Lúcio, M. (2021). A biophysical insight of camptothecin biodistribution: Towards a molecular understanding of its pharmacokinetic issues. *Pharmaceutics*, *13*(6), 869. DOI: 10.3390/pharmaceutics13060869 PMID: 34204692

Almontasser, A., Al-Mufti, S. M. S., & Arya, R. K. (2025). Nanofillers in Drug Delivery Industry: Applications. In Mallakpour, S., & Hussain, C. M. (Eds.), *Handbook of Nanofillers* (pp. 1479–1519). Springer Nature Singapore., DOI: 10.1007/978-981-96-2407-2_78

Al-Omari, A., Altalla, B., Alayyan, O., Abdel-Razeq, H., & Mansour, A. (2025). *Barriers to Cancer Research in Low- and Middle-Income Countries: Findings from the Arab Region and Beyond*. DOI: 10.21203/rs.3.rs-7184575/v1

Alsamhary, K. I. (2020). Eco-friendly synthesis of silver nanoparticles by Bacillus subtilis and their antibacterial activity. *Saudi Journal of Biological Sciences*, *27*(8), 2185–2191. DOI: 10.1016/j.sjbs.2020.04.026 PMID: 32714045

Alsawaftah, N., Awad, N. S., Pitt, W., & Husseini, G. (2022). pH-Responsive Nanocarriers in Cancer Therapy. *Polymers*, *14*(5), 936. Advance online publication. DOI: 10.3390/polym14050936 PMID: 35267759

Alsehli, M. (2020). Polymeric nanocarriers as stimuli-responsive systems for targeted tumor (cancer) therapy: Recent advances in drug delivery. *Saudi Pharmaceutical Journal : SPJ : The Official Publication of the Saudi Pharmaceutical Society*, *28*(3), 255–265. https://doi.org/https://doi.org/10.1016/j.jsps.2020.01.004. DOI: 10.1016/j.jsps.2020.01.004 PMID: 32194326

Al-Shadidi, J. R. M. H., Al-Shammari, S., Al-Mutairi, D., Alkhudhair, D., Thu, H. E., & Hussain, Z. (2024). Chitosan nanoparticles for targeted cancer therapy: A review of stimuli-responsive, passive, and active targeting strategies. *International Journal of Nanomedicine*, *19*, 8373–8400. DOI: 10.2147/IJN.S472433 PMID: 39161363

Alsheleh, T., Zraikat, M., Daoud, F., Alqudah, D., Abdelghany, S., Siniyeh, A., & Alshaer, W. (2025). In vitro cellular interaction of drug-loaded liposomes with 2D and 3D cell culture. *PLoS One*, *20*(3), e0320374. Advance online publication. DOI: 10.1371/journal.pone.0320374 PMID: 40131912

Alsubaie, F., Alshammari, B., Grira, S., Mohammad Sayem Mozumder, A. S., & Mourad, A.-H. I. (2026). Recent advances, methods, and applications of copper-mediated Atom Transfer Radical Polymerization (ATRP) for precise block copolymers synthesis. *Materials Today. Communications*, *53*, 115385. https://doi.org/https://doi.org/10.1016/j.mtcomm.2026.115385. DOI: 10.1016/j.mtcomm.2026.115385

Alvandi, H., Shafie, A., Najafi, F., Sabzini, M., Mashayekhi, M., Omami, S. H., Eskandarisani, M. M., Dashti, S., Javanmard, A., Tajik, M., Bräse, S., & Farmani, A. R. (2025). Carbon-based nanostructure drug delivery systems and their biologic applications–a review. *Carbon Letters*, *35*(4), 1529–1587. DOI: 10.1007/s42823-025-00927-1

Alvarez, N., & Sevilla, A. (2024). Current advances in photodynamic therapy (PDT) and the future potential of PDT-combinatorial cancer therapies. *International Journal of Molecular Sciences*, *25*(2), 1023. DOI: 10.3390/ijms25021023 PMID: 38256096

Amaral, M. N., Kumar, P., Faísca, P., Ferreira, H. A., Coelho, J. M. P., Gaspar, M. M., & Reis, C. P. (2025). Gold nanoparticle-mediated photothermal therapy: Expanding the frontiers of cancer treatment and theragnostics. *Biomedicine & Pharmacotherapy = Biomédecine & Pharmacothérapie*, *190*, 118399. DOI: 10.1016/j.biopha.2025.118399 PMID: 40784123

Amarante-Mendes, G. P., Rana, A., Datoguia, T. S., Hamerschlak, N., & Brumatti, G. (2022). BCR-ABL1 Tyrosine Kinase Complex Signaling Transduction: Challenges to Overcome Resistance in Chronic Myeloid Leukemia. *Pharmaceutics*, *14*(1), 215. DOI: 10.3390/pharmaceutics14010215 PMID: 35057108

Ambekar, R. S., Choudhary, M., & Kandasubramanian, B. (2020). Recent advances in dendrimer-based nanoplatform for cancer treatment: A review. *European Polymer Journal*, *126*, 109546. DOI: 10.1016/j.eurpolymj.2020.109546

Ameen, F., Almalki, N. S., Alshalan, R., & Sakayanathan, P. (2025). Green Synthesis of Selenium Nanoparticles Utilizing *Drimia indica*: Insights Into Anticancer and Antimicrobial Activities. *Microscopy Research and Technique*, *88*(3), 749–760. DOI: 10.1002/jemt.24726 PMID: 39558605

Amin, M., Huang, W., Seynhaeve, A. L. B., & ten Hagen, T. L. M. (2020). Hyperthermia and Temperature-Sensitive Nanomaterials for Spatiotemporal Drug Delivery to Solid Tumors. *Pharmaceutics*, *12*(11), 1007. DOI: 10.3390/pharmaceutics12111007 PMID: 33105816

Amirinezhadfard, E., Niazi Tabar, A., Xia, L.-J., & Yang, W.-C. (2025). Plant-derived nanomaterials in cancer therapy: Advances from green synthesis to application of PANoptosis-mediated tumor suppression. *Industrial Crops and Products*, *236*, 121850. DOI: 10.1016/j.indcrop.2025.121850

Ammar, M. M., Ali, R., Abd Elaziz, N. A., Habib, H., Abbas, F. M., Yassin, M. T., Maniah, K., & Abdelaziz, R. (2025). Nanotechnology in oncology: Advances in biosynthesis, drug delivery, and theranostics. *Discover. Oncology*, *16*(1), 1172. DOI: 10.1007/s12672-025-02664-3 PMID: 40542991

Anand, U., Dey, A., Chandel, A. K. S., Sanyal, R., Mishra, A., Pandey, D. K., De Falco, V., Upadhyay, A., Kandimalla, R., Chaudhary, A., Dhanjal, J. K., Dewanjee, S., Vallamkondu, J., & Pérez de la Lastra, J. M. (2023). Cancer chemotherapy and beyond: Current status, drug candidates, associated risks and progress in targeted therapeutics. *Genes & Diseases*, *10*(4), 1367–1401. 10.1016/j.gendis.2022.02.007. DOI: 10.1016/j.gendis.2022.02.007 PMID: 37397557

Anarjan, F. S. (2019). Active targeting drug delivery nanocarriers: Ligands. *Nano-Structures & Nano-Objects*, *19*, 100370. DOI: 10.1016/j.nanoso.2019.100370

Anas, M., Khan, A., Falak, A., Shinwari, Z. K., Mohamed, H. E. A., Hkiri, K., Maaza, M., Fahad, S., & Quraishi, U. M. (2025). Harnessing Biomedical Applications of Ascorbic Acid, Berberine, and Glutathione-Capped Quantum Dots Through Multifunctional Physico-chemical, Biochemical, and Anti-microbial Assays. *BioNanoScience*, *15*(2), 259. DOI: 10.1007/s12668-025-01864-z

Andrés, C. M., Pérez de la Lastra, J. M., Munguira, E. B., Andrés Juan, C., & Pérez-Lebeña, E. (2024). Dual-Action Therapeutics: DNA Alkylation and Antimicrobial Peptides for Cancer Therapy. *Cancers (Basel)*, *16*(18), 3123. DOI: 10.3390/cancers16183123 PMID: 39335095

Ang, W. H., Khalaila, I., Allardyce, C. S., Juillerat-Jeanneret, L., & Dyson, P. J. (2005). Rational Design of Platinum(IV) Compounds to Overcome Glutathione-S-Transferase Mediated Drug Resistance. *Journal of the American Chemical Society*, *127*(5), 1382–1383. DOI: 10.1021/ja0432618 PMID: 15686364

Anjum, S., Hashim, M., Imran, M., Babur, S., Adnan, S., Hano, C., & Ibrahim, W. N. (2025). Selenium Nanoparticles in Cancer Therapy: Unveiling Cytotoxic Mechanisms and Therapeutic Potential. *Cancer Reports*, *8*(6), e70210. DOI: 10.1002/cnr2.70210 PMID: 40452566

Anzalone, A., Koblan, L., & Liu, D. (2020). Genome editing with CRISPR–Cas nucleases, base editors, transposases and prime editors. *Nature Biotechnology*, *38*(7), 824–844. DOI: 10.1038/s41587-020-0561-9 PMID: 32572269

Appidi, T.. (2025). Nanoparticle-based triple therapy: A three-in-one approach for cancer treatment. *Molecular Cancer*, *24*, 119. DOI: 10.1186/s12943-025-02266-1 PMID: 40050802

Arif, U., Haider, S., Haider, A., Khan, N., Alghyamah, A. A., Jamila, N., Khan, M. I., Almasry, W. A., & Kang, I.-K. (2019). Biocompatible polymers and their potential biomedical applications: A review. *Current Pharmaceutical Design*, *25*(34), 3608–3619. DOI: 10.2174/1381612825999191011105148 PMID: 31604409

Arif, Z. U., Khalid, M., Noroozi, R., Hossain, M., Shi, H., Tariq, A., Ramakrishna, S., & Umer, R. (2023). Additive manufacturing of sustainable biomaterials for biomedical applications. *Asian Journal of Pharmaceutical Sciences*, *18*(3), 100812. Advance online publication. DOI: 10.1016/j.ajps.2023.100812 PMID: 37274921

Armenia, I., Ayllón, C. C., Herrero, B. T., Bussolari, F., Alfranca, G., Grazú, V., & De La Fuente, M. J. (2022). Photonic and magnetic materials for on-demand local drug delivery. *Advanced Drug Delivery Reviews*, *114584*, 114584. Advance online publication. DOI: 10.1016/j.addr.2022.114584 PMID: 36273514

Arpicco, S., Bartkowski, M., Barge, A., Zonari, D., Serpe, L., Milla, P., Dosio, F., Stella, B., & Giordani, S. (2020). Effects of the molecular weight of hyaluronic acid in a carbon nanotube drug delivery conjugate. *Frontiers in Chemistry*, *8*, 578008. DOI: 10.3389/fchem.2020.578008 PMID: 33381490

Arshad, R., Hassan, D., Sani, A., Mustafa, G., Rahdar, A., Fathi-karkan, S., Kharaba, Z., Medina, D. I., & Pandey, S. (2024). Nano-engineered solutions for ibuprofen therapy: Unveiling advanced co-delivery strategies and nanoparticle systems. *Journal of Drug Delivery Science and Technology*, *98*, 105815. DOI: 10.1016/j.jddst.2024.105815

Arunkumar, P., Vedagiri, H., & Premkumar, K. (2013). Rapid bioreduction of trivalent aurum using banana stem powder and its cytotoxicity against MCF-7 and HEK-293 cell lines. *Journal of Nanoparticle Research, 15*(3), 1481. DOI: 10.1007/s11051-013-1481-0

Asaadi, Y., Jouneghani, F. F., Janani, S., & Rahbarizadeh, F. (2021). A comprehensive comparison between camelid nanobodies and single chain variable fragments. *Biomarker Research, 9*(1), 87. DOI: 10.1186/s40364-021-00332-6 PMID: 34863296

Ashique, S., Garg, A., Mishra, N., Raina, N., Ming, L. C., Tulli, H. S., Behl, T., Rani, R., & Gupta, M. (2023). Nano-mediated strategy for targeting and treatment of non-small cell lung cancer (NSCLC). *Naunyn-Schmiedeberg's Archives of Pharmacology, 396*(11), 2769–2792. DOI: 10.1007/s00210-023-02522-5 PMID: 37219615

Ashrafizadeh, M., Zarrabi, A., Bigham, A., Taheriazam, A., Saghari, Y., Mirzaei, S., Hashemi, M., Hushmandi, K., Karimi-Maleh, H., Zare, E. N., Sharifi, E., Ertas, Y. N., Rabiee, N., Sethi, G., & Shen, M. (2023). (Nano)platforms in breast cancer therapy: Drug/gene delivery, advanced nanocarriers and immunotherapy. *Medicinal Research Reviews, 43*(6), 2115–2176. DOI: 10.1002/med.21971 PMID: 37165896

Attri, N., Das, S., Banerjee, J., Shamsuddin, S. H., Dash, S. K., & Pramanik, A. (2024). Liposomes to cubosomes: The evolution of lipidic nanocarriers and their cutting-edge biomedical applications. *ACS Applied Bio Materials, 7*(5), 2677–2694. DOI: 10.1021/acsabm.4c00153 PMID: 38613498

Avramović, N., Mandić, B., Savić-Radojević, A., & Simić, T. (2020). Polymeric Nanocarriers of Drug Delivery Systems in Cancer Therapy. *Pharmaceutics, 12*(4), 298. Advance online publication. DOI: 10.3390/pharmaceutics12040298 PMID: 32218326

Ayub, A., Wani, A. K., Malik, S. M., Ayub, M., Singh, R., Chopra, C., & Malik, T. (2025). Green nanoscience for healthcare: Advancing biomedical innovation through eco-synthesized nanoparticle. *Biotechnology Reports (Amsterdam, Netherlands), 47*, e00913. 10.1016/j.btre.2025.e00913. DOI: 10.1016/j.btre.2025.e00913 PMID: 40895589

Azimizonuzi, H., Ghayourvahdat, A., Ahmed, M. H., Kareem, R. A., Zrzor, A. J., Mansoor, A. S., Athab, Z. H., & Kalavi, S. (2025). A state-of-the-art review of the recent advances of theranostic liposome hybrid nanoparticles in cancer treatment and diagnosis. *Cancer Cell International, 25*(1), 26. DOI: 10.1186/s12935-024-03610-z PMID: 39871316

Azmi, F., Xu, X., Duong, H., Ye, P., Chen, T., Li, H., Chen, J., Ardekani, S. M., Dehghani, A., Zheng, G., Harris, D., Lu, H., Wang, Y., & Cao, Q. (2025). Renal clearable sucrose carbon dots for doxorubicin delivery to treat renal carcinoma. *Nanoscale Advances*, *7*(9), 2751–2760. DOI: 10.1039/D4NA01082E PMID: 40160256

Badgugar, L., Maru, S., Sisodiya, N., Maru, R., & Sugunan, S. (2026). Chapter Thirteen - Safety, ethical, and regulatory considerations in nanotechnology for skin cancer. In Kumar Singh, A., Kaur, R., Pareek, A., Maru, S., & Prakash, A. (Eds.), *Advances in Cancer Research* (Vol. 169, pp. 381–415). Academic Press., ,,10.1016/bs.acr.2026.02.008

Bagheri, B., Surwase, S. S., Lee, S. S., Park, H., Rad, Z. F., Trevaskis, N. L., & Kim, Y.-C. (2022). Carbon-based nanostructures for cancer therapy and drug delivery applications. *Journal of Materials Chemistry. B*, *10*(48), 9944–9967. DOI: 10.1039/D2TB01741E PMID: 36415922

Baharara, J., Ramezani, T., Divsalar, A., Mousavi, M., & Seyedarabi, A. (2016). *Induction of Apoptosis by Green Synthesized Gold Nanoparticles Through Activation of Caspase-3 and 9 in Human Cervical Cancer Cells.*

Bahuguna, S., Kumar, M., Sharma, G., Kumar, R., Singh, B., & Raza, K. (2018). Fullerenol-based intracellular delivery of methotrexate: A water-soluble nanoconjugate for enhanced cytotoxicity and improved pharmacokinetics. *AAPS PharmSciTech*, *19*(3), 1084–1092. DOI: 10.1208/s12249-017-0920-0 PMID: 29159749

Baig, M. M. F. A., Chien, W. T., & Chair, S. Y. (2025). Nanotechnological approaches for the targeted delivery of CRISPR-Cas systems for genomic modifications, biomolecular sensing, and precision medicine. *Biomaterials Science*, *13*(17), 4597–4638. DOI: 10.1039/D5BM00711A PMID: 40704512

Bai, X., Smith, Z. L., Wang, Y., Butterworth, S., & Tirella, A. (2022). Sustained Drug Release from Smart Nanoparticles in Cancer Therapy: A Comprehensive Review. *Micromachines*, *13*(10), 1623. Advance online publication. DOI: 10.3390/mi13101623 PMID: 36295976

Bajracharya, R., Song, J. G., Patil, B. R., Lee, S. H., Noh, H. M., Kim, D. H., Kim, G. L., Seo, S. H., Park, J. W., Jeong, S. H., Lee, C. H., & Han, H. K. (2022). Functional ligands for improving anticancer drug therapy: Current status and applications to drug delivery systems. *Drug Delivery*, *29*(1), 1959–1970. DOI: 10.1080/10717544.2022.2089296 PMID: 35762636

Balasubramanian, A., Veluswami, K., Rao, S., Aggarwal, S., & Mani, S. (2024). Exploring Clustered Regularly Interspaced Short Palindromic Repeats-CRISPR-Associated Protein 9 (CRISPR-Cas9) as a therapeutic modality for cancer: A scoping review. *Cureus, 16.* Advance online publication. DOI: 10.7759/cureus.64324 PMID: 39130943

Bala, V. M., Lampropoulou, D. I., Grammatikaki, S., Spanoudaki, M., Zambouli, D., Magkouta, S., Lagopati, N., Gazouli, M., & Efstathopoulos, E. P. (2023). Nanoparticle-mediated hyperthermia and cytotoxicity mechanisms in cancer. *International Journal of Molecular Sciences*, *25*(1), 296. DOI: 10.3390/ijms25010296 PMID: 38203467

Baldea, I., Iacovi ă, C., Gurgu, R. A., Vizitiu, A. S., Râzniceanu, V., & Mitrea, D. R. (2025). Magnetic hyperthermia with iron oxide nanoparticles: From toxicity challenges to cancer applications. *Nanomaterials (Basel, Switzerland)*, *15*(19), 1519. DOI: 10.3390/nano15191519 PMID: 41090863

Banda, A., Impomeni, O., Singh, A., Baloch, A. R., Hu, W., & Jaijyan, D. (2024). Precision in action: The role of Clustered Regularly Interspaced Short Palindromic Repeats/Cas in gene therapies. *Vaccines*, *12*(6), 636. DOI: 10.3390/vaccines12060636 PMID: 38932365

Bang, J. H., & Suslick, K. S. (2010). Applications of ultrasound to the synthesis of nanostructured materials. *Advanced Materials*, *22*(10), 1039–1059. DOI: 10.1002/adma.200904093 PMID: 20401929

Bansal, D. (2025). Personalized medicine in oncology: advancements, difficulties, and prospects. *International journal of therapeutic innovation, 3*(4).

Bartkowski, M., & Giordani, S. (2021). Carbon nano-onions as potential nanocarriers for drug delivery. *Dalton Transactions : An International Journal of Inorganic Chemistry*, *50*(7), 2300–2309. DOI: 10.1039/D0DT04093B PMID: 33471000

Basety, S., Gudepu, R., & Velidandi, A. (2026). Smart Nanoformulations for Oncology: A Review on Overcoming Biological Barriers with Active Targeting, Stimuli-Responsive, and Controlled Release for Effective Drug Delivery. *Pharmaceutics*, *18*(2), 196. Advance online publication. DOI: 10.3390/pharmaceutics18020196 PMID: 41754938

Batool, S., Sohail, S., ud Din, F., Alamri, A. H., Alqahtani, A. S., Alshahrani, M. A., Alshehri, M. A., & Choi, H. G. (2023). A detailed insight of the tumor targeting using nanocarrier drug delivery system. *Drug Delivery*, *30*(1), 2183815. DOI: 10.1080/10717544.2023.2183815 PMID: 36866455

Baust, J. G., Santucci, K. L., Snyder, K. K., Robilotto, A., & Baust, J. M. (2023). Mechanisms of Tissue Injury in Cryosurgery. In *The Application of Heat in Oncology* (pp. 45-71). 10.1002/9781119799627.ch4

Bazak, R., Houri, M., El Achy, S., Hussein, W., & Refaat, T. (2014). Passive targeting of nanoparticles to cancer: A comprehensive review of the literature. *Molecular and Clinical Oncology*, *2*(6), 904–908. DOI: 10.3892/mco.2014.356 PMID: 25279172

Beach, E. S., Cui, Z., & Anastas, P. T. (2009). Green Chemistry: A design framework for sustainability. *Energy & Environmental Science*, *2*(10), 1038–1049. DOI: 10.1039/b904997p

Beach, M. A., Nayanathara, U., Gao, Y., Zhang, C., Xiong, Y., Wang, Y., & Such, G. K. (2024). Polymeric Nanoparticles for Drug Delivery. *Chemical Reviews*, *124*(9), 5505–5616. DOI: 10.1021/acs.chemrev.3c00705 PMID: 38626459

Beh, C. Y., Prajnamitra, R. P., Chen, L. L., & Hsieh, P. C. (2021). Advances in Biomimetic Nanoparticles for Targeted Cancer Therapy and Diagnosis. *Molecules (Basel, Switzerland)*, *26*(16), 5052. Advance online publication. DOI: 10.3390/molecules26165052 PMID: 34443638

Bell, B. I., Vercellino, J., Brodin, N. P., Velten, C., Nanduri, L. S. Y., Nagesh, P. K. B., Tanaka, K. E., Fang, Y., Wang, Y., Macedo, R., English, J., Schumacher, M. M., Duddempudi, P. K., Asp, P., Koba, W., Shajahan, S., Liu, L., Tomé, W. A., Yang, W.-L., & Guha, C. (2022). Orthovoltage X-Rays Exhibit Increased Efficacy Compared with γ-Rays in Preclinical Irradiation. *Cancer Research*, *82*(15), 2678–2691. DOI: 10.1158/0008-5472.CAN-22-0656 PMID: 35919990

Beltrán-Gracia, E., López-Camacho, A., Higuera-Ciapara, I., Velázquez-Fernández, J. B., & Vallejo-Cardona, A. A. (2019). Nanomedicine review: Clinical developments in liposomal applications. *Cancer Nanotechnology*, *10*(1), 11. DOI: 10.1186/s12645-019-0055-y

Bendre, A., Bhat, M. P., Lee, K.-H., Altalhi, T., Alruqi, M. A., & Kurkuri, M. (2022). Recent developments in microfluidic technology for synthesis and toxicity-efficiency studies of biomedical nanomaterials. *Materials Today. Advances*, *13*, 100205. DOI: 10.1016/j.mtadv.2022.100205

Betke, A., & Kickelbick, G. (2014). Bottom-Up, Wet Chemical Technique for the Continuous Synthesis of Inorganic Nanoparticles. *Inorganics*, *2*(1), 1–15. DOI: 10.3390/inorganics2010001

Bevilacqua, G., Corvino, R., Capriotti, A. L., Montone, C. M., Moriconi, M., Salciccia, S., Brunelli, V., Santarelli, V., Sciarra, B., Laganà, A., Santini, D., Sciarra, A., & Gentilucci, A. (2024). The protein corona on nanoparticles for tumor targeting in prostate cancer: A review of the literature and experimental trial protocol. *Biology (Basel), 13*(12), 1024. DOI: 10.3390/biology13121024 PMID: 39765691

Bezelya, A., Kucukturkmen, B., & bozkır, A. (2023). *The Use of Microfluidic Systems for Nanoparticle Production*. ,DOI: 10.20944/preprints202309.1452.v1

Bhalani, D. V., Nutan, B., Kumar, A., & Singh Chandel, A. K. (2022). Bioavailability enhancement techniques for poorly aqueous soluble drugs and therapeutics. *Biomedicines, 10*(9), 2055. DOI: 10.3390/biomedicines10092055 PMID: 36140156

Bharadwaj, K., Rabha, B., Pati, S., Sarkar, T., Choudhury, B., Barman, A., Bhattacharjya, D., Srivastava, A., Baishya, D., Edinur, H. A., Abdul Kari, Z., & Mohd Noor, N. H. (2021). Green synthesis of gold nanoparticles using plant extracts as beneficial prospect for cancer theranostics. *Molecules (Basel, Switzerland), 26*(21), 6389. Advance online publication. DOI: 10.3390/molecules26216389 PMID: 34770796

Bhatia, S. N., & Ingber, D. E. (2014). Microfluidic organs-on-chips. *Nature Biotechnology*, *32*(8), 760–772. DOI: 10.1038/nbt.2989 PMID: 25093883

Bhattacharya, T., Preetam, S., Ghosh, B., Chakrabarti, T., Chakrabarti, P., Samal, S. K., & Thorat, N. (2023). Advancement in biopolymer-assisted cancer theranostics. *ACS Applied Bio Materials*, *6*(9), 3488–3510. DOI: 10.1021/acsabm.3c00458 PMID: 37699558

Bhilare, N. V., Marulkar, V. S., Chatap, V. K., Kulkarni, A. S., Dombe, S. A., & Shirote, P. J. (2024). Biodegradability and Biodegradable Polymers: An Overview. *Handbook of Biodegradable Polymers*, 1-38.

Biju, V. (2014). Chemical modifications and bioconjugate reactions of nanomaterials for sensing, imaging, drug delivery and therapy. *Chemical Society Reviews*, *43*(3), 744–764. DOI: 10.1039/C3CS60273G PMID: 24220322

Bindhu, M. R., & Umadevi, M. (2014). Silver and gold nanoparticles for sensor and antibacterial applications. *Spectrochimica Acta. Part A, Molecular and Biomolecular Spectroscopy*, *128*, 37–45. DOI: 10.1016/j.saa.2014.02.119 PMID: 24657466

Bizuayehu, H., Ahmed, K., Kibret, G., Dadi, A., Belachew, S., Bagade, T., Tegegne, T., Venchiarutti, R., Kibret, K., Hailegebireal, A., Assefa, Y., Khan, M., Abajobir, A., Alene, K., Mengesha, Z., Erku, D., Enquobahrie, D., Minas, T., Misgan, E., & Ross, A. (2024). Global disparities of cancer and its projected burden in 2050. *JAMA Network Open*, *7*(4), e2443198. DOI: 10.1001/jamanetworkopen.2024.43198 PMID: 39499513

Björnsson, B., Borrebaeck, C., Elander, N., Gasslander, T., Gawel, D. R., Gustafsson, M., Jörnsten, R., Lee, E. J., Li, X., Lilja, S., Martínez-Enguita, D., Matussek, A., Sandström, P., Schäfer, S., Stenmarker, M., Sun, X. F., Sysoev, O., Zhang, H., & Benson, M.Swedish Digital Twin. (2019). Digital twins to personalize medicine. *Genome Medicine*, *12*(1), 4. DOI: 10.1186/s13073-019-0701-3 PMID: 31892363

Blackwell, J., Kraśny, M. J., O'Brien, A., Ashkan, K., Galligan, J., Destrade, M., & Colgan, N. (2022). Proton Resonance Frequency Shift Thermometry: A Review of Modern Clinical Practices. *Journal of Magnetic Resonance Imaging*, *55*(2), 389–403. DOI: 10.1002/jmri.27446 PMID: 33217099

Blagojevic, L., & Kamaly, N. (2025). Nanogels: A chemically versatile drug delivery platform. *Nano Today*, *61*, 102645. https://doi.org/https://doi.org/10.1016/j.nantod.2025.102645. DOI: 10.1016/j.nantod.2025.102645

Blank, C., Gajewski, T. F., & Mackensen, A. (2005). Interaction of PD-L1 on tumor cells with PD-1 on tumor-specific T cells as a mechanism of immune evasion: Implications for tumor immunotherapy. *Cancer Immunology, Immunotherapy : CII*, *54*(4), 307–314. DOI: 10.1007/s00262-004-0593-x PMID: 15599732

Boix-Montesinos, P., Soriano-Teruel, P., Armiñán, A., Orzáez, M., & Vicent, M. (2021). The past, present, and future of breast cancer models for nanomedicine development. *Advanced Drug Delivery Reviews*, *173*, 306–330. DOI: 10.1016/j.addr.2021.03.018 PMID: 33798642

Boros, E., Dyson, P. J., & Gasser, G. (2020). Classification of Metal-Based Drugs according to Their Mechanisms of Action. *Chem*, *6*(1), 41–60. DOI: 10.1016/j.chempr.2019.10.013 PMID: 32864503

Boulaiz, H., Alvarez, P. J., Ramirez, A., Marchal, J. A., Prados, J., Rodríguez-Serrano, F., Perán, M., Melguizo, C., & Aranega, A. (2011). Nanomedicine: Application areas and development prospects. *International Journal of Molecular Sciences*, *12*(5), 3303–3321. DOI: 10.3390/ijms12053303 PMID: 21686186

Bousbaa, H. (2021). Novel Anticancer Strategies. *Pharmaceutics*, *13*(2), 275. Advance online publication. DOI: 10.3390/pharmaceutics13020275 PMID: 33670469

Bozzer, S., Dal Bo, M., Grimaldi, M. C., Toffoli, G., & Macor, P. (2022). Nanocarriers as a Delivery Platform for Anticancer Treatment: Biological Limits and Perspectives in B-Cell Malignancies. *Pharmaceutics, 14*(9), 1965. https://www.mdpi.com/1999-4923/14/9/1965. DOI: 10.3390/pharmaceutics14091965 PMID: 36145713

Bravo, M., Fortuni, B., Mulvaney, P., Hofkens, J., Uji-i, H., Rocha, S., & Hutchison, J. A. (2024). Nanoparticle-mediated thermal cancer therapies: Strategies to improve clinical translatability. *Journal of Controlled Release : Official Journal of the Controlled Release Society, 372*, 751–777. DOI: 10.1016/j.jconrel.2024.06.055 PMID: 38909701

Bray, F., Jemal, A., Grey, N., Ferlay, J., & Forman, D. (2012). Global cancer transitions according to the Human Development Index (2008–2030): A population-based study. *The Lancet. Oncology, 13*(8), 790–801. DOI: 10.1016/S1470-2045(12)70211-5 PMID: 22658655

Bray, F., Laversanne, M., Sung, H., Ferlay, J., Siegel, R. L., Soerjomataram, I., & Jemal, A. (2024). Global cancer statistics 2022: GLOBOCAN estimates of incidence and mortality worldwide for 36 cancers in 185 countries. *CA: a Cancer Journal for Clinicians, 74*(3), 229–263. DOI: 10.3322/caac.21834 PMID: 38572751

Brown, J. S., Amend, S. R., Austin, R. H., Gatenby, R. A., Hammarlund, E. U., & Pienta, K. J. (2023). Updating the Definition of Cancer. *Molecular Cancer Research, 21*(11), 1142–1147. DOI: 10.1158/1541-7786.MCR-23-0411 PMID: 37409952

Bufalini, C., & Campardelli, R. (2025). Versatile Emulsion-Based Encapsulation System Production Processes: A Review. *Processes (Basel, Switzerland), 13*(5), 1409. DOI: 10.3390/pr13051409

Cai, R., Lv, R., Shi, X., Yang, G., & Jin, J. (2023). CRISPR/dCas9 tools: Epigenetic mechanism and application in gene transcriptional regulation. *International Journal of Molecular Sciences, 24*(19), 14865. DOI: 10.3390/ijms241914865 PMID: 37834313

Chaffer, C. L., San Juan, B. P., Lim, E., & Weinberg, R. A. (2016). EMT, cell plasticity and metastasis. *Cancer Metastasis Reviews, 35*(4), 645–654. DOI: 10.1007/s10555-016-9648-7 PMID: 27878502

Chai, M., Wang, S., Chen, Y., Pei, X., & Zhen, X. (2025). Targeted and intelligent nano-drug delivery systems for colorectal cancer treatment. *Frontiers in Bioengineering and Biotechnology, 13*, 1582659. DOI: 10.3389/fbioe.2025.1582659 PMID: 40352359

Chakraborty, S., Hosen, M. I., Ahmed, M., & Shekhar, H. U. (2018). Onco-multi-OMICS approach: A new frontier in cancer research. *BioMed Research International*, *2018*(1), 9836256. DOI: 10.1155/2018/9836256 PMID: 30402498

Chalamala, S. R., Kummari, N. K., Singh, A. K., Saibewar, A., & Chalavadi, K. M. (2022). Federated learning to comply with data protection regulations. *CSI Transactions on ICT*, *10*(1), 47–60. DOI: 10.1007/s40012-022-00351-0

Chand, B., & Phogat, P. Shreya, Singh, S., & Jha, R. (2025). Regulatory Framework and Standards on Bionanocomposites Manufacturing and Applications: Social Impact and Ethical Considerations. In *Advanced Bionanocomposite Materials: Innovations for Sustainable Development* (pp. 625-667). Springer.

Chandrakala, V., Aruna, V., & Angajala, G. (2022). Review on metal nanoparticles as nanocarriers: Current challenges and perspectives in drug delivery systems. *Emergent Materials*, *5*(6), 1593–1615. DOI: 10.1007/s42247-021-00335-x PMID: 35005431

Chandran, S. P., Chaudhary, M., Pasricha, R., Ahmad, A., & Sastry, M. (2006). Synthesis of Gold Nanotriangles and Silver Nanoparticles Using Aloe vera Plant Extract. *Biotechnology Progress*, *22*(2), 577–583. DOI: 10.1021/bp0501423 PMID: 16599579

Chand, S., Kumar, R., Thakur, N., Kumar, K., Umar, A., Almas, T., & Baskoutas, S. (2025). Sustainable synthesis and multifunctional applications of biowaste-derived carbon nanomaterials and metal oxide composites: A review. *Chemosphere*, *385*, 144540. DOI: 10.1016/j.chemosphere.2025.144540 PMID: 40609513

Chaudhary, S., Jain, V. P., & Jaiswar, G. (2022). The composition of polysaccharides: monosaccharides and binding, group decorating, polysaccharides chains. In *Innovation in nano-polysaccharides for eco-sustainability* (pp. 83–118). Elsevier. DOI: 10.1016/B978-0-12-823439-6.00005-2

Chauhan, A., Arya, A., & Kumar, V.. (2025). The evolution of integrated magnetic hyperthermia and chemodynamic therapy: Multifunctional nanomaterials as precision tools in cancer treatment. *Nanoscale Advances*, *7*, 2139–2157. DOI: 10.1039/D4NA01004C PMID: 40708892

Chauhan, D. S., Prasad, R., Dhanka, M., Kaur, N., Vithalani, H., Liyanapathirana, K., Hebbandi Nanjundappa, R., Gao, H., & Umeshappa, C. S. (2026). Nanoparticles-based phototherapy systems: Molecular mechanisms and clinical applications. *Signal Transduction and Targeted Therapy*, *11*(1), 143. DOI: 10.1038/s41392-025-02536-w PMID: 41839827

Chehelgerdi, M., Chehelgerdi, M., Allela, O. Q. B., Pecho, R. D. C., Jayasankar, N., Rao, D. P., Thamaraikani, T., Vasanthan, M., Viktor, P., Lakshmaiya, N., Saadh, M. J., Amajd, A., Abo-Zaid, M. A., Castillo-Acobo, R. Y., Ismail, A. H., Amin, A. H., & Akhavan-Sigari, R. (2023). Progressing nanotechnology to improve targeted cancer treatment: Overcoming hurdles in its clinical implementation. *Molecular Cancer*, *22*(1), 169. DOI: 10.1186/s12943-023-01865-0 PMID: 37814270

Chehelgerdi, M., Chehelgerdi, M., Khorramian-Ghahfarokhi, M., Shafieizadeh, M., Mahmoudi, E., Eskandari, F., Rashidi, M., Arshi, A., & Mokhtari-Farsani, A. (2024). Comprehensive review of CRISPR-based gene editing: Mechanisms, challenges, and applications in cancer therapy. *Molecular Cancer*, *23*(1), 25. DOI: 10.1186/s12943-023-01925-5 PMID: 38195537

Chen, C., Wang, Z.-K., & Qin, Y. (2023). CRISPR/Cas9 system: Recent applications in immuno-oncology and cancer immunotherapy. *Experimental Hematology & Oncology*, *12*(1), 76. DOI: 10.1186/s40164-023-00457-4 PMID: 37964355

Chen, D., Liu, X. W., Lu, X., & Tian, J. (2023). Nanoparticle drug delivery systems for synergistic delivery of tumor therapy. *Frontiers in Pharmacology*, *14*, 1111991. Advance online publication. DOI: 10.3389/fphar.2023.1111991 PMID: 36874010

Chen, D., Sarkar, S., Candia, J., Florczyk, S. J., Bodhak, S., Driscoll, M. K., Simon, C. G.Jr, Dunkers, J. P., & Losert, W. (2016). Machine learning based methodology to identify cell shape phenotypes associated with microenvironmental cues. *Biomaterials*, *104*, 104–118. DOI: 10.1016/j.biomaterials.2016.06.040 PMID: 27449947

Cheng, Y., Zhao, Y., Chen, C., & Zhang, F. (2025). Heavy Metals Toxicity: Mechanism, Health Effects, and Therapeutic Interventions. *MedComm*, *6*(9), e70241. 10.1002/mco2.70241. DOI: 10.1002/mco2.70241 PMID: 40843132

Cheng, Z., Huang, H., Yin, M., & Liu, H. (2025). Applications of liposomes and lipid nanoparticles in cancer therapy: Current advances and prospects. *Experimental Hematology & Oncology*, *14*(1), 11. DOI: 10.1186/s40164-025-00602-1 PMID: 39891180

Cheng, Z., Li, M., Dey, R., & Chen, Y. (2021). Nanomaterials for cancer therapy. *Journal of Hematology & Oncology*, *14*(1), 85. Advance online publication. DOI: 10.1186/s13045-021-01096-0 PMID: 34059100

Chen, H., Zhang, L., Liu, M., Li, Y., & Rao, P. (2025). Multidimensional perspectives on cancer therapy-induced cardiotoxicity: Characteristics, mechanisms, biomarkers, psychological stress effects, and treatment progress. *Cancer Pathogenesis and Therapy*, *3*, E036–E135. DOI: 10.1016/j.cpt.2025.12.003

Chen, J., Larsson, L., Swarbrick, A., & Lundeberg, J. (2024). Spatial landscapes of cancers: Insights and opportunities. *Nature Reviews. Clinical Oncology*, *21*(9), 660–674. DOI: 10.1038/s41571-024-00926-7 PMID: 39043872

Chen, R., Chen, Y., Liao, M., Lee, Y., Chen, Z., Yan, S., Yeh, Y., Yang, L., Lee, Y., Wu, Y., & Wang, Y. (2020). Autophagy in nanomaterial toxicity. *International Journal of Molecular Sciences*, *21*(7), 2387. Advance online publication. DOI: 10.3390/ijms21072387 PMID: 32235610

Chen, T., Wong, Y.-S., Zheng, W., Bai, Y., & Huang, L. (2008). Selenium nanoparticles fabricated in Undaria pinnatifida polysaccharide solutions induce mitochondria-mediated apoptosis in A375 human melanoma cells. *Colloids and Surfaces. B, Biointerfaces*, *67*(1), 26–31. DOI: 10.1016/j.colsurfb.2008.07.010 PMID: 18805679

Chen, X., Wu, D., & Chen, Z. (2024). Biomedical applications of stimuli-responsive nanomaterials. *MedComm*, *5*(8), e643. Advance online publication. DOI: 10.1002/mco2.643 PMID: 39036340

Chen, Y.-S., Jin, E., & Day, P. J. (2024). Use of drug sensitisers to improve therapeutic index in cancer. *Pharmaceutics*, *16*(7), 928. DOI: 10.3390/pharmaceutics16070928 PMID: 39065625

Chen, Y., Sun, H., Li, Y., Han, X., Yang, Y., Chen, Z., Zhao, X., Qian, Y., Liu, X., Zhou, F., Bai, J., & Qiao, Y. (2025). Magnetic nanomaterials for hyperthermia-based therapy and controlled drug delivery. *Bioactive Materials*, *53*, 591–629. DOI: 10.1016/j.bioactmat.2025.07.033 PMID: 40761549

Chen, Y., Wang, S., & Zhang, F. (2023). Near-infrared luminescence high-contrast in vivo biomedical imaging. *Nature Reviews Bioengineering*, *1*(1), 60–78. DOI: 10.1038/s44222-022-00002-8

Chen, Z. (2010). Small-molecule delivery by nanoparticles for anticancer therapy. *Trends in Molecular Medicine*, *16*(12), 594–602. DOI: 10.1016/j.molmed.2010.08.001 PMID: 20846905

Chen, Z., Wang, X., Zhao, N., Chen, H., & Guo, G. (2023). Advancements in pH-responsive nanocarriers: Enhancing drug delivery for tumor therapy. *Expert Opinion on Drug Delivery*, *20*(11), 1623–1642. DOI: 10.1080/17425247.2023.2292678 PMID: 38059646

Chifiriuc, M. C., Filip, R., Constantin, M., Pircalabioru, G. G., Bleotu, C., Burlibasa, L., Ionica, E., Corcionivoschi, N., & Mihaescu, G. (2022). Common themes in antimicrobial and anticancer drug resistance [Review]. *Frontiers in Microbiology, Volume 13 - 2022*. DOI: 10.3389/fmicb.2022.960693

Chopra, H., Bibi, S., Singh, I., Hasan, M. M., Khan, M. S., Yousafi, Q., Baig, A. A., Rahman, M., Islam, F., Emran, T. B., & Cavalu, S. (2022). Green Metallic Nanoparticles: Biosynthesis to Applications. *Frontiers in Bioengineering and Biotechnology*, *10*, 874742. DOI: 10.3389/fbioe.2022.874742 PMID: 35464722

Cho, S., Shin, J., & Cho, B.-K. (2018). Applications of CRISPR/Cas System to Bacterial Metabolic Engineering. *International Journal of Molecular Sciences*, *19*(4), 1089. DOI: 10.3390/ijms19041089 PMID: 29621180

Chou, W.-C., & Lin, Z. (2024). Impact of protein coronas on nanoparticle interactions with tissues and targeted delivery. *Current Opinion in Biotechnology*, *85*, 103046. DOI: 10.1016/j.copbio.2023.103046 PMID: 38103519

Chuprin, J., Buettner, H., Seedhom, M., Greiner, D., Keck, J., Ishikawa, F., Shultz, L., & Brehm, M. (2023). Humanized mouse models for immuno-oncology research. *Nature Reviews. Clinical Oncology*, *20*(3), 192–206. DOI: 10.1038/s41571-022-00721-2 PMID: 36635480

Chu, S., Shi, X., Tian, Y., & Gao, F. (2022). pH-responsive polymer nanomaterials for tumor therapy. *Frontiers in Oncology*, *12*, 855019. DOI: 10.3389/fonc.2022.855019 PMID: 35392227

Ciftci, F., Özarslan, A. C., Kantarci, İ. C., Yelkenci, A., Tavukcuoglu, O., & Ghorbanpour, M. (2025). Advances in drug targeting, drug delivery, and nanotechnology applications: Therapeutic significance in cancer treatment. *Pharmaceutics*, *17*(1), 121. DOI: 10.3390/pharmaceutics17010121 PMID: 39861768

Cisneros, E. P., Morse, B. A., Savk, A., Malik, K., Peppas, N. A., & Lanier, O. L. (2024). The role of patient-specific variables in protein corona formation and therapeutic efficacy in nanomedicine. *Journal of Nanobiotechnology*, *22*(1), 714. DOI: 10.1186/s12951-024-02954-y PMID: 39548452

Coene, A., & Leliaert, J. (2022). Magnetic nanoparticles in theranostic applications. *Journal of Applied Physics*, *131*(16), 160902. DOI: 10.1063/5.0085202

Colonna, G. (2025). Overcoming Barriers in Cancer Biology Research: Current Limitations and Solutions. *Cancers (Basel)*, *17*(13), 2102. https://www.mdpi.com/2072-6694/17/13/2102. DOI: 10.3390/cancers17132102 PMID: 40647400

Coppola, P., Andersson, A., & Cole, S. (2019). The importance of the human mass balance study in regulatory submissions. *CPT: Pharmacometrics & Systems Pharmacology*, *8*(11), 792–804. DOI: 10.1002/psp4.12466 PMID: 31515957

Corral-Acero, J., Margara, F., Marciniak, M., Rodero, C., Loncaric, F., Feng, Y., Gilbert, A., Fernandes, J. F., Bukhari, H. A., Wajdan, A., Martinez, M. V., Santos, M. S., Shamohammdi, M., Luo, H., Westphal, P., Leeson, P., DiAchille, P., Gurev, V., Mayr, M., & Lamata, P. (2020). The 'Digital Twin' to enable the vision of precision cardiology. *European Heart Journal*, *41*(48), 4556–4564. DOI: 10.1093/eurheartj/ehaa159 PMID: 32128588

Costechareyre, C., Teinturier, R., Loskutov, J., Mongellaz, M., Nadolny, M., Lacourrège, M., Jarrosson, L., Delloye-Bourgeois, C., Wedeken, L., & Castellani, V. (2025). Integrating AVI-PDX and patient-derived organoid technologies for enhanced preclinical modeling in immuno-oncology. *Cancer Research*, *85*(8_Supplement_1), 6491. Advance online publication. DOI: 10.1158/1538-7445.AM2025-6491

Crawford, J., Herndon, D., Gmitter, K., & Weiss, J. (2024). The impact of myelosuppression on quality of life of patients treated with chemotherapy. *Future Oncology*, *20*(21), 1515–1530. DOI: 10.2217/fon-2023-0513 PMID: 38587388

Crecente-Campo, J., & Alonso, M. J. (2019). Engineering, on-demand manufacturing, and scaling-up of polymeric nanocapsules. *Bioengineering & Translational Medicine*, *4*(1), 38–50. DOI: 10.1002/btm2.10118 PMID: 30680317

Cui, X., Ruan, Q., Zhuo, X., Xia, X., Hu, J., Fu, R., Li, Y., Wang, J., & Xu, H. (2023). Photothermal nanomaterials: A powerful light-to-heat converter. *Chemical Reviews*, *123*(11), 6891–6952. DOI: 10.1021/acs.chemrev.3c00159 PMID: 37133878

da Costa, R., Passos, G. F., Quintão, N. L. M., Fernandes, E. S., Maia, J. R. L. C. B., Campos, M. M., & Calixto, J. B. (2020). Taxane-induced neurotoxicity: Pathophysiology and therapeutic perspectives. *British Journal of Pharmacology*, *177*(14), 3127–3146. 10.1111/bph.15086. DOI: 10.1111/bph.15086 PMID: 32352155

da Silva Correia, A. B. A., de Souza Melo, B. M., da Silva, B. A., Maciel, G. C., Cavalcante, G. M. S. M., Almeida dos Santos, J. P., Silva, L. M. B., Rocha, L. F. P., Brêda, M. B. C., de Oliveira, M. E. T., Lima, K. F. B., de Oliveira Wanderley Nepomuceno, L. B.-H., da Silva, T. H., Rodrigues, G. H. B., dos Santos, M. L., da Silva Fernandes Duarte, A. K., Gomes, E. S. B., de Araújo, S. T., de França-Júnior, R. R., & de Carvalho Fraga, C. A. (2025). Physical Interactions between Cancer Cells and Tumor Microenvironment. *Cell Biochemistry and Biophysics*, *84*(1), 1–16. DOI: 10.1007/s12013-025-01900-5 PMID: 40956489

Dai, X., Dai, Y., Zheng, Y., & Lv, Y. (2024). Magnetic nanoparticles and possible synergies with cold atmospheric plasma for cancer treatment. *RSC Advances*, *14*(40), 29039–29051. DOI: 10.1039/D4RA03837A PMID: 39282063

Dakal, T. C., Kumar, A., Majumdar, R. S., & Yadav, V. (2016). Mechanistic Basis of Antimicrobial Actions of Silver Nanoparticles. *Frontiers in Microbiology, 7*. Advance online publication. DOI: 10.3389/fmicb.2016.01831 PMID: 27899918

Dalbanjan, N. P., Korgaonkar, K., Eelager, M. P., Gonal, B. N., Kadapure, A. J., Arakera, S. B., & Kumar, S. K. P. (2025). In-silico strategies in nano-drug design: Bridging nanomaterials and pharmacological applications. *Nano TransMed, 4*, 100091. https://doi.org/https://doi.org/10.1016/j.ntm.2025.100091

Damasco, J., Ravi, S., Perez, J., Hagaman, D., & Melancon, M. (2020). Safe-by-design nanomedicine. *Nanomaterials (Basel, Switzerland), 10*. Advance online publication. DOI: 10.3390/nano10112186 PMID: 33147800

Damavandi, S. E., Banihashemi Jozdani, S. M., Elyasigorji, Z., & Vosough, M. (2026). Bridging cancer therapies: The role of magnetic nanoparticles in combination cancer therapy. *Discover Nano, 21*(1), 49. DOI: 10.1186/s11671-026-04439-3 PMID: 41712065

Daniels, B., Gräff, T., Kotschik, P., & Walter-Rohde, S. (2025). High time to Update Statistical Guidance in Ecotoxicology—A Workshop Synthesis on the Revision of OECD Document No.54. *Integrated Environmental Assessment and Management*. DOI: 10.1093/inteam/vjaf113

Darbandi, A., Asadi, A., Mahdizade Ari, M., Ohadi, E., Talebi, M., Halaj Zadeh, M., Darb Emamie, A., Ghanavati, R., & Kakanj, M. (2022). Bacteriocins: Properties and potential use as antimicrobials. *Journal of Clinical Laboratory Analysis, 36*(1), e24093. DOI: 10.1002/jcla.24093 PMID: 34851542

Das, K., & J, C. (2023). Nanoparticles and convergence of artificial intelligence for targeted drug delivery for cancer therapy: Current progress and challenges. *Frontiers in Medical Technology, 4*, 1067144. Advance online publication. DOI: 10.3389/fmedt.2022.1067144 PMID: 36688144

Das, S. S., Bharadwaj, P., Bilal, M., Barani, M., Rahdar, A., Taboada, P., Bungău, S., & Kyzas, G. (2020). Stimuli-Responsive Polymeric Nanocarriers for Drug Delivery, Imaging, and Theragnosis. *Polymers, 12*(6), 1397. Advance online publication. DOI: 10.3390/polym12061397 PMID: 32580366

de Almeida Campos, L. A., de Souza, J. B., de Queiroz Macêdo, H. L. R., Borges, J. C., de Oliveira, D. N., & Cavalcanti, I. M. F. (2024). Synthesis of polymeric nanoparticles by double emulsion and pH-driven: Encapsulation of antibiotics and natural products for combating Escherichia coli infections. *Applied Microbiology and Biotechnology, 108*(1), 351. DOI: 10.1007/s00253-024-13114-5 PMID: 38819646

De Jong, W. H., Geertsma, R. E., & Borchard, G. (2022). Regulatory safety evaluation of nanomedical products: Key issues to refine. *Drug Delivery and Translational Research, 12*(9), 2042–2047. DOI: 10.1007/s13346-022-01208-4 PMID: 35908133

De Roode, K., Hashemi, K., Verdurmen, W., & Brock, R. (2024). Tumor-on-a-chip models for predicting nanoparticle behavior. *Small, 20*(35), 2402311. Advance online publication. DOI: 10.1002/smll.202402311 PMID: 38700060

Deivayanai, V. C., Thamarai, P., Karishma, S., Saravanan, A., Yaashikaa, P. R., Vickram, A. S., Hemavathy, R. V., Rohith Kumar, R., Rishikesavan, S., & Shruthi, S. (2025). Advances in nanoparticle-mediated cancer therapeutics: Current research and future perspectives. *Cancer Pathogenesis and Therapy*, *3*(4), 293–308. DOI: 10.1016/j.cpt.2024.11.002 PMID: 40697517

Deka, H., Choudhury, A., & Dey, B. K. (2022). An Overview on Plant Derived Phenolic Compounds and Their Role in Treatment and Management of Diabetes. *Journal of Pharmacopuncture*, *25*(3), 199–208. DOI: 10.3831/KPI.2022.25.3.199 PMID: 36186092

Delgado, J. L., Hsieh, C.-M., Chan, N.-L., & Hiasa, H. (2018). Topoisomerases as anticancer targets. *The Biochemical Journal*, *475*(2), 373–398. DOI: 10.1042/BCJ20160583 PMID: 29363591

Delgado-Pujol, E. J., Martínez, G., Casado-Jurado, D., Vázquez, J., León-Barberena, J., Rodríguez-Lucena, D., Torres, Y., Alcudia, A., & Begines, B. (2025). Hydrogels and Nanogels: Pioneering the Future of Advanced Drug Delivery Systems. *Pharmaceutics*, *17*(2), 215. Advance online publication. DOI: 10.3390/pharmaceutics17020215 PMID: 40006582

Deng, H., Huang, W., & Zhang, Z. (2019). Nanotechnology based CRISPR/Cas9 system delivery for genome editing: Progress and prospect. *Nano Research, 12*(10), 2437–2450. DOI: 10.1007/s12274-019-2465-x

Deng, X., Zhang, Z., Ren, T., & Chen, L. (2025). Regulation of oxidative stress and inflammation caused by drug accumulation in the TME based on EPR-passive strategy and active targeting. *Cancer Nanotechnology*, *16*(1), 40. DOI: 10.1186/s12645-025-00342-1

De, R., Mahata, M. K., & Kim, K.-T. (2022). Structure-Based Varieties of Polymeric Nanocarriers and Influences of Their Physicochemical Properties on Drug Delivery Profiles. *Advanced Science (Weinheim, Baden-Württemberg, Germany)*, *9*(10), 2105373. https://doi.org/https://doi.org/10.1002/advs.202105373. DOI: 10.1002/advs.202105373 PMID: 35112798

Desai, N., Rana, D., Patel, M., Bajwa, N., Prasad, R., & Vora, L. K. (2025). Nanoparticle Therapeutics in Clinical Perspective: Classification, Marketed Products, and Regulatory Landscape. *Small*, *21*(29), 2502315. DOI: 10.1002/smll.202502315 PMID: 40454890

Devnarain, N., Osman, N., Fasiku, V., Makhathini, S., Salih, M., Ibrahim, U. H., & Govender, T. (2020). Intrinsic stimuli-responsive nanocarriers for smart drug delivery of antibacterial agents-An in-depth review of the last two decades. *Wiley Interdisciplinary Reviews. Nanomedicine and Nanobiotechnology*, *1664*. Advance online publication. DOI: 10.1002/wnan.1664 PMID: 32808486

Dhahri, S., Shall, H., & Thabet Mliki, N. (2024). Towards sustainable Nanomaterials: Exploring green synthesis methods and their impact on electrical properties. *Inorganic Chemistry Communications*, *168*, 112872. DOI: 10.1016/j.inoche.2024.112872

Dhanapal, Y., Sridhar, D., Khute, S., Subash, P., & AnandaThangadurai, S. (2026). Artificial intelligence assisted design of nanomedicines for colorectal cancer diagnosis and therapy: Advances, challenges, and future directions. Intelligent Pharmacy. DOI: 10.1016/j.ipha.2026.01.001

Dheyab, M. A., Abdullah, W., Aziz, A. A., Alanezi, S. T., Oladzadabbasabadi, N., Khaniabadi, P. M., Jameel, M. S., Braim, F. S., Al-Jawarneh, M. A., Mohammad, A. I. J. M., Rahman, A. A., & Mehrdel, B. (2025). Mechanistic insights and advances in polyphenol- and flavonoid-mediated sustainable synthesis of gold nanoparticles from agricultural waste: A review. *International Journal of Biological Macromolecules*, *320*, 145978. DOI: 10.1016/j.ijbiomac.2025.145978 PMID: 40669619

Dhumal, P., Chakraborty, S., Ibrahim, B., Kaur, M., & Valsami-Jones, E. (2024). Green-synthesised carbon nanodots: A SWOT analysis for their safe and sustainable innovation. *Journal of Cleaner Production*, *480*, 144115. DOI: 10.1016/j.jclepro.2024.144115

Dhuri, A., Kanp, T., Rode, K., Gupta, U., Guru, S. K., & Singh, P. K. (2024). Current advances in non-viral nanoparticle-based gene therapy for effective management of cancer. *Journal of Drug Delivery Science and Technology*, *100*, 106083. DOI: 10.1016/j.jddst.2024.106083

Dikshit, P. K., Kumar, J., Das, A. K., Sadhu, S., Sharma, S., Singh, S., Gupta, P. K., & Kim, B. S. (2021). Green Synthesis of Metallic Nanoparticles: Applications and Limitations. *Catalysts*, *11*(8), 902. DOI: 10.3390/catal11080902

Ding, S., Hsu, C., Wang, Z., Natesh, N., Millen, R., Negrete, M., Giroux, N., Rivera, G., Dohlman, A., Bose, S., Rotstein, T., Spiller, K., Yeung, A., Sun, Z., Jiang, C., Xi, R., Wilkin, B., Randon, P., Williamson, I., & Shen, X. (2022). Patient-derived micro-organospheres enable clinical precision oncology. *Cell Stem Cell*, *29*(6), 905–917.e6. Advance online publication. DOI: 10.1016/j.stem.2022.04.006 PMID: 35508177

Ding, S., Liu, J., Han, X., & Tang, M. (2023). CRISPR/Cas9-mediated genome editing in cancer therapy. *International Journal of Molecular Sciences*, *24*(22), 16325. DOI: 10.3390/ijms242216325 PMID: 38003514

Ding, W., Zhang, Y., & Shi, S. (2020). Development and Application of CRISPR/Cas in Microbial Biotechnology. *Frontiers in Bioengineering and Biotechnology*, *8*, 711. DOI: 10.3389/fbioe.2020.00711 PMID: 32695770

Dogbey, D. M., Torres, V. E. S., Fajemisin, E., Mpondo, L., Ngwenya, T., Akinrinmade, O. A., Perriman, A. W., & Barth, S. (2023). Technological advances in the use of viral and non-viral vectors for delivering genetic and non-genetic cargos for cancer therapy. *Drug Delivery and Translational Research*, *13*(11), 2719–2738. DOI: 10.1007/s13346-023-01362-3 PMID: 37301780

Đorđević, S., Gonzalez, M. M., Conejos-Sánchez, I., Carreira, B., Pozzi, S., Acúrcio, R. C., Satchi-Fainaro, R., Florindo, H. F., & Vicent, M. J. (2022). Current hurdles to the translation of nanomedicines from bench to the clinic. *Drug Delivery and Translational Research*, *12*(3), 500–525. DOI: 10.1007/s13346-021-01024-2 PMID: 34302274

Dri, D. A., Rinaldi, F., Carafa, M., & Marianecci, C. (2023). Nanomedicines and nanocarriers in clinical trials: Surfing through regulatory requirements and physico-chemical critical quality attributes. *Drug Delivery and Translational Research*, *13*(3), 757–769. DOI: 10.1007/s13346-022-01262-y PMID: 36450964

Duan, L., Kan, O., Xu, X., Xu, L., Wen, C., Zhou, X., Qin, Z., Xu, Z., Sun, W., & Liang, Y. (2021). Nanoparticle delivery of CRISPR/Cas9 for genome editing. *Frontiers in Genetics*, *12*, 673286. DOI: 10.3389/fgene.2021.673286 PMID: 34054927

Duan, Q., Zhao, Z., Zhang, Y.-J., Fu, L., Yuan, Y., Du, J.-Z., & Wang, J. (2023). Activatable fluorescent probes for real-time imaging-guided tumor therapy. *Advanced Drug Delivery Reviews*, *114793*, 114793. Advance online publication. DOI: 10.1016/j.addr.2023.114793 PMID: 36963569

Dubey, R., Dutta, D., Sarkar, A., & Chattopadhyay, P. (2021). Functionalized carbon nanotubes: Synthesis, properties and applications in water purification, drug delivery, and material and biomedical sciences. *Nanoscale Advances*, *3*(20), 5722–5744. DOI: 10.1039/D1NA00293G PMID: 36132675

Dudhwala, Y. D., Mehta, R. K., Kapoor, D. U., Shah, D. P., Shah, K. V., & Saini, V. (2026). Ligand-functionalized nanocarriers for targeted breast cancer therapy: Advances in receptor-mediated drug delivery. *Health Nanotechnology*, *2*(1), 7. DOI: 10.1186/s44301-026-00029-5

Eaton, P., Quaresma, P., Soares, C., Neves, C., De Almeida, M. P., Pereira, E., & West, P. (2017). A direct comparison of experimental methods to measure dimensions of synthetic nanoparticles. *Ultramicroscopy*, *182*, 179–190. DOI: 10.1016/j.ultramic.2017.07.001 PMID: 28692935

Eder, K. M., Marzi, A., Wågbø, A. M., Vermeulen, J. P., de la Fonteyne-Blankestijn, L. J. J., Rösslein, M., Ossig, R., Klinkenberg, G., Vandebriel, R. J., & Schnekenburger, J. (2022). Standardization of an in vitro assay matrix to assess cytotoxicity of organic nanocarriers: A pilot interlaboratory comparison. *Drug Delivery and Translational Research*, *12*(9), 2187–2206. DOI: 10.1007/s13346-022-01203-9 PMID: 35794354

Egbe Vydaline, A. O., Rozhkov, S., Sosa, G., & Mallikaratchy, P. (2025). Aptamers as target-specific recognition elements in drug delivery. *Advanced Drug Delivery Reviews*, *226*, 115685. https://doi.org/https://doi.org/10.1016/j.addr.2025.115685. DOI: 10.1016/j.addr.2025.115685 PMID: 40925520

Ege, E., Briggi, D., Vu, P., Cheng, J., Lin, F., & Xu, J. (2024). Targeting dorsal root ganglia for chemotherapy-induced peripheral neuropathy: From bench to bedside. *Therapeutic Advances in Neurological Disorders*, *17*, 17562864241252718. DOI: 10.1177/17562864241252718 PMID: 39318973

Einolghasi, F. B., Zeinali, B., Vafai, K., & Mojra, A. (2025). Development of a prediction model for hyperthermia-enhanced drug delivery using thermosensitive nanoparticles. *Journal of Thermal Biology*, *129*, 104124. DOI: 10.1016/j.jtherbio.2025.104124 PMID: 40300402

Ejigah, V., Owoseni, O., Bataille-Backer, P., Ogundipe, O. D., Fisusi, F., & Adesina, S. (2022). Approaches to Improve Macromolecule and Nanoparticle Accumulation in the Tumor Microenvironment by the Enhanced Permeability and Retention Effect. *Polymers*, *14*(13), 2601. Advance online publication. DOI: 10.3390/polym14132601 PMID: 35808648

El-Awady, S. M. M. (2023). Overview of Failure Mode and Effects Analysis (FMEA): A Patient Safety Tool. *Global Journal on Quality and Safety in Healthcare*, *6*(1), 24–26. DOI: 10.36401/JQSH-23-X2 PMID: 37260856

Eljack, S., David, S., Faggad, A., Chourpa, I., & Allard-Vannier, E. (2022). Nanoparticles design considerations to co-deliver nucleic acids and anti-cancer drugs for chemoresistance reversal. *International Journal of Pharmaceutics: X*, *4*, 100126. DOI: 10.1016/j.ijpx.2022.100126 PMID: 36147518

El-Khawaga, A. M.. (2026). Advances in green-synthesized magnetic nanoparticles for cancer diagnosis and therapy. *Frontiers in Bioengineering and Biotechnology*, *14*, 1655936. DOI: 10.3389/fbioe.2026.1655936

El-Khawaga, A. M., Youssef, O. I., El-Dydamoni, O. A., Ragab, W. M., & Mohammed, R. A. E. (2026). Advances in green-synthesized magnetic nanoparticles for targeted cancer therapy: Mechanisms, applications, and future perspectives. *European Journal of Medical Research*, *31*(1), 347. DOI: 10.1186/s40001-025-03804-9 PMID: 41654877

El-Saadony, M., Saad, A., Sitohy, M., Alkafaas, S. S., Dladla, M., Ghosh, S., Mohammed, D. M., Soliman, T. N., Ibrahim, E. H., Fahmy, M. A., Qamar, J. A. S., & Tarabily, K. E. A. (2025). Chitosan nanoparticles: Green synthesis, biological activities, and sustainable frontiers in targeted drug delivery and cancer nanomedicine – A comprehensive review. *Materials Today. Bio*, *35*, 102358. Advance online publication. DOI: 10.1016/j.mtbio.2025.102358 PMID: 41215768

El-Seedi, H. R., El-Shabasy, R. M., Khalifa, S. A. M., Saeed, A., Shah, A., Shah, R., Iftikhar, F. J., Abdel-Daim, M. M., Omri, A., Hajrahand, N. H., Sabir, J. S. M., Zou, X., Halabi, M. F., Sarhan, W., & Guo, W. (2019). Metal nanoparticles fabricated by green chemistry using natural extracts: Biosynthesis, mechanisms, and applications. *RSC Advances*, *9*(42), 24539–24559. DOI: 10.1039/C9RA02225B PMID: 35527869

El-Shabasy, R. M., Zayed, A., Farag, M. A., & Shoueir, K. R. (2026). Green synthesis of relevant and sustainable bio-applications of few-layer graphene: A multifaceted review and future perspectives. *Materials Today Sustainability, 33*, 101259. 10.1016/j.mtsust.2025.101259

EMA. (2025). Nanotechnology-based medicinal products for human use: EU-IN Horizon Scanning Report. European Medicines Agency, EMA/20989/2025/Rev.1. https://www.ema.europa.eu/

Erhirhie, E. O., Ihekwereme, C. P., & Ilodigwe, E. E. (2018). Advances in acute toxicity testing: Strengths, weaknesses and regulatory acceptance. *Interdisciplinary Toxicology*, *11*(1), 5–12. DOI: 10.2478/intox-2018-0001 PMID: 30181707

Esfahani, K., Elkrief, A., Calabrese, C., Lapointe, R., Hudson, M., Routy, B., Miller, W. H.Jr, & Calabrese, L. (2020). Moving towards personalized treatments of immune-related adverse events. *Nature Reviews. Clinical Oncology, 17*(8), 504–515. DOI: 10.1038/s41571-020-0352-8 PMID: 32246128

Esmaeili, Y., Ghaheh, H. S., Ghasemi, F., Shariati, L., Rafienia, M., Bidram, E., & Zarrabi, A. (2022). Graphene oxide quantum dot-chitosan nanotheranostic platform as a pH-responsive carrier for improving curcumin uptake internalization: In vitro & in silico study. *Biomaterials Advances, 139*, 213017. DOI: 10.1016/j.bioadv.2022.213017 PMID: 35882115

Esposito, A., Ferraresi, A., & Vallino, L.. (2024). 3D cell cultures as alternatives in cancer research. *International Journal of Biological Sciences, 20*, 5293–5311. DOI: 10.7150/ijbs.96469 PMID: 39430243

European Medicines Agency. (2025). Nanotechnology-based medicinal products for human use: EU-IN Horizon Scanning Report. EMA/20989/2025/Rev.1. https://www.ema.europa.eu/

Ewii, U. E., Attama, A. A., Olorunsola, E. O., Onugwu, A. L., Nwakpa, F. U., Anyiam, C., Chijioke, C., & Ogbulie, T. (2025). Nanoparticles for drug delivery: Insight into in vitro and in vivo drug release from nanomedicines. *Nano TransMed, 4*, 100083. https://doi.org/https://doi.org/10.1016/j.ntm.2025.100083

Ezike, T. C., Okpala, U. S., Onoja, U. L., Nwike, C. P., Ezeako, E. C., Okpara, O. J., Okoroafor, C. C., Eze, S. C., Kalu, O. L., Odoh, E. C., Nwadike, U. G., Ogbodo, J. O., Umeh, B. U., Ossai, E. C., & Nwanguma, B. C. (2023). Advances in drug delivery systems, challenges and future directions. *Heliyon, 9*(6), e17488. DOI: 10.1016/j.heliyon.2023.e17488 PMID: 37416680

Fadaei, M. R., Fadaei, M. S., Kheirieh, A. E., Hatami, H., Rahmanian-Devin, P., Tayebi-Khorrami, V., Fathabadi, M. F., Baradaran Rahimi, V., & Askari, V. R. (2025). Overview of dendrimers as promising drug delivery systems with insight into anticancer and anti-microbial applications. *International Journal of Pharmaceutics: X, 10*, 100390. https://doi.org/https://doi.org/10.1016/j.ijpx.2025.100390. DOI: 10.1016/j.ijpx.2025.100390 PMID: 41497304

Fadeel, B. (2022). Understanding the immunological interactions of engineered nanomaterials: Role of the bio-corona. *Wiley Interdisciplinary Reviews. Nanomedicine and Nanobiotechnology, 14*(6), e1798. 10.1002/wnan.1798. DOI: 10.1002/wnan.1798 PMID: 36416023

Fadeel, B., Bussy, C., Merino, S., Vázquez, E., Flahaut, E., Mouchet, F., Evariste, L., Gauthier, L., Koivisto, A. J., Vogel, U., Martín, C., Delogu, L. G., Buerki-Thurnherr, T., Wick, P., Beloin-Saint-Pierre, D., Hischier, R., Pelin, M., Candotto Carniel, F., Tretiach, M., & Bianco, A. (2018). Safety Assessment of Graphene-Based Materials: Focus on Human Health and the Environment. *ACS Nano*, *12*(11), 10582–10620. DOI: 10.1021/acsnano.8b04758 PMID: 30387986

Fakhradini, S. S., Mosharaf-Dehkordi, M., & Ahmadikia, H. (2024). Enhancing liver cancer treatment: Exploring the frequency effects of magnetic nanoparticles for heat-based tumor therapy with microwaves. *International Journal of Thermal Sciences*, *204*, 109154. DOI: 10.1016/j.ijthermalsci.2024.109154

Falak, A., Anas, M., Khan, A., Hayat, A., Shaheen, Z., Saleem, M. H., Fahad, S., & Quraishi, U. M. (2025). Efficacy of ascorbic acid coated quantum dots in alleviating lead-induced oxidative damage and enhancing growth parameters in rice (Oryza sativa L.) for sustainable cultivation. *Journal of Trace Elements in Medicine and Biology : Organ of the Society for Minerals and Trace Elements (GMS)*, *88*, 127603. DOI: 10.1016/j.jtemb.2025.127603 PMID: 39847985

Falsini, S., Bardi, U., Abou-Hassan, A., & Ristori, S. (2018). Sustainable strategies for large-scale nanotechnology manufacturing in the biomedical field. *Green Chemistry*, *20*(17), 3897–3907. DOI: 10.1039/C8GC01248B

Falzone, L., Salomone, S., & Libra, M. (2018). Evolution of cancer pharmacological treatments at the turn of the third millennium. *Frontiers in Pharmacology*, *9*, 421926. DOI: 10.3389/fphar.2018.01300 PMID: 30483135

Fam, S. Y., Chee, C. F., Yong, C. Y., Ho, K. L., Mariatulqabtiah, A. R., & Tan, W. S. (2020). Stealth coating of nanoparticles in drug-delivery systems. *Nanomaterials (Basel, Switzerland)*, *10*(4), 787. DOI: 10.3390/nano10040787 PMID: 32325941

Fan, D., Cao, Y., Cao, M., Wang, Y., Cao, Y., & Gong, T. (2023). Nanomedicine in cancer therapy. *Signal Transduction and Targeted Therapy*, *8*(1), 293. DOI: 10.1038/s41392-023-01536-y PMID: 37544972

Fang, R. H., Gao, W., & Zhang, L. (2022). Targeting drugs to tumours using cell membrane-coated nanoparticles. *Nature Reviews. Clinical Oncology*, *20*(1), 33–48. DOI: 10.1038/s41571-022-00699-x PMID: 36307534

Farsani, A. M., Mokhtari, N., Nooraei, S., Bahrulolum, H., Akbari, A., Farsani, Z. M., Khatami, S., Ebadi, M. S., & Ahmadian, G. (2024). Lipid nanoparticles: The game-changer in CRISPR-Cas9 genome editing. *Heliyon*, *10*(2), e24606. DOI: 10.1016/j.heliyon.2024.e24606 PMID: 38288017

Fatima, I., Rahdar, A., Sargazi, S., Barani, M., Hassanisaadi, M., & Thakur, V. K. (2021). Quantum dots: Synthesis, antibody conjugation, and HER2-receptor targeting for breast cancer therapy. *Journal of Functional Biomaterials*, *12*(4), 75. DOI: 10.3390/jfb12040075 PMID: 34940554

Fazal, T., Murtaza, B. N., Shah, M., Iqbal, S., Rehman, M., Jaber, F., Dera, A. A., Awwad, N. S., & Ibrahium, H. A. (2023). Recent developments in natural biopolymer based drug delivery systems. *RSC Advances*, *13*(33), 23087–23121. DOI: 10.1039/D3RA03369D PMID: 37529365

Feng, X., Li, Z., Liu, Y., Chen, D., & Zhou, Z. (2024). CRISPR/Cas9 technology for advancements in cancer immunotherapy: From uncovering regulatory mechanisms to therapeutic applications. *Experimental Hematology & Oncology*, *13*(1), 58. DOI: 10.1186/s40164-024-00570-y PMID: 39427211

Fernandes, D. A.. (2023). Review on metal-based theranostic nanoparticles for cancer therapy and imaging. *Technology in Cancer Research & Treatment*, *22*, 15330338231191493. DOI: 10.1177/15330338231191493 PMID: 37642945

Fernandez Ainaga, D. L., Roncal-Herrero, T., Ilett, M., Aslam, Z., Cheng, C., Hitchcock, J. P., Cayre, O. J., & Hondow, N. (2025). Native state structural and chemical characterisation of Pickering emulsions: A cryo-electron microscopy study. *Journal of Microscopy*, *298*(1), 92–105. DOI: 10.1111/jmi.13391 PMID: 39887717

Fernandez-Fernandez, A., Manchanda, R., & McGoron, A. J. (2011). Theranostic applications of nanomaterials in cancer: Drug delivery, image-guided therapy, and multifunctional platforms. *Applied Biochemistry and Biotechnology*, *165*(7), 1628–1651. DOI: 10.1007/s12010-011-9383-z PMID: 21947761

Ferroudj, A., Csík, A., & Prokisch, J. (2026). Allotrope-dependent physicochemical and optical properties of red and grey selenium nanoparticles. *RSC Advances*, *16*(22), 20391–20398. DOI: 10.1039/D6RA01409G PMID: 42017156

Firdhouse, J., & Lalitha, P. (2015). Apoptotic efficacy of biogenic silver nanoparticles on human breast cancer MCF-7 cell lines. *Progress in Biomaterials*, *4*(2–4), 113–121. DOI: 10.1007/s40204-015-0042-2 PMID: 26566469

Firoozi, A. A., Firoozi, A. A., & Maghami, M. R. (2025). Transformative Impacts of Nanotechnology on Sustainable Construction: A Comprehensive Review. *Results in Engineering*, *26*, 104973. DOI: 10.1016/j.rineng.2025.104973

Firouzpour, H., Najafi, F., Mohammadgholi, A., Alizadegan, D., Mahmoudi Beram, F., Akhtari, N., Zoghi, M., Ansar, N., & Tavakkoli Yaraki, M. (2026). Nanoparticle-based drug delivery systems for effective cancer treatment: Mechanisms and applications. *Next Nanotechnology, 9*, 100372. 10.1016/j.nxnano.2026.100372

Flieger, J., Franus, W., Panek, R., Szymańska-Chargot, M., Flieger, W., Flieger, M., & Kołodziej, P. (2021). Green Synthesis of Silver Nanoparticles Using Natural Extracts with Proven Antioxidant Activity. *Molecules (Basel, Switzerland), 26*(16), 4986. DOI: 10.3390/molecules26164986 PMID: 34443574

Fojo, T. (2007). Multiple paths to a drug resistance phenotype: Mutations, translocations, deletions and amplification of coding genes or promoter regions, epigenetic changes and microRNAs. *Drug Resistance Updates, 10*(1-2), 59–67. DOI: 10.1016/j.drup.2007.02.002 PMID: 17350322

Forghani, R., Savadjiev, P., Chatterjee, A., Muthukrishnan, N., Reinhold, C., & Forghani, B. (2019). Radiomics and artificial intelligence for biomarker and prediction model development in oncology. *Computational and Structural Biotechnology Journal, 17*, 995–1008. DOI: 10.1016/j.csbj.2019.07.001 PMID: 31388413

Fouda, A., Al-Otaibi, W. A., Saber, T., AlMotwaa, S. M., Alshallash, K. S., Elhady, M., Badr, N. F., & Abdel-Rahman, M. A. (2022). Antimicrobial, Antiviral, and In-Vitro Cytotoxicity and Mosquitocidal Activities of Portulaca oleracea-Based Green Synthesis of Selenium Nanoparticles. *Journal of Functional Biomaterials, 13*(3), 157. DOI: 10.3390/jfb13030157 PMID: 36135592

Freitas, S. C., Sanderson, D., Caspani, S., Magalhães, R., Cortés-Llanos, B., Granja, A., Reis, S., Belo, J. H., Azevedo, J., Gómez-Gaviro, M. V., & Sousa, C. T. (2023). New frontiers in colorectal cancer treatment combining nanotechnology with photo- and radiotherapy. *Cancers (Basel), 15*(2), 383. DOI: 10.3390/cancers15020383 PMID: 36672333

Fu, S., Rempson, C. M., Puche, V., Zhao, B., & Zhang, F. (2022). Construction of disulfide containing redox-responsive polymeric nanomedicine. *Methods : A Companion to Methods in Enzymology, 199*, 67–79. DOI: 10.1016/j.ymeth.2021.12.011 PMID: 34971759

Gacem, M. A., & Abd-Elsalam, K. A. (2022). Strategies for scaling up of green-synthesized nanomaterials: Challenges and future trends. In *Green Synthesis of Silver Nanomaterials* (pp. 669–698). Elsevier., DOI: 10.1016/B978-0-12-824508-8.00008-3

Gaddimath, S., Payamalle, S., Puttaningaiah, K. P. C. H., & Hur, J.-K. (2024). Recent Advances in pH and Redox Responsive Polymer Nanocomposites for Cancer Therapy. *Journal of Composites Science*, *8*(1), 28. Advance online publication. DOI: 10.3390/jcs8010028

Gaebler, D., Hachey, S., & Hughes, C. (2024). Tumor microenvironment in chip systems. *Frontiers in Bioengineering and Biotechnology*, •••, 12. PMID: 39386043

Gagliardi, A., Giuliano, E., Venkateswararao, E., Fresta, M., Bulotta, S., Awasthi, V., & Cosco, D. (2021). Biodegradable Polymeric Nanoparticles for Drug Delivery to Solid Tumors. *Frontiers in Pharmacology*, *12*, 601626. DOI: 10.3389/fphar.2021.601626 PMID: 33613290

Galluzzi, L., Buqué, A., Kepp, O., Zitvogel, L., & Kroemer, G. (2017). Immunogenic cell death in cancer and infectious disease. *Nature Reviews. Immunology*, *17*(2), 97–111. DOI: 10.1038/nri.2016.107 PMID: 27748397

Gameel, M., Elsayed, D. A., Shehab, W. S., Magdy, N., & Al-Erian, M. A., Emwas, A.-H. M., Jaremko, M., & Elmansy, M. (2025). Enzymes reloaded: Nanotechnology's role in the next generation of cancer therapies. *Results in Chemistry, 18*, 102889. 10.1016/j.rechem.2025.102889

Gao, Q., Gao, J., Ding, C., Li, S., Deng, L., & Kong, Y. (2021). Construction of a pH-and near-infrared irradiation-responsive nanoplatform for chemo-photothermal therapy. *International Journal of Pharmaceutics*, *593*, 120112. DOI: 10.1016/j.ijpharm.2020.120112 PMID: 33259903

Gao, Z., Du, S., Song, J., Gao, Y., Peng, X., Lin, X., e, S., Zhao, Y., & Zhang, S. (2026). Aptamer-liposome targeted nanotherapeutics for cancer therapy: Bibliometric analysis, recent developments and future perspectives. *Materials Today. Bio*, *36*, 102766. DOI: 10.1016/j.mtbio.2026.102766 PMID: 41560835

Gárate-Vélez, L., & Quintana, M. (2025). Carbon-based nanomaterials: Interactions with cells, brain therapies, and neural sensing. *Journal of Materials Science. Materials in Engineering*, *20*(1), 23. DOI: 10.1186/s40712-025-00236-5

García-Quintero, A., & Palencia, M. (2021). A critical analysis of environmental sustainability metrics applied to green synthesis of nanomaterials and the assessment of environmental risks associated with the nanotechnology. *The Science of the Total Environment*, *793*, 148524. DOI: 10.1016/j.scitotenv.2021.148524 PMID: 34182452

Gardea-Torresdey, J. L., Gomez, E., Peralta-Videa, J. R., Parsons, J. G., Troiani, H., & Jose-Yacaman, M. (2003). Alfalfa Sprouts: A Natural Source for the Synthesis of Silver Nanoparticles. *Langmuir : The ACS Journal of Surfaces and Colloids*, *19*(4), 1357–1361. DOI: 10.1021/la020835i

Garg, P., Krishna, M., Horne, D., Salgia, R., & Singhal, S. S. (2025). Hyperthermia-induced apoptosis: The role of magnetic nanoparticles in targeted delivery for breast cancer treatment. Biochimica et Biophysica Acta (BBA) -. *Revisiones en Cáncer*, *1880*(6), 189490. DOI: 10.1016/j.bbcan.2025.189490 PMID: 41173154

Gavas, S., Quazi, S., & Karpiński, T. (2021). Nanoparticles for Cancer Therapy: Current Progress and Challenges. *Nanoscale Research Letters*, *16*(1), 173. Advance online publication. DOI: 10.1186/s11671-021-03628-6 PMID: 34866166

Gavilán, H., Gallo-Cordova, A., Rábade-Chediak, M., Páez-Rodríguez, A., Morales, M. P., & Gutiérrez, L. (2025). Magnetic hyperthermia in focus: Emerging non-cancer applications of magnetic nanoparticles. *Nanoscale*, *17*(48), 27734–27758. DOI: 10.1039/D5NR03329B PMID: 41292379

Gavrila , S. (2025). Nanomembranes as Eco-Friendly Instruments for Modern Food Processing, from Filtration to Packaging. *Membranes (Basel)*, *15*(6), 167. Advance online publication. DOI: 10.3390/membranes15060167 PMID: 40559345

Gawne, P. J., Ferreira, M., Papaluca, M., Grimm, J., & Decuzzi, P. (2023). New opportunities and old challenges in the clinical translation of nanotheranostics. *Nature Reviews. Materials*, *8*(12), 783–798. DOI: 10.1038/s41578-023-00581-x PMID: 39022623

Gencheva, M., Hare, I., Kurian, S., Fortney, J., Piktel, D., Wysolmerski, R., & Gibson, L. F. (2013). Bone marrow osteoblast vulnerability to chemotherapy. *European Journal of Haematology*, *90*(6), 469–478. 10.1111/ejh.12109. DOI: 10.1111/ejh.12109 PMID: 23551534

Gerzon, G., Sheng, Y., & Kirkitadze, M. (2021). Process Analytical Technologies-Advances in Bioprocess Integration and Future Perspectives. *Journal of Pharmaceutical and Biomedical Analysis*, *207*, 114379. DOI: 10.1016/j.jpba.2021.114379 PMID: 34607168

Ghosh, A., Ghosh, A. K., Chowdhury, M., & Das, P. K. (2022). Folic acid-functionalized carbon dot-enabled starvation therapy in synergism with paclitaxel against breast cancer. *ACS Applied Bio Materials*, *5*(5), 2389–2402. DOI: 10.1021/acsabm.2c00235 PMID: 35452214

Ghosh, R., Kumar, M., Kumar, S., Komal, K., Sharma, R., & Kurmi, B. D. (2025). Small molecule therapeutics for receptor-mediated targeting through liposomes in breast cancer treatment: A comprehensive review. *Bioorganic Chemistry*, *160*, 108442. DOI: 10.1016/j.bioorg.2025.108442 PMID: 40199009

Ghosh, S., Sarkar, B., Kaushik, A., & Mostafavi, E. (2022). Nanobiotechnological prospects of probiotic microflora: Synthesis, mechanism, and applications. *The Science of the Total Environment*, *838*, 156212. DOI: 10.1016/j.scitotenv.2022.156212 PMID: 35623529

Ghuttora, N. (2016). Increase the usage of biopolymers and biodegradable polymers for sustainable environment.

Gijsen, M., Vlasselaers, D., Spriet, I., & Allegaert, K. (2021). Pharmacokinetics of Antibiotics in Pediatric Intensive Care: Fostering Variability to Attain Precision Medicine. *Antibiotics (Basel, Switzerland)*, *10*(10), 1182. https://www.mdpi.com/2079-6382/10/10/1182. DOI: 10.3390/antibiotics10101182 PMID: 34680763

Gilbertson, L. M., Zimmerman, J. B., Plata, D. L., Hutchison, J. E., & Anastas, P. T. (2015). Designing nanomaterials to maximize performance and minimize undesirable implications guided by the Principles of Green Chemistry. *Chemical Society Reviews*, *44*(16), 5758–5777. DOI: 10.1039/C4CS00445K PMID: 25955514

Gilson, P., Merlin, J.-L., & Harlé, A. (2022). Deciphering tumour heterogeneity: From tissue to liquid biopsy. *Cancers (Basel)*, *14*(6), 1384. DOI: 10.3390/cancers14061384 PMID: 35326534

Gimondi, S., Ferreira, H., Reis, R. L., & Neves, N. M. (2023). Microfluidic Devices: A Tool for Nanoparticle Synthesis and Performance Evaluation. *ACS Nano*, *17*(15), 14205–14228. DOI: 10.1021/acsnano.3c01117 PMID: 37498731

Giri, P., Banerjee, A., & Layek, B. (2023). A recent review on cancer nanomedicine. *Cancers (Basel)*, *15*(8), 2256. DOI: 10.3390/cancers15082256 PMID: 37190185

Goldstein, J. I., Newbury, D. E., Michael, J. R., Ritchie, N. W. M., Scott, J. H. J., & Joy, D. C. (2018). *Scanning Electron Microscopy and X-Ray Microanalysis*. Springer New York., DOI: 10.1007/978-1-4939-6676-9

Gong, N., Zhong, W., Alameh, M.-G., Han, X., Xue, L., El-Mayta, R., Zhao, G., Vaughan, A. E., Qin, Z., Xu, F., Hamilton, A. G., Kim, D., Xu, J., Kim, J., Teng, X., Li, J., Liang, X.-J., Weissman, D., Guo, W., & Mitchell, M. J. (2024). Tumour-derived small extracellular vesicles act as a barrier to therapeutic nanoparticle delivery. *Nature Materials*, *23*(12), 1736–1747. DOI: 10.1038/s41563-024-01961-6 PMID: 39223270

Gonzalez-Ibarra, F. P., Varon, J., & Lopez-Meza, E. G. (2011). Therapeutic Hypothermia: Critical Review of the Molecular Mechanisms of Action [Review]. *Frontiers in Neurology, volume 2 - 2011*. DOI: 10.3389/fneur.2011.00004

Gorgieva, S., & Kokol, V. (2011). Collagen-vs. gelatine-based biomaterials and their biocompatibility: review and perspectives. *Biomaterials applications for nanomedicine, 2*(2011), 17-52.

Gottimukkala, K. S. V., Lane, D. D., Cunningham, R., Malik, H. S., Jwa, Y., Cassidy, M. E., Castelli, J. M. P., Enstrom, M. R., Poljakov, K., Gastelum, G., Ho, S. H., Tassa, C., & Adair, J. E. (2026). CRISPR-AuNP: Physicochemical optimization of a gold nanoparticle platform for cost-effective and modular non-viral gene editing in HSPCs. *Gene Therapy, 33*(2), 188–202. DOI: 10.1038/s41434-025-00591-0 PMID: 41535609

Graham, W., Torbett, B. E., & Dougherty, J. P. (2025). Magnetic nanoparticles and drug delivery systems for anti-cancer therapy. *Nanomaterials (Basel, Switzerland), 15*(4), 285. DOI: 10.3390/nano15040285 PMID: 39997849

Graván, P., Marchal, J. A., & Galisteo-González, F. (2025). Improving tumor treatment: Cell membrane-coated nanoparticles for targeted therapies. *Materials Today. Bio, 32*, 101716. DOI: 10.1016/j.mtbio.2025.101716 PMID: 40391023

Gressler, S., Hipfinger, C., Part, F., Pavlicek, A., Zafiu, C., & Giese, B. (2025). A systematic review of nanocarriers used in medicine and beyond - definition and categorization framework. *Journal of Nanobiotechnology, 23*(1), 90. DOI: 10.1186/s12951-025-03113-7 PMID: 39920688

Gronnow, M. J., White, R. J., Clark, J. H., & Macquarrie, D. J. (2005). Energy efficiency in chemical reactions: A comparative study of different reaction techniques. *Organic Process Research & Development, 9*(4), 516–518. DOI: 10.1021/op0498060

Gu, A., Li, J., Li, M., & Liu, Y. (2025). Patient-derived xenograft model in cancer: Establishment and applications. *MedComm, 6*(2), e70059. Advance online publication. DOI: 10.1002/mco2.70059 PMID: 39830019

Gui, R., & Jin, H. (2025). Synthesis, characterizations, properties, modifications and applications of zero-dimensional monoelemental selenium nanomaterials. *Coordination Chemistry Reviews, 535*, 216626. DOI: 10.1016/j.ccr.2025.216626

Gujjar, S., Kukal, S., Jayabal, P., Balaji, N., Sainger, S., Roy, S., Rallapalli, S., Mahadevappa, R., Minocha, S., Kumar, S., & Mathapati, S. (2025). Nanomaterials for biomedical applications: Addressing regulatory hurdles and strategic solutions. *Nano Trends, 11*, 100127. 10.1016/j.nwnano.2025.100127

Gul, A., Fozia, , Shaheen, A., Ahmad, I., Khattak, B., Ahmad, M., Ullah, R., Bari, A., Ali, S. S., Alobaid, A., Asmari, M. M., & Mahmood, H. M. (2021). Green Synthesis, Characterization, Enzyme Inhibition, Antimicrobial Potential, and Cytotoxic Activity of Plant Mediated Silver Nanoparticle Using Ricinus communis Leaf and Root Extracts. *Biomolecules*, *11*(2), 206. DOI: 10.3390/biom11020206 PMID: 33540690

Gull, S. A., Khan[1], M. A., Hashmi, S., & Noor, T. (2025). A COMPREHENSIVE REVIEW OF TREATMENT APPROACHES FOR CONGENITAL AND HEREDITARY ANOMALIES. survival, 3(2). https://doi.org/DOI: 10.5281/zenodo.15286622

Gupta, D., Boora, A., Thakur, A., & Gupta, T. K. (2023). Green and sustainable synthesis of nanomaterials: Recent advancements and limitations. *Environmental Research*, *231*, 116316. DOI: 10.1016/j.envres.2023.116316 PMID: 37270084

Gupta, D., Gupta, T. K., Roy, P., Sharma, R., Kasana, R., & Rathore, P. (2024). Recent nanotheranostic approaches in cancer research. *Clinical and Experimental Medicine*, *24*(1), 8. DOI: 10.1007/s10238-023-01262-3 PMID: 38240834

Gupta, V., Kumar, D., Gupta, S., Tanwar, R., Jaiswal, N. K., Islam, M. M., Singh, S., Choudhary, N., Gowri, S., Webster, T. J., & Faiyazuddin, M. (2026). Nanotechnology-driven cancer therapies for precision oncology: Advances and clinical outlook. *International Journal of Nanomedicine*, *21*, 568254. DOI: 10.2147/IJN.S568254 PMID: 41869396

Gustafson, H. H., Holt-Casper, D., Grainger, D. W., & Ghandehari, H. (2015). Nanoparticle uptake: The phagocyte problem. *Nano Today*, *10*(4), 487–510. 10.1016/j.nantod.2015.06.006. DOI: 10.1016/j.nantod.2015.06.006 PMID: 26640510

Gutiérrez Coronado, O., Sandoval Salazar, C., Muñoz Carrillo, J. L., Gutiérrez Villalobos, O. A., Miranda Beltrán, M. L., Soriano Hernández, A. D., Beltrán Campos, V., & Villalobos Gutiérrez, P. T. (2025). Functionalized nanomaterials in cancer treatment: A review. *International Journal of Molecular Sciences*, *26*(6), 2633. DOI: 10.3390/ijms26062633 PMID: 40141274

Guzmán Rodríguez, A., Sablón Carrazana, M., Rodríguez Tanty, C., Malessy, M. J. A., Fuentes, G., & Cruz, L. J. (2022). Smart Polymeric Micelles for Anticancer Hydrophobic Drugs. *Cancers (Basel)*, *15*(1), 4. Advance online publication. DOI: 10.3390/cancers15010004 PMID: 36612002

Guzmán-Sastoque, P., Monsalve, M. C., Rodríguez, C. F., Castellanos, S., Cruz, J. C., & Reyes, L. H. Delivery of Nucleic Acids Using Nanocarriers: siRNA and miRNA Delivery, mRNA and DNA Delivery, CRISPR-Cas Systems. In *Nanocarriers for Nucleic Acids and Proteins* (pp. 248–291). CRC Press.

Habanjar, O., Diab-Assaf, M., Caldefie-Chézet, F., & Delort, L. (2021). 3D cell culture systems in cancer. *International Journal of Molecular Sciences*, •••, 22.

Habeeb, M., Deepthi, K. L., Vengateswaran, H. T., Arsey, S., You, H. W., & Aher, K. B. (2024). Innovations in Metal-Based Nanomedicines to Interfere with the Signaling Communications of Prostate Cancer. *BioNanoScience*, *15*(1), 123. DOI: 10.1007/s12668-024-01723-3

Haider, M., Zaki, K. Z., El Hamshary, M. R., Hussain, Z., Orive, G., & Ibrahim, H. O. (2022). Polymeric nanocarriers: A promising tool for early diagnosis and efficient treatment of colorectal cancer. *Journal of Advanced Research*, *39*, 237–255. https://doi.org/https://doi.org/10.1016/j.jare.2021.11.008. DOI: 10.1016/j.jare.2021.11.008 PMID: 35777911

Haider, T., Pandey, V., Banjare, N., Gupta, P. N., & Soni, V. (2020). Drug resistance in cancer: Mechanisms and tackling strategies. *Pharmacological Reports*, *72*(5), 1125–1151. DOI: 10.1007/s43440-020-00138-7 PMID: 32700248

Halder, A. K., Biswas, A., Saha, M., & Sadi, S. M. (2025). *Machine learning-based prediction of acquired antimicrobial resistance in multiple bacterial species using K-mer analysis, mutation detection, and AMR gene profiling*. Brac University.

Haleem, A., Javaid, M., Singh, R. P., Rab, S., & Suman, R. (2023). Applications of nanotechnology in medical field: A brief review. *Global Health Journal (Amsterdam, Netherlands)*, *7*(2), 70–77. DOI: 10.1016/j.glohj.2023.02.008

Halwani, A. A. (2022). Development of Pharmaceutical Nanomedicines: From the Bench to the Market. *Pharmaceutics*, *14*(1), 106. Advance online publication. DOI: 10.3390/pharmaceutics14010106 PMID: 35057002

Hamouda, R. A., Hussein, M. H., Abo-elmagd, R. A., & Bawazir, S. S. (2019). Synthesis and biological characterization of silver nanoparticles derived from the cyanobacterium Oscillatoria limnetica. *Scientific Reports*, *9*(1), 13071. DOI: 10.1038/s41598-019-49444-y PMID: 31506473

Hamzah, H. (2026). Targeted therapy and biomarker-guided applications of ecofriendly silver nanoparticles in precision oncology [Review]. *Journal of Pharmacy & Pharmaceutical Sciences, Volume 28 - 2025*. DOI: 10.3389/jpps.2025.15403

Han, X., Feng, X., & Zhou, H.. (2025). Functional nanomaterials for enhanced tumor photothermal therapy-The mechanisms and applications. *Frontiers in Pharmacology*, *16*, 1604965. DOI: 10.3389/fphar.2025.1604965 PMID: 40727107

Harle, S. (2025). Integrative approaches to tackle multidisciplinary challenges: A review of multi-science problem analysis. *Current Materials Science*, *18*(6), 703–728. DOI: 10.2174/0126661454279515231225151023

Hassan, D., Khalil, A. T., Saleem, J., Diallo, A., Khamlich, S., Shinwari, Z. K., & Maaza, M. (2018). Biosynthesis of pure hematite phase magnetic iron oxide nanoparticles using floral extracts of Callistemon viminalis (bottlebrush): their physical properties and novel biological applications. *Artificial Cells, Nanomedicine, and Biotechnology, 46*(sup1), 693–707. DOI: 10.1080/21691401.2018.1434534

Hassan, D., Sani, A., Chanihoon, G. Q., Antonio Pérez, A., Ehsan, M., & Torres Huerta, A. L. (2024). Environmentally Sustainable and Green Polymeric Method for Chitosan (CH) Film Synthesis Using Natural Acids and Impact of Zinc Ferrite Nanoparticles (NPs) on Water Solubility (WS) and Physical Properties. *Polymers*, *16*(24), 3466. DOI: 10.3390/polym16243466 PMID: 39771318

Hassan, D., Sani, A., & Medina, D. I. (2023). Limitations of Nanocarriers Such as Cell and Tissue Toxicity, Genotoxicity, Scale-Up of Nanomaterials. In Khan, F. A. (Ed.), *Nano Drug Delivery for Cancer Therapy: Principles and Practices* (pp. 149–171). Springer Nature Singapore., DOI: 10.1007/978-981-99-6940-1_8

Hawthorne, D., Pannala, A., Sandeman, S., & Lloyd, A. (2022). Sustained and targeted delivery of hydrophilic drug compounds: A review of existing and novel technologies from bench to bedside. *Journal of Drug Delivery Science and Technology*, *78*, 103936. https://doi.org/https://doi.org/10.1016/j.jddst.2022.103936. DOI: 10.1016/j.jddst.2022.103936

Heinzelmann, E., Piraino, F., Costa, M., Roch, A., Norkin, M., Garnier, V., Homicsko, K., & Brandenberg, N. (2024). iPSC-derived and Patient-Derived Organoids: Applications and challenges in scalability and reproducibility as pre-clinical models. *Current Research in Toxicology*, *7*, 100197. Advance online publication. DOI: 10.1016/j.crtox.2024.100197 PMID: 40276485

Hemlata, M., Meena, P. R., Singh, A. P., & Tejavath, K. K. (2020). Biosynthesis of Silver Nanoparticles Using *Cucumis prophetarum* Aqueous Leaf Extract and Their Antibacterial and Antiproliferative Activity Against Cancer Cell Lines. *ACS Omega*, *5*(10), 5520–5528. DOI: 10.1021/acsomega.0c00155 PMID: 32201844

Herdiana, Y., Wathoni, N., Shamsuddin, S., & Muchtaridi, M. (2022). Drug release study of the chitosan-based nanoparticles. *Heliyon*, *8*(1), e08674. https://doi.org/https://doi.org/10.1016/j.heliyon.2021.e08674. DOI: 10.1016/j.heliyon.2021.e08674 PMID: 35028457

Herdiana, Y., Wathoni, N., Shamsuddin, S., & Muchtaridi, M. (2022). Scale-up polymeric-based nanoparticles drug delivery systems: Development and challenges. *OpenNano*, *7*, 100048. DOI: 10.1016/j.onano.2022.100048

Herrmann, H. (2003). Kinetics of aqueous phase reactions relevant for atmospheric chemistry. *Chemical Reviews*, *103*(12), 4691–4716. DOI: 10.1021/cr020658q PMID: 14664629

Hertig, J. B., Shah, V. P., Flühmann, B., Mühlebach, S., Stemer, G., Surugue, J., Moss, R., & Di Francesco, T. (2021). Tackling the challenges of nanomedicines: Are we ready? *American Journal of Health-System Pharmacy*, *78*(12), 1047–1056. DOI: 10.1093/ajhp/zxab048 PMID: 33599767

Hidalgo, M., Amant, F., Biankin, A. V., Budinská, E., Byrne, A. T., Caldas, C., Clarke, R. B., de Jong, S., Jonkers, J., Mælandsmo, G. M., Roman-Roman, S., Seoane, J., Trusolino, L., & Villanueva, A. (2014). Patient-derived xenograft models: An emerging platform for translational cancer research. *Cancer Discovery*, *4*(9), 998–1013. DOI: 10.1158/2159-8290.CD-14-0001 PMID: 25185190

Hoang, C. N. M., Nguyen, S. H., & Tran, M. T. (2025). Nanoparticles in cancer therapy: Strategies to penetrate and modulate the tumor microenvironment - A review. *Smart Materials in Medicine*, *6*(2), 270–284. DOI: 10.1016/j.smaim.2025.07.004

Hocaoğlu, R., Zıkşahna, K., & Ihlamur, M. (2026). Selecting nanocarrier platforms for drug delivery: A criteria-based, translational review integrating CQA/CMC and immune interactions. *Health Nanotechnology*, *2*(1), 5. DOI: 10.1186/s44301-026-00026-8

Hojjat-Farsangi, M. (2014). Small-Molecule Inhibitors of the Receptor Tyrosine Kinases: Promising Tools for Targeted Cancer Therapies. *International Journal of Molecular Sciences*, *15*(8), 13768–13801. DOI: 10.3390/ijms150813768 PMID: 25110867

Holghoomi, R., Kharab, Z., Rahdar, A., Pandey, S., & Ferreira, L. (2024). Harnessing the power of green synthesis of nanomaterials for anticancer applications. *Coordination Chemistry Reviews*, *513*, 215903. Advance online publication. DOI: 10.1016/j.ccr.2024.215903

Hong, L., Li, W., Li, Y., & Yin, S. (2023). Nanoparticle-based drug delivery systems targeting cancer cell surfaces. *RSC Advances*, *13*(31), 21365–21382. DOI: 10.1039/D3RA02969G PMID: 37465582

Hossain, A., Manik, M. H., Rakib, S., Mahmud, N., Khan, S., Ahsan, Z., Islam, M. S., Hossain, N., & Akter, M. A. (2025). Green Nanotechnology for Implantable Biosensors: Biocompatibility and Functional Integration in Medical Applications. *Biosensors & Bioelectronics: X*, *27*, 100678. Advance online publication. DOI: 10.1016/j.biosx.2025.100678

Hossam Abdelmonem, B., Abdelaal, N. M., Anwer, E. K. E., Rashwan, A. A., Hussein, M. A., Ahmed, Y. F., Khashana, R., Hanna, M. M., & Abdelnaser, A. (2024). Decoding the Role of CYP450 Enzymes in Metabolism and Disease: A Comprehensive Review. *Biomedicines*, *12*(7), 1467. DOI: 10.3390/biomedicines12071467 PMID: 39062040

Hosseini, S. M., Mohammadnejad, J., Salamat, S., Beiram Zadeh, Z., Tanhaei, M., & Ramakrishna, S. (2023). Theranostic polymeric nanoparticles as a new approach in cancer therapy and diagnosis: A review. *Materials Today. Chemistry*, *29*, 101400. https://doi.org/https://doi.org/10.1016/j.mtchem.2023.101400. DOI: 10.1016/j.mtchem.2023.101400

Hou, J., He, Z., Liu, T., Chen, D., Wang, B., Wen, Q., & Zheng, X. (2022). Evolution of molecular targeted cancer therapy: Mechanisms of drug resistance and novel opportunities identified by CRISPR-Cas9 screening. *Frontiers in Oncology*, *12*, 755053. DOI: 10.3389/fonc.2022.755053 PMID: 35372044

Hou, Y., Zhu, C., Shen, Z., Xu, Y., Zhou, S., & Zhou, X. (2025). Carbon Nanomaterials in Biomedicine: Opportunities and Toxicological Concerns. *International Journal of Nanomedicine*, *20*, 14707–14730. DOI: 10.2147/IJN.S552319 PMID: 41393084

Huang, H., Zheng, Y., Chang, M., Song, J., Xia, L., Wu, C., Jia, W., Ren, H., Feng, W., & Chen, Y. (2024). Ultrasound-based micro-/nanosystems for biomedical applications. *Chemical Reviews*, *124*(13), 8307–8472. DOI: 10.1021/acs.chemrev.4c00009 PMID: 38924776

Huber, M. J., Ivleva, N. P., Booth, A. M., Beer, I., Bianchi, I., Drexel, R., Geiss, O., Mehn, D., Meier, F., Molska, A., Parot, J., Sørensen, L., Vella, G., Prina-Mello, A., Vogel, R., & Caputo, F. (2023). Physicochemical characterization and quantification of nanoplastics: Applicability, limitations and complementarity of batch and fractionation methods. *Analytical and Bioanalytical Chemistry*, *415*(15), 3007–3031. DOI: 10.1007/s00216-023-04689-5 PMID: 37106123

Hughes, D., & Andersson, D. I. (2015). Evolutionary consequences of drug resistance: Shared principles across diverse targets and organisms. *Nature Reviews. Genetics*, *16*(8), 459–471. DOI: 10.1038/nrg3922 PMID: 26149714

Hulkoti, N. I., & Taranath, T. C. (2014a). Biosynthesis of nanoparticles using microbes—A review. *Colloids and Surfaces. B, Biointerfaces*, *121*, 474–483. DOI: 10.1016/j.colsurfb.2014.05.027 PMID: 25001188

Hussain, S. M., Braydich-Stolle, L. K., Schrand, A. M., Murdock, R. C., Yu, K. O., Mattie, D. M., Schlager, J. J., & Terrones, M. (2009). Toxicity evaluation for safe use of nanomaterials: Recent achievements and technical challenges. *Advanced Materials*, *21*(16), 1549–1559. DOI: 10.1002/adma.200801395

Hussen, B., Rasul, M. F., Abdullah, S. R., Hidayat, H. J., Faraj, G. S. H., Ali, F. A., Salihi, A., Baniahmad, A., Ghafouri-Fard, S., Rahman, M., Glassy, M. C., Branicki, W., & Taheri, M. (2023). Targeting miRNA by CRISPR/Cas in cancer: Advantages and challenges. *Military Medical Research*, *10*(1), 47. DOI: 10.1186/s40779-023-00468-6 PMID: 37460924

Huston, M., DeBella, M., DiBella, M., & Gupta, A. (2021). Green Synthesis of Nanomaterials. *Nanomaterials (Basel, Switzerland)*, *11*(8), 2130. DOI: 10.3390/nano11082130 PMID: 34443960

Hwang, J., Kang, J., Oh, N., Li, C., Yoo, C.-Y., & Kwak, M. (2025). Effect of Surface Charge of Nanoparticles on Cellular Uptake in HeLa Cells. *Langmuir : The ACS Journal of Surfaces and Colloids*, *41*(21), 13192–13198. DOI: 10.1021/acs.langmuir.5c00832 PMID: 40384298

Hwang, J., & Oh, J.. (2026). Nanoparticle-enabled photothermal therapy integrated with gene delivery, immunotherapy, and chemotherapy: A review of recent advances and future perspectives. *International Journal of Nanomedicine*, *21*, 101–132. DOI: 10.2147/IJN.S573093 PMID: 41869393

Ijaz, A., & Javed, M. R. (2026). Synthesis, characterization, antifungal and SERS analysis of mycosynthesized silver nanoparticles against Aspergillus parasiticus. *Process Biochemistry (Barking, London, England)*, *162*, 170–182. DOI: 10.1016/j.procbio.2026.01.017

Ijaz, M., Aslam, B., Hasan, I., Ullah, Z., Roy, S., & Guo, B. (2024). Cell membrane-coated biomimetic nanomedicines: Productive cancer theranostic tools. *Biomaterials Science*, *12*(4), 863–891. DOI: 10.1039/D3BM01552A PMID: 38230669

Imani, S., Moradi, S., Faraj, T. A., Hassanpoor, P., Musapour, N., Najmaldin, S. K., Abdulhamd, A. H., Mohammadi, A. T., Taha, C. H., & Aminnezhad, S. (2025). Nanoparticle technologies in precision oncology and personalized vaccine development: Challenges and advances. *International Journal of Pharmaceutics: X*, *10*, 100353. DOI: 10.1016/j.ijpx.2025.100353 PMID: 40688030

Imani, Z., Amini, M., Ghahremani, M. H., Khonsari, F., Shahbaz, S., Atyabi, F., & Dinarvand, R. (2025). Docetaxel-dimer prodrug as a glutathione-responsive nano drug delivery system for improved chemotherapy. *Journal of Pharmaceutical Sciences, 114*(9), 103874. DOI: 10.1016/j.xphs.2025.103874 PMID: 40541678

Ingber, D. (2022). Human organs-on-chips for disease modelling, drug development and personalized medicine. *Nature Reviews. Genetics, 23*(8), 467–491. DOI: 10.1038/s41576-022-00466-9 PMID: 35338360

Invrea, F., Rovito, R., Torchiaro, E., Petti, C., Isella, C., & Medico, E. (2020). Patient-derived xenografts (PDXs) as model systems for human cancer. *Current Opinion in Biotechnology, 63*, 151–156. DOI: 10.1016/j.copbio.2020.01.003 PMID: 32070860

Iqbal, I., Wilairatana, P., Saqib, F., Nasir, B., Wahid, M., Latif, M. F., Iqbal, A., Naz, R., & Mubarak, M. S. (2023). Plant Polyphenols and Their Potential Benefits on Cardiovascular Health: A Review. *Molecules (Basel, Switzerland), 28*(17), 6403. DOI: 10.3390/molecules28176403 PMID: 37687232

Iqbal, N., & Iqbal, N. (2014). Imatinib: A Breakthrough of Targeted Therapy in Cancer. *Chemotherapy Research and Practice, 2014*(1), 357027. 10.1155/2014/357027. DOI: 10.1155/2014/357027 PMID: 24963404

Iram, A., Dong, Y., & Ignea, C. (2024). Synthetic biology advances towards a bio-based society in the era of artificial intelligence. *Current Opinion in Biotechnology, 87*, 103143. DOI: 10.1016/j.copbio.2024.103143 PMID: 38781699

Iravani, S., & Varma, R. (2022). Nanosponges for Drug Delivery and Cancer Therapy: Recent Advances. *Nanomaterials (Basel, Switzerland), 12*(14), 2440. Advance online publication. DOI: 10.3390/nano12142440 PMID: 35889665

Ishola, B. O., Rahaman, K. A., Razzak, S. A., Rumon, M. M. H., Shakil, M. S., & Uddin, S. (2026). Advanced mechanisms of polymer-based drug delivery systems for clinical applications. *RSC Pharmaceutics*.

Islam, S., Ahmed, M. M., Islam, M., Hossain, N., & Chowdhury, M. (2025). Advances in Nanoparticles in Targeted Drug Delivery- A Review. *Results in Surfaces and Interfaces, 19*, 100529. DOI: 10.1016/j.rsurfi.2025.100529

Jacob, S., Varkey, N. R., Boddu, S. H. S., Gorain, B., Rao, R., & Nair, A. B. (2025). Advances in Lipid-Polymer Hybrid Nanoparticles: Design Strategies, Functionalization, Oncological and Non-Oncological Clinical Prospects. *Pharmaceuticals (Basel, Switzerland), 18*(12), 1772. https://www.mdpi.com/1424-8247/18/12/1772. DOI: 10.3390/ph18121772 PMID: 41471261

Jaffee, E. M., Van Dang, C., Agus, D. B., Alexander, B. M., Anderson, K. C., Ashworth, A., Barker, A. D., Bastani, R., Bhatia, S., & Bluestone, J. A. (2017). Future cancer research priorities in the USA: A Lancet Oncology Commission. *The Lancet. Oncology*, *18*(11), e653–e706. DOI: 10.1016/S1470-2045(17)30698-8 PMID: 29208398

Jagadeesh, P., Rangappa, S. M., & Siengchin, S. (2024). Advanced characterization techniques for nanostructured materials in biomedical applications. *Advanced Industrial and Engineering Polymer Research*, *7*(1), 122–143. DOI: 10.1016/j.aiepr.2023.03.002

Jain, A., & Bhise, K. (2025). Nano-pharmacokinetics and pharmacodynamics of green-synthesized ZnO nanoparticles: A pathway to safer therapeutic applications. *Xenobiotica; The Fate of Foreign Compounds in Biological Systems*, *55*(4), 265–276. DOI: 10.1080/00498254.2025.2505062 PMID: 40380770

Jain, S., & Mehata, M. S. (2017). Medicinal Plant Leaf Extract and Pure Flavonoid Mediated Green Synthesis of Silver Nanoparticles and their Enhanced Antibacterial Property. *Scientific Reports*, *7*(1), 15867. DOI: 10.1038/s41598-017-15724-8 PMID: 29158537

Janik, E., Niemcewicz, M., Ceremuga, M., Krzowski, L., Saluk-Bijak, J., & Bijak, M. (2020). Various aspects of a gene editing system—CRISPR–Cas9. *International Journal of Molecular Sciences*, *21*(24), 9604. DOI: 10.3390/ijms21249604 PMID: 33339441

Janitri, V., Aruljothi, K., Mythili, V., Singh, S., Prasher, P., Gupta, G., Dua, K., Hanumanthappa, R., Karthikeyan, K., & Anand, K. (2024). The roles of patient-derived xenograft models and artificial intelligence toward precision medicine. *MedComm*, *5*(10), e745. Advance online publication. DOI: 10.1002/mco2.745 PMID: 39329017

Javid, H., Oryani, M. A., Rezagholinejad, N., Esparham, A., Tajaldini, M., & Karimi-Shahri, M. (2024). RGD peptide in cancer targeting: Benefits, challenges, solutions, and possible integrin-RGD interactions. *Cancer Medicine*, *13*(2), e6800. DOI: 10.1002/cam4.6800 PMID: 38349028

Jayaraj, S., Jayaprakash, N., Devaraji, M., Murugavel, S., Robinson, A. X., & Murugesan, L. (2026). Nanocarrier-based strategies for advanced cancer drug delivery and therapeutics: Design principles, biological interactions, and clinical potential. *Nano Trends, 14*, 100203. 10.1016/j.nwnano.2026.100203

Jayaraj, S., Jayaprakash, N., Devaraji, M., Murugavel, S., Robinson, A. X., & Murugesan, L. (2026). Nanocarrier-based strategies for advanced cancer drug delivery and therapeutics: Design principles, biological interactions, and clinical potential. *Nano Trends, 14*, 100203. https://doi.org/https://doi.org/10.1016/j.nwnano.2026.100203

Jeevanandam, J., Kiew, S. F., Boakye-Ansah, S., Lau, S. Y., Barhoum, A., Danquah, M. K., & Rodrigues, J. (2022). Green approaches for the synthesis of metal and metal oxide nanoparticles using microbial and plant extracts. *Nanoscale, 14*(7), 2534–2571. DOI: 10.1039/D1NR08144F PMID: 35133391

Jeong, W., Yu, U., & Lee, S. H. (2025). Global decomposition of networks into multiple cores formed by local hubs. *Physical Review. E, 111*(5), 054302. DOI: 10.1103/PhysRevE.111.054302 PMID: 40534039

Jiang, J., Cui, X., Huang, Y., Yan, D., Wang, B., Yang, Z., Chen, M., Wang, J., Zhang, Y., Liu, G., Zhou, C., Cui, S., Ni, J., Yang, F., & Cui, D. (2024). Advances and prospects in integrated nano-oncology. *Nano Biomedicine and Engineering, 16*(2), 152–187. Advance online publication. DOI: 10.26599/NBE.2024.9290060

Jiang, M.. (2025). Nanomaterials in gene therapy and genome editing: Challenges and emerging directions. *Journal of Nanobiotechnology*, •••, 24. DOI: 10.1186/s12951-025-03124-4 PMID: 41402770

Jiang, T., Gonzalez, K. M., Cordova, L., & Lu, J. (2023). Nanotechnology-enabled gene delivery for cancer and other genetic diseases. *Expert Opinion on Drug Delivery, 20*(4), 523–540. DOI: 10.1080/17425247.2023.2200246 PMID: 37017558

Jiao, W., Dai, L., Yan, B., Lyu, Y., Fan, H., & Liu, X. (2024). Heating up the immune battle: Magnetic hyperthermia against cancer. *Fundamental Research (Beijing)*. Advance online publication. DOI: 10.1016/j.fmre.2024.08.006 PMID: 41466973

Jia, Y., Chen, S., Wang, C., Sun, T., & Yang, L. (2022). Hyaluronic acid-based nano drug delivery systems for breast cancer. *Frontiers in Bioengineering and Biotechnology, 10*, 990145. DOI: 10.3389/fbioe.2022.990145 PMID: 36091467

Ji, J., Li, L., Guo, W., Zhang, J., Yao, Y., Chen, H., Liao, F., Jin, Z., Liu, L., Ouyang, J., & Liang, X.-J. (2025). Nanoscale synthetic biology with innovative medicinal applications. *Fundamental Research (Beijing), 5*(5), 1889–1910. DOI: 10.1016/j.fmre.2024.11.006 PMID: 41613444

Jin, J., Yoshimura, K., Sewastjanow-Silva, M., Song, S., & Ajani, J. (2023). Challenges and Prospects of Patient-Derived Xenografts for Cancer Research. *Cancers (Basel), 15*(17), 4352. Advance online publication. DOI: 10.3390/cancers15174352 PMID: 37686627

Jin, Z., & Amili, H.. (2024). Tumor microenvironment-responsive drug delivery based on micelles. *Biomedicines*, *12*(2), 417. DOI: 10.3390/biomedicines12020417 PMID: 38398021

John, P. M., Emmanuel, M., Beegum, J., John, F., & George, J. (2023). Nano-drug delivery systems for tumour-targeting: overcoming the limitations of chemotherapy. In *Targeted Cancer Therapy in Biomedical Engineering* (pp. 487–521). Springer. DOI: 10.1007/978-981-19-9786-0_14

Joudeh, N., & Linke, D. (2022). Nanoparticle classification, physicochemical properties, characterization, and applications: A comprehensive review for biologists. *Journal of Nanobiotechnology*, *20*(1), 262. DOI: 10.1186/s12951-022-01477-8 PMID: 35672712

Jouybar, M., de Winde, C. M., Wolf, K., Friedl, P., Mebius, R. E., & den Toonder, J. M. (2024). Cancer-on-chip models for metastasis: Importance of the tumor microenvironment. *Trends in Biotechnology*, *42*(4), 431–448. DOI: 10.1016/j.tibtech.2023.10.001 PMID: 37914546

Joyce, K., Fabra, G. T., Bozkurt, Y., & Pandit, A. (2021). Bioactive potential of natural biomaterials: Identification, retention and assessment of biological properties. *Signal Transduction and Targeted Therapy*, *6*(1), 122. DOI: 10.1038/s41392-021-00512-8 PMID: 33737507

Joye, I. J., & McClements, D. J. (2014). Biopolymer-based nanoparticles and microparticles: Fabrication, characterization, and application. *Current Opinion in Colloid & Interface Science*, *19*(5), 417–427. DOI: 10.1016/j.cocis.2014.07.002

Jubelin, C., Muñoz-García, J., & Griscom, L.. (2022). 3D in vitro models in oncology. *Cell & Bioscience*, •••, 12.

Jumadi, J., Harun, W. S. W., Kadirgama, K., Samylingam, L., Aslfattahi, N., S, C. K., Kok, C. K., & Kiai, M. S. (2025). A Comparative Review of Polylactic Acid and Poly(Lactic-co-Glycolic Acid) Biomaterials: Optimizing Drug Delivery. *BioNanoScience*, *16*(1), 19. DOI: 10.1007/s12668-025-02233-6

Jung, G. T., Kim, K.-P., & Kim, K. (2020). How to interpret and integrate multi-omics data at systems level. *Animal Cells and Systems*, *24*(1), 1–7. DOI: 10.1080/19768354.2020.1721321 PMID: 32158610

Jurczyk, M., Jelonek, K., Musiał-Kulik, M., Beberok, A., Wrześniok, D., & Kasperczyk, J. (2021). Single- versus Dual-Targeted Nanoparticles with Folic Acid and Biotin for Anticancer Drug Delivery. *Pharmaceutics*, *13*(3), 326. Advance online publication. DOI: 10.3390/pharmaceutics13030326 PMID: 33802531

Jurkovicova, D., Neophytou, C. M., Gašparović, A. Č., & Gonçalves, A. C. (2022). DNA damage response in cancer therapy and resistance: Challenges and opportunities. *International Journal of Molecular Sciences*, *23*(23), 14672. DOI: 10.3390/ijms232314672 PMID: 36499000

Kalimuthu, K., Suresh Babu, R., Venkataraman, D., Bilal, M., & Gurunathan, S. (2008). Biosynthesis of silver nanocrystals by Bacillus licheniformis. *Colloids and Surfaces. B, Biointerfaces*, *65*(1), 150–153. DOI: 10.1016/j.colsurfb.2008.02.018 PMID: 18406112

Kalishwaralal, K., Deepak, V., Ram Kumar Pandian, S., Kottaisamy, M., BarathManiKanth, S., Kartikeyan, B., & Gurunathan, S. (2010). Biosynthesis of silver and gold nanoparticles using Brevibacterium casei. *Colloids and Surfaces. B, Biointerfaces*, *77*(2), 257–262. DOI: 10.1016/j.colsurfb.2010.02.007 PMID: 20197229

Kalita, I., Bhattacharjee, S., & Saharia, M. (2025). Advancements in medicinal plant research: Harnessing artificial intelligence, machine learning, deep learning, and bioinformatics. In *Biotechnology, Multiple Omics, and Precision Breeding in Medicinal Plants* (pp. 135-145). CRC Press.

Kalpage, C., & Isuranga, T. (2025). Supercritical Fluid Technologies for Nanoparticle Production. In (pp. 215-246). DOI: 10.1007/978-981-96-9176-0_11

Kanaoujiya, R., Saroj, S. K., Rajput, V. D., Alimuddin, , Srivastava, S., Minkina, T., Igwegbe, C. A., Singh, M., & Kumar, A. (2023). Emerging application of nanotechnology for mankind. *Emergent Materials*, *6*(2), 439–452. DOI: 10.1007/s42247-023-00461-8 PMID: 36743193

Kang, C. S., Zhang, S., Wang, H., Liu, Y., Ren, S., Chen, Y., Li, J., Bandara, N., Rogachev, A. Y., Rogers, B. E., & Chong, H.-S. (2022). Novel Chelating Agents for Zirconium-89-Positron Emission Tomography (PET) Imaging: Synthesis, DFT Calculation, Radiolabeling, and In Vitro and In Vivo Complex Stability. *ACS Omega*, *7*(42), 37229–37236. DOI: 10.1021/acsomega.2c03478 PMID: 36312419

Kang, T.-H. (2023). *DNA damage, repair, and cancer metabolism* (Vol. 24). MDPI.

Kantor, A., McClements, M. E., & MacLaren, R. E. (2020). CRISPR-Cas9 DNA base-editing and prime-editing. *International Journal of Molecular Sciences*, *21*(17), 6240. DOI: 10.3390/ijms21176240 PMID: 32872311

Kant, S., Deepika, , & Roy, S. (2025). Integrative Multi-Omics and Artificial Intelligence: A New Paradigm for Systems Biology. *Omics : A Journal of Integrative Biology*, *29*(12), 576–587. DOI: 10.1177/15578100251392371 PMID: 41203247

Kara, A., Ongoren, B., Anaya, B. J., Lalatsa, A., & Serrano, D. R. (2025). Continuous manufacturing of nanomedicines using 3D-printed microfluidic devices. *Applied Materials Today*, *43*, 102672. DOI: 10.1016/j.apmt.2025.102672

Kardamakis, D., Baatout, S., Bourguignon, M., Foray, N., & Socol, Y. (2023). History of Radiation Biology. In Baatout, S. (Ed.), *Radiobiology Textbook* (pp. 1–24). Springer International Publishing., DOI: 10.1007/978-3-031-18810-7_1

Karimi, S., Bakhshali, R., Bolandi, S., Zahed, Z., Zadeh, S. S. M., Zenjanab, M. K., & Esfahlan, R. J. (2025). For and against tumor microenvironment: Nanoparticle-based strategies for active cancer therapy. *Materials Today. Bio*, *31*, 101626. Advance online publication. DOI: 10.1016/j.mtbio.2025.101626 PMID: 40124335

Karnwal, A., Jassim, A. Y., Mohammed, A. A., Sharma, V., Al-Tawaha, A. R., & Sivanesan, I. (2024). Nanotechnology for Healthcare: Plant-Derived Nanoparticles in Disease Treatment and Regenerative Medicine. *Pharmaceuticals (Basel, Switzerland)*, *17*(12), 1711. Advance online publication. DOI: 10.3390/ph17121711 PMID: 39770553

Kashapov, R., Gaynanova, G., Gabdrakhmanov, D., Kuznetsov, D., Pavlov, R., Petrov, K., Zakharova, L., & Sinyashin, O. (2020). Self-Assembly of Amphiphilic Compounds as a Versatile Tool for Construction of Nanoscale Drug Carriers. *International Journal of Molecular Sciences*, *21*(18), 6961. Advance online publication. DOI: 10.3390/ijms21186961 PMID: 32971917

Kashinath, K. P., Sanjay, L. R., Ashokbhai, M. K., Roy, S., Sardar, M. S., & Kaity, S. (2025). Continuous manufacturing based paradigm shift in pharmaceuticals production and current regulatory framework. *Chemical Engineering Research & Design*, *215*, 1–22. DOI: 10.1016/j.cherd.2025.01.003

Kashkooli, F. M., Soltani, M., & Momeni, M. M. (2023). Integrating therapeutic ultrasound with nanosized drug delivery systems in the battle against cancer. *Technology in Cancer Research & Treatment*, *22*, 15330338231211472. DOI: 10.1177/15330338231211472 PMID: 37946517

Kashyap, B. K., Singh, V. V., Solanki, M. K., Kumar, A., Ruokolainen, J., & Kesari, K. K. (2023). Smart nanomaterials in cancer theranostics: Challenges and opportunities. *ACS Omega*, *8*(16), 14290–14320. DOI: 10.1021/acsomega.2c07840 PMID: 37125102

Kaur, R., Bhardwaj, A., & Gupta, S. (2023). Cancer treatment therapies: Traditional to modern approaches to combat cancers. *Molecular Biology Reports*, *50*(11), 9663–9676. DOI: 10.1007/s11033-023-08809-3 PMID: 37828275

Kaushik, N., Borkar, S. B., Nandanwar, S. K., Panda, P. K., Choi, E. H., & Kaushik, N. K. (2022). Nanocarrier cancer therapeutics with functional stimuli-responsive mechanisms. *Journal of Nanobiotechnology*, *20*(1), 152. DOI: 10.1186/s12951-022-01364-2 PMID: 35331246

Kaye, A. D., Shah, S. S., Johnson, C. D., De Witt, A. S., Thomassen, A. S., Daniel, C. P., Ahmadzadeh, S., Tirumala, S., Bembenick, K. N., Kaye, A. M., & Shekoohi, S. (2025). Tacrolimus- and Mycophenolate-Mediated Toxicity: Clinical Considerations and Options in Management of Post-Transplant Patients. *Current Issues in Molecular Biology*, *47*(1), 2. DOI: 10.3390/cimb47010002 PMID: 39852117

Kazemi Siyanaki, H. S., Rezaei, F., Kabiri, S., & Ashrafkhani, B. (2025). Computational modeling of combined laser-ultrasound hyperthermia therapy for breast cancer using gold nanoparticles: A simulation study. *Lasers in Medical Science*, *40*(1), 430. DOI: 10.1007/s10103-025-04657-8 PMID: 41081940

Kazemian, P., Yu, S.-Y., Thomson, S. B., Birkenshaw, A., Leavitt, B., & Ross, C. J. D. (2022). Lipid-nanoparticle-based delivery of CRISPR/Cas9 genome-editing components. *Molecular Pharmaceutics*, *19*(5), 1669–1686. DOI: 10.1021/acs.molpharmaceut.1c00916 PMID: 35594500

Kazi, R. N. A., Hasani, I. W., Khafaga, D. S. R., Kabba, S., Farhan, M., Aatif, M., Muteeb, G., & Fahim, Y. A. (2025). Nanomedicine: The Effective Role of Nanomaterials in Healthcare from Diagnosis to Therapy. *Pharmaceutics*, *17*(8), 987. Advance online publication. DOI: 10.3390/pharmaceutics17080987 PMID: 40871010

Keitel, U., Scheel, A., Thomale, J., Halpape, R., Kaulfuß, S., Scheel, C., & Dobbelstein, M. (2014). BclxL mediates therapeutic resistance of a mesenchymal breast cancer cell subpopulation. *Oncotarget*, *5*(23), 11778–11791. DOI: 10.18632/oncotarget.2634 PMID: 25473892

Keklikcioglu Cakmak, N., & Eroglu, A. (2023). Doxorubicin and tamoxifen loaded graphene oxide nanoparticle functionalized with chitosan and folic acid for anticancer drug delivery. *Polymer Bulletin*, *80*(2), 2171–2185. DOI: 10.1007/s00289-022-04549-9

Kesharwani, P., Kumar, V., Goh, K. W., Gupta, G., Alsayari, A., Wahab, S., & Sahebkar, A. (2025). PEGylated PLGA nanoparticles: Unlocking advanced strategies for cancer therapy. *Molecular Cancer*, *24*(1), 205. DOI: 10.1186/s12943-025-02410-x PMID: 40702523

Khalaf, K., Hana, D., Chou, J. T.-T., Singh, C., Mackiewicz, A., & Kaczmarek, M. (2021). Aspects of the tumor microenvironment involved in immune resistance and drug resistance. *Frontiers in Immunology, 12*, 656364. DOI: 10.3389/fimmu.2021.656364 PMID: 34122412

Khalbas, A. H., Albayati, T. M., Ali, N. S., & Salih, I. K. (2024). Drug loading methods and kinetic release models using of mesoporous silica nanoparticles as a drug delivery system: A review. *South African Journal of Chemical Engineering, 50*, 261–280. DOI: 10.1016/j.sajce.2024.08.013

Khalid-Salako, F., Salimi Khaligh, S., Fathi, F., Demirci, O. C., Öncer, N., Kurt, H., & Yüce, M. (2025). The Nanocarrier Landscape—Evaluating Key Drug Delivery Vehicles and Their Capabilities: A Translational Perspective. *ACS Applied Materials & Interfaces, 17*(26), 37383–37403. DOI: 10.1021/acsami.5c07366 PMID: 40526827

Khalighi, S., Reddy, K., Midya, A., Pandav, K. B., Madabhushi, A., & Abedalthagafi, M. (2024). Artificial intelligence in neuro-oncology: Advances and challenges in brain tumor diagnosis, prognosis, and precision treatment. *NPJ Precision Oncology, 8*(1), 80. DOI: 10.1038/s41698-024-00575-0 PMID: 38553633

Khan, A., Khan, R. A., Anas, M., Qadri, T. A., Ashiq, B., Younis, T., Ullah, N., Khan, A. M., Nasreen, Z., & Shinwari, Z. K. (2026). Biogenic Nanoparticles and Green Nanocomposites as Sustainable Antimicrobial Strategies Against Bacteria, Fungi and Viruses. *Nano-Horizons: Journal of Nanosciences and Nanotechnologies, 5*, 31 pages. DOI: 10.25159/3005-2602/20247

Khan, I. A., Yu, T., Yang, M., Liu, J., & Chen, Z. (2025). A systematic review of toxicity, biodistribution, and biosafety in upconversion nanomaterials: Critical insights into toxicity mitigation strategies and future directions for safe applications. *BME frontiers, 6*, 0120.

Khan, M. Z., Naeem, M. S., Khalid, S., Gill, Z., Ali, H., Iqbal, S., Shahid, S., Sulman, A., Awan, A., & Kharl, H. A. A. (2026). AI-GUIDED QSAR, MOLECULAR DYNAMICS, AND PHARMACOPHORE MODELING. *PharmA AI: Revolutionizing Drug Discovery and Healthcare with Artificial Intelligence*.

Khan, A., Anas, M., Qadri, T. A., Algopishi, U. B., Fayad, E., Binjawhar, D. N., Khan, R. A., Maaza, M., & Shinwari, Z. K. (2025). Eco-friendly synthesis of copper oxide-silver bimetallic nanoparticles using Withania coagulans: Characterization and biomedical potential through antioxidant, antibacterial, and catalytic activities. *Inorganic Chemistry Communications, 181*, 115131. 10.1016/j.inoche.2025.115131. DOI: 10.1016/j.inoche.2025.115131

Khan, A., Anas, M., Ullah, N., Khattak, W. A., Ali, M., Maaza, M., & Shinwari, Z. K. (2025). Harnessing Green Nanotechnology: Exploring the Potential of Silver, Copper, and their Nanocomposites for Biomedical and Agricultural Applications. *Pakistan Journal of Botany*, *57*(6), 6. DOI: 10.30848/PJB2025-6(5)

Khan, A., Ashiq, B., Qadri, T. A., Khattak, W. A., Anas, M., Khan, R. A., Abbas, S., Shinwari, Z. K., & Maaza, M. (2025). Nanotechnology Solutions For Advancing Sustainable Development Goals (SDGs): Integrating Nanomaterials For Global Sustainability. *BioNanoScience*, *15*(4), 590. DOI: 10.1007/s12668-025-02214-9

Khan, A., Qadri, T. A., Khan, R. A., Hassan, D., Ashiq, B., Sani, A., Shinwari, Z. K., & Maaza, M. (2026). Sustainable green synthesis of CeO 2–ZnO nanocomposites using Rhazya stricta: Enhanced synergistic solutions for biomedical challenges and environmental remediation. *RSC Advances*, *16*(26), 23370–23390. DOI: 10.1039/D6RA02267G PMID: 42100458

Khan, A., Younis, T., Anas, M., Ali, M., Shinwari, Z. K., Khalil, A. T., Munawar, K. S., Mohamed, H. E. A., Hkiri, K., Maaza, M., Seleiman, M. F., & Khan, N. (2025). Withania coagulans-mediated green synthesis of silver nanoparticles: Characterization and assessment of their phytochemical, antioxidant, toxicity, and antimicrobial activities. *BMC Plant Biology*, *25*(1), 574. DOI: 10.1186/s12870-025-06533-7 PMID: 40316892

Khan, F., Shariq, M., Asif, M., Siddiqui, M. A., Malan, P., & Ahmad, F. (2022). Green Nanotechnology: Plant-Mediated Nanoparticle Synthesis and Application. *Nanomaterials (Basel, Switzerland)*, *12*(4), 673. DOI: 10.3390/nano12040673 PMID: 35215000

Khan, I. U., Khan, R. U., Asif, H., Khalid, S. H., Asghar, S., Saleem, M., Shah, K. U., Shah, S. U., Rizvi, S. A., & Shahzad, Y. (2017). Co-delivery strategies to overcome multidrug resistance in ovarian cancer. *International Journal of Pharmaceutics*, *533*(1), 111–124. DOI: 10.1016/j.ijpharm.2017.09.060 PMID: 28947245

Khan, M. (2024). Polymers as Efficient Non-Viral Gene Delivery Vectors: The Role of the Chemical and Physical Architecture of Macromolecules. *Polymers*, *16*(18), 2629. Advance online publication. DOI: 10.3390/polym16182629 PMID: 39339093

Khanna, P., Ong, C., Bay, B. H., & Baeg, G. H. (2015). Nanotoxicity: An Interplay of Oxidative Stress, Inflammation and Cell Death. *Nanomaterials (Basel, Switzerland)*, *5*(3), 1163–1180. DOI: 10.3390/nano5031163 PMID: 28347058

Kharl, H. A. A. (2026). *PharmA AI: Revolutionizing Drug Discovery and Healthcare with Artificial Intelligence*. IBH Publisher.

Khoobchandani, M., Katti, K., Karikachery, A., Thipe, V., Srisrimal, D., & Mohandoss, D.. (2020). New approaches in breast cancer therapy through green nanotechnology and nano-Ayurvedic medicine. *International Journal of Nanomedicine*, *15*, 181–197. DOI: 10.2147/IJN.S219042 PMID: 32021173

Khorsand Zak, A., & Hashim, A. M. (2025). Advanced XRD peak broadening analysis of gallium-doped ZnO nanoparticles for crystallite size evaluation. *Scientific Reports*, *16*(1), 1717. DOI: 10.1038/s41598-025-31317-2 PMID: 41372362

Khorsheed, K., Asadi, F., Asadi, R., Aydoğmuş, E., Erzen, B., Karataş, M., Deniz, Ş., & Orhan, R. (2025). Process Optimization with Response Surface Methodology: A Comprehensive Approach by Experimental Design Types. *International Journal of Advanced Natural Sciences and Engineering Researches*, *9*, 546–558.

Khursheed, R., Dua, K., Vishwas, S., Gulati, M., Jha, N. K., Aldhafeeri, G. M., Alanazi, F. G., Goh, B. H., Gupta, G., Paudel, K. R., Hansbro, P. M., Chellappan, D. K., & Singh, S. K. (2022). Biomedical applications of metallic nanoparticles in cancer: Current status and future perspectives. *Biomedicine & Pharmacotherapy = Biomédecine & Pharmacothérapie*, *150*, 112951. DOI: 10.1016/j.biopha.2022.112951 PMID: 35447546

Kiattisewee, C., Dong, C., Fontana, J., Sugianto, W., Peralta-Yahya, P., Carothers, J. M., & Zalatan, J. G. (2021). Portable bacterial CRISPR transcriptional activation enables metabolic engineering in Pseudomonas putida. *Metabolic Engineering*, *66*, 283–295. DOI: 10.1016/j.ymben.2021.04.002 PMID: 33930546

Ki, M.-R., Kim, D. H., Abdelhamid, M. A. A., & Pack, S. P. (2025). Cancer and Aging Biomarkers: Classification, Early Detection Technologies and Emerging Research Trends. *Biosensors (Basel)*, *15*(11), 737. DOI: 10.3390/bios15110737 PMID: 41294748

Kim, D. J. (2026). Molecular and Immune Mechanisms Governing Cancer Metastasis, Including Dormancy, Microenvironmental Niches, and Tumor-Specific Programs. *International Journal of Molecular Sciences*, *27*(2), 875. DOI: 10.3390/ijms27020875 PMID: 41596524

Kim, K., Narsinh, K., & Ozhinsky, E. (2024). Technical advances in motion-robust MR thermometry. *Magnetic Resonance in Medicine*, *92*(1), 15–27. DOI: 10.1002/mrm.30057 PMID: 38501903

Kim, M., Hwang, Y., Lim, S., Jang, H.-K., & Kim, H.-O. (2024). Advances in nanoparticles as non-viral vectors for efficient delivery of CRISPR/Cas9. *Pharmaceutics*, *16*(9), 1197. DOI: 10.3390/pharmaceutics16091197 PMID: 39339233

Koç, M. M.. (2025). Nanoparticles in photothermal therapy-based medical and biological applications. *The European Physical Journal Plus*, *140*, 496. DOI: 10.1140/epjp/s13360-025-06452-4

Kok, H. P., Cressman, E. N., Ceelen, W., Brace, C. L., Ivkov, R., Grüll, H., Ter Haar, G., Wust, P., & Crezee, J. (2020). Heating technology for malignant tumors: A review. *International Journal of Hyperthermia*, *37*(1), 711–741. DOI: 10.1080/02656736.2020.1779357 PMID: 32579419

Kong, X., Gao, P., Wang, J., Fang, Y., & Hwang, K. (2023). Advances of medical nanorobots for future cancer treatments. *Journal of Hematology & Oncology*, *16*(1), 74. Advance online publication. DOI: 10.1186/s13045-023-01463-z PMID: 37452423

Kovalev, I. S., Zyryanov, G. V., Santra, S., Majee, A., Varaksin, M. V., & Charushin, V. N. (2022). Folic Acid Antimetabolites (Antifolates): A Brief Review on Synthetic Strategies and Application Opportunities. *Molecules (Basel, Switzerland)*, *27*(19), 6229. DOI: 10.3390/molecules27196229 PMID: 36234766

Kowalski, G., Witczak, M., & Kuterasiński, Ł. (2024). Structure Effects on Swelling Properties of Hydrogels Based on Sodium Alginate and Acrylic Polymers. *Molecules (Basel, Switzerland)*, *29*(9), 1937. Advance online publication. DOI: 10.3390/molecules29091937 PMID: 38731429

Krasnovskaya, O. O., Abramchuck, D., Erofeev, A., Gorelkin, P., Kuznetsov, A., Shemukhin, A., & Beloglazkina, E. K. (2023). Recent Advances in 64Cu/67Cu-Based Radiopharmaceuticals. *International Journal of Molecular Sciences*, *24*(11), 9154. DOI: 10.3390/ijms24119154 PMID: 37298101

Kraus, S., Arbib, S., Ventura-Bixenshpaner, H., Rukenstein, P., Hirsh-Kovacs, M., Shalev, B., Eltanani, M., Ron, U., Gumina, O., Shalev, O., & Geschwind, J. (2026). Case report: Treatment with magnetic nanoparticle-based hyperthermia stabilizes metastatic disease in stage IV primary liver cancer patients. *Frontiers in Oncology*, *15*, 1742249. DOI: 10.3389/fonc.2025.1742249 PMID: 41626176

Kučuk, N., Primožič, M., Knez, Ž., & Leitgeb, M. (2023). Sustainable biodegradable biopolymer-based nanoparticles for healthcare applications. *International Journal of Molecular Sciences*, *24*(4), 3188. DOI: 10.3390/ijms24043188 PMID: 36834596

Kumari, R., Xu, X., & Li, H. (2022). Translational and Clinical Relevance of PDX-Derived Organoid Models in Oncology Drug Discovery and Development. *Current Protocols*, *2*(7), e431. Advance online publication. DOI: 10.1002/cpz1.431 PMID: 35789132

Kumar, N., Chamoli, P., Misra, M., Manoj, M., & Sharma, A. (2022). Advanced metal and carbon nanostructures for medical, drug delivery and bio-imaging applications. *Nanoscale*, *14*(11), 3987–4017. DOI: 10.1039/D1NR07643D PMID: 35244647

Kumar, N., Singh, S., Sharma, P., Kumar, B., & Kumar, A. (2024). Single-, Dual-, and Multi-Stimuli-Responsive Nanogels for Biomedical Applications. *Gels (Basel, Switzerland)*, *10*(1), 61. Advance online publication. DOI: 10.3390/gels10010061 PMID: 38247784

Kumar, S., Banerjee, S., Ghosh, A., Pramanik, C., Dey, I., Choudhury, N., Sarkar, S., Das, A., Ghosh, S., Samanta, S., Das, S., Chattopadhyay, N., Bag, M., Talukdar, P., & Das, S. (2026). Biological synergies for heavy metal bioremediation across microbial, algal, and phytoremediation perspectives. *Discover Green Chemistry*, *1*(1), 3. DOI: 10.1007/s44509-026-00003-5

Kunjiappan, S., Pavadai, P., Vellaichamy, S., Ram Kumar Pandian, S., Ravishankar, V., Palanisamy, P., Govindaraj, S., Srinivasan, G., Premanand, A., Sankaranarayanan, M., & Theivendren, P. (2021). Surface receptor-mediated targeted drug delivery systems for enhanced cancer treatment: A state-of-the-art review. *Drug Development Research*, *82*(3), 309–340. DOI: 10.1002/ddr.21758 PMID: 33170541

Kurul, F., Turkmen, H., Cetin, A. E., & Topkaya, S. N. (2025). Nanomedicine: How nanomaterials are transforming drug delivery, bio-imaging, and diagnosis. *Next Nanotechnology, 7*, 100129. 10.1016/j.nxnano.2024.100129

Kurul, F., Turkmen, H., Cetin, A. E., & Topkaya, S. N. (2025). Nanomedicine: How nanomaterials are transforming drug delivery, bio-imaging, and diagnosis. *Next Nanotechnology, 7*, 100129. https://doi.org/https://doi.org/10.1016/j.nxnano.2024.100129

Kuwahata, A.. (2025). Magnetic hyperthermia using iron oxide nanoparticles via lymphatic drug delivery for metastatic lymph nodes. *Scientific Reports*, *15*, 40658. DOI: 10.1038/s41598-025-25808-5 PMID: 41290791

Kwak, K., Yu, B., Lewandowski, R. J., & Kim, D. H. (2022). Recent progress in cryoablation cancer therapy and nanoparticles mediated cryoablation. *Theranostics*, *12*(5), 2175–2204. DOI: 10.7150/thno.67530 PMID: 35265206

La Monica, G., Bono, A., Alamia, F., Lauria, A., & Martorana, A. (2026). Spatio-temporal control of prodrug activation through external stimuli for effective and safe on-site release in solid tumors: From current advances to future perspectives. *Drug Discovery Today*, *31*(1), 104595. https://doi.org/https://doi.org/10.1016/j.drudis.2025.104595. DOI: 10.1016/j.drudis.2025.104595 PMID: 41494612

La Rosa, R., Johansen, H. K., & Molin, S. (2022). Persistent bacterial infections, antibiotic treatment failure, and microbial adaptive evolution. *Antibiotics (Basel, Switzerland), 11*(3), 419. DOI: 10.3390/antibiotics11030419 PMID: 35326882

Lammers, T. (2024). Nanomedicine tumor targeting. *Advanced Materials, 36*(26), e2312169. DOI: 10.1002/adma.202312169 PMID: 38361435

Lee, B. J., Kanaan, M. H. G., Abdullah, S. S., & Ghasemian, A. (2025). Plant-derived nanoparticles in cancer therapy: A comprehensive review of recent advances and future prospects. *OBM Genetics, 9*(3), 308. DOI: 10.21926/obm.genet.2503308

Lee, H., Rho, W., Kim, Y.-H., Chang, H., & Jun, B.-H. (2025). CRISPR-Cas9 gene therapy: Non-viral delivery and stimuli-responsive nanoformulations. *Molecules (Basel, Switzerland), 30*(3), 542. DOI: 10.3390/molecules30030542 PMID: 39942646

Lee, J., Kim, J., Cho, E., Jeong, J., Song, J., & Kang, M. (2026). Nanotechnology-Enhanced Approaches for Early Cancer Diagnosis. *Nano Select, 7*(1), e70105. DOI: 10.1002/nano.70105

Lee, S. H., & Jun, B.-H. (2019). Silver Nanoparticles: Synthesis and Application for Nanomedicine. *International Journal of Molecular Sciences, 20*(4), 865. DOI: 10.3390/ijms20040865 PMID: 30781560

Lee, Y. J., & Kim, M. S. (2025). Advances in drug-loaded microspheres for targeted, controlled, and sustained drug delivery: Potential, applications, and future directions. *Biomedicine & Pharmacotherapy = Biomédecine & Pharmacothérapie, 189*, 118244. DOI: 10.1016/j.biopha.2025.118244 PMID: 40516334

Lei, T., Wang, Y., Zhang, Y., Yang, Y., Cao, J., Huang, J., Chen, J., Chen, H., Zhang, J., Wang, L., Xu, X., Gale, R. P., & Wang, L. (2024). Leveraging CRISPR gene editing technology to optimize the efficacy, safety and accessibility of CAR T-cell therapy. *Leukemia, 38*(12), 2517–2543. DOI: 10.1038/s41375-024-02444-y PMID: 39455854

Lei, Z. N., Teng, Q. X., Wu, Z. X., Ping, F. F., Song, P., Wurpel, J. N., & Chen, Z. S. (2021). Overcoming multidrug resistance by knockout of ABCB1 gene using CRISPR/Cas9 system in SW620/Ad300 colorectal cancer cells. *MedComm, 2*(4), 765–777. DOI: 10.1002/mco2.106 PMID: 34977876

Li, A., Zhao, Y., Li, Y., Jiang, L., Gu, Y., & Liu, J. (2021). Cell-derived biomimetic nanocarriers for targeted cancer therapy: Cell membranes and extracellular vesicles. *Drug Delivery, 28*(1), 1237–1255. DOI: 10.1080/10717544.2021.1938757 PMID: 34142930

Lian, S., Diko, C. S., Yan, Y., Li, Z., Zhang, H., Ma, Q., & Qu, Y. (2019). Characterization of biogenic selenium nanoparticles derived from cell-free extracts of a novel yeast Magnusiomyces ingens. *3 Biotech*, *9*(6), 221. DOI: 10.1007/s13205-019-1748-y

Li, C., Brant, E. J., Budak, H., & Zhang, B. (2021). CRISPR/Cas: A Nobel Prize award-winning precise genome editing technology for gene therapy and crop improvement. *Journal of Zhejiang University. Science. B*, *22*(3), 253–284. DOI: 10.1631/jzus.B2100009 PMID: 33835761

Li, C., Yang, T., Weng, Y., Zhang, M., Zhao, D., Guo, S., Hu, B., Shao, W., Wang, X., Hussain, A., Liang, X., & Huang, Y. (2021). Ionizable lipid-assisted efficient hepatic delivery of gene editing elements for oncotherapy. *Bioactive Materials*, *9*, 590–601. DOI: 10.1016/j.bioactmat.2021.05.051 PMID: 34853819

Li, H., Qiao, R., Davis, T. P., & Tang, S.-Y. (2020). Biomedical Applications of Liquid Metal Nanoparticles: A Critical Review. *Biosensors (Basel)*, *10*(12), 196. DOI: 10.3390/bios10120196 PMID: 33266097

Li, J., & Burgess, D. (2020). Nanomedicine-based drug delivery towards tumor biological and immunological microenvironment. *Acta Pharmaceutica Sinica. B*, *10*(11), 2110–2124. DOI: 10.1016/j.apsb.2020.05.008 PMID: 33304781

Li, J., Gu, A., Tang, N., Zengin, G., Li, M., & Liu, Y. (2025). Patient-derived xenograft models in pan-cancer: From bench to clinic. *Interdisciplinary Medicine*, *3*(5), e20250016. Advance online publication. DOI: 10.1002/INMD.20250016

Li, J., Wang, S., Fontana, F., Tapeinos, C., Shahbazi, M.-A., Han, H., & Santos, H. A. (2023). Nanoparticles-based phototherapy systems for cancer treatment: Current status and clinical potential. *Bioactive Materials*, *23*, 471–507. DOI: 10.1016/j.bioactmat.2022.11.013 PMID: 36514388

Li, L., Lee, C., Cruz, D. F., Krovi, S. A., Hudgens, M. G., Cottrell, M. L., & Johnson, L. M. (2022). Reservoir-Style Polymeric Drug Delivery Systems: Empirical and Predictive Models for Implant Design. *Pharmaceuticals (Basel, Switzerland)*, *15*(10), 1226. Advance online publication. DOI: 10.3390/ph15101226 PMID: 36297338

Lima, B. S., & Videira, M. A. (2018). Toxicology and biodistribution: The clinical value of animal biodistribution studies. *Molecular Therapy. Methods & Clinical Development*, *8*, 183–197. DOI: 10.1016/j.omtm.2018.01.003 PMID: 29541646

Lim, Z.-F. (2024). *Personalized Molecular Therapies for Advanced Non-Small Cell Lung Cancer: Overcoming Heterogeneity to Optimize Treatment Response and Clinical Outcomes*. West Virginia University.

Lin, J., Jiao, G., & Kermanshahi-Pour, A. (2022). Algal polysaccharides-based hydrogels: Extraction, synthesis, characterization, and applications. *Marine Drugs*, *20*(5), 306. DOI: 10.3390/md20050306 PMID: 35621958

Li, S., Wang, L., Wang, Y., Zhang, C., Hong, Z., & Han, Z. (2022). The synthetic lethality of targeting cell cycle checkpoints and PARPs in cancer treatment. *Journal of Hematology & Oncology*, *15*(1), 147. DOI: 10.1186/s13045-022-01360-x PMID: 36253861

Li, T., Yang, Y., Qi, H., Cui, W., Zhang, L., Fu, X., He, X., Liu, M., Li, P., & Yu, T. (2023). CRISPR/Cas9 therapeutics: Progress and prospects. *Signal Transduction and Targeted Therapy*, *8*(1), 36. DOI: 10.1038/s41392-023-01309-7 PMID: 36646687

Liu, B., Zhou, H., Tan, L., Siu, K., & Guan, X. (2024). Exploring treatment options in cancer: Tumor treatment strategies. *Signal Transduction and Targeted Therapy*, *9*(1), 1856. Advance online publication. DOI: 10.1038/s41392-024-01856-7 PMID: 39013849

Liu, G., Lovell, J. F., Zhang, L., & Zhang, Y. (2020). Stimulus-Responsive Nanomedicines for Disease Diagnosis and Treatment. *International Journal of Molecular Sciences*, *21*(17), 6380. Advance online publication. DOI: 10.3390/ijms21176380 PMID: 32887466

Liu, H., Lu, H.-H., Alp, Y., Wu, R., & Thayumanavan, S. (2023). Structural Determinants of Stimuli-Responsiveness in Amphiphilic Macromolecular Nano-assemblies. *Progress in Polymer Science*, *148*, 101765. Advance online publication. DOI: 10.1016/j.progpolymsci.2023.101765 PMID: 38476148

Liu, J., Zhao, C., Chen, W. R., & Zhou, B. (2022). Recent progress in two-dimensional nanomaterials for cancer theranostics. *Coordination Chemistry Reviews*, *469*, 214654. Advance online publication. DOI: 10.1016/j.ccr.2022.214654

Liu, M., & Yang, X. (2025). Patient-derived xenograft models: Current status, challenges, and innovations in cancer research. *Genes & Diseases*, *12*(5), 101520. Advance online publication. DOI: 10.1016/j.gendis.2025.101520 PMID: 40548062

Liu, Q., Chen, H., Mi, R., Min, X., Fang, M., Wu, X., Huang, Z., & Liu, Y. (2025). Biomass-derived carbon dots: Preparation, properties, and applications. *Nanomaterials (Basel, Switzerland)*, *15*(16), 1279. DOI: 10.3390/nano15161279 PMID: 40863859

Liu, X., Tang, I., Wainberg, Z., & Meng, H. (2020). Safety considerations in cancer nanomedicine. *Small*, *16*(36), 2000673. Advance online publication. DOI: 10.1002/smll.202000673 PMID: 32406992

Liu, Y., Li, C., Yang, X., Yang, B., & Fu, Q. (2024). Stimuli-responsive polymer-based nanosystems for cardiovascular disease theranostics. *Biomaterials Science*, *12*(15), 3805–3825. Advance online publication. DOI: 10.1039/D4BM00415A PMID: 38967109

Liu, Y., Liu, Y., Zang, J., Abdullah, A. A. I., Li, Y., & Dong, H. (2020). Design Strategies and Applications of ROS-Responsive Phenylborate Ester-Based Nanomedicine. *ACS Biomaterials Science & Engineering*, *6*(12), 6510–6527. DOI: 10.1021/acsbiomaterials.0c01190 PMID: 33320631

Liu, Y., Si, L., Jiang, Y., Jiang, S., Zhang, X., Li, S., Chen, J., & Hu, J. (2025). Design of pH-Responsive Nanomaterials Based on the Tumor Microenvironment. *International Journal of Nanomedicine*, *20*, 705–721. DOI: 10.2147/IJN.S504629 PMID: 39845771

Liu, Y., Wu, W., Cai, C., Zhang, H., Shen, H., & Han, Y. (2023). Patient-derived xenograft models in cancer therapy: Technologies and applications. *Signal Transduction and Targeted Therapy*, *8*(1), 160. Advance online publication. DOI: 10.1038/s41392-023-01419-2 PMID: 37045827

Liu, Y., Zhang, Y., Li, H., & Hu, T. Y. (2025). Recent advances in the bench-to-bedside translation of cancer nanomedicines. *Acta Pharmaceutica Sinica. B*, *15*(1), 97–122. DOI: 10.1016/j.apsb.2024.12.007 PMID: 40041906

Liu, Z., Dong, H., Cui, Y., Cong, L., & Zhang, D. (2020). Application of different types of CRISPR/Cas-based systems in bacteria. *Microbial Cell Factories*, *19*(1), 172. DOI: 10.1186/s12934-020-01431-z PMID: 32883277

Li, Y. R., Halladay, T., & Yang, L. (2024). Immune evasion in cell-based immunotherapy: Unraveling challenges and novel strategies. *Journal of Biomedical Science*, *31*(1), 5. DOI: 10.1186/s12929-024-00998-8 PMID: 38217016

Li, Y.. (2024). Nanomaterial-mediated photothermal therapy modulates tumor-associated macrophages for cancer treatment. *Journal of Materials Chemistry. B*, *12*, 11732–11749. DOI: 10.1039/D4TB01928H PMID: 39501854

Li, Z., Ning, F., Wang, C., Yu, H., Ma, Q., & Sun, Y. (2021). Normalization of the tumor microvasculature based on targeting and modulation of the tumor microenvironment. *Nanoscale*, *13*(41), 17254–17271. DOI: 10.1039/D1NR03387E PMID: 34651623

Lodi, M. B., Possidente, R., Melis, A., Di Meglio, A., Fanti, A., & Baccoli, R. (2025). Thermal behavior of magnetic scaffolds for RF-induced hyperthermia. *Applied Sciences (Basel, Switzerland)*, *15*(17), 9782. DOI: 10.3390/app15179782

Lombardi, L., Genio, V. D., Albericio, F., & Williams, D. R. (2025). Advances in peptidomimetics for next-generation therapeutics: Strategies, modifications, and applications. *Chemical Reviews*, *125*(15), 7099–7166. DOI: 10.1021/acs.chemrev.4c00989 PMID: 40698392

Long, X., Zhang, X., Chen, Q., Liu, M., Xiang, Y., Yang, Y., Xiao, Z., Huang, J., Wang, X., Liu, C., Nan, Y., & Huang, Q. (2022). Nucleus-targeting phototherapy nanodrugs for high-efficiency cancer treatment. *Frontiers in Pharmacology*, *13*, 905375. DOI: 10.3389/fphar.2022.905375 PMID: 35645841

Lopatkin, A. J., & Collins, J. J. (2020). Predictive biology: Modelling, understanding and harnessing microbial complexity. *Nature Reviews. Microbiology*, *18*(9), 507–520. DOI: 10.1038/s41579-020-0372-5 PMID: 32472051

Lotfi, M., Nejib, M., & Naceur, M. (2013). Cell adhesion to biomaterials: Concept of biocompatibility. *Advances in biomaterials science and biomedical applications, 8*, 208-240.

Lovatt, C., O'Sullivan, T., Ortega-de San Luis, C., Ryan, T. J., & Frank, R. A. W. (2026). Memory engram synapse 3D macromolecular architecture visualized by cryoCLEM-guided cryoET. *Structure (London, England)*, *34*(1), 100–112.e103. DOI: 10.1016/j.str.2025.10.014 PMID: 41240908

Luce, E., & Duclos-Vallée, J. (2025). Stem Cells and Organoids: A Paradigm Shift in Preclinical Models Toward Personalized Medicine. *Pharmaceuticals (Basel, Switzerland)*, *18*(7), 992. Advance online publication. DOI: 10.3390/ph18070992 PMID: 40732281

Lu, H., Zhang, S., Wang, J., & Chen, Q. (2021). A review on polymer and lipid-based nanocarriers and its application to nano-pharmaceutical and food-based systems. *Frontiers in Nutrition*, *8*, 783831. DOI: 10.3389/fnut.2021.783831 PMID: 34926557

Luiz, M. T., Dutra, J. A. P., Viegas, J. S. R., de Araújo, J. T. C., Tavares, A. G.Junior, & Chorilli, M. (2023). Hybrid magnetic lipid-based nanoparticles for cancer therapy. *Pharmaceutics*, *15*(3), 751. DOI: 10.3390/pharmaceutics15030751 PMID: 36986612

Lukyanov, A. N., & Torchilin, V. P. (2004). Micelles from lipid derivatives of water-soluble polymers as delivery systems for poorly soluble drugs. *Advanced Drug Delivery Reviews*, *56*(9), 1273–1289. DOI: 10.1016/j.addr.2003.12.004 PMID: 15109769

Luo, D., Yan, C., & Wang, T. (2015). Interparticle forces underlying nanoparticle self-assemblies. *Small*, *11*(45), 5984–6008. DOI: 10.1002/smll.201501783 PMID: 26436692

Lu, Y., Zhang, B., Wang, Z., He, Y., Ge, H., Ma, H., & Cui, P. (2026). From Host-Derived Pressures to the Environmental Anti-Antimicrobial Peptides Resistome: Mechanisms, Reservoirs and Implications for Therapeutic Peptide Design. *Marine Drugs*, *24*(2), 76. DOI: 10.3390/md24020076 PMID: 41745479

Ly, P. D., Ly, K. N., Phan, H. L., Nguyen, H. H. T., Duong, V. A., & Nguyen, H. V. (2024). Recent advances in surface decoration of nanoparticles in drug delivery. *Frontiers in Nanotechnology*, *6*, 1456939. DOI: 10.3389/fnano.2024.1456939

Lytle, N. K., Barber, A. G., & Reya, T. (2018). Stem cell fate in cancer growth, progression and therapy resistance. *Nature Reviews. Cancer*, *18*(11), 669–680. DOI: 10.1038/s41568-018-0056-x PMID: 30228301

Ma, B., Teixeira, W., & Jiang, L. (2024). Cryoablation techniques in bladder cancer [Frigid Zone Medicine]. *RE:view*, *4*(2), 72–77. doi:10.1515/fzm-2024-0008

Maboudi, A. H., Lotfipour, M. H., Rasouli, M., Azhdari, M., Macloughlin, R., Bekeschus, S., & Doroudian, M. (2024). Micelle-based nanoparticles with stimuli-responsive properties for drug delivery. *Nanotechnology Reviews*, *13*(1), 20230218. Advance online publication. DOI: 10.1515/ntrev-2023-0218

Madani, S. Y., Naderi, N., Dissanayake, O., Tan, A., & Seifalian, A. M. (2011). A new era of cancer treatment: Carbon nanotubes as drug delivery tools. *International Journal of Nanomedicine*, •••, 2963–2979. PMID: 22162655

Maimaitijiang, A., He, D., Li, D., Li, W., Su, Z., Fan, Z., & Li, J. (2024). Progress in Research of Nanotherapeutics for Overcoming Multidrug Resistance in Cancer. *International Journal of Molecular Sciences*, *25*(18), 9973. Advance online publication. DOI: 10.3390/ijms25189973 PMID: 39337463

Majhi, K. C., Karfa, P., Kumar, S., & Madhuri, R. (2019). Water as the green solvent in organic synthesis. *Ind. Appl. Green Solvents*, *2*, 182–201.

Makharadze, D., del Valle, L. J., Katsarava, R., & Puiggalí, J. (2025). The Art of PEGylation: From Simple Polymer to Sophisticated Drug Delivery System. *International Journal of Molecular Sciences*, *26*(7), 3102. DOI: 10.3390/ijms26073102 PMID: 40243857

Makhoba, X. H., Viegas, C.Jr, Mosa, R. A., Viegas, F. P., & Pooe, O. J. (2020). Potential impact of the multi-target drug approach in the treatment of some complex diseases. *Drug Design, Development and Therapy*, *14*, 3235–3249. DOI: 10.2147/DDDT.S257494 PMID: 32884235

Mak, I. W., Evaniew, N., & Ghert, M. (2014). Lost in translation: Animal models and clinical trials in cancer treatment. *American Journal of Translational Research, 6*(2), 114. PMID: 24489990

Malik, S., Muhammad, K., & Waheed, Y. (2023). Nanotechnology: A Revolution in Modern Industry. *Molecules (Basel, Switzerland), 28*(2), 661. DOI: 10.3390/molecules28020661 PMID: 36677717

Malinowska, J. M., & Whelan, M. (2025). The important role of standards for the uptake of transcriptomics and metabolomics based in vitro methods in regulatory toxicology. *Archives of Toxicology, 99*(10), 3865–3875. DOI: 10.1007/s00204-025-04119-8 PMID: 40739409

Manfioletti, G., & Fedele, M. (2023). *Epithelial–mesenchymal transition (EMT)* (Vol. 24). MDPI.

Mangu, K., Gudeli, R., & Rizwanullah, M. (2025). Recent advancement in polymeric nanoparticles for oral chemotherapy: Transforming cancer treatment. *BioImpacts, 15*, 31117. DOI: 10.34172/bi.31117 PMID: 41185868

Manzari-Tavakoli, A., Babajani, A., Tavakoli, M. M., Safaeinejad, F., & Jafari, A. (2024). Integrating natural compounds and nanoparticle-based drug delivery systems: A novel strategy for enhanced efficacy and selectivity in cancer therapy. *Cancer Medicine, 13*(5), e7010. DOI: 10.1002/cam4.7010 PMID: 38491817

Mao, Y., Xie, J., Yang, F., Luo, Y., Du, J., & Xiang, H. (2024). Advances and prospects of precision nanomedicine in personalized tumor theranostics [Review]. *Frontiers in Cell and Developmental Biology, Volume 12 - 2024*. DOI: 10.3389/fcell.2024.1514399

Mao, Y., Shangguan, D., Huang, Q., Xiao, L., Cao, D., Zhou, H., & Wang, Y.-K. (2025). Emerging artificial intelligence-driven precision therapies in tumor drug resistance: Recent advances, opportunities, and challenges. *Molecular Cancer, 24*(1), 123. DOI: 10.1186/s12943-025-02321-x PMID: 40269930

Ma, P., & Mumper, R. J. (2013). Paclitaxel nano-delivery systems: A comprehensive review. *Journal of Nanomedicine & Nanotechnology, 4*(2), 1000164. DOI: 10.4172/2157-7439.1000164 PMID: 24163786

Marinho, A., Nunes, C., & Reis, S. (2021). Hyaluronic acid: A key ingredient in the therapy of inflammation. *Biomolecules, 11*(10), 1518. DOI: 10.3390/biom11101518 PMID: 34680150

Marquez, J. M. (2026). William Halsted's Radical Mastectomy Patients: A Historical Analysis Through Patient Narratives. *Annals of Surgery Open : Perspectives of Surgical History, Education, and Clinical Approaches*, *7*(1), e651. DOI: 10.1097/AS9.0000000000000651 PMID: 41890517

Martins, G., Luiz Gogola, J., Gevaerd, A., Kalinke, C., de Oliveira, P. R., Caetano, F. R., Janegitz, B. C., Bergamini, M. F., & Marcolino-Junior, L. H. (2026). Exploring the Potential of Zero-Dimensional Carbon Nanomaterials in Photoluminescent, Electrochemiluminescent and Electrochemical Sensors. *Advanced Materials Interfaces*, *13*(11), 01018. DOI: 10.1002/admi.202501018

Marusyk, A., Almendro, V., & Polyak, K. (2012). Intra-tumour heterogeneity: A looking glass for cancer? *Nature Reviews. Cancer*, *12*(5), 323–334. DOI: 10.1038/nrc3261 PMID: 22513401

Masimirembwa, C. M., Bredberg, U., & Andersson, T. B. (2003). Metabolic stability for drug discovery and development: Pharmacokinetic and biochemical challenges. *Clinical Pharmacokinetics*, *42*(6), 515–528. DOI: 10.2165/00003088-200342060-00002 PMID: 12793837

Mavila, S., Eivgi, O., Berkovich, I., & Lemcoff, N. G. (2016). Intramolecular cross-linking methodologies for the synthesis of polymer nanoparticles. *Chemical Reviews*, *116*(3), 878–961. DOI: 10.1021/acs.chemrev.5b00290 PMID: 26270435

Ma, X., Shu, C., Guo, J., Pang, L., Su, L., Fu, D., & Zhong, W. (2014). Targeted cancer therapy based on single-wall carbon nanohorns with doxorubicin in vitro and in vivo. *Journal of Nanoparticle Research*, *16*(7), 2497. DOI: 10.1007/s11051-014-2497-9

Ma, X., Tian, Y., Yang, R., Wang, H., Allahou, L. W., Chang, J., Williams, G., Knowles, J. C., & Poma, A. (2024). Nanotechnology in healthcare, and its safety and environmental risks. *Journal of Nanobiotechnology*, *22*(1), 715. DOI: 10.1186/s12951-024-02901-x PMID: 39548502

Meng, W., Huang, L., Guo, J., Xin, Q., Liu, J., & Hu, Y. (2024). Innovative Nanomedicine Delivery: Targeting Tumor Microenvironment to Defeat Drug Resistance. *Pharmaceutics*, *16*(12), 1549. Advance online publication. DOI: 10.3390/pharmaceutics16121549 PMID: 39771528

Menon, G., Anandan, T., Kolage, S. M., Shenoy, S. C., Lohakare, O. K., Ahire, A. S., & Ravindran, D. (2025). Theranostics in the management of colorectal cancer. *Discover Nano*, *20*(1), 214. DOI: 10.1186/s11671-025-04390-9 PMID: 41276733

Menon, S., & Shanmugam, V. (2020). Cytotoxicity Analysis of Biosynthesized Selenium Nanoparticles Towards A549 Lung Cancer Cell Line. *Journal of Inorganic and Organometallic Polymers and Materials*, *30*(5), 1852–1864. DOI: 10.1007/s10904-019-01409-4

Meraj, A., Singh, S. P., Jawaid, M., Nasef, M. M., Alomar, T. S., & AlMasoud, N. (2023). A review on eco-friendly isolation of lignin by natural deep eutectic solvents from agricultural wastes. *Journal of Polymers and the Environment*, *31*(8), 3283–3316. DOI: 10.1007/s10924-023-02817-x

Metselaar, J. M., & Lammers, T. (2020). Challenges in nanomedicine clinical translation. *Drug Delivery and Translational Research*, *10*(3), 721–725. DOI: 10.1007/s13346-020-00740-5 PMID: 32166632

Miao, G., Zhou, L., Zhu, Y., & Zhang, Y. (2023). Accelerated blood clearance of PEGylated nanoparticles in cancer therapy: Mechanisms and mitigation strategies. *Acta Biomaterialia*, *170*, 1–16. DOI: 10.1016/j.actbio.2023.08.040

Mikušová, V., & Mikuš, P. (2021). Advances in chitosan-based nanoparticles for drug delivery. *International Journal of Molecular Sciences*, *22*(17), 9652. DOI: 10.3390/ijms22179652 PMID: 34502560

Mills, J. A., Liu, F., Jarrett, T. R., Fletcher, N. L., & Thurecht, K. J. (2022). Nanoparticle based medicines: approaches for evading and manipulating the mononuclear phagocyte system and potential for clinical translation []. *Biomaterials Science*, *10*(12), 3029–3053. DOI: 10.1039/D2BM00181K

Mi, P. (2020). Stimuli-responsive nanocarriers for drug delivery, tumor imaging, therapy and theranostics. *Theranostics*, *10*(10), 4557–4588. DOI: 10.7150/thno.38069 PMID: 32292515

Mishra, M., Ahmed, R., Das, D. K., Pramanik, D. D., Dash, S. K., & Pramanik, A. (2024). Recent advancements in the application of circulating tumor DNA as biomarkers for early detection of cancers. *ACS Biomaterials Science & Engineering*, *10*(8), 4740–4756. DOI: 10.1021/acsbiomaterials.4c00606 PMID: 38950521

Mishra, P., Ray, S., Sinha, S., Das, B., Khan, M., Behera, S. K., Yun, S.-I., Tripathy, S. K., & Mishra, A. (2016). Facile bio-synthesis of gold nanoparticles by using extract of Hibiscus sabdariffa and evaluation of its cytotoxicity against U87 glioblastoma cells under hyperglycemic condition. *Biochemical Engineering Journal*, *105*, 264–272. DOI: 10.1016/j.bej.2015.09.021

Mitchell, M. J., Billingsley, M. M., Haley, R. M., Wechsler, M. E., Peppas, N. A., & Langer, R. (2021). Engineering precision nanoparticles for drug delivery. *Nature Reviews. Drug Discovery*, *20*(2), 101–124. DOI: 10.1038/s41573-020-0090-8 PMID: 33277608

Mittal, A. K., Chisti, Y., & Banerjee, U. C. (2013). Synthesis of metallic nanoparticles using plant extracts. *Biotechnology Advances*, *31*(2), 346–356. DOI: 10.1016/j.biotechadv.2013.01.003 PMID: 23318667

Moacă, E. A.. (2022). Biologic impact of green synthesized magnetic iron oxide nanoparticles. *Pharmaceutics*, *15*(1), 95. DOI: 10.3390/pharmaceutics15010002 PMID: 36678723

Mobarak, M. H., & Al Mahmud, M., Hossain, A., Abir, Md. A. H., Hossain, N., & Chowdhury, M. A. (2026). Advances of Green-Synthesized Nanoparticles for Biomedical Applications. *BioMed Research International*, *9207147*(1), •••. DOI: 10.1155/bmri/9207147 PMID: 41925124

Moghimi, N., Hosseini, S., Dalan, A., Mohammadizadeh, D., Goldman, A., & Kohandel, M. (2023). Controlled tumor heterogeneity in a co-culture system by 3D bio-printed tumor-on-chip model. *Scientific Reports*, *13*(1), 13648. Advance online publication. DOI: 10.1038/s41598-023-40680-x PMID: 37607994

Mohamed, R. G. A., Ali, S. M., Ahmed, I. S., Rawas-Qalaji, M., & Hussain, Z. (2025). Next-generation nanocarriers for colorectal cancer: Passive, active, and stimuli-responsive strategies for precision therapy. *Biomaterials Science*, *13*(20), 5626–5664. DOI: 10.1039/D5BM01176K PMID: 40928395

Mohammed Fayaz, A., Balaji, K., Kalaichelvan, P. T., & Venkatesan, R. (2009). Fungal based synthesis of silver nanoparticles—An effect of temperature on the size of particles. *Colloids and Surfaces. B, Biointerfaces*, *74*(1), 123–126. DOI: 10.1016/j.colsurfb.2009.07.002 PMID: 19674875

Mohammed, A. E., & Al-Megrin, W. A. (2021). Biological Potential of Silver Nanoparticles Mediated by Leucophyllum frutescens and Russelia equisetiformis Extracts. *Nanomaterials (Basel, Switzerland)*, *11*(8), 2098. DOI: 10.3390/nano11082098 PMID: 34443930

Mohan, H., Bincoletto, V., Arpicco, S., & Giordani, S. (2022). Supramolecular functionalisation of B/N co-doped carbon nano-onions for novel nanocarrier systems. *Materials (Basel)*, *15*(17), 5987. DOI: 10.3390/ma15175987 PMID: 36079368

Molaei, M. J.. (2024). Magnetic hyperthermia in cancer therapy, mechanisms, challenges, and perspectives. *Journal of Biomaterials Applications*, *39*, 786–808. DOI: 10.1177/08853282241244707 PMID: 38606627

Molla, G., & Bitew, M. (2025). The Future of Cancer Diagnosis and Treatment: Unlocking the Power of Biomarkers and Personalized Molecular-Targeted Therapies. *Journal of Molecular Pathology*, *6*(3), 20. DOI: 10.3390/jmp6030020

Molla, K. A., & Yang, Y. (2019). CRISPR/Cas-mediated base editing: Technical considerations and practical applications. *Trends in Biotechnology*, *37*(10), 1121–1142. DOI: 10.1016/j.tibtech.2019.03.008 PMID: 30995964

Monica Ahmad, N., Husaini Mohamed, A., Hasan, N., Zainal- Abidin, N., Zaini Nawahwi, M., & Mohamad Azzeme, A. (2024). Effect of optimisation variable and the role of plant extract in the synthesis of nanoparticles using plant-mediated synthesis approaches. *Inorganic Chemistry Communications*, *161*, 111839. DOI: 10.1016/j.inoche.2023.111839

Monti, C., Zilocchi, M., Colugnat, I., & Alberio, T. (2019). Proteomics turns functional. *Journal of Proteomics*, *198*, 36–44. DOI: 10.1016/j.jprot.2018.12.012 PMID: 30553948

More, P. R., Pandit, S., Filippis, A. D., Franci, G., Mijakovic, I., & Galdiero, M. (2023). Silver Nanoparticles: Bactericidal and Mechanistic Approach against Drug Resistant Pathogens. *Microorganisms*, *11*(2), 369. DOI: 10.3390/microorganisms11020369 PMID: 36838334

Mostafavi, E., Iravani, S., Varma, R. S., Khatami, M., & Rahbarizadeh, F. (2022). Eco-friendly synthesis of carbon nanotubes and their cancer theranostic applications. *Materials Advances*, *3*(12), 4765–4782. DOI: 10.1039/D2MA00341D PMID: 35812837

Mougiakos, I., Bosma, E. F., Ganguly, J., Van Der Oost, J., & Van Kranenburg, R. (2018). Hijacking CRISPR-Cas for high-throughput bacterial metabolic engineering: Advances and prospects. *Current Opinion in Biotechnology*, *50*, 146–157. DOI: 10.1016/j.copbio.2018.01.002 PMID: 29414054

Mourdikoudis, S., Pallares, R. M., & Thanh, N. T. K. (2018). Characterization techniques for nanoparticles: Comparison and complementarity upon studying nanoparticle properties. *Nanoscale*, *10*(27), 12871–12934. DOI: 10.1039/C8NR02278J PMID: 29926865

Mousavi-Kiasary, S. M. S., Senabreh, A., Zandi, A., Pena, R., Cruz, F., Adibi, A., & Hooshmand, N. (2025). Synergistic cancer therapies enhanced by nanoparticles: Advancing nanomedicine through multimodal strategies. *Pharmaceutics*, *17*(6), 682. DOI: 10.3390/pharmaceutics17060682 PMID: 40573996

Muhammad, A. I., Dalia, A. A. M., Hemly, N. I. M., Zainudin, N. N., & Samsudin, A. A. (2025). Biofortified Bacteria: The Role of Selenium-Enriched Microorganisms in Enhancing Animal Selenium Uptake—A Review. *Journal of Animal Physiology and Animal Nutrition*, *109*(6), 1298–1320. DOI: 10.1111/jpn.70001 PMID: 40674621

Mukherjee, P., Ahmad, A., Mandal, D., Senapati, S., Sainkar, S. R., Khan, M. I., Parishcha, R., Ajaykumar, P. V., Alam, M., Kumar, R., & Sastry, M. (2001). Fungus-Mediated Synthesis of Silver Nanoparticles and Their Immobilization in the Mycelial Matrix: A Novel Biological Approach to Nanoparticle Synthesis. *Nano Letters*, *1*(10), 515–519. DOI: 10.1021/nl0155274

Mukherjee, P., Bhattacharya, R., Bone, N., Lee, Y. K., Patra, C., Wang, S., Lu, L., Secreto, C., Banerjee, P. C., Yaszemski, M. J., Kay, N. E., & Mukhopadhyay, D. (2013). Correction: Potential therapeutic application of gold nanoparticles in B-chronic lymphocytic leukemia (BCLL): enhancing apoptosis. *Journal of Nanobiotechnology*, *11*(1), 23. DOI: 10.1186/1477-3155-11-23

Mundekkad, D., & Cho, W. (2022). Nanoparticles in Clinical Translation for Cancer Therapy. *International Journal of Molecular Sciences*, *23*(3), 1685. Advance online publication. DOI: 10.3390/ijms23031685 PMID: 35163607

Murar, M., Pujals, S., & Albertazzi, L. (2023). Multivalent effect of peptide functionalized polymeric nanoparticles towards selective prostate cancer targeting. *Nanoscale Advances*, *5*(5), 1378–1385. DOI: 10.1039/D2NA00601D PMID: 36866255

Murray, C. J., & Lopez, A. D. (1997). Mortality by cause for eight regions of the world: Global Burden of Disease Study. *Lancet*, *349*(9061), 1269–1276. DOI: 10.1016/S0140-6736(96)07493-4 PMID: 9142060

Mustapha, S., Ndamitso, M. M., Abdulkareem, A. S., Tijani, J. O., Shuaib, D. T., Mohammed, A. K., & Sumaila, A. (2019). Comparative study of crystallite size using Williamson-Hall and Debye-Scherrer plots for ZnO nanoparticles. *Advances in Natural Sciences: Nanoscience and Nanotechnology*, *10*(4), 045013. DOI: 10.1088/2043-6254/ab52f7

N v, N. K., Mathew, A. A., Jafri, R. I., & Balachandran, M.NV. (2026). Biomass-derived Carbon Dots for Fluorescence-based Pollutant Sensing, Cellular Imaging, Antibacterial and Anticancer Therapy. *BioNanoScience*, *16*(4), 291. DOI: 10.1007/s12668-026-02527-3

Nadeem-Tariq, A., Rafanan, J. R., Kang, N., Zhang, S., Kanniyappan, H., & Merchant, A. (2026). Electrochemical Detection of Cancer Biomarkers: From Molecular Sensing to Clinical Translation. *Biosensors (Basel), 16*(1), 44. DOI: 10.3390/bios16010044 PMID: 41590296

Naderinezhad, S., Amoabediny, G., & Haghiralsadat, F. (2017). Co-delivery of hydrophilic and hydrophobic anticancer drugs using biocompatible pH-sensitive lipid-based nano-carriers for multidrug-resistant cancers. *RSC Advances, 7*(48), 30008–30019. DOI: 10.1039/C7RA01736G

Naeem, M. S., Ahmed, S., Sulman, A., Usman, S., Farooq, M. A., Talha, M., Awan, A., Afzal, Z., Khan, M. Z., & Hanif, A. (2025). Role of Medicinal Plants in the Management of Metabolic Disorders: Mechanisms, Efficacy, and Therapeutic Prospects.

Nagasawa, S., Kashima, Y., Suzuki, A., & Suzuki, Y. (2021). Single-cell and spatial analyses of cancer cells: Toward elucidating the molecular mechanisms of clonal evolution and drug resistance acquisition. *Inflammation and Regeneration, 41*(1), 22. DOI: 10.1186/s41232-021-00170-x PMID: 34271973

Nagaya, T., Nakamura, Y. A., Choyke, P. L., & Kobayashi, H. (2017). Fluorescence-Guided Surgery [Review]. *Frontiers in Oncology, Volume 7 - 2017*. DOI: 10.3389/fonc.2017.00314

Nair, D. G., & Weiskirchen, R. (2024). Advanced In Vitro Models for Preclinical Drug Safety: Recent Progress and Prospects. *Current Issues in Molecular Biology, 47*(1), 7. Advance online publication. DOI: 10.3390/cimb47010007 PMID: 39852122

Najdian, A., Beiki, D., Abbasi, M., Gholamrezanezhad, A., Ahmadzadehfar, H., Amani, A. M., Shafiee Ardestani, M., & Assadi, M. (2024). Exploring innovative strides in radiolabeled nanoparticle progress for multimodality cancer imaging and theranostic applications. *Cancer Imaging; the Official Publication of the International Cancer Imaging Society, 24*(1), 127. DOI: 10.1186/s40644-024-00762-z PMID: 39304961

Nalla, B. T., Choudhary, B. K., Sunil Kumar, M., Ramachandra, C. G., Vaghela, D. R., Pradhan, P., Smerat, A., & Kamakshi Priya, K. (2025). Nanocatalyst-enabled waste-to-energy systems: from material innovation to techno-economic and sustainability pathways. *Reviews in Inorganic Chemistry*(0).

Narwade, M., Shaikh, A., Gajbhiye, K. R., Kesharwani, P., & Gajbhiye, V. (2023). Advanced cancer targeting using aptamer functionalized nanocarriers for site-specific cargo delivery. *Biomaterials Research, 27*(1), 42. DOI: 10.1186/s40824-023-00365-y PMID: 37149607

Nascimento, A. T. D., Stoddart, P. R., Goris, T., Kael, M., Manasseh, R., Alt, K., Tashkandi, J., Kim, B. C., & Moulton, S. E. (2025). Stimuli-Responsive Materials for Biomedical Applications. *Advanced Materials*, *37*(36), e07559. Advance online publication. DOI: 10.1002/adma.202507559 PMID: 40801175

Natarajan, J. V., Darwitan, A., Barathi, V. A., Ang, M., Htoon, H. M., Boey, F., Tam, K. C., Wong, T. T., & Venkatraman, S. S. (2014). Sustained drug release in nanomedicine: A long-acting nanocarrier-based formulation for glaucoma. *ACS Nano*, *8*(1), 419–429. DOI: 10.1021/nn4046024 PMID: 24392729

Naveenkumar, S., Alagumanikumaran, N., Kaviyarasu, K., & Muthukumaran, A. (2024). Influence of encapsulated sodium alginates and pectin on selenium nanoparticles and efficient cardioprotective effect in C2C12 cell line. *Journal of Nanoparticle Research*, *26*(3), 52. DOI: 10.1007/s11051-024-05956-x

Nayak, N., & Sahoo, T. R. (2025). Sustainable Nanomaterials Production Using Bio-based Polymers. *Green Chemistry Methods for Producing Biopolymeric Nanomaterials*, 133-160.

Nayak, P., Bentivoglio, V., Varani, M., & Signore, A. (2023). Tumor spheroid models for anticancer evaluation. *Cancers (Basel)*, •••, 15. PMID: 37835541

Németh, Z., Csóka, I., Semnani Jazani, R., Sipos, B., Haspel, H., Kozma, G., Kónya, Z., & Dobó, D. G. (2022). Quality by Design-Driven Zeta Potential Optimisation Study of Liposomes with Charge Imparting Membrane Additives. *Pharmaceutics*, *14*(9), 1798. Advance online publication. DOI: 10.3390/pharmaceutics14091798 PMID: 36145546

Nevedomskaya, E., & Haendler, B. (2022). From omics to multi-omics approaches for in-depth analysis of the molecular mechanisms of prostate cancer. *International Journal of Molecular Sciences*, *23*(11), 6281. DOI: 10.3390/ijms23116281 PMID: 35682963

Nidhi, S., Anand, U., Oleksak, P., Tripathi, P., Lal, J. A., Thomas, G., Kuca, K., & Tripathi, V. (2021). Novel CRISPR–Cas systems: An updated review of the current achievements, applications, and future research perspectives. *International Journal of Molecular Sciences*, *22*(7), 3327. DOI: 10.3390/ijms22073327 PMID: 33805113

Niemeyer, C. M. (2001). Nanoparticles, proteins, and nucleic acids: Biotechnology meets materials science. *Angewandte Chemie International Edition*, *40*(22), 4128–4158. DOI: 10.1002/1521-3773(20011119)40:22<4128::AID-ANIE4128>3.0.CO;2-S PMID: 29712109

Nikhar, S., Jahan, A., Mahato, T., & Giri, A. S. (2026). Peptide and Antibody Functionalized Nanoparticles: Active Targeting Strategies for Next-Generation Contrast Agents and Drug Delivery Systems. *ACS Applied Nano Materials*, *9*(23), 10448–10477. Advance online publication. DOI: 10.1021/acsanm.5c05449

Nirmala, M., Kizhuveetil, U., Johnson, A., G, B., Nagarajan, R., & Muthuvijayan, V. (2023). Cancer nanomedicine: A review of nano-therapeutics and challenges ahead. *RSC Advances*, *13*(13), 8606–8629. DOI: 10.1039/D2RA07863E PMID: 36926304

Nisha, N.. (2024). Plant-mediated gold nanoparticles in cancer therapy: Green synthesis, mechanisms and applications. *Frontiers in Nanotechnology*, *6*, 1490980. DOI: 10.3389/fnano.2024.1490980

Nugent, P., Duncan, J. N., & Colagiovanni, D. (2017). Preparation of a Preclinical Dossier to Support an Investigational New Drug (IND) Application and First-In-Human Clinical Trial. In (pp. 189-213). DOI: 10.1016/B978-0-12-803620-4.00008-6

Nunziata, G., Borroni, A., & Rossi, F. (2025). Advanced microfluidic strategies for core-shell nanoparticles: The next-generation of polymeric and lipid-based drug nanocarriers. *Chemical Engineering Journal Advances*, *22*, 100759. https://doi .org/https://doi.org/10.1016/j.ceja.2025.100759. DOI: 10.1016/j.ceja.2025.100759

Obalola, A. A., Abrahamse, H., & Dhilip Kumar, S. S. (2025). Recent advances in gold nanoparticle-based targeted photodynamic and photothermal therapies. *Nanoscale Advances*, *7*(23), 5853–5875. DOI: 10.1039/D5NA00767D PMID: 41159015

Okesola, B. O., & Smith, D. K. (2016). Applying low-molecular weight supramolecular gelators in an environmental setting–self-assembled gels as smart materials for pollutant removal. *Chemical Society Reviews*, *45*(15), 4226–4251. DOI: 10.1039/C6CS00124F PMID: 27241027

Olędzka, A. J., & Czerwińska, M. E. (2023). Role of Plant-Derived Compounds in the Molecular Pathways Related to Inflammation. *International Journal of Molecular Sciences*, *24*(5), 4666. DOI: 10.3390/ijms24054666 PMID: 36902097

Oliveira, B., Fernandes, A., & Baptista, P. (2024). Gene silencing in 2D vs 3D cultures. *Frontiers in Bioengineering and Biotechnology*, •••, 12.

Oliveira, T. P. D. S., Lima, A. K. O., Gonçalves, T. P., Florêncio, I., Báo, S. N., Silva, N. O. D., Albuquerque, P., Silva-Pereira, I., & Muehlmann, L. A. (2026). Synthesis of Selenium Nanoparticles: Influence of Reaction Parameters on Physicochemical, Morphological, and Biological Properties. *Drugs and Drug Candidates*, *5*(1), 22. DOI: 10.3390/ddc5010022

Omidian, H., Cubeddu, L. X., & Wilson, R. L. (2025). Peptide-functionalized nanomedicine: Advancements in drug delivery, diagnostics, and biomedical applications. *Molecules (Basel, Switzerland)*, *30*(7), 1572. DOI: 10.3390/molecules30071572 PMID: 40286158

Omran, B., & Baek, K.-H. (2021). Nanoantioxidants: Pioneer Types, Advantages, Limitations, and Future Insights. *Molecules (Basel, Switzerland)*, *26*(22), 7031. DOI: 10.3390/molecules26227031 PMID: 34834124

Ong, L., Chia, S., Wong, S., Zhang, X., Chua, H., Loo, J., Chua, W., Chua, C., Tan, E., Hentze, H., Tan, I., Dasgupta, R., & Toh, Y. (2022). A comparative study of tumour-on-chip models with patient-derived xenografts for predicting chemotherapy efficacy in colorectal cancer patients. *Frontiers in Bioengineering and Biotechnology*, *10*, 952726. Advance online publication. DOI: 10.3389/fbioe.2022.952726 PMID: 36147524

Ortiz-Tirado, A., Medina-Ganem, L. F., Bandala, E. R., Conejo-Davila, A. S., Vega-Rios, A., Goonetilleke, A., & Rodriguez-Narvaez, O. M. (2025). Biogenic nanoparticles: Synthesis, characterization, applications and scaling up limitations in water treatment. *Journal of Environmental Chemical Engineering*, *13*(5), 117730. DOI: 10.1016/j.jece.2025.117730

Osataphan, N., Phrommintikul, A., Chattipakorn, S. C., & Chattipakorn, N. (2020). Effects of doxorubicin-induced cardiotoxicity on cardiac mitochondrial dynamics and mitochondrial function: Insights for future interventions. *Journal of Cellular and Molecular Medicine*, *24*(12), 6534–6557. DOI: 10.1111/jcmm.15305 PMID: 32336039

Osman, A. I., Zhang, Y., Farghali, M., Rashwan, A. K., Eltaweil, A. S., Abd El-Monaem, E. M., Mohamed, I. M. A., Badr, M. M., Ihara, I., Rooney, D. W., & Yap, P.-S. (2024). Synthesis of green nanoparticles for energy, biomedical, environmental, agricultural, and food applications: A review. *Environmental Chemistry Letters*, *22*(2), 841–887. DOI: 10.1007/s10311-023-01682-3

Ou, W., Stewart, S., White, A., Kwizera, E. A., Xu, J., Fang, Y., Shamul, J. G., Xie, C., Nurudeen, S., Tirada, N. P., Lu, X., Tkaczuk, K. H. R., & He, X. (2023). In-situ cryo-immune engineering of tumor microenvironment with cold-responsive nanotechnology for cancer immunotherapy. *Nature Communications*, *14*(1), 392. DOI: 10.1038/s41467-023-36045-7 PMID: 36693842

Overchuk, M., Weersink, R. A., Wilson, B. C., & Zheng, G. (2023). Photodynamic and photothermal therapies: Synergy opportunities for nanomedicine. *ACS Nano*, *17*(9), 7979–8003. DOI: 10.1021/acsnano.3c00891 PMID: 37129253

Padrela, L., Mouras, R., & Killackey, D. (2025). Innovation testbeds as enabling ecosystems driving nanopharmaceuticals to market. *Drug Discovery Today*, *30*(8), 104433. DOI: 10.1016/j.drudis.2025.104433 PMID: 40675439

Pallares, R. M., Barmin, R. A., Wang, A., Kiessling, F., & Lammers, T. (2025). Clinical cancer nanomedicines. *Journal of Controlled Release : Official Journal of the Controlled Release Society*, *385*, 113991. DOI: 10.1016/j.jconrel.2025.113991 PMID: 40582644

Pandey, S., Awasthee, N., Shekher, A., Rai, L. C., Gupta, S. C., & Dubey, S. K. (2021). Biogenic synthesis and characterization of selenium nanoparticles and their applications with special reference to antibacterial, antioxidant, anticancer and photocatalytic activity. *Bioprocess and Biosystems Engineering*, *44*(12), 2679–2696. DOI: 10.1007/s00449-021-02637-0 PMID: 34599397

Panzarini, E., Mariano, S., Carata, E., Mura, F., Rossi, M., & Dini, L. (2018). Intracellular Transport of Silver and Gold Nanoparticles and Biological Responses: An Update. *International Journal of Molecular Sciences*, *19*(5), 1305. DOI: 10.3390/ijms19051305 PMID: 29702561

Papa, S., Ortolani, C., Fernández, P., & O'Connor, J. E. (2023). Flow Cytometry and Its Applications to Molecular Biology and Diagnosis 2.0. *International Journal of Molecular Sciences*, *24*(22), 16215. Advance online publication. DOI: 10.3390/ijms242216215 PMID: 38003405

Parashar, A. K., Saraogi, G. K., Jain, P. K., Kurmi, B., Shrivastava, V., & Arora, V. (2024). Polymer-drug conjugates: Revolutionizing nanotheranostic agents for diagnosis and therapy. *Discover. Oncology*, *15*(1), 641. DOI: 10.1007/s12672-024-01509-9 PMID: 39527173

Park, S., Lu, G., Zheng, Y., Davison, E. K., & Li, Y. (2025). Nanoparticle-Based Delivery Strategies for Combating Drug Resistance in Cancer Therapeutics. *Cancers (Basel)*, *17*(16), 2628. Advance online publication. DOI: 10.3390/cancers17162628 PMID: 40867257

Park, Y. (2014). A New Paradigm Shift for the Green Synthesis of Antibacterial Silver Nanoparticles Utilizing Plant Extracts. *Toxicological Research*, *30*(3), 169–178. DOI: 10.5487/TR.2014.30.3.169 PMID: 25343010

Parthasarathy, N.. (2025). Recent progress in multifunctional theranostic inorganic nanoparticles for cancer therapy. *Smart Materials in Medicine*, *6*, 801–825. DOI: 10.1016/j.smaim.2025.07.006

Parvizpour, S., Beyrampour-Basmenj, H., Razmara, J., Farhadi, F., & Shamsir, M. S. (2024). Cancer treatment comes to age: From one-size-fits-all to next-generation sequencing (NGS) technologies. *BioImpacts*, *14*(4), 29957–29957. DOI: 10.34172/bi.2023.29957 PMID: 39104623

Pashkina, E., Bykova, M., Berishvili, M., Lazarev, Y., & Kozlov, V. (2025). Hyaluronic acid-based drug delivery systems for cancer therapy. *Cells*, *14*(2), 61. DOI: 10.3390/cells14020061 PMID: 39851489

Pasieczna-Patkowska, S., Cichy, M., & Flieger, J. (2025). Application of Fourier Transform Infrared (FTIR) Spectroscopy in Characterization of Green Synthesized Nanoparticles. *Molecules (Basel, Switzerland)*, *30*(3), 684. DOI: 10.3390/molecules30030684 PMID: 39942788

Patel, R., & Patel, A. (2024). Revolutionizing drug development: AI-driven predictive modeling for accelerated small molecule and biologic therapeutics. *Well Testing Journal*, *33*(S2), 668–691.

Patil, R. M., Deshpande, P. P., Aalhate, M., Gananadhamu, S., & Singh, P. K. (2022). An Update on Sophisticated and Advanced Analytical Tools for Surface Characterization of Nanoparticles. *Surfaces and Interfaces*, *33*, 102165. DOI: 10.1016/j.surfin.2022.102165

Patiño-Ruiz, D. A., Meramo-Hurtado, S. I., González-Delgado, Á. D., & Herrera, A. (2021). Environmental Sustainability Evaluation of Iron Oxide Nanoparticles Synthesized via Green Synthesis and the Coprecipitation Method: A Comparative Life Cycle Assessment Study. *ACS Omega*, *6*(19), 12410–12423. DOI: 10.1021/acsomega.0c05246 PMID: 34056392

Patnaik, C., Rath, S., & De, A. (2025). Avenues for integrating photothermal therapy in cancer clinic. *Communications Chemistry*, *8*(1), 225. DOI: 10.1038/s42004-025-01705-w PMID: 41188487

Patra, S., Mukherjee, S., Barui, A. K., Ganguly, A., Sreedhar, B., & Patra, C. R. (2015). Green synthesis, characterization of gold and silver nanoparticles and their potential application for cancer therapeutics. *Materials Science and Engineering C*, *53*, 298–309. DOI: 10.1016/j.msec.2015.04.048 PMID: 26042718

Paul, B., & Montoya, G. (2020). CRISPR-Cas12a: Functional overview and applications. *Biomedical Journal*, *43*(1), 8–17. DOI: 10.1016/j.bj.2019.10.005 PMID: 32200959

Paveethra, S., Manisekaran, H., & Sasidharan, S. (2024). Medicinal plants derived green carbon dots: Synthesis, characterization and their potential applications in cancer therapy. *Asian Pacific Journal of Cancer Prevention : APJCP*, *25*(10), 3393–3411. DOI: 10.31557/APJCP.2024.25.10.3393 PMID: 39471005

Pawar, P., & Prabhu, A. (2025). Smart SPIONs for multimodal cancer theranostics: A review. *Molecular Pharmaceutics*, *22*(5), 2372–2391. DOI: 10.1021/acs.molpharmaceut.5c00411 PMID: 40223773

Peer, D., Karp, J. M., Hong, S., Farokhzad, O. C., Margalit, R., & Langer, R. (2020). Nanocarriers as an emerging platform for cancer therapy. *Nano-enabled medical applications*, 61-91.

Peng, C., Huang, Y., & Zheng, J. (2020). Renal clearable nanocarriers: Overcoming the physiological barriers for precise drug delivery and clearance. *Journal of Controlled Release : Official Journal of the Controlled Release Society*, *322*, 64–80. DOI: 10.1016/j.jconrel.2020.03.020 PMID: 32194171

Peng, Q., Qiu, X., Zhang, Z., Zhang, S., Zhang, Y., Liang, Y., Guo, J., Peng, H., Chen, M., Fu, Y.-X., & Tang, H. (2020). PD-L1 on dendritic cells attenuates T cell activation and regulates response to immune checkpoint blockade. *Nature Communications*, *11*(1), 4835. DOI: 10.1038/s41467-020-18570-x PMID: 32973173

Peng, X., Zhang, M., Gao, Y., Zhao, C., Li, S., Liao, S., Zhao, X., Huang, R., Wang, J., & Zeng, L. (2025). Targeted endocytosis: Strategies in drug delivery systems for enhancing antitumor efficacy. *Colloids and Surfaces. B, Biointerfaces*, •••, 115276. PMID: 41242027

Peng, Y., Yang, Z., Li, J., & Liu, S. (2024). Research progress on nanotechnology of traditional Chinese medicine to enhance the therapeutic effect of osteoarthritis. *Drug Delivery and Translational Research*, *14*(6), 1517–1534. DOI: 10.1007/s13346-024-01517-w PMID: 38225521

Peng, Y., Yu, M., Li, B., Zhang, S., Cheng, J., Wu, F., Du, S., Miao, J., Hu, B., Olkhovsky, I., & Li, S. (2025). Advances in nanocarrier-mediated cancer therapy: Progress in immunotherapy, chemotherapy, and radiotherapy. *Chinese Medical Journal*, *138*(16), 1927–1944. DOI: 10.1097/CM9.0000000000003703 PMID: 40545573

Perron, U., Grassi, E., Chatzipli, A., Viviani, M., Karakoc, E., Trastulla, L., Brochier, L., Isella, C., Zanella, E., Klett, H., Molineris, I., Schueler, J., Esteller, M., Médico, E., Conte, N., McDermott, U., Trusolino, L., Bertotti, A., & Iorio, F. (2024). Integrative ensemble modelling of cetuximab sensitivity in colorectal cancer patient-derived xenografts. *Nature Communications*, *15*(1), 9139. Advance online publication. DOI: 10.1038/s41467-024-53163-y PMID: 39528460

Petrović, S., Bita, B., & Barbinta-Patrascu, M. (2024). Nanoformulations in Pharmaceutical and Biomedical Applications: Green Perspectives. *International Journal of Molecular Sciences*, *25*(11), 5842. Advance online publication. DOI: 10.3390/ijms25115842 PMID: 38892030

Pettini, F., Visibelli, A., Cicaloni, V., Iovinelli, D., & Spiga, O. (2021). Multi-omics model applied to cancer genetics. *International Journal of Molecular Sciences*, *22*(11), 5751. DOI: 10.3390/ijms22115751 PMID: 34072237

Pham, S. H., Choi, Y., & Choi, J. (2020). Stimuli-Responsive Nanomaterials for Application in Antitumor Therapy and Drug Delivery. *Pharmaceutics*, *12*(7), 630. Advance online publication. DOI: 10.3390/pharmaceutics12070630 PMID: 32635539

Pickar-Oliver, A., & Gersbach, C. A. (2019). The next generation of CRISPR–Cas technologies and applications. *Nature Reviews. Molecular Cell Biology*, *20*(8), 490–507. DOI: 10.1038/s41580-019-0131-5 PMID: 31147612

Pinto, B., Henriques, A., Silva, P., & Bousbaa, H. (2020). 3D spheroids as preclinical models. *Pharmaceutics*, •••, 12.

Pinto, G. A., Durao, F. O., Fiúza, A. M., Guimaraes, M. M., & Madureira, C. N. (2004). Design optimisation study of solvent extraction: Chemical reaction, mass transfer and mixer–settler hydrodynamics. *Hydrometallurgy*, *74*(1-2), 131–147. DOI: 10.1016/j.hydromet.2004.02.002

Pourmadadi, M., Aslani, A., Hassan, D., Sani, A., Rahdar, A., & Medina, I. D., Abdouss, M., & Romanholo Ferreira, L. F. (2023). Recent advancements in the targeted delivery of Gemcitabine: Harnessing nanomedicine for enhanced cancer therapy. *OpenNano, 13*, 100177. 10.1016/j.onano.2023.100177

Pourmadadi, M., Shabestari, S. M., Abdouss, H., Rahdar, A., Fathi-Karkan, S., & Pandey, S. (2025). Artificial intelligence-driven intelligent nanocarriers for cancer theranostics: A paradigm shift with focus on brain tumors. *Seminars in Oncology*, *52*(6), 152429. DOI: 10.1016/j.seminoncol.2025.152429 PMID: 41218468

Prasad, R., Peng, B., Mendes, B. B., Kilian, H. I., Gorain, M., Zhang, H., Kundu, G., Xia, J., Lovell, J. F., & Conde, J. (2024). Biomimetic bright optotheranostics for metastasis monitoring and multimodal image-guided breast cancer therapeutics. *Journal of Controlled Release : Official Journal of the Controlled Release Society*, *367*, 300–315. Advance online publication. DOI: 10.1016/j.jconrel.2024.01.056 PMID: 38281670

Procopio, A., Lagreca, E., Jamaledin, R., La Manna, S., Corrado, B., Di Natale, C., & Onesto, V. (2022). Recent Fabrication Methods to Produce Polymer-Based Drug Delivery Matrices (Experimental and In Silico Approaches). *Pharmaceutics, 14*(4), 872. Advance online publication. DOI: 10.3390/pharmaceutics14040872 PMID: 35456704

Pujitha, K., Harichandhana, N., Suchitra, M., & Chandra, Y. (2025). The Role of Stability Studies in Ensuring Pharmaceutical Product Quality. *International Journal of Allied Medical Sciences and Clinical Research, 13*(4), 660–675. DOI: 10.61096/ijamscr.v13.iss4.2025.660-675

Puri, S., Mazza, M., Roy, G., England, R. M., Zhou, L., Nourian, S., & Subramony, A. J. (2023). Evolution of Nanomedicine Formulations for Targeted Delivery and Controlled Release. *Advanced Drug Delivery Reviews, 114962*, 114962. Advance online publication. DOI: 10.1016/j.addr.2023.114962 PMID: 37321376

Rabiee, N. (2023). Natural components as surface engineering agents for CRISPR delivery. *Environmental Research, 232*, 116333. DOI: 10.1016/j.envres.2023.116333 PMID: 37286127

Rai, M., Wypij, M., Ingle, A. P., Trzcińska-Wencel, J., & Golińska, P. (2021). Emerging trends in pullulan-based antimicrobial systems for various applications. *International Journal of Molecular Sciences, 22*(24), 13596. DOI: 10.3390/ijms222413596 PMID: 34948392

Rajan, A., Laha, S. S., Sahu, N. K., Thorat, N. D., & Shankar, B. (2024). Recent advancements and clinical aspects of engineered iron oxide nanoplatforms for magnetic hyperthermia-induced cancer therapy. *Materials Today. Bio, 29*, 101348. DOI: 10.1016/j.mtbio.2024.101348 PMID: 39669801

Rajanathadurai, J., Perumal, E., & Sindya, J. (2024). Advances in targeting cancer epigenetics using CRISPR-dCas9 technology: A comprehensive review and future prospects. *Functional & Integrative Genomics, 24*(5), 164. DOI: 10.1007/s10142-024-01455-3 PMID: 39292321

Rajpali, R., Teotia, A., Thota, P., Khanna, D., Samaniya, V., Kumar, P., Pandey, M., Lagashetti, A., Kalaiselvan, V., & Raghuvanshi, R. (2025). A comparative analysis of pharmacopeial quality standards for antibiotics with respect to bacterial endotoxin limits. *Journal of Biological Methods, 13*(1), e99010088. DOI: 10.14440/jbm.0127

Raj, S., Khurana, S., Choudhari, R., Kesari, K. K., Kamal, M., Garg, N., Ruokolainen, J., Das, B., & Kumar, D. (2019). Specific targeting cancer cells with nanoparticles and drug delivery in cancer therapy. *Seminars in Cancer Biology*. Advance online publication. DOI: 10.1016/j.semcancer.2019.11.002 PMID: 31715247

Ramos, A. P., Cruz, M. A. E., Tovani, C. B., & Ciancaglini, P. (2017). Biomedical applications of nanotechnology. *Biophysical Reviews*, *9*(2), 79–89. DOI: 10.1007/s12551-016-0246-2 PMID: 28510082

Ramos, T. I., Villacis-Aguirre, C. A., López-Aguilar, K. V., Santiago Padilla, L., Altamirano, C., Toledo, J. R., & Santiago Vispo, N. (2022). The Hitchhiker's Guide to Human Therapeutic Nanoparticle Development. *Pharmaceutics*, *14*(2), 247. Advance online publication. DOI: 10.3390/pharmaceutics14020247 PMID: 35213980

Randviir, E. P., Brownson, D. A. C., & Banks, C. E. (2014). A decade of graphene research: Production, applications and outlook. *Materials Today*, *17*(9), 426–432. DOI: 10.1016/j.mattod.2014.06.001

Rani, M. Keshu, & Shanker, U. (2022). Chapter 3 - Green nanomaterials: An overview. In U. Shanker, C. M. Hussain, & M. Rani (Eds.), *Green Functionalized Nanomaterials for Environmental Applications* (pp. 43–80). Elsevier. 10.1016/B978-0-12-823137-1.00026-9

Rasool, M., Malik, A., Waquar, S., Arooj, M., Zahid, S., Asif, M., Shaheen, S., Hussain, A., Ullah, H., & Gan, S. H. (2022). New challenges in the use of nanomedicine in cancer therapy. *Bioengineered*, *13*(1), 759–773. DOI: 10.1080/21655979.2021.2012907 PMID: 34856849

Rauf, M. A., Rao, A., Sivasoorian, S. S., & Iyer, A. K. (2025). Nanotechnology-based delivery of CRISPR/Cas9 for cancer treatment: A comprehensive review. *Cells*, *14*(15), 1136. DOI: 10.3390/cells14151136 PMID: 40801569

Riaz, M. U., Naeem, M. S., Zohaib, M., Ahmad, B., Ijaz, M. T., Tahir, M. U., Sulman, A., Awan, A., & Kharl, H. A. A. (2026). AI IN TARGET IDENTIFICATION AND VALIDATION. *PharmA AI: Revolutionizing Drug Discovery and Healthcare with Artificial Intelligence*.

Rid, A., Aguilera, B., Banda, C., Divi, R., Harris, M., Kim, A., Ossandon, M., Zervos, J., & Rowthorn, V. (2024). Global health reciprocal innovation: Ethical, legal and regulatory considerations. *BMJ Global Health*, *8*(Suppl 7), e014693. Advance online publication. DOI: 10.1136/bmjgh-2023-014693 PMID: 38821558

Rieke, V., & Butts Pauly, K. (2008). MR thermometry. *Journal of Magnetic Resonance Imaging*, *27*(2), 376–390. DOI: 10.1002/jmri.21265 PMID: 18219673

Rizawan, A., Rehman, U., Gupta, G., Alsayari, A., Wahab, S., & Kesharwani, P. (2025). Polyglutamic acid in cancer nanomedicine. *International Journal of Pharmaceutics*, *676*, 125623. Advance online publication. DOI: 10.1016/j.ijpharm.2025.125623 PMID: 40254191

Rizo-Téllez, S. A., Sekheri, M., & Filep, J. G. (2022). Myeloperoxidase: Regulation of neutrophil function and target for therapy. *Antioxidants, 11*(11), 2302. DOI: 10.3390/antiox11112302 PMID: 36421487

Rodrigues, A. S., Batista, J. G. S., Rodrigues, M. Á. V., Thipe, V. C., Minarini, L. A. R., Lopes, P. S., & Lugão, A. B. (2024). Advances in silver nanoparticles: A comprehensive review on their potential as antimicrobial agents and their mechanisms of action elucidated by proteomics. *Frontiers in Microbiology, 15*, 1440065. DOI: 10.3389/fmicb.2024.1440065 PMID: 39149204

Rodríguez, F., Caruana, P., De La Fuente, N., Español, P., Gamez, M., Balart, J., Llurba, E., Rovira, R., Ruiz, R., Martín-Lorente, C., Corchero, J., & Céspedes, M. V. (2022). Nano-Based Approved Pharmaceuticals for Cancer Treatment: Present and Future Challenges. *Biomolecules, 12*(6), 784. Advance online publication. DOI: 10.3390/biom12060784 PMID: 35740909

Rodríguez-Gómez, F. D., Monferrer, D., Penon, O., & Rivera-Gil, P. (2025). Regulatory pathways and guidelines for nanotechnology-enabled health products: A comparative review of EU and US frameworks. *Frontiers in Medicine, 12*, 1544393. DOI: 10.3389/fmed.2025.1544393 PMID: 40109724

Rodríguez-Rojas, M. D. P., Bustos-Terrones, V., Díaz-Cárdenas, M. Y., Vázquez-Vélez, E., & Martínez, H. (2024). Life Cycle Assessment of Green Synthesis of TiO2 Nanoparticles vs. Chemical Synthesis. *Sustainability (Basel), 16*(17), 7751. DOI: 10.3390/su16177751

Rojas Burbano, J. C., Ruiz, N. I., Rojas Burbano, G. A., Guacho Inca, J. S., Arragan Lezama, C. A., & González, M. S. (2025). Robot-Assisted Surgery: Current Applications and Future Trends in General Surgery. *Cureus, 17*(4), e82318. DOI: 10.7759/cureus.82318 PMID: 40376327

Roma-Rodrigues, C., Rivas-García, L., Baptista, P. V., & Fernandes, A. R. (2020). Gene therapy in cancer treatment: Why go nano? *Pharmaceutics, 12*(3), 233. DOI: 10.3390/pharmaceutics12030233 PMID: 32151052

Rónavári, A., Igaz, N., Adamecz, D. I., Szerencsés, B., Molnar, C., Kónya, Z., Pfeiffer, I., & Kiricsi, M. (2021). Green Silver and Gold Nanoparticles: Biological Synthesis Approaches and Potentials for Biomedical Applications. *Molecules (Basel, Switzerland), 26*(4), 844. DOI: 10.3390/molecules26040844 PMID: 33562781

Rozhin, P., Abdel Monem Gamal, J., Giordani, S., & Marchesan, S. (2022). Carbon nanomaterials (CNMs) and enzymes: From nanozymes to cnm-enzyme conjugates and biodegradation. *Materials (Basel), 15*(3), 1037. DOI: 10.3390/ma15031037 PMID: 35160982

Rulten, S. L., Grose, R. P., Gatz, S. A., Jones, J. L., & Cameron, A. J. M. (2023). The Future of Precision Oncology. *International Journal of Molecular Sciences, 24*(16), 12613. https://www.mdpi.com/1422-0067/24/16/12613. DOI: 10.3390/ijms241612613 PMID: 37628794

Russo, M., Sogari, A., & Bardelli, A. (2021). Adaptive evolution: How bacteria and cancer cells survive stressful conditions and drug treatment. *Cancer Discovery, 11*(8), 1886–1895. DOI: 10.1158/2159-8290.CD-20-1588 PMID: 33952585

Sabir, S., Thani, A. S. B., & Abbas, Q. (2025). Nanotechnology in cancer treatment: Revolutionizing strategies against drug resistance. *Frontiers in Bioengineering and Biotechnology, 13*, 1548588. DOI: 10.3389/fbioe.2025.1548588 PMID: 40370595

Sabit, H., Pawlik, T., Radwan, F. F. Y., Abdel-Hakeem, M. A., Abdel-Ghany, S., Wadan, A.-H. S., Elzawahri, M., El-Hashash, A., & Arneth, B. (2025). Precision nanomedicine: Navigating the tumor microenvironment for enhanced cancer immunotherapy and targeted drug delivery. *Molecular Cancer, 24*(1), 160. Advance online publication. DOI: 10.1186/s12943-025-02357-z PMID: 40457437

Sahel, D. K., Vora, L. K., Saraswat, A., Sharma, S., Monpara, J., D'Souza, A. A., Mishra, D., Tryphena, K. P., Kawakita, S., Khan, S., Azhar, M., Khatri, D. K., Patel, K., & Singh Thakur, R. R. (2023). CRISPR/Cas9 genome editing for tissue-specific in vivo targeting: Nanomaterials and translational perspective. *Advanced Science (Weinheim, Baden-Württemberg, Germany), 10*(19), 2207512. DOI: 10.1002/advs.202207512 PMID: 37166046

Sahu, S., & Tiwari, S. (2023). Carbon allotropes: Fundamental, synthesis, characterization, and properties functionalization. In *Carbon allotropes* (pp. 41–70). CRC Press. DOI: 10.1201/9781003323976-5

Sajjadi, M., Nasrollahzadeh, M., Jaleh, B., Soufi, G. J., & Iravani, S. (2021). Carbon-based nanomaterials for targeted cancer nanotherapy: Recent trends and future prospects. *Journal of Drug Targeting, 29*(7), 716–741. DOI: 10.1080/1061186X.2021.1886301 PMID: 33566719

Salavatipour, M. S., Poursalehi, Z., Rouzbahani, N. H., Mohammadyar, S., & Vasei, M. (2024). CRISPR-Cas9 in basic and translational aspects of cancer therapy. *BioImpacts, 14*(2), 237–252. DOI: 10.34172/bi.2024.30087 PMID: 39493894

Salem, S. S., & Fouda, A. (2021). Green Synthesis of Metallic Nanoparticles and Their Prospective Biotechnological Applications: An Overview. *Biological Trace Element Research, 199*(1), 344–370. DOI: 10.1007/s12011-020-02138-3 PMID: 32377944

Salgueiro, M. J., & Zubillaga, M. (2025). Theranostic nanoplatforms in nuclear medicine: Current advances, emerging trends, and perspectives for personalized oncology. *Journal of Nanotheranostics*, *6*(4), 27. DOI: 10.3390/jnt6040027

Salinas, Y., Chauhan, S. C., & Bandyopadhyay, D. (2025). Small-Molecule Mitotic Inhibitors as Anticancer Agents: Discovery, Classification, Mechanisms of Action, and Clinical Trials. *International Journal of Molecular Sciences*, *26*(7), 3279. DOI: 10.3390/ijms26073279 PMID: 40244152

Salmaso, S., & Caliceti, P. (2013). Stealth Properties to Improve Therapeutic Efficacy of Drug Nanocarriers. *Journal of Drug Delivery*, *2013*(1), 374252. 10.1155/2013/374252. DOI: 10.1155/2013/374252 PMID: 23533769

Samathoti, P., Kumarachari, R. K., Bukke, S. P. N., Rajasekhar, E. S. K., Jaiswal, A. A., & Eftekhari, Z. (2025). The role of nanomedicine and artificial intelligence in cancer health care: Individual applications and emerging integrations-a narrative review. *Discover. Oncology*, *16*(1), 697. DOI: 10.1007/s12672-025-02469-4 PMID: 40338421

Sander, J. D., & Joung, J. K. (2014). CRISPR-Cas systems for genome editing, regulation and targeting. *Nature Biotechnology*, *32*(4), 347–355. DOI: 10.1038/nbt.2842 PMID: 24584096

Sanjeev, N. O., Vallabha, M. S., & Valsan, A. E. (2024). Azadirachta indica leaf extract based green-synthesized ZnO nanoparticles coated on spent tea waste activated carbon for pharmaceuticals and personal care products removal. *Environmental Research*, *252*, 119047. DOI: 10.1016/j.envres.2024.119047 PMID: 38704006

Santhamoorthy, M., Asaithambi, P., Ramkumar, V., Elangovan, N., Perumal, I., & Kim, S.-C. (2025). A Review on the Recent Advancements of Polymer-Modified Mesoporous Silica Nanoparticles for Drug Delivery Under Stimuli-Trigger. *Polymers*, *17*(12), 1640. Advance online publication. DOI: 10.3390/polym17121640 PMID: 40574168

Saratale, R. G., Cho, S. K., Bharagava, R. N., Vivekanand, V., Bhatia, S. K., Romanholo Ferreira, L. F., Shin, H. S., Jeon, B.-H., Kumar, R., & Saratale, G. D. (2025). Biorefinery of avocado seed waste for sustainable production of bioplastics and functionalized silver nanocomposites. *International Journal of Biological Macromolecules*, *323*, 147042. DOI: 10.1016/j.ijbiomac.2025.147042 PMID: 40848789

Saratale, R. G., Karuppusamy, I., Saratale, G. D., Pugazhendhi, A., Kumar, G., Park, Y., Ghodake, G. S., Bharagava, R. N., Banu, J. R., & Shin, H. S. (2018). A comprehensive review on green nanomaterials using biological systems: Recent perception and their future applications. *Colloids and Surfaces. B, Biointerfaces*, *170*, 20–35. DOI: 10.1016/j.colsurfb.2018.05.045 PMID: 29860217

Saroj, N., Dholaniya, P. S., Alvi, S. B., Sridharan, D., Soni, N., Ashraf, S. A., Choudhry, A., Ashraf, Y. A., Mikula, S. K., Singla, D. K., & Khan, M. (2025). SiRNA-mediated knockdown of TOP2B protects hiPSC-derived cardiomyocytes from doxorubicin-induced toxicity. *Life Sciences*, *371*, 123595. DOI: 10.1016/j.lfs.2025.123595 PMID: 40158615

Sasidharan, S., Kumari, K., & Ramakrishnan, V. (2024). Aromatic π–π Interactions in Self-assembled Nanostructured Materials. Conference on Research and Industrial Conclave-Integration, Sergeeva, O. V., Luo, L., & Guiseppi-Elie, A. (2025). Cancer theragnostics: Closing the loop for advanced personalized cancer treatment through the platform integration of therapeutics and diagnostics. *Frontiers in Bioengineering and Biotechnology*, *12*, 1499474.

Sathuvan, M., Narayanan, K., Cheong, K.-L., & Thangam, R. (2026). Nanoplatforms for multimodal imaging and targeted cancer therapy: Recent advances and future perspectives. *Bioengineering (Basel, Switzerland)*, *13*(2), 174. DOI: 10.3390/bioengineering13020174 PMID: 41749714

Satyavani, K., Gurudeeban, S., Ramanathan, T., & Balasubramanian, T. (2011). Biomedical potential of silver nanoparticles synthesized from calli cells of Citrullus colocynthis (L.) Schrad. *Journal ofNanobiotechnology*, *9*(1), 43. DOI: 10.1186/1477-3155-9-43 PMID: 21943321

Saw, W. S., Anasamy, T., Foo, Y. Y., Kwa, Y. C., Kue, C. S., Yeong, C. H., Kiew, L. V., Lee, H. B., & Chung, L. Y. (2021). Delivery of Nanoconstructs in Cancer Therapy: Challenges and Therapeutic Opportunities. *Advanced Therapeutics*, *4*(3), 2000206. DOI: 10.1002/adtp.202000206

Saxena, R., Kotnala, S., Bhatt, S. C., Uniyal, M., Rawat, B. S., Negi, P., & Riyal, M. K. (2025). A review on green synthesis of nanoparticles toward sustainable environment. *Sustainable Chemistry for Climate Action*, *6*, 100071. DOI: 10.1016/j.scca.2025.100071

Sayed, Z. S., Ghiaty, A., & Elbaz, N. M. (2024). Cancer treatment approaches within the frame of nanotechnology. *RSC Advances*, *14*, 33645–33669. DOI: 10.1039/D4RA06992G PMID: 39670162

Schreiner, O. D., Schreiner, T. G., Miron, L., & Ciobanu, R. C. (2026). Two-stage microwave hyperthermia using magnetic nanoparticles for optimal chemotherapy activation in liver cancer: Concept and preliminary tests on Wistar rat model. *Cancers (Basel)*, *18*(2), 330. DOI: 10.3390/cancers18020330 PMID: 41595249

Sebastián, V. (2026). Microfluidic reactors for the synthesis of inorganic and hybrid nanoparticles for drug delivery. *Advanced Drug Delivery Reviews*, *234*, 115872. DOI: 10.1016/j.addr.2026.115872 PMID: 41951054

Sehgal, K., Portell, A., Ivanova, E. V., Lizotte, P. H., Mahadevan, N. R., Greene, J. R., Vajdi, A., Gurjao, C., Teceno, T., Taus, L. J., Thai, T. C., Kitajima, S., Liu, D., Tani, T., Noureddine, M., Lau, C. J., Kirschmeier, P. T., Liu, D., Giannakis, M., & Barbie, D. A. (2021). Dynamic single-cell RNA sequencing identifies immunotherapy persister cells following PD-1 blockade. *The Journal of Clinical Investigation*, *131*(2), e135038. DOI: 10.1172/JCI135038 PMID: 33151910

Selewski, D. T., Shah, G. V., Mody, R. J., Rajdev, P. A., & Mukherji, S. K. (2010). Rituximab (Rituxan). *AJNR. American Journal of Neuroradiology*, *31*(7), 1178–1180. DOI: 10.3174/ajnr.A2142 PMID: 20448016

Sell, M., Lopes, A. R., Escudeiro, M., Esteves, B., Monteiro, A. R., Trindade, T., & Cruz-Lopes, L. (2023). Application of Nanoparticles in Cancer Treatment: A Concise Review. *Nanomaterials (Basel, Switzerland)*, *13*(21), 2887. https://www.mdpi.com/2079-4991/13/21/2887. DOI: 10.3390/nano13212887 PMID: 37947732

Selvaraj, N., Swaroop, A. K., Nidamanuri, B. S. S., Kumar, R. R., Natarajan, J., & Selvaraj, J. (2022). Network-based drug repurposing: A critical review. *Current Drug Research Reviews*, *14*(2), 116–131. DOI: 10.2174/258997751466622021420403 PMID: 35156575

Şengül, H., Theis, T. L., & Ghosh, S. (2008). Toward sustainable nanoproducts: An overview of nanomanufacturing methods. *Journal of Industrial Ecology*, *12*(3), 329–359. DOI: 10.1111/j.1530-9290.2008.00046.x

Serdar, G. (2024). Biogenic Synthesis and Characterization of Gold Nanoparticles from Morus alba L. Leaves by Microwave Extraction Method. *Journal of the Turkish Chemical Society Section A: Chemistry*, *11*(2), 615–622. DOI: 10.18596/jotcsa.1372302

Serras, A., Faustino, C., & Pinheiro, L. (2024). Functionalized polymeric micelles for targeted cancer therapy: Steps from conceptualization to clinical trials. *Pharmaceutics*, *16*(8), 1047. DOI: 10.3390/pharmaceutics16081047 PMID: 39204392

Sezer, E., Ulucan Karnak, F., & Akgöl, S. (2025). Green carbon dots in the era of AI: Sustainable synthesis, intelligent drug delivery, advanced diagnostics, and bioimaging. *Turkish Journal of Biology = Türk Biyoloji Dergisi, 49*(5), 498–533. DOI: 10.55730/1300-0152.2762 PMID: 41246234

Shafaei, N., Khorshidi, S., & Karkhaneh, A. (2023). The immune-stealth polymeric coating on drug delivery nanocarriers: In vitro engineering and in vivo fate. *Journal of Biomaterials Applications, 38*(2), 159–178. DOI: 10.1177/08853282231185352 PMID: 37480331

Shah, M., Prakashan, D., & Gandhi, S. (2024). Microfluidics, organs-on-a-chip, and 3D printing. In Mohanan, P. V. (Ed.), *Human Organs-on-a-Chip Technology* (pp. 91–112). Academic Press., DOI: 10.1016/B978-0-443-13782-2.00008-5

Shah, S., & D'Souza, G. (2025). Tumor microenvironment modeling in vitro. *Cells*, •••, 14. PMID: 40422235

Shahverdi, A. R., Minaeian, S., Shahverdi, H. R., Jamalifar, H., & Nohi, A.-A. (2007). Rapid synthesis of silver nanoparticles using culture supernatants of Enterobacteria: A novel biological approach. *Process Biochemistry (Barking, London, England), 42*(5), 919–923. DOI: 10.1016/j.procbio.2007.02.005

Shakeri-Zadeh, A., & Bulte, J. W. M. (2025). Imaging-guided precision hyperthermia with magnetic nanoparticles. *Nature Reviews Bioengineering, 3*(3), 245–260. DOI: 10.1038/s44222-024-00257-3 PMID: 40260131

Shamsian, S., Siddique, A. B., Kordzadeh-Kermani, V., de la Vega Tejuca, L., Falcone, F., Ray, M., Ashrafizadeh, S. N., Chapa, S. O. M., Madou, M. J., & Madadelahi, M. (2025). Nanomaterials in PCR: Exploring light-to-heat conversion mechanisms and microfluidic integration. *Microsystems & Nanoengineering, 11*(1), 127. DOI: 10.1038/s41378-025-00898-3 PMID: 40533488

Shanahan, K., Coen, D., & Nafo, W. (2025). Polymer-based nanoparticles for cancer theranostics: Advances, challenges, and future perspectives. *Exploration of Biomat-X, 2*, 101342. DOI: 10.37349/ebmx.2025.101342

Shang, Y., Yi, X., Xiang, D., & Zhou, L. (2025). Nanoagent-mediated photothermal therapy: From delivery system design to synergistic theranostic applications. *International Journal of Nanomedicine, 20*, 6891–6927. DOI: 10.2147/IJN.S522736 PMID: 40458746

Sharaf, M., Alhamad, A. A., Ltaief, O. O., & Amor, I. B. (2023). Challenges of nanomaterials-based cancer therapy: A future destination. *International Journal of Surgery*, *109*(6), 1819–1820. DOI: 10.1097/JS9.0000000000000412 PMID: 37072147

Sharifi, S., Mahmoud, N. N., Voke, E., Landry, M. P., & Mahmoudi, M. (2022). Importance of Standardizing Analytical Characterization Methodology for Improved Reliability of the Nanomedicine Literature. *Nano-Micro Letters*, *14*(1), 172. DOI: 10.1007/s40820-022-00922-5 PMID: 35987931

Sharma, A., Tyagi, V. V., Chen, C. R., & Buddhi, D. (2009). Review on thermal energy storage with phase change materials and applications. *Renewable & Sustainable Energy Reviews*, *13*(2), 318–345. DOI: 10.1016/j.rser.2007.10.005

Sharma, M., Tellili, N., Kacem, I., & Rouissi, T. (2024). Microbial biopolymers: From production to environmental applications—a review. *Applied Sciences (Basel, Switzerland)*, *14*(12), 5081. DOI: 10.3390/app14125081

Sharma, R., Borah, S. J., Bhawna, , Kumar, S., Gupta, A., Singh, P., Goel, V. K., Kumar, R., & Kumar, V. (2022). Functionalized peptide-based nanoparticles for targeted cancer nanotherapeutics. *ACS Omega*, *7*(41), 37362–37389. DOI: 10.1021/acsomega.2c03974 PMID: 36278104

Sharom, F. J. (2014). Complex interplay between the P-glycoprotein multidrug efflux pump and the membrane: Its role in modulating protein function. *Frontiers in Oncology*, *4*, 41. DOI: 10.3389/fonc.2014.00041 PMID: 24624364

Sheibani, S., Basu, K., Farnudi, A., Ashkarran, A., Ichikawa, M., Presley, J. F., Bui, K. H., Ejtehadi, M. R., Vali, H., & Mahmoudi, M. (2021). Nanoscale characterization of the biomolecular corona by cryo-electron microscopy, cryo-electron tomography, and image simulation. *Nature Communications*, *12*(1), 573. DOI: 10.1038/s41467-020-20884-9 PMID: 33495475

Sheldon, R. A. (2005). Green solvents for sustainable organic synthesis: State of the art. *Green Chemistry*, *7*(5), 267–278. DOI: 10.1039/b418069k

Shen, X., Pan, D., Gong, Q., Gu, Z., & Luo, K. (2023). Nanomedicine in solid tumors. *Bioactive Materials*, *32*, 445–472. DOI: 10.1016/j.bioactmat.2023.10.017 PMID: 37965242

Sheth, P., Sandhu, H., Singhal, D., Malick, W., Shah, N., & Serpil Kislalioglu, M. (2012). Nanoparticles in the pharmaceutical industry and the use of supercritical fluid technologies for nanoparticle production. *Current Drug Delivery*, *9*(3), 269–284. DOI: 10.2174/156720112800389052 PMID: 22283656

Shinde, V. R., Revi, N., Murugappan, S., Singh, S. P., & Rengan, A. K. (2022). Enhanced permeability and retention effect: A key facilitator for solid tumor targeting by nanoparticles. *Photodiagnosis and Photodynamic Therapy*, *39*, 102915. DOI: 10.1016/j.pdpdt.2022.102915 PMID: 35597441

Shi, P., Cheng, Z., Zhao, K., Chen, Y., Zhang, A., Gan, W., & Zhang, Y. (2023). Active targeting schemes for nano-drug delivery systems in osteosarcoma therapeutics. *Journal of Nanobiotechnology*, *21*(1), 103. DOI: 10.1186/s12951-023-01826-1 PMID: 36944946

Shirzad, M.. (2025). Artificial intelligence-assisted design of nanomedicines for cancer diagnosis and therapy. *Drug Delivery and Translational Research*, *15*, 4127–4150. DOI: 10.1007/s12668-025-01980-w

Shleeva, M. O., Kondratieva, D. A., & Kaprelyants, A. S. (2023). Bacillus licheniformis: A Producer of Antimicrobial Substances, including Antimycobacterials, Which Are Feasible for Medical Applications. *Pharmaceutics*, *15*(7), 1893. DOI: 10.3390/pharmaceutics15071893 PMID: 37514078

Siddique, S., & Chow, J. (2022). Recent Advances in Functionalized Nanoparticles in Cancer Theranostics. *Nanomaterials (Basel, Switzerland)*, *12*(16), 2826. Advance online publication. DOI: 10.3390/nano12162826 PMID: 36014691

Sidorenko, V., Tobi, A., Sugahara, K. N., & Teesalu, T. (2025). Peptide-targeted nanoparticles for tumor therapy. *Journal of Controlled Release : Official Journal of the Controlled Release Society*, *387*, 114195. DOI: 10.1016/j.jconrel.2025.114195 PMID: 40925419

Simelane, N., & Abrahamse, H. (2024). Photodynamic therapy in 2D and 3D systems. *Frontiers in Molecular Biosciences*.

Simões, A., Veiga, F., & Vitorino, C. (2024). Question-based review for pharmaceutical development: An enhanced quality approach. *European Journal of Pharmaceutics and Biopharmaceutics : Official Journal of Arbeitsgemeinschaft für Pharmazeutische Verfahrenstechnik e.V*, *195*, 114174. DOI: 10.1016/j.ejpb.2023.114174 PMID: 38160986

Simonsen, T. G., Gaustad, J.-V., Leinaas, M. N., & Rofstad, E. K. (2012). High interstitial fluid pressure is associated with tumor-line specific vascular abnormalities in human melanoma xenografts. *PLoS One*, *7*(6), e40006. DOI: 10.1371/journal.pone.0040006 PMID: 22768196

Singh, D., Rathod, V., Ninganagouda, S., Hiremath, J., Singh, A. K., & Mathew, J. (2014). Optimization and Characterization of Silver Nanoparticle by Endophytic Fungi *Penicillium* sp. Isolated from *Curcuma longa* (Turmeric) and Application Studies against MDR *E. coli* and *S. aureus*. *Bioinorganic Chemistry and Applications*, *2014*, 1–8. DOI: 10.1155/2014/408021 PMID: 24639625

Singh, D., Rathod, V., Singh, A. K., Ul Haq, M., Mathew, J., & Kulkarni, P. (2016). A Study on Extracellular Synthesis of Silver Nanoparticles from Endophytic Fungi, Isolated from Ethanomedicinal Plants *Curcuma longa* and *Catharanthus roseus*. *International Letters of Natural Sciences*, *57*, 58–66. DOI: 10.56431/p-sx5071

Singh, H., Desimone, M. F., Pandya, S., Jasani, S., George, N., Adnan, M., Aldarhami, A., Bazaid, A. S., & Alderhami, S. A. (2023). Revisiting the Green Synthesis of Nanoparticles: Uncovering Influences of Plant Extracts as Reducing Agents for Enhanced Synthesis Efficiency and Its Biomedical Applications. *International Journal of Nanomedicine*, *18*, 4727–4750. DOI: 10.2147/IJN.S419369 PMID: 37621852

Singh, H., Mijaković, I., & Singh, P. (2025). Advances in precision oncology using patient-derived organoids and functional biomaterials. *Frontiers in Cell and Developmental Biology*, *13*, 1670328. Advance online publication. DOI: 10.3389/fcell.2025.1670328 PMID: 41098309

Singh, N. A. (2017). Nanotechnology innovations, industrial applications and patents. *Environmental Chemistry Letters*, *15*(2), 185–191. DOI: 10.1007/s10311-017-0612-8

Singh, N., Kim, J., Kim, J., Lee, K., Zunbul, Z., Lee, I., Kim, E., Chi, S.-G., & Kim, J. S. (2022). Covalent organic framework nanomedicines. *Bioactive Materials*, *21*, 358–380. DOI: 10.1016/j.bioactmat.2022.08.016 PMID: 36185736

Singh, R., Sharma, A., Saji, J., Umapathi, A., Kumar, S., & Daima, H. (2022). Smart nanomaterials for cancer diagnosis and treatment. *Nano Convergence*, *9*(1), 21. Advance online publication. DOI: 10.1186/s40580-022-00313-x PMID: 35569081

Singh, R., Shedbalkar, U. U., Wadhwani, S. A., & Chopade, B. A. (2015). Bacteriagenic silver nanoparticles: Synthesis, mechanism, and applications. *Applied Microbiology and Biotechnology*, *99*(11), 4579–4593. DOI: 10.1007/s00253-015-6622-1 PMID: 25952110

Singh, S., Vidyarthi, A. S., Nigam, V. K., & Dev, A. (2014). Extracellular facile biosynthesis, characterization and stability of gold nanoparticles by *Bacillus licheniformis*. *Artificial Cells, Nanomedicine, and Biotechnology*, *42*(1), 6–12. DOI: 10.3109/21691401.2012.759122 PMID: 23438180

Skotland, T., Iversen, T. G., Llorente, A., & Sandvig, K. (2022). Biodistribution, pharmacokinetics and excretion studies of intravenously injected nanoparticles and extracellular vesicles: Possibilities and challenges. *Advanced Drug Delivery Reviews*, *186*, 114326. DOI: 10.1016/j.addr.2022.114326 PMID: 35588953

Slater, T. J. A., Janssen, A., Camargo, P. H. C., Burke, M. G., Zaluzec, N. J., & Haigh, S. J. (2016). STEM-EDX tomography of bimetallic nanoparticles: A methodological investigation. *Ultramicroscopy*, *162*, 61–73. DOI: 10.1016/j.ultramic.2015.10.007 PMID: 26780684

Sliem, A. R. (2024). Green Synthesis of Antimicrobial Silver Nanoparticles: Exploring Plant-Derived Reducing Agents from Camellia Sinensis, Syzygium aromaticum, and Citrus Sinensis. *Al-Azhar Journal of Agricultural Research*, *49*(1), 81–94. DOI: 10.21608/ajar.2024.269482.1329

Smirnov, I., Kaszkur, Z., & Hoell, A. (2023). Development of nanoparticle bulk morphology analysis: A multidomain XRD approach. *Nanoscale*, *15*(19), 8633–8642. DOI: 10.1039/D3NR00456B PMID: 37073880

Soetaert, F., Korangath, P., Serantes, D., Fiering, S., & Ivkov, R. (2020). Cancer therapy with iron oxide nanoparticles: Agents of thermal and immune therapies. *Advanced Drug Delivery Reviews*, *163–164*, 65–83. DOI: 10.1016/j.addr.2020.06.025 PMID: 32603814

Song, R., Sang, G., Geng, M., & Jiang, H. (2019). China's reform of the regulatory system for medical products and its impact. *National Science Review*, *6*(1), 1. DOI: 10.1093/nsr/nwz001 PMID: 34691816

Song, Y., & Prather, K. L. J. (2024). Strategies in engineering sustainable biochemical synthesis through microbial systems. *Current Opinion in Chemical Biology*, *81*, 102493. DOI: 10.1016/j.cbpa.2024.102493 PMID: 38971129

Song, Z., Tao, Y., Liu, Y., & Li, J. (2024). Advances in delivery systems for CRISPR/Cas-mediated cancer treatment: A focus on viral vectors and extracellular vesicles. *Frontiers in Immunology*, *15*, 1444437. DOI: 10.3389/fimmu.2024.1444437 PMID: 39281673

Soni, A. K., & Jha, R. K. (2024). Nanotechnology's Applications and Potential in Various Fields. *Cureus*. Advance online publication. DOI: 10.7759/cureus.59234 PMID: 38813303

Soni, G., Kale, K., Shetty, S., Gupta, M. K., & Yadav, K. S. (2020). Quality by design (QbD) approach in processing polymeric nanoparticles loading anticancer drugs by high pressure homogenizer. *Heliyon*, *6*(4), e03846. DOI: 10.1016/j.heliyon.2020.e03846 PMID: 32373744

Soomro, R.. (2025). A novel plant-based approach for synthesis of iron oxide nanoparticles in cancer theranostics. *Nano Biomedicine and Engineering*, *17*, 241–259. DOI: 10.26599/NBE.2024.9290097

Soukar, J., & Gaharwar, A. K. (2025). Organelle-targeting nanoparticles. *Advanced Science (Weinheim, Baden-Württemberg, Germany)*, *12*(7), 2411720. DOI: 10.1002/advs.202411720 PMID: 39806939

Souri, M., Soltani, M., Kashkooli, F. M., Shahvandi, M. K., Chiani, M., Shariati, F. S., Mehrabi, M., & Munn, L. (2022). Towards principled design of cancer nanomedicine to accelerate clinical translation. *Materials Today. Bio*, *13*, 100208. Advance online publication. DOI: 10.1016/j.mtbio.2022.100208 PMID: 35198957

Souto, E. B., Silva, G. F., Dias-Ferreira, J., Zielinska, A., Ventura, F., Durazzo, A., Lucarini, M., Novellino, E., & Santini, A. (2020). Nanopharmaceutics: Part I—Clinical trials legislation and good manufacturing practices (GMP) of nanotherapeutics in the EU. *Pharmaceutics*, *12*(2), 146. DOI: 10.3390/pharmaceutics12020146 PMID: 32053962

Spada, A., & Gerber-Lemaire, S. (2025). Surface functionalization of nanocarriers with anti-EGFR ligands for cancer active targeting. *Nanomaterials (Basel, Switzerland)*, *15*(3), 158. DOI: 10.3390/nano15030158 PMID: 39940134

Stevanović, M., & Filipović, N. (2024). A review of recent developments in biopolymer nano-based drug delivery systems with antioxidative properties: Insights into the last five years. *Pharmaceutics*, *16*(5), 670. DOI: 10.3390/pharmaceutics16050670 PMID: 38794332

Stojanov, J., Grozdanić, M., & Ljujić, M.. (2025). Cancer 3D models and drug resistance. *Oncology Research*, •••, 33. PMID: 41050078

Suárez Priede, A., Gómez-Sánchez, M., García-Cancela, P., Bettmer, J., & Díez, P. (2026). Deciphering the formation of biogenic nanoparticles and their protein corona: State-of-the-art and analytical challenges. *Analytical and Bioanalytical Chemistry*, *418*(2), 415–435. DOI: 10.1007/s00216-025-06144-z PMID: 41076454

Subhan, M. A., Yalamarty, S. S. K., Filipczak, N., Parveen, F., & Torchilin, V. P. (2021). Recent Advances in Tumor Targeting via EPR Effect for Cancer Treatment. *Journal of Personalized Medicine*, *11*(6), 571. Advance online publication. DOI: 10.3390/jpm11060571 PMID: 34207137

Sugrue, I., Ross, R. P., & Hill, C. (2024). Bacteriocin diversity, function, discovery and application as antimicrobials. *Nature Reviews. Microbiology*, *22*(9), 556–571. DOI: 10.1038/s41579-024-01045-x PMID: 38730101

Sukri, S., Shameli, K., Teow, S., Chew, J., Ooi, L., & Soon, M.. (2023). Enhanced antibacterial and anticancer activities of green synthesized nanoparticles. *Frontiers in Microbiology*, *14*. Advance online publication. DOI: 10.3389/fmicb.2023.1194292 PMID: 37577438

Sulman, A., Farooq, M. A., Naseer, R. D., Hanif, A., Afzal, Z., Khan, M. Z., Naeem, M. S., Ahmad, H. M., Mahmood, S., & Kharl, H. A. A. (2025). Neurobiology and Management of Drug Addiction and Dependence: A Multidisciplinary Review. *Journal of Pharma and Biomedics*, *3*(1), 191–207.

Sun, B., Wang, W., & Sain, M. (2022). Carbonaceous nanocomposites for biomedical applications as high-drug loading nanocarriers for sustained delivery: A Review. *Journal of Composites Science*, *6*(12), 379. DOI: 10.3390/jcs6120379

Sung, H., Ferlay, J., Siegel, R. L., Laversanne, M., Soerjomataram, I., Jemal, A., & Bray, F. (2021). Global cancer statistics 2020: GLOBOCAN estimates of incidence and mortality worldwide for 36 cancers in 185 countries. *CA: a Cancer Journal for Clinicians*, *71*(3), 209–249. DOI: 10.3322/caac.21660 PMID: 33538338

Sung, Y. K., & Kim, S. W. (2019). Recent advances in the development of gene delivery systems. *Biomaterials Research*, *23*(1), 8. DOI: 10.1186/s40824-019-0156-z PMID: 30915230

Sun, H., Li, Y., Xue, M., & Feng, D. (2025). Tumor Microenvironment-Responsive Nanoparticles: Promising Cancer PTT Carriers. *International Journal of Nanomedicine*, *20*, 7987–8001. DOI: 10.2147/IJN.S526497 PMID: 40584786

Sun, L., Liu, H., Ye, Y., Lei, Y., Islam, R., Tan, S., Tong, R., Miao, Y., & Cai, L. (2023). Smart nanoparticles for cancer therapy. *Signal Transduction and Targeted Therapy*, *8*(1), 418. Advance online publication. DOI: 10.1038/s41392-023-01642-x PMID: 37919282

Sun, Q., Wang, Z., Liu, B., He, F., Gai, S., Yang, P., Yang, D., Li, C., & Lin, J. (2022). Recent advances on endogenous/exogenous stimuli-triggered nanoplatforms for enhanced chemodynamic therapy. *Coordination Chemistry Reviews*, *451*, 214267. Advance online publication. DOI: 10.1016/j.ccr.2021.214267

Sun, Y., & Davis, E. W. (2021). Nanoplatforms for Targeted Stimuli-Responsive Drug Delivery: A Review of Platform Materials and Stimuli-Responsive Release and Targeting Mechanisms. *Nanomaterials (Basel, Switzerland)*, *11*(3), 746. Advance online publication. DOI: 10.3390/nano11030746 PMID: 33809633

Sun, Y., Zhou, Y., Rehman, M., Wang, Y.-F., & Guo, S. (2024). Protein corona of nanoparticles: Isolation and analysis. *Chemical & Biomedical Imaging*, *2*(9), 728–744. DOI: 10.1021/cbe.4c00105 PMID: 39974182

Supple, W. F. (2026). *Cancer Is a Parasite: Kill it with the Safe, Over-the-Counter Antiparasitic Fenbendazole*. Simon and Schuster.

Sushnitha, M., Evangelopoulos, M., Tasciotti, E., & Taraballi, F. (2020). Cell Membrane-Based Biomimetic Nanoparticles and the Immune System: Immunomodulatory Interactions to Therapeutic Applications [Review]. *Frontiers in Bioengineering and Biotechnology, Volume 8 - 2020*. DOI: 10.3389/fbioe.2020.00627

Su, Y., Wang, N., Liu, B., Du, Y., Li, R., Meng, Y., Feng, Y., Shan, Z., & Meng, S. (2020). A phototheranostic nanoparticle for cancer therapy fabricated by BODIPY and graphene to realize photo-chemo synergistic therapy and fluorescence/photothermal imaging. *Dyes and Pigments*, *177*, 108262. DOI: 10.1016/j.dyepig.2020.108262

Syed, S. M., Kulkarni, S., Patil, M., & Satpute, K. (2026). A comprehensive review of green synthesis methods and applications of nanoparticles derived from plant extracts and microorganisms. *Discover Green Chemistry*, *1*(1), 6. DOI: 10.1007/s44509-026-00006-2

Sysak, S., Czarczynska-Goslinska, B., Szyk, P., Koczorowski, T., Mlynarczyk, D. T., Szczolko, W., Lesyk, R., & Goslinski, T. (2023). Metal Nanoparticle-Flavonoid Connections: Synthesis, Physicochemical and Biological Properties, as Well as Potential Applications in Medicine. *Nanomaterials (Basel, Switzerland)*, *13*(9), 1531. DOI: 10.3390/nano13091531 PMID: 37177076

Szczyglewska, P., Feliczak-Guzik, A., & Nowak, I. (2023). Nanotechnology–General Aspects: A Chemical Reduction Approach to the Synthesis of Nanoparticles. *Molecules (Basel, Switzerland)*, *28*(13), 4932. DOI: 10.3390/molecules28134932 PMID: 37446593

Szwed, M., & Marczak, A. (2024). Application of nanoparticles for magnetic hyperthermia for cancer treatment-The current state of knowledge. *Cancers (Basel), 16*(6), 1156. DOI: 10.3390/cancers16061156 PMID: 38539491

Taghavimandi, F., Kim, M. G., Lee, M., & Shin, K. (2025). Beyond PEGylation: Nanoparticle surface modulation for enhanced cancer therapy. *Health Nanotechnology, 1*(1), 13. DOI: 10.1186/s44301-025-00014-4

Taha, B., Sulaiman, G. M., Addie, A. J., Khalil, K. A., Ahmed, E. M., Chaudhary, V., & Arsad, N. (2025). Multifunctional nanophotonic photoacoustic biosensors: A new era in molecular imaging-guided deep-tissue cancer monitoring. *Methods : A Companion to Methods in Enzymology, 242*, 1–23. Advance online publication. DOI: 10.1016/j.ymeth.2025.06.005 PMID: 40516638

Tai, L.-H., Zhang, J., & Auer, R. C. (2013). Preventing surgery-induced NK cell dysfunction and cancer metastases with influenza vaccination. *OncoImmunology, 2*(11), e26618. DOI: 10.4161/onci.26618 PMID: 24404430

Takechi-Haraya, Y. (2025). Cryogenic transmission electron microscopy for the evaluation of lipid-based nanomedicines: Principles, applications, and challenges. *Biological & Pharmaceutical Bulletin, 48*(11), 1641–1651. DOI: 10.1248/bpb.b25-00158 PMID: 41183888

Tamariz, E., & Rios-Ramírez, A. (2013). Biodegradation of medical purpose polymeric materials and their impact on biocompatibility. *Biodegradation-Life of Science Croatia: Intech*, 1-29.

Tang, L., Li, J., Pan, T., Yin, Y., Mei, Y., Xiao, Q., Wang, R., Yan, Z., & Wang, W. (2022). Versatile carbon nanoplatforms for cancer treatment and diagnosis: Strategies, applications and future perspectives. *Theranostics, 12*(5), 2290–2321. DOI: 10.7150/thno.69628 PMID: 35265211

Tang, S., Chen, X., Tong, X., & Zhu, L. (2025). Overcoming the delivery challenges in CRISPR/Cas9 gene editing for effective cancer treatment: A review of delivery systems. *International Journal of Medical Sciences, 22*(14), 3625–3649. DOI: 10.7150/ijms.112724 PMID: 40959562

Tang, S.-Y., Tabor, C., Kalantar-Zadeh, K., & Dickey, M. D. (2021). Gallium liquid metal: The devil's elixir. *Annual Review of Materials Research, 51*(1), 381–408. DOI: 10.1146/annurev-matsci-080819-125403

Tan, K., Zhang, H., Yang, J., Wang, H., Li, Y., Ding, G., Gu, P., Yang, S., Li, J., & Fan, X. (2025). Organelle-oriented nanomedicines in tumor therapy: Targeting, escaping, or collaborating? *Bioactive Materials, 49*, 291–339. DOI: 10.1016/j.bioactmat.2025.02.040 PMID: 40161442

Tan, P., Cai, H., Wei, Q., Tang, X., Zhang, Q., Kopytynski, M., Yang, J., Yi, Y., Zhang, H., Gong, Q., Gu, Z., Chen, R., & Luo, K. (2021). Enhanced chemo-photodynamic therapy of an enzyme-responsive prodrug in bladder cancer patient-derived xenograft models. *Biomaterials, 277*, 121061. DOI: 10.1016/j.biomaterials.2021.121061 PMID: 34508957

Thang, N. H., Chien, T. B., & Cuong, D. X. (2023). Polymer-Based Hydrogels Applied in Drug Delivery: An Overview. *Gels (Basel, Switzerland), 9*(7), 523. Advance online publication. DOI: 10.3390/gels9070523 PMID: 37504402

Thapa, R., & Kim, J. O. (2022). Nanomedicine-based commercial formulations: Current developments and future prospects. *Journal of Pharmaceutical Investigation, 53*(1), 19–33. DOI: 10.1007/s40005-022-00607-6 PMID: 36568502

Thipe, V. C., Karikachery, A. R., Çakılkaya, P., Farooq, U., Genedy, H. H., Kaeokhamloed, N., Phan, D.-H., Rezwan, R., Tezcan, G., Roger, E., & Katti, K. V. (2022). Green nanotechnology—An innovative pathway towards biocompatible and medically relevant gold nanoparticles. *Journal of Drug Delivery Science and Technology, 70*, 103256. DOI: 10.1016/j.jddst.2022.103256

Thölke, D. J. C. (2025). *Diagnostic and Prognostic Role of Circulating Hsp70 Levels and Lymphocyte Subpopulation Frequencies as Tumor Biomarkers in Patients with Glioblastoma*. Technische Universität München.

Thomas, O. S., & Weber, W. (2019). Overcoming Physiological Barriers to Nanoparticle Delivery—Are We There Yet? [Review]. *Frontiers in Bioengineering and Biotechnology, Volume 7 - 2019*. DOI: 10.3389/fbioe.2019.00415

Thomas, R., Surendran, S. P., & Jeong, Y. (2020). Tumor Microenvironment-Stimuli Responsive Nanoparticles for Anticancer Therapy. *Frontiers in Molecular Biosciences, 7*, 610533. Advance online publication. DOI: 10.3389/fmolb.2020.610533 PMID: 33392264

Thongpon, P., Tang, M., & Cong, Z. (2025). Peptide-based nanoparticle for tumor therapy. *Biomedicines, 13*(6), 1415. DOI: 10.3390/biomedicines13061415 PMID: 40564134

Thorel, L., Perréard, M., Florent, R., Divoux, J., Coffy, S., Vincent, A., Gaggioli, C., Guasch, G., Gidrol, X., Weiswald, L., & Poulain, L. (2024). Patient-derived tumor organoids: A new avenue for preclinical research and precision medicine in oncology. *Experimental & Molecular Medicine*, *56*(7), 1531–1551. DOI: 10.1038/s12276-024-01272-5 PMID: 38945959

Tian, H., Zhang, T., Qin, S., Huang, Z., Zhou, L., Shi, J., Nice, E. C., Xie, N., Huang, C., & Shen, Z. (2022). Enhancing the therapeutic efficacy of nanoparticles for cancer treatment using versatile targeted strategies. *Journal of Hematology & Oncology*, *15*(1), 132. DOI: 10.1186/s13045-022-01320-5 PMID: 36096856

Tian, Y., Lei, Y., Wang, Y., Lai, J., Wang, J., & Xia, F. (2023). Mechanism of multidrug resistance to chemotherapy mediated by P-glycoprotein. *International Journal of Oncology*, *63*(5), 119. DOI: 10.3892/ijo.2023.5567 PMID: 37654171

Tian, Y., Sun, M., Song, R., Yang, Z., & Zhang, H. (2025). DNA dendrimer-based nanocarriers for targeted Co-delivery and controlled release of multiple chemotherapeutic drugs. *RSC Advances*, *15*(4), 2981–2987. DOI: 10.1039/D4RA07839J PMID: 39882009

Tia, S. T., Luo, M., & Fan, W. (2025). Mapping the Role of P-gp in Multidrug Resistance: Insights from Recent Structural Studies. *International Journal of Molecular Sciences*, *26*(9), 4179. DOI: 10.3390/ijms26094179 PMID: 40362415

Tincu, C.-E., Andri oiu, C. V., Popa, M., & Ochiuz, L. (2023). Recent advancements and strategies for overcoming the blood–brain barrier using albumin-based drug delivery systems to treat brain cancer, with a focus on glioblastoma. *Polymers*, *15*(19), 3969. DOI: 10.3390/polym15193969 PMID: 37836018

Tiwari, A., Mishra, S., & Kuo, T.-R. (2025). Current AI technologies in cancer diagnostics and treatment. *Molecular Cancer*, *24*(1), 159. DOI: 10.1186/s12943-025-02369-9 PMID: 40457408

Torchilin, V. (2008). Multifunctional pharmaceutical nanocarriers: development of the concept. In *Multifunctional pharmaceutical nanocarriers* (pp. 1–32). Springer. DOI: 10.1007/978-0-387-76554-9_1

Torchilin, V. P. (2014). Multifunctional, stimuli-sensitive nanoparticulate systems for drug delivery. *Nature Reviews. Drug Discovery*, *13*(11), 813–827. DOI: 10.1038/nrd4333 PMID: 25287120

Torres-Cavanillas, R., & Forment-Aliaga, A. (2024). Design of stimuli-responsive transition metal dichalcogenides. *Communications Chemistry*, *7*(1), 241. Advance online publication. DOI: 10.1038/s42004-024-01322-z PMID: 39462088

Torres, J., Valenzuela Oses, J. K., Rabasco-Álvarez, A. M., González-Rodríguez, M. L., & García, M. C. (2025). Innovations in cancer therapy: Endogenous stimuli-responsive liposomal nanocarriers. *Pharmaceutics*, *17*(2), 245. DOI: 10.3390/pharmaceutics17020245 PMID: 40006612

Torres-Saavedra, P. A., & Winter, K. A. (2022). An Overview of Phase 2 Clinical Trial Designs. *International Journal of Radiation Oncology, Biology, Physics*, *112*(1), 22–29. DOI: 10.1016/j.ijrobp.2021.07.1700 PMID: 34363901

Tosca, E., Ronchi, D., Facciolo, D., & Magni, P. (2023). 3D models in drug development. *Biomedicines*, •••, 11. PMID: 37189676

Tran, S., DeGiovanni, P. J., Piel, B., & Rai, P. (2017). Cancer nanomedicine: A review of recent success in drug delivery. *Clinical and Translational Medicine*, *6*(1), 44. DOI: 10.1186/s40169-017-0175-0 PMID: 29230567

Trédan, O., Galmarini, C. M., Patel, K., & Tannock, I. F. (2007). Drug resistance and the solid tumor microenvironment. *Journal of the National Cancer Institute*, *99*(19), 1441–1454. DOI: 10.1093/jnci/djm135 PMID: 17895480

Triantafyllopoulou, E., Selianitis, D., Chountoulesi, M., Karayianni, M., Valsami, G., Pispas, S., & Pippa, N. (2026). Lipid/copolymer hybrid drug delivery systems: From design to performance. *Advances in Colloid and Interface Science*, *353*, 103884. https://doi.org/https://doi.org/10.1016/j.cis.2026.103884. DOI: 10.1016/j.cis.2026.103884 PMID: 41923680

Tripathi, D., Pandey, P., Sharma, S., Rai, A. K., & Prabhu, B. H. M. (2025). Advances in nanomaterials for precision drug delivery: Insights into pharmacokinetics and toxicity. *BioImpacts*, *15*(1), 30573–30573. DOI: 10.34172/bi.30573 PMID: 40256227

Trofin, A.-M., Scripcariu, D. V., Filipiuc, S.-I., Neagu, A.-N., Filipiuc, L.-E., Tamba, B.-I., Palaghia, M. M., & Uritu, C. M. (2025). From Nature to Nanomedicine: Enhancing the Antitumor Efficacy of Rhein, Curcumin, and Resveratrol. *Medicina (Kaunas, Lithuania)*, *61*(6), 981. DOI: 10.3390/medicina61060981 PMID: 40572668

Trombino, S., Sole, R., Di Gioia, M. L., Procopio, D., Curcio, F., & Cassano, R. (2023). Green chemistry principles for nano-and micro-sized hydrogel synthesis. *Molecules (Basel, Switzerland)*, *28*(5), 2107. DOI: 10.3390/molecules28052107 PMID: 36903352

Truong-Dinh Tran, T., Ha-Lien Tran, P., Tu Nguyen, K., & Tran, V.-T. (2016). Nano-precipitation: Preparation and application in the field of pharmacy. *Current Pharmaceutical Design*, *22*(20), 2997–3006. DOI: 10.2174/1381612822666160408151702 PMID: 27055935

Truong, T. T., Mondal, S., Doan, V. H. M., Tak, S., Choi, J., Oh, H., Nguyen, T. D., Misra, M., Lee, B., & Oh, J. (2024). Precision-engineered metal and metal-oxide nanoparticles for biomedical imaging and healthcare applications. *Advances in Colloid and Interface Science*, *332*, 103263. DOI: 10.1016/j.cis.2024.103263 PMID: 39121830

Tsekhmistrenko, S. I., Bityutskyy, V. S., Tsekhmistrenko, O. S., Horalskyi, L. P., Tymoshok, N. O., & Spivak, M. Y. (2020). Bacterial synthesis of nanoparticles: A green approach. *Biosystems Diversity*, *28*(1), 9–17. DOI: 10.15421/012002

Tsiaxerli, A., Vardaka, E., Moutroupidis, C., Taylor, K. M. G., Kachrimanis, K., & Malamatari, M. (2024). Updates on the conversion of nanosuspensions to solid oral dosage forms. *Journal of Food and Drug Analysis*, *32*(4), 398–411. DOI: 10.38212/2224-6614.3525 PMID: 39752860

Tsiklis, D. S. (2012). *Handbook of techniques in high-pressure research and engineering*. Springer Science & Business Media.

Tumilaar, S. G., Hardianto, A., Dohi, H., & Kurnia, D. (2024). A Comprehensive Review of Free Radicals, Oxidative Stress, and Antioxidants: Overview, Clinical Applications, Global Perspectives, Future Directions, and Mechanisms of Antioxidant Activity of Flavonoid Compounds. *Journal of Chemistry*, *2024*, 1–21. DOI: 10.1155/2024/5594386

Ullah, N., Khan, A., Maqbool, H., Ali, M., Munawar, K. S., Fallatah, A. M., Mohamed, H. E. A., Hkiri, K., Maaza, M., & Shinwari, Z. K. (2026). Green synthesis and biomedical evaluation of graphene oxide nanoparticles using Monotheca buxifolia extract. *3 Biotech, 16*(4), 144. DOI: 10.1007/s13205-026-04758-3

Use, I. C. H. T. R. R. P. H. (2009). ICH harmonised tripartite guideline: Pharmaceutical development Q8 (R2). In: International Council for Harmonisation of Technical Requirements for….

Uvarov, V., & Popov, I. (2013). Metrological characterization of X-ray diffraction methods at different acquisition geometries for determination of crystallite size in nano-scale materials. *Materials Characterization*, *85*, 111–123. DOI: 10.1016/j.matchar.2013.09.002

Vagena, I.-A., Malapani, C., Gatou, M.-A., Lagopati, N., & Pavlatou, E. A. (2025). Enhancement of EPR Effect for Passive Tumor Targeting: Current Status and Future Perspectives. *Applied Sciences (Basel, Switzerland)*, *15*(6), 3189. https://www.mdpi.com/2076-3417/15/6/3189. DOI: 10.3390/app15063189

Vaghari-Tabari, M., Hassanpour, P., Sadeghsoltani, F., Malakoti, F., Alemi, F., Qujeq, D., Asemi, Z., & Yousefi, B. (2022). CRISPR/Cas9 gene editing: A new approach for overcoming drug resistance in cancer. *Cellular & Molecular Biology Letters*, *27*(1), 16. DOI: 10.1186/s11658-022-00348-2 PMID: 35715750

Vakhshiteh, F., Bagheri, Z., & Soleimani, M.. (2023). Heterotypic tumor spheroids for nanomedicine. *Journal of Nanobiotechnology*, •••, 21. PMID: 37533100

Valentino, A., Yazdanpanah, S., Conte, R., Calarco, A., & Peluso, G. (2024). Smart Nanocomposite Hydrogels as Next-Generation Therapeutic and Diagnostic Solutions. *Gels (Basel, Switzerland)*, *10*(11), 689. DOI: 10.3390/gels10110689 PMID: 39590045

van der Meel, R., Sulheim, E., Shi, Y., Kiessling, F., Mulder, W. J. M., & Lammers, T. (2019). Smart cancer nanomedicine. *Nature Nanotechnology*, *14*(11), 1007–1017. DOI: 10.1038/s41565-019-0567-y PMID: 31695150

VanDuyse, S. A., Fulford, M. J., & Bartlett, M. G. (2021). ICH Q10 Pharmaceutical Quality System Guidance: Understanding Its Impact on Pharmaceutical Quality. *The AAPS Journal*, *23*(6), 117. DOI: 10.1208/s12248-021-00657-y PMID: 34773177

Varma, S., Dey, S., & Palanisamy, D. (2022). Cellular uptake pathways of nanoparticles: Process of endocytosis and factors affecting their fate. *Current Pharmaceutical Biotechnology*, *23*(5), 679–706. DOI: 10.2174/13892010226662107141453 56 PMID: 34264182

Veleva, V., Hart, M., Greiner, T., & Crumbley, C. (2003). Indicators for measuring environmental sustainability: A case study of the pharmaceutical industry. *Benchmarking*, *10*(2), 107–119. DOI: 10.1108/14635770310469644

Venkataraman, S., Rajendran, D. S., Kumar, P. S., Rangasamy, G., & Vaidyanathan, V. K. (2024). A comprehensive review on the refinery of citrus peel towards the production of bioenergy, biochemical and biobased value-added products: Present insights and futuristic challenges. *Waste and Biomass Valorization*, *15*(11), 6491–6512. DOI: 10.1007/s12649-024-02557-6

Venturini, J.. (2025). Developments in nanotechnology approaches for the treatment of cancer. *Cancer Nanotechnology*, *16*, 32. DOI: 10.1186/s12645-025-00315-7

Venturini, J., Chakraborty, A., Baysal, M. A., & Tsimberidou, A. M. (2025). Developments in nanotechnology approaches for the treatment of solid tumors. *Experimental Hematology & Oncology*, *14*(1), 76. Advance online publication. DOI: 10.1186/s40164-025-00656-1 PMID: 40390104

Verma, J., Warsame, C., Seenivasagam, R. K., Katiyar, N. K., Aleem, E., & Goel, S. (2023). Nanoparticle-mediated cancer cell therapy: Basic science to clinical applications. *Cancer Metastasis Reviews*, *42*(3), 601–627. DOI: 10.1007/s10555-023-10086-2 PMID: 36826760

Verza, F. A., Silva, G. C. D., & Nishimura, F. G. (2025). The impact of oxidative stress and the NRF2-KEAP1-ARE signaling pathway on anticancer drug resistance. *Oncology Research*, *33*(8), 1819–1834. Advance online publication. DOI: 10.32604/or.2025.065755 PMID: 40746900

Vicente, T. T., Arsalani, S., Quiel, M. S., Fernandes, G. S., da Silva, K. R., Fukada, S. Y., Gualdi, A. J., Guidelli, É. J., Baffa, O., Carneiro, A. A. O., Ramos, A. P., & Pavan, T. Z. (2024). Improving the theranostic potential of magnetic nanoparticles by coating with natural rubber latex for ultrasound, photoacoustic imaging, and magnetic hyperthermia: An in vitro study. *Pharmaceutics*, *16*(11), 1474. DOI: 10.3390/pharmaceutics16111474 PMID: 39598597

Viegas, C., Patrício, A. B., Prata, J. M., Nadhman, A., Chintamaneni, P. K., & Fonte, P. (2023). Solid Lipid Nanoparticles vs. Nanostructured Lipid Carriers: A Comparative Review. *Pharmaceutics*, *15*(6), 1593. Advance online publication. DOI: 10.3390/pharmaceutics15061593 PMID: 37376042

Vijayaram, S., Razafindralambo, H., Sun, Y.-Z., Vasantharaj, S., Ghafarifarsani, H., Hoseinifar, S. H., & Raeeszadeh, M. (2024). Applications of Green Synthesized Metal Nanoparticles—A Review. *Biological Trace Element Research*, *202*(1), 360–386. DOI: 10.1007/s12011-023-03645-9 PMID: 37046039

Villiger, L., Joung, J., Koblan, L., Weissman, J. S., Abudayyeh, O. O., & Gootenberg, J. S. (2024). CRISPR technologies for genome, epigenome and transcriptome editing. *Nature Reviews. Molecular Cell Biology*, *25*(6), 464–487. DOI: 10.1038/s41580-023-00697-6 PMID: 38308006

Visan, A. I., Popescu-Pelin, G., & Socol, G. (2021). Degradation Behavior of Polymers Used as Coating Materials for Drug Delivery—A Basic Review. *Polymers*, *13*(8), 1272. https://www.mdpi.com/2073-4360/13/8/1272. DOI: 10.3390/polym13081272 PMID: 33919820

Vlastou, E., Zolota, V., Koukourakis, M. I., & Koukourakis, G. V. (2025). The emerging role of nanoparticles combined with either radiotherapy or immunotherapy in head and neck cancer. *Cancers (Basel)*, *17*(5), 899. DOI: 10.3390/cancers17050899 PMID: 40075746

Voss, M., Valle, C., Calcio Gaudino, E., Tabasso, S., Forte, C., & Cravotto, G. (2024). Unlocking the potential of agrifood waste for sustainable innovation in agriculture. *Recycling*, *9*(2), 25. DOI: 10.3390/recycling9020025

Vyawhare, N. (2025). An FMEA-driven holistic approach to quality risk management for new product introduction in pharmaceutical commercial plant: Navigating global regulatory landscapes. *International Journal of Research in Pharmaceutical Sciences*, *16*(3), 24–45. DOI: 10.26452/ijrps.v16i3.4788

Wagh, H., & Bhattacharya, S. (2025). Therapeutic potential, formulation strategies and cancer treatment applications of ripe papaya pectin (Carica papaya L.): A comprehensive review. *Carbohydrate Polymer Technologies and Applications*, *11*, 100942. https://doi.org/https://doi.org/10.1016/j.carpta.2025.100942. DOI: 10.1016/j.carpta.2025.100942

Wagner, B., Ettner-Sitter, A., Ihlo, N., Behr, M., Koelbl, S., Brunner, S., Weber, F., Rau, B., Schlitt, H., Brochhausen, C., Schoenmehl, R., Artinger, A., Schott, D., Pizon, M., Pachmann, K., Aung, T., Haerteis, S., & Hackl, C. (2025). Patient-derived xenografts from circulating cancer stem cells as a preclinical model for personalized pancreatic cancer research. *Scientific Reports*, *15*(1), 87054. Advance online publication. DOI: 10.1038/s41598-025-87054-z PMID: 39843495

Wahab, W., Alshamsi, R., Alharsousi, B., Alnuaimi, M., Alhammadi, Z., & Al-Zaitone, B. (2025). Recent Developments in Pharmaceutical Spray Drying: Modeling, Process Optimization, and Emerging Trends with Machine Learning. *Pharmaceutics*, *17*(12), 1605. Advance online publication. DOI: 10.3390/pharmaceutics17121605 PMID: 41471119

Waheed, I., Ali, A., Tabassum, H., Khatoon, N., Lai, W. F., & Zhou, X. (2024). Lipid-based nanoparticles as drug delivery carriers for cancer therapy. *Frontiers in Oncology*, *14*, 1296091. DOI: 10.3389/fonc.2024.1296091 PMID: 38660132

Waitz, R., Solomon, S. B., Petre, E. N., Trumble, A. E., Fassò, M., Norton, L., & Allison, J. P. (2012). Potent induction of tumor immunity by combining tumor cryoablation with anti–CTLA-4 therapy. *Cancer Research*, *72*(2), 430–439. DOI: 10.1158/0008-5472.CAN-11-1782 PMID: 22108823

Wang, B., Hu, S., Teng, Y., Chen, J., Wang, H., Xu, Y., Wang, K., Xu, J., Cheng, Y., & Gao, X. (2024). Current advance of nanotechnology in diagnosis and treatment for malignant tumors. *Signal Transduction and Targeted Therapy*, *9*(1), 200. Advance online publication. DOI: 10.1038/s41392-024-01889-y PMID: 39128942

Wang, H., Agarwal, P., Liang, Y., Xu, J., Zhao, G., Tkaczuk, K. H. R., Lu, X., & He, X. (2018). Enhanced cancer therapy with cold-controlled drug release and photothermal warming enabled by one nanoplatform. *Biomaterials*, *180*, 265–278. DOI: 10.1016/j.biomaterials.2018.07.021 PMID: 30055400

Wang, J. (2025). The Application of Nanotechnology. *Applied and Computational Engineering*, *205*(1), 81–87. DOI: 10.54254/2755-2721/2025.TJ23325

Wang, M., Liu, H., Huang, J., Cai, T., Xu, Z. P., & Zhang, L. (2025). Advancing cancer gene therapy: The emerging role of nanoparticle delivery systems. *Journal of Nanobiotechnology*, *23*(1), 362. DOI: 10.1186/s12951-025-03433-8 PMID: 40394591

Wang, R., Hu, B., Pan, Z., Mo, C., Zhao, X., Liu, G., Hou, P., Cui, Q., Xu, Z., Wang, W., Yu, Z., Zhao, L., He, M., Wang, Y., Fu, C., Wei, M., & Yu, L. (2025). Antibody–Drug Conjugates (ADCs): Current and future biopharmaceuticals. *Journal of Hematology & Oncology*, *18*(1), 51. DOI: 10.1186/s13045-025-01704-3 PMID: 40307936

Wang, S.-W., Gao, C., Zheng, Y.-M., Yi, L., Lu, J.-C., Huang, X.-Y., Cai, J.-B., Zhang, P.-F., Cui, Y.-H., & Ke, A.-W. (2022). Current applications and future perspective of CRISPR/Cas9 gene editing in cancer. *Molecular Cancer*, *21*(1), 57. DOI: 10.1186/s12943-022-01518-8 PMID: 35189910

Wang, X., Li, C., Wang, Y., Chen, H., Zhang, X., Luo, C., Zhou, W., Li, L., Teng, L., Yu, H., & Wang, J. (2022). Smart drug delivery systems for precise cancer therapy. *Acta Pharmaceutica Sinica. B*, *12*(11), 4098–4121. DOI: 10.1016/j.apsb.2022.08.013 PMID: 36386470

Wang, X., Xiong, T., Cui, M., Li, N., Li, Q., Zhu, L., Duan, S., Wang, Y., & Guo, Y. (2021). A novel targeted co-delivery nanosystem for enhanced ovarian cancer treatment via multidrug resistance reversion and mTOR-mediated signaling pathway. *Journal of Nanobiotechnology*, *19*(1), 444. Advance online publication. DOI: 10.1186/s12951-021-01139-1 PMID: 34949180

Wang, X., & Zhang, Y. (2022). The Janus of protein corona on nanoparticles for tumor targeting, immunotherapy and diagnosis. *Journal of Controlled Release : Official Journal of the Controlled Release Society*, *345*, 832–850. DOI: 10.1016/j.jconrel.2022.03.056 PMID: 35367478

Wang, Y., Chang, J., & Liu, J. (2026). Molecular engineering approaches to half-life extension of therapeutic biomolecules. *Frontiers in Pharmacology*, *17*, 1778569. DOI: 10.3389/fphar.2026.1778569 PMID: 42038289

Wang, Y.-H., Zhan, J., Huang, J., Wang, X., Chen, Z., Yang, Z., & Li, J. (2023). Dynamic responsiveness of self-assembling peptide-based nano-drug systems. *Interdisciplinary Medicine*, *1*(1), e20220005. Advance online publication. DOI: 10.1002/INMD.20220005

Wang, Z., Hasan, I., Zhang, Y., Peng, T., & Guo, B. (2025). Photothermal combination therapy for metastatic breast cancer: A new strategy and future perspectives. *Biomedicines*, *13*(10), 2558. DOI: 10.3390/biomedicines13102558 PMID: 41153838

Wang, Z., Li, N., Zhou, X., Wei, S., Zhu, Y., Li, M., Gong, J., He, Y., Dong, X., Gao, C., & Cheng, S. (2024). Optimization of fermentation parameters to improve the biosynthesis of selenium nanoparticles by Bacillus licheniformis F1 and its comprehensive application. *BMC Microbiology*, *24*(1), 271. DOI: 10.1186/s12866-024-03410-5 PMID: 39033096

Watson, J. L., Juergens, D., Bennett, N. R., Trippe, B. L., Yim, J., Eisenach, H. E., Ahern, W., Borst, A. J., Ragotte, R. J., Milles, L. F., Wicky, B. I. M., Hanikel, N., Pellock, S. J., Courbet, A., Sheffler, W., Wang, J., Venkatesh, P., Sappington, I., Torres, S. V., & Baker, D. (2023). De novo design of protein structure and function with RFdiffusion. *Nature*, *620*(7976), 1089–1100. DOI: 10.1038/s41586-023-06415-8 PMID: 37433327

Wei, D., Sun, Y., Zhu, H., & Fu, Q. (2023). Stimuli-Responsive Polymer-Based Nanosystems for Cancer Theranostics. *ACS Nano*, *17*(23), 23223–23261. Advance online publication. DOI: 10.1021/acsnano.3c06019 PMID: 38041800

Wei, H.-X., Liu, M.-H., Wang, T.-Y., Shih, M.-H., Yu, J., & Yeh, Y.-C. (2025). Fabrication of pH- and Ultrasound-Responsive Polymeric Micelles: The Effect of Amphiphilic Block Copolymers with Different Hydrophilic/Hydrophobic Block Ratios for Self-Assembly and Controlled Drug Release. *Biomacromolecules*, *26*(4), 2116–2130. Advance online publication. DOI: 10.1021/acs.biomac.4c01202 PMID: 40067950

Wei, X., Song, M., Li, W., Huang, J., Yang, G., & Wang, Y. (2021). Multifunctional nanoplatforms co-delivering combinatorial dual-drug for eliminating cancer multidrug resistance. *Theranostics*, *11*(13), 6334–6354. DOI: 10.7150/thno.59342 PMID: 33995661

Wei, Z., Zhou, Y., Wang, R., Wang, J., & Chen, Z. (2022). Aptamers as Smart Ligands for Targeted Drug Delivery in Cancer Therapy. *Pharmaceutics*, *14*(12), 2561. DOI: 10.3390/pharmaceutics14122561 PMID: 36559056

Wen, F., Wang, L., Li, X., Zhao, J., Xu, T., Zhu, J., Ma, L., & Wang, X. (2025). Precision nanomedicine for cancer: Innovations, strategies, and translational challenges. *OncoTargets and Therapy*, *18*, 1125–1148. DOI: 10.2147/OTT.S550104 PMID: 41098351

Weng, Z., Xu, Y., Gao, J., & Wang, X. (2022). Research progress of stimuli-responsive ZnO-based nanomaterials in biomedical applications. *Biomaterials Science*, *11*(1), 76–95. Advance online publication. DOI: 10.1039/D2BM01460B PMID: 36385188

Wen, Z., Liu, C., Teng, Z., Jin, Q., Liao, Z., Zhu, X., & Huo, S. (2023). Ultrasound meets the cell membrane: For enhanced endocytosis and drug delivery. *Nanoscale*, *15*(33), 13532–13545. DOI: 10.1039/D3NR02562D PMID: 37548587

Winkler, J. S., Barai, M., & Tomassone, M. S. (2019). Dual drug-loaded biodegradable Janus particles for simultaneous co-delivery of hydrophobic and hydrophilic compounds. *Experimental Biology and Medicine (Maywood, N.J.)*, *244*(14), 1162–1177. DOI: 10.1177/1535370219876554 PMID: 31617755

Wolfram, J., & Ferrari, M. (2019). Clinical cancer nanomedicine. *Nano Today*, *25*, 85–98. DOI: 10.1016/j.nantod.2019.02.005 PMID: 31360214

Wong, K. Y.. (2024). Aptamer-functionalized liposomes for targeted drug delivery. *Biomaterials Advances*, *156*, 213716. DOI: 10.1016/j.bioadv.2023.213716

Wozeak, D. R., Pereira, I. L., Cardoso, T. L., Panagio, L. A., Nakazato, G., Acosta, I. B., Varela, A. S., & Hartwig, D. D. (2026). Biogenic Silver Nanoparticles: An Antibacterial and Antibiofilm Approach to Control Carbapenem-Resistant Klebsiella pneumoniae. *Current Microbiology*, *83*(4), 232. DOI: 10.1007/s00284-026-04747-y PMID: 41801463

Wu, J., Zhou, X., Tsang, C. Y., Mei, Q., & Zhang, Y. (2025). Bioengineered nanomaterials for dynamic diagnostics in vivo. *Chemical Society Reviews*, *54*(11), 5470–5515. DOI: 10.1039/D5CS00136F PMID: 40289891

Wu, Q., Chen, L., Huang, X., Lin, J., Gao, J., Yang, G., Wu, Y., Wang, C., Kang, X., Yao, Y., Wang, Y., Xue, M., Luan, X., Chen, X., Zhang, Z., & Sun, S. (2023). A biomimetic nanoplatform for customized photothermal therapy of HNSCC evaluated on patient-derived xenograft models. *International Journal of Oral Science*, *15*(1), 9. Advance online publication. DOI: 10.1038/s41368-022-00211-2 PMID: 36765028

Wu, X., Xin, Y., Zhang, H., Quan, L., & Ao, Q. (2024). Biopolymer-based nanomedicine for cancer therapy: Opportunities and challenges. *International Journal of Nanomedicine*, *19*, 7415–7471. DOI: 10.2147/IJN.S460047 PMID: 39071502

Wu, Y., Song, Y., Wang, R., & Wang, T. (2023). Molecular mechanisms of tumor resistance to radiotherapy. *Molecular Cancer, 22*(1), 96. DOI: 10.1186/s12943-023-01801-2 PMID: 37322433

Xiao, X., Teng, F., Shi, C., Chen, J., Wu, S., Wang, B., Meng, X., Essiet Imeh, A., & Li, W. (2022). Polymeric nanoparticles—Promising carriers for cancer therapy. *Frontiers in Bioengineering and Biotechnology, 10*, 1024143. DOI: 10.3389/fbioe.2022.1024143 PMID: 36277396

Xia, W., Tao, Z., Zhu, B., Zhang, W., Liu, C., Chen, S., & Song, M. (2021). Targeted Delivery of Drugs and Genes Using Polymer Nanocarriers for Cancer Therapy. *International Journal of Molecular Sciences, 22*(17), 9118. Advance online publication. DOI: 10.3390/ijms22179118 PMID: 34502028

Xie, Y., Zou, Y., Lyu, L., Yang, M., Liu, Y.-N., Zeng, K., & Chen, W. (2025). Nanomaterials for Cryomedicine: Classification, Mechanisms, and Applications. *Advanced Functional Materials, 35*(41), 2506488. DOI: 10.1002/adfm.202506488

Xiong, L., Liang, B., & Yu, K. (2025). Magnetic hyperthermia in oncology: Nanomaterials-driven combinatorial strategies for synergistic therapeutic gains. *Materials Today. Bio, 33*, 102070. DOI: 10.1016/j.mtbio.2025.102070 PMID: 40688660

Xuan, L., Ju, Z., Skonieczna, M., Zhou, P., & Huang, R. (2023). Nanoparticle toxicity and evaluation models. *MedComm, 4*. Advance online publication. DOI: 10.1002/mco2.327 PMID: 37457660

Xu, H., Wen, J., Yang, J., Zhou, S., Li, Y., Xu, K., Li, W., & Li, S. (2024). Tumor-microenvironment-on-a-chip: The construction and application. *Cell Communication and Signaling : CCS, 22*(1), 515. DOI: 10.1186/s12964-024-01884-4 PMID: 39438954

Xu, L., Wang, Y.-Y., Huang, J., Chen, C.-Y., Wang, Z.-X., & Xie, H. (2020). Silver nanoparticles: Synthesis, medical applications and biosafety. *Theranostics, 10*(20), 8996–9031. DOI: 10.7150/thno.45413 PMID: 32802176

Xu, R., Wang, S., Guo, Q., Zhong, R., Chen, X., & Xia, X. (2025). Anti-tumor strategies of photothermal therapy combined with other therapies using nanoplatforms. *Pharmaceutics, 17*(3), 306. DOI: 10.3390/pharmaceutics17030306 PMID: 40142970

Xu, X., Gao, Y., Dai, J., Wang, Q., Wang, Z., Liang, W., Zhang, Q., Liu, Z., Luo, H., Qiao, Z., Li, L., Wang, Z., Chen, L., Zhang, Y., & Xiong, Z. (2024). Gastric cancer assembloids derived from patient-derived xenografts: A preclinical model for therapeutic drug screening. *Small Methods, 8*(9), e2400204. DOI: 10.1002/smtd.202400204 PMID: 38948952

Xu, X., Liu, C., Wang, Y., Koivisto, O., Zhou, J.-N., Shu, Y., & Zhang, H. (2021). Nanotechnology-based delivery of CRISPR/Cas9 for cancer treatment. *Advanced Drug Delivery Reviews, 168*, 113891. DOI: 10.1016/j.addr.2021.113891 PMID: 34324887

Yadav, A., & Yadav, K. (2015). Nanomedicines in the treatment of methicillin-resistant Staphylococcus aureus. Academia Drug Development and Pharmacotherapy 2026; 2. In (Vol. 33, pp. 941-951).

Yang, H., & Patel, D. J. (2024). Structures, mechanisms and applications of RNA-centric CRISPR-Cas13. *Nature Chemical Biology*, *20*(6), 673–688. DOI: 10.1038/s41589-024-01593-6 PMID: 38702571

Yang, M., Ou, J., Yan, H., He, Y., Gan, L., & Zhang, S.-L. (2025). Triphenyl-phosphonium-based mitochondrial targeting in cancer therapy: Mechanisms, progress, and perspectives. *Chemical Communications (Cambridge)*, *61*(96), 18937–18950. DOI: 10.1039/D5CC05709D PMID: 41236308

Yang, S., Hu, X., Zhu, J., Zheng, B., Bi, W., Wang, X., Wu, J., Mi, Z., & Wu, Y. (2025). Aspects and Implementation of Pharmaceutical Quality by Design from Conceptual Frameworks to Industrial Applications. *Pharmaceutics*, *17*(5), 623. Advance online publication. DOI: 10.3390/pharmaceutics17050623 PMID: 40430914

Yang, S., Liu, J., Ping, Y., Wang, Z., Zhang, J., Zhang, L., Cui, L., Xiao, Y., & Qu, L. (2022). Multi-functionalized single-walled carbon nanotubes as delivery carriers: Promote the targeting uptake and antitumor efficacy of doxorubicin. *Journal of Inclusion Phenomena and Macrocyclic Chemistry*, *102*(9), 801–817. DOI: 10.1007/s10847-022-01163-0

Yang, T., Lou, Y., & Ying, Z. (2025). Nanoparticle-ultrasound synergy: An emerging theranostic paradigm for breast and gynecologic cancers. *Frontiers in Oncology*, *15*, 1617939. DOI: 10.3389/fonc.2025.1617939 PMID: 41426324

Yang, T., Shi, Y., Yang, Q., Xin, Y., Gu, Z., & Zhang, L. (2021). A Novel Regulator Participating in Nitrogen Removal Process of Bacillus subtilis JD-014. *International Journal of Molecular Sciences*, *22*(12), 6543. DOI: 10.3390/ijms22126543 PMID: 34207153

Yang, X., Ren, H., Guo, X., Hu, C., & Fu, J. (2020). Radiation-induced skin injury: Pathogenesis, treatment, and management. *Aging*, *12*(22), 23379–23393. DOI: 10.18632/aging.103932 PMID: 33202382

Yan, L., Duan, B., Wang, T., Wang, T., Chi, H., & Gao, H. (2025). Omics-enhanced nanomedicine: Integrating multi-omics for precision cancer diagnosis and therapy. *Discover. Oncology*, *17*(1), 97. DOI: 10.1007/s12672-025-04088-5 PMID: 41379246

Yan, S., Na, J., Liu, X., & Wu, P. (2024). Different Targeting Ligands-Mediated Drug Delivery Systems for Tumor Therapy. *Pharmaceutics*, *16*(2), 248. Advance online publication. DOI: 10.3390/pharmaceutics16020248 PMID: 38399302

Yao, R., Liu, D., Jia, X., Zheng, Y., Liu, W., & Xiao, Y. (2018). CRISPR-Cas9/Cas12a biotechnology and application in bacteria. *Synthetic and Systems Biotechnology*, *3*(3), 135–149. DOI: 10.1016/j.synbio.2018.09.004 PMID: 30345399

Yao, Y., Zhou, Y., Liu, L., Xu, Y., Chen, Q., Wang, Y., Wu, S., Deng, Y., Zhang, J., & Shao, A. (2020). Nanoparticle-based drug delivery in cancer therapy and its role in overcoming drug resistance. *Frontiers in Molecular Biosciences*, *7*, 193. DOI: 10.3389/fmolb.2020.00193 PMID: 32974385

Yarahmadi, R., Damiri, Z., Darvishi, M.-M., Alipoor, A., Pourbabaki, R., Moridi, P., Sharifi, J., & Jebeli, M. (2026). Evaluating Clean Room Performance in the Pharmaceutical Industry: A Numerical and Experimental Investigation Based on ISO 14644–3 Standards. *Aerosol and Air Quality Research*, *26*(3), 25. Advance online publication. DOI: 10.1007/s44408-026-00101-w

Yaramiri, A., Asalh, R. A., Asalh, M. A., AlSawaftah, N., Abuwatfa, W. H., & Husseini, G. A. (2025). A Comprehensive Review of Smart Thermosensitive Nanocarriers for Precision Cancer Therapy. *International Journal of Molecular Sciences*, *26*(15), 7322. DOI: 10.3390/ijms26157322 PMID: 40806454

Yasir, M., Mishra, R., Tripathi, A. S., Maurya, R. K., shahi, A., Zaki, M. E. A., Al Hussain, S. A., & Masand, V. H. (2024). Theranostics: A multifaceted approach utilizing nano-biomaterials. *Discover Nano*, *19*(1), 35. DOI: 10.1186/s11671-024-03979-w PMID: 38407670

Yetisgin, A. A., Cetinel, S., Zuvin, M., Kosar, A., & Kutlu, O. (2020). Therapeutic Nanoparticles and Their Targeted Delivery Applications. *Molecules (Basel, Switzerland)*, *25*(9), 2193. Advance online publication. DOI: 10.3390/molecules25092193 PMID: 32397080

Ye, W., Han, H., Li, H., Jin, Q., Wu, Y., Chakrabortty, S., Weil, T., & Ji, J. (2020). Polymer coated nanodiamonds as gemcitabine prodrug with enzymatic sensitivity for pancreatic cancer treatment. *Progress in Natural Science*, *30*(5), 711–717. DOI: 10.1016/j.pnsc.2020.10.011

Ying, H., Zhang, K., Chen, Y., Deng, Q., Mao, C., & Li, J. (2026). Recent development on hyperthermia: An effective cotreatment improving radiotherapy outcome. *Oncology Letters*, *31*(3), 112. DOI: 10.3892/ol.2026.15465 PMID: 41613581

Yoo, J., Park, C., Yi, G., Lee, D., & Koo, H. (2019). Active Targeting Strategies Using Biological Ligands for Nanoparticle Drug Delivery Systems. *Cancers (Basel)*, *11*(5), 640. Advance online publication. DOI: 10.3390/cancers11050640 PMID: 31072061

Yoshida, G. (2020). Applications of patient-derived tumor xenograft models and tumor organoids. *Journal of Hematology & Oncology*, *13*(1), 4. Advance online publication. DOI: 10.1186/s13045-019-0829-z PMID: 31910904

Younas, A., Wang, S., Asad, M., Al Mamun, A., Majeed, S., Sharif, A., Zhou, Q., Liu, Y., Geng, P., Shao, C., & Xiao, J. (2025). Recent advances in cancer nanomedicine: From smart targeting to personalized therapeutics-pioneering a new era in precision oncology. *Materials Today. Bio*, *31*, 101232. DOI: 10.1016/j.mtbio.2025.101232 PMID: 41560793

Younas, A., Wang, S., Asad, M., Mamun, A. A., Majeed, S., Sharif, A., Zhou, Q., Liu, Y., Geng, P., Shao, C., & Xiao, J. (2025). Recent advances in cancer nanomedicine: From smart targeting to personalized therapeutics - pioneering a new era in precision oncology. *Materials Today. Bio*, *36*, 102660. Advance online publication. DOI: 10.1016/j.mtbio.2025.102660 PMID: 41560793

Youssef, E., Fletcher, B., & Palmer, D. (2025). Enhancing precision in cancer treatment: The role of gene therapy and immune modulation in oncology. *Frontiers in Medicine*, *11*, 1527600. DOI: 10.3389/fmed.2024.1527600 PMID: 39871848

Yu, J. B., Cramer, L. D., Herrin, J., Soulos, P. R., Potosky, A. L., & Gross, C. P. (2014). Stereotactic Body Radiation Therapy Versus Intensity-Modulated Radiation Therapy for Prostate Cancer: Comparison of Toxicity. *Journal of Clinical Oncology : Official Journal of the American Society of Clinical Oncology*, *32*(12), 1195–1201. DOI: 10.1200/JCO.2013.53.8652 PMID: 24616315

Yuksekdag, Z., Kilickaya, R., Kara, F., & Acar, B. C. (2025). Biogenic-Synthesized Silver Nanoparticles Using the Ligilactobacillus salivarius KC27L Postbiotic: Antimicrobial, Anti-Biofilm, and Antioxidant Activity and Cytotoxic Effects. *Probiotics and Antimicrobial Proteins*, *17*(5), 3685–3699. DOI: 10.1007/s12602-025-10481-x PMID: 40011382

Yu, Y.-P., Wang, Q., Liu, Y.-C., & Xie, Y. (2014). Molecular basis for the targeted binding of RGD-containing peptide to integrin αVβ3. *Biomaterials*, *35*(5), 1667–1675. 10.1016/j.biomaterials.2013.10.072. DOI: 10.1016/j.biomaterials.2013.10.072 PMID: 24268666

Yu, Y., Song, M., Chen, C., Du, Y., Li, C., Han, Y., Yan, F., Shi, Z., & Feng, S. (2020). Bortezomib-encapsulated CuS/carbon dot nanocomposites for enhanced photothermal therapy via stabilization of polyubiquitinated substrates in the proteasomal degradation pathway. *ACS Nano*, *14*(8), 10688–10703. DOI: 10.1021/acsnano.0c05332 PMID: 32790339

Zafar, A., Khatoon, S., Khan, M. J., Abu, J., & Naeem, A. (2025). Advancements and limitations in traditional anti-cancer therapies: A comprehensive review of surgery, chemotherapy, radiation therapy, and hormonal therapy. *Discover. Oncology*, *16*(1), 607. DOI: 10.1007/s12672-025-02198-8 PMID: 40272602

Zaiki, Y., Iskandar, A., & Wong, T. W. (2023). Functionalized chitosan for cancer nano drug delivery. *Biotechnology Advances*, *67*, 108200. 10.1016/j.biotechadv.2023.108200. DOI: 10.1016/j.biotechadv.2023.108200 PMID: 37331671

Zalba, S., & Garrido, M. J. (2022). Stealth nanoparticles in oncology: Facing the PEG dilemma. *Journal of Controlled Release : Official Journal of the Controlled Release Society*, *351*, 22–36. DOI: 10.1016/j.jconrel.2022.09.002 PMID: 36087801

Zanbili, F., & Marjani, A. P. (2025). Innovative green and bio-based approaches for photosensitive nanoparticle synthesis: A review on methodologies, characterization, and applications. *Micro and Nano Systems Letters*, *13*(1), 3. Advance online publication. DOI: 10.1186/s40486-025-00223-7

Zanella, E., Grassi, E., & Trusolino, L. (2022). Towards precision oncology with patient-derived xenografts. *Nature Reviews. Clinical Oncology*, *19*(11), 719–732. DOI: 10.1038/s41571-022-00682-6 PMID: 36151307

Zeinali, R., Zaeifi, D., Zolfaghari-Moghaddam, S. Y., Paul, M., & Biazar, E. (2025). Current advances in nanocarriers for cancer therapy. *International Journal of Nanomedicine*, *20*, 12217–12262. DOI: 10.2147/IJN.S548088 PMID: 41078404

Zeng, M., Ruan, Z., Tang, J., Liu, M., Hu, C., Fan, P., & Dai, X. (2023). Generation, evolution, interfering factors, applications, and challenges of patient-derived xenograft models in immunodeficient mice. *Cancer Cell International*, *23*(1), 2953. Advance online publication. DOI: 10.1186/s12935-023-02953-3 PMID: 37344821

Zhang, D., Liu, X., Li, X., Cai, X., Diao, Z., Qiu, L., Chen, X., Liu, Y., Sun, J., Cui, D., Ye, Q., & Yin, T. (2024). A Multifunctional Low-Temperature Photothermal Nanomedicine for Melanoma Treatment via the Oxidative Stress Pathway Therapy. *International Journal of Nanomedicine*, *19*(0), 11671–11688. DOI: 10.2147/IJN.S487683 PMID: 39553457

Zhang, G., Chang, H., Amatore, C., Chen, Y., Jiang, H., & Wang, X. (2013). Apoptosis induction and inhibition of drug resistant tumor growth in vivo involving daunorubicin-loaded graphene–gold composites. *Journal of Materials Chemistry. B, 1*(4), 493–499. DOI: 10.1039/C2TB00378C PMID: 32260820

Zhang, H., Li, S., & Ma, X. (2024). Transforming Healthcare with Nanomedicine: A SWOT Analysis of Drug Delivery Innovation. *Drug Design, Development and Therapy, 18*, 3499–3521. DOI: 10.2147/DDDT.S470210 PMID: 39132625

Zhang, H., Liu, Y., & Qu, S. (2024). Recent advances in photo-responsive carbon dots for tumor therapy. *Responsive Materials, 2*(2), e20240012. DOI: 10.1002/rpm.20240012

Zhang, H., Qin, C., An, C., Zheng, X., Wen, S., Chen, W., Liu, X., Lv, Z., Yang, P., Xu, W., Gao, W., & Wu, Y. (2021). Application of the CRISPR/Cas9-based gene editing technique in basic research, diagnosis, and therapy of cancer. *Molecular Cancer, 20*(1), 126. DOI: 10.1186/s12943-021-01431-6 PMID: 34598686

Zhang, J.. (2025). Targeted cancer treatment using folic acid-functionalized carbohydrate polymer nanocarriers. *Industrial Crops and Products, 226*, 120653. DOI: 10.1016/j.indcrop.2025.120653

Zhang, N., Xiong, G., & Liu, Z. (2022). Toxicity of metal-based nanoparticles: Challenges in the nano era. *Frontiers in Bioengineering and Biotechnology, 10*, 1001572. DOI: 10.3389/fbioe.2022.1001572 PMID: 36619393

Zhang, N., Yuan, Y., Cao, X., Du, Y., Zhang, Z., & Gui, Y. (2018). Latent Heat Thermal Energy Storage Systems with Solid–Liquid Phase Change Materials: A Review. *Advanced Engineering Materials, 20*(6), 1700753. DOI: 10.1002/adem.201700753

Zhang, Q., Kuang, G., Li, W., Wang, J., Ren, H., & Zhao, Y. (2023). Stimuli-responsive gene delivery nanocarriers for cancer therapy. *Nano-Micro Letters, 15*(1), 44. DOI: 10.1007/s40820-023-01018-4 PMID: 36752939

Zhang, R. X., Ahmed, T., Li, L. Y., Li, J., Abbasi, A. Z., & Wu, X. Y. (2017). Design of nanocarriers for nanoscale drug delivery to enhance cancer treatment using hybrid polymer and lipid building blocks. *Nanoscale, 9*(4), 1334–1355. DOI: 10.1039/C6NR08486A PMID: 27973629

Zhang, S., Zhang, X., Gao, H., Zhang, X., Sun, L., Huang, Y., Zhang, J., & Ding, B. (2024). Cell membrane-coated biomimetic nanoparticles in cancer treatment. *Pharmaceutics, 16*(4), 531. DOI: 10.3390/pharmaceutics16040531 PMID: 38675192

Zhang, X., Chan, H. W., Shao, Z., Wang, Q., Chow, S., & Chow, S. F. (2025). Navigating translational research in nanomedicine: A strategic guide to formulation and manufacturing. *International Journal of Pharmaceutics*, *671*, 125202. DOI: 10.1016/j.ijpharm.2025.125202 PMID: 39799998

Zhang, X., Heinonen, S., & Levänen, E. (2014). Applications of supercritical carbon dioxide in materials processing and synthesis. *RSC Advances*, *4*(105), 61137–61152. DOI: 10.1039/C4RA10662H

Zhang, X., Jing, H., Wang, Y., Mu, S., Wang, X., Zheng, H., Chen, Y., Wang, K., Sun, J., Bi, J., & Zhang, G. (2026). Membrane-derived biomimetic nanovesicles in anti-tumor immunotherapy: Advances and outlook. *Acta Pharmaceutica Sinica. B*, *16*(3), 1388–1420. DOI: 10.1016/j.apsb.2026.01.019 PMID: 41909733

Zhang, X., Wang, S., Cheng, G., Yu, P., Chang, J., & Chen, X. (2021). Cascade Drug-Release Strategy for Enhanced Anticancer Therapy. *Matter*, *4*(1), 26–53. DOI: 10.1016/j.matt.2020.10.002 PMID: 33718863

Zhang, X., Xu, P., Lou, Y., Liu, Y., Shan, Q., Xiong, Y., Wei, H., & Song, J. (2024). Characteristics of Novel Heterotrophic Nitrification–Aerobic Denitrification Bacteria Bacillus subtilis F4 and Alcaligenes faecalis P4 Isolated from Landfill Leachate Biochemical Treatment System. *Water (Basel)*, *16*(14), 1993. DOI: 10.3390/w16141993

Zhang, X., Yin, Z., Xiang, S., Yan, H., & Tian, H. (2024). Degradation of polymer materials in the environment and its impact on the health of experimental animals: A review. *Polymers*, *16*(19), 2807. DOI: 10.3390/polym16192807 PMID: 39408516

Zhang, Y. F., Wang, Y., Zhi, D., Du, Z., & Zhou, B. (2024). Advances in magnetic induction hyperthermia. *Frontiers in Bioengineering and Biotechnology*, *12*, 1432189. DOI: 10.3389/fbioe.2024.1432189 PMID: 39161353

Zhang, Y. Q., Guo, R. R., Chen, Y. H., Li, T. C., Du, W. Z., Xiang, R. W., Guan, J. B., Li, Y. P., Huang, Y. Y., Yu, Z. Q., Cai, Y., Zhang, P., & Ling, G. X. (2023). Ionizable drug delivery systems for efficient and selective gene therapy. *Military Medical Research*, *10*(1), 9. DOI: 10.1186/s40779-023-00445-z PMID: 36843103

Zhang, Y., Li, Y., Yang, K., Liu, W., Zhao, S., Tang, Y., & Chen, Z. (2025). Synergistic innovations of nanomedicine in lymphoma treatment: A systematic review. *Journal of Translational Medicine*, *23*(1), 1389. Advance online publication. DOI: 10.1186/s12967-025-07249-w PMID: 41366795

Zhang, Y., Paraskeva, C., Chen, Q., Maisuradze, A., Ansari, S. R., Sarkar, T., Koliaraki, V., & Teleki, A. (2025). Flame-made nanoparticles for magnetic hyperthermia and MRI in colorectal cancer theranostics. *Nanoscale Advances*, *7*(17), 5284–5299. DOI: 10.1039/D5NA00603A PMID: 40704284

Zhang, Y., Payab, N., Weigelin, B., & Schürch, C. (2026). Evolution of cancer models from 2D to 3D systems. *Theranostics*, *16*, 4042–4057. DOI: 10.7150/thno.127053 PMID: 41695491

Zhang, Y., Wang, Q., Zhao, X. Y., Zhang, H., & Pan, G. (2023). Molecularly Imprinted Nanomaterials with Stimuli Responsiveness for Applications in Biomedicine. *Molecules (Basel, Switzerland)*, *28*(3), 918. Advance online publication. DOI: 10.3390/molecules28030918 PMID: 36770595

Zhang, Y., & Wu, B. M. (2023). Current Advances in Stimuli-Responsive Hydrogels as Smart Drug Delivery Carriers. *Gels (Basel, Switzerland)*, *9*(10), 838. Advance online publication. DOI: 10.3390/gels9100838 PMID: 37888411

Zhao, C., Kang, J., Wang, Y., Wang, Y., Tang, X., & Jiang, Z. (2023). Carbon-Based Stimuli-Responsive Nanomaterials: Classification and Application. *Cyborg and Bionic Systems (Washington, D.C.)*, *4*, 0022. Advance online publication. DOI: 10.34133/cbsystems.0022 PMID: 37223546

Zhao, J., Li, X., Ma, T., Chang, B., Zhang, B., & Fang, J. (2024). Glutathione-triggered prodrugs: Design strategies, potential applications, and perspectives. *Medicinal Research Reviews*, *44*(3), 1013–1054. DOI: 10.1002/med.22007 PMID: 38140851

Zhao, L., Zhou, Y., Zhang, J., Liang, H., Chen, X., & Tan, H. (2023). Natural Polymer-Based Hydrogels: From Polymer to Biomedical Applications. *Pharmaceutics*, *15*(10), 2514. Advance online publication. DOI: 10.3390/pharmaceutics15102514 PMID: 37896274

Zhao, M., Ma, J., Li, M., Zhang, Y., Jiang, B., Zhao, X., Huai, C., Shen, L., Zhang, N., He, L., & Qin, S. (2021). Cytochrome P450 enzymes and drug metabolism in humans. *International Journal of Molecular Sciences*, *22*(23), 12808. DOI: 10.3390/ijms222312808 PMID: 34884615

Zhao, N., Shi, Y., Liu, P., & Lv, C. (2025). pH-responsive carbohydrate polymer-based nanoparticles in cancer therapy. *International Journal of Biological Macromolecules*, *306*, 141236. DOI: 10.1016/j.ijbiomac.2025.141236 PMID: 39978518

Zhao, R., Feng, Q., Xia, Y., Liao, L., & Xie, S. (2025). Iteration of Tumor Organoids in Drug Development: Simplification and Integration. *Pharmaceuticals (Basel, Switzerland), 18*(10), 1540. Advance online publication. DOI: 10.3390/ph18101540 PMID: 41155654

Zhao, X., Chen, W., Wu, J., Shen, Y., Xu, B., Chen, Z., & Sun, Y. (2025). Application of Biomimetic Cell Membrane-Coated Nanocarriers in Cardiovascular Diseases. *International Journal of Nanomedicine, 20*, 8249–8289. DOI: 10.2147/IJN.S531558 PMID: 40589723

Zheng, J., Sun, Y., Shen, Y., & Zhou, Z. (2025). Surface engineering of nanoparticles for precision medicine. *Precision Medicine and Engineering, 2*(3), 100037. DOI: 10.1016/j.preme.2025.100037

Zheng, L.. (2026). Engineering biopolymer nanoparticles for targeted cancer therapy. *Frontiers in Nutrition, 13*, 1778133. DOI: 10.3389/fnut.2026.1778133 PMID: 42146093

Zhong, Z., Deng, W., Wu, J., Shang, H., Tong, Y., He, Y., Huang, Q., Ba, X., Chen, Z., & Tang, K. (2024). Cell membrane coated nanoparticles as a biomimetic drug delivery platform for enhancing cancer immunotherapy. *Nanoscale, 16*(18), 8708–8738. DOI: 10.1039/D4NR00284A PMID: 38634521

Zhou, W., Jia, Y., Liu, Y., Chen, Y., & Zhao, P. (2022). Tumor Microenvironment-Based Stimuli-Responsive Nanoparticles for Controlled Release of Drugs in Cancer Therapy. *Pharmaceutics, 14*(11), 2346. Advance online publication. DOI: 10.3390/pharmaceutics14112346 PMID: 36365164

Zhou, Y., Chen, X., Cao, J., & Gao, H. (2020). Overcoming the biological barriers in the tumor microenvironment for improving drug delivery and efficacy. *Journal of Materials Chemistry. B, 8*(31), 6765–6781. Advance online publication. DOI: 10.1039/D0TB00649A PMID: 32315375

Zhou, Y., Li, Q., Wu, Y., Li, X., Zhou, Y., Wang, Z., Liang, H., Ding, F., Hong, S., Steinmetz, N., & Cai, H. (2023). Molecularly Stimuli-Responsive Self-Assembled Peptide Nanoparticles for Targeted Imaging and Therapy. *ACS Nano, 17*(9), 8004–8025. Advance online publication. DOI: 10.1021/acsnano.3c01452 PMID: 37079378

Zhuang, L., Lian, Y., & Zhu, T. (2025). Multifunctional gold nanoparticles in cancer. *Molecular Cancer, 24*. Advance online publication. DOI: 10.1186/s12943-025-02431-6 PMID: 40993729

Zhu, J., Lee, H., Huang, R., Zhou, J., Zhang, J., Yang, X., Zhou, W., Jiang, W., & Chen, S. (2025). Harnessing nanotechnology for cancer treatment. *Frontiers in Bioengineering and Biotechnology*, *12*, 1514890. Advance online publication. DOI: 10.3389/fbioe.2024.1514890 PMID: 39902172

Zhu, L., Jiang, Y., Tian, H., Yu, Y., Gan, Y., Li, H., Yuan, M., Huang, X., & Liu, X. (2025). Recent advancement in MRI-based nanotheranostic agents for tumor diagnosis and treatment. *International Journal of Nanomedicine*, *20*, 7251–7284. DOI: 10.2147/IJN.S529003 PMID: 40904647

Zhu, Q., Liu, S., Sun, J., Liu, J., Kirubaharan, C. J., Chen, H., Xu, W.-H., & Wang, Q. (2020). Stimuli-responsive cellulose nanomaterials for smart applications. *Carbohydrate Polymers*, *235*, 115933. DOI: 10.1016/j.carbpol.2020.115933 PMID: 32122477

Zhu, S., Xu, R., Xu, H., Wang, Y., Wang, L., He, N., Wang, H.-H., & Ren, X. (2026). The emerging role of aptamers in targeted cancer immunotherapy. *Molecular Therapy. Oncology*, *34*(1), 201110. DOI: 10.1016/j.omton.2025.201110 PMID: 41659919

Zhu, W., Fan, W., Zhang, X., & Gao, M. (2021). Sustained-Release Solid Dispersion of High-Melting-Point and Insoluble Resveratrol Prepared through Hot Melt Extrusion to Improve Its Solubility and Bioavailability. *Molecules (Basel, Switzerland)*, *26*(16), 4982. DOI: 10.3390/molecules26164982 PMID: 34443569

Zou, J., Jiang, C., Hu, Q., Jia, X., Wang, S., Wan, S., Mao, Y., Zhang, D., Zhang, P., Dai, B., & Li, Y. (2025). Tumor microenvironment-responsive engineered hybrid nanomedicine for photodynamic-immunotherapy via multi-pronged amplification of reactive oxygen species. *Nature Communications*, *16*(1), 424. Advance online publication. DOI: 10.1038/s41467-024-55658-0 PMID: 39762214

Zou, J., Xing, X., Teng, C., Zhao, Q., He, W., Wu, X., & Xia, Y. (2024). Cocrystal@ protein-anchoring nanococktail for combinatorially treating multidrug-resistant cancer. *Acta Pharmaceutica Sinica. B*, *14*(10), 4509–4525. DOI: 10.1016/j.apsb.2024.08.014 PMID: 39525582

Zuhrotun, A., Oktaviani, D. J., & Hasanah, A. N. (2023). Biosynthesis of Gold and Silver Nanoparticles Using Phytochemical Compounds. *Molecules (Basel, Switzerland)*, *28*(7), 3240. DOI: 10.3390/molecules28073240 PMID: 37050004

# About the Contributors

**Ahmed El-Mallul** is a distinguished surgeon and academic with a PhD in Surgery from the Medical University of Łódź. He has led surgical departments in Poland and Libya and is currently a professor at the University of Radom, Department of Surgery. His research focuses on nanomedicine, oncology, and surgical innovations, with numerous publications, book chapters, and patents. He is the editor of Nanotechnology to Nanomedicine: From Beaker to Body and continues to advance biomedical research globally.

**Amjid Khan** is an interdisciplinary scientist and nanomaterials expert currently serving as a Postdoctoral Research Fellow under the prestigious UNESCO-UNISA Africa Chair in Nanoscience and Nanotechnology, College of Graduate Studies, at the University of South Africa, South Africa. Dr. Khan is a Senior Fellow under the prestigious UNESCO-UNISA-iTLABS/NRF Africa Chair in Nanoscience and Nanotechnology in Pretoria, Republic of South Africa. Dr. Khan's interdisciplinary research bridges advanced nanotechnology applications across biomedicine and sustainable agriculture. In biomedicine, his work focuses on the green and biomimetic synthesis of functional nanomaterials, including plasmonic nanoparticles, quantum dots, and metal-oxide nanocomposites, tailored for diagnostic nanoprobes, bio-imaging, targeted drug delivery, and precision therapeutics. In sustainable agriculture, he specializes in smart nanobiotechnology, abiotic stress mitigation, and the development of next-generation nano-biofertilizers to enhance crop resilience, soil health, and sustainable yield. A highly productive scholar, Dr. Khan has authored over 50 SCI-indexed peer-reviewed articles. Beyond his research appointments, he serves as an Advisory and Affiliate Member of the Paschim Medinipur Microbiological Innovation and Technological Advancement Society (PMMITAS). Demonstrating global editorial leadership, Dr. Khan serves as an Associate Editor for *Frontiers in Systems Biology*, a Reviewing Editor for the Springer Nature Reviewer Community, and a Lead Volume Editor for academic books with premier publishers, including IGI Global, John Wiley & Sons, and CRC Press/Taylor & Francis.

***

**Muhammad Ali** is an MPhil graduate in Biological Sciences with extensive experience in teaching, research, and laboratory practice. His research interests encompass in silico vaccine design, computer-aided drug design, and molecular biology, with a particular focus on intellectual disabilities. During his postgraduate training, he developed strong expertise in advanced molecular and computational techniques, including next-generation sequencing (NGS) and RNA sequencing (RNA-seq) data analysis, alongside a broad range of core methodologies in cell and molecular biology.

**Uswa Amir** worked as a volunteer with Shaukat Khanum and gained valuable experience in community service. Alongside this, I have actively participated as a general member in different clubs, where I contributed to various activities and team projects. These experiences have helped me develop strong communication, teamwork, and organizational skills, as well as a sense of responsibility toward society.

**Farhad Badshah** is affiliated with the Chinese Academy of Agricultural Sciences, China, including the Agricultural Genomics Institute at Shenzhen and the Institute of Animal Sciences, Beijing. He is associated with research platforms focused on livestock and poultry multi-omics, agricultural genomics, and animal biotech breeding. He is an author of the book chapter titled "Lab-to-Clinic: Scale-Up and Global Regulatory Frameworks.

**Sheikh Ahmed Gull** Department of Zoology, Faculty of Biological Sciences, Quaid-i-Azam University, Islamabad, Pakistan Email: ahmedgull@bs.qau.edu.pk

**Mujtaba ul Hassan** is affiliated with the Department of Biosciences, COMSATS University Islamabad, Pakistan. He is an author of the book chapter titled "Lab-to-Clinic: Scale-Up and Global Regulatory Frameworks." His affiliation supports the biosciences and translational context of the chapter.

**Ayesha Khan** is a passionate educator dedicated to helping students build confidence in learning. With a focus on clarity and patience, she breaks down complex topics into simple, practical steps.

**Bilal Khan** Department of Bioscience Technology, Chung Yuan Christian University, Taoyuan City, Taiwan Email: bilalkhancycu@gmail.com

**Muhammad Zakria Iqbal Khan** is affiliated with the Department of Biosciences, COMSATS University Islamabad, Pakistan. He is an author of the book chapter titled "Lab-to-Clinic: Scale-Up and Global Regulatory Frameworks." His academic affiliation is within biosciences, supporting the chapter's focus on laboratory-to-clinical translation and regulatory frameworks.

**Andrew Tapiwa Kugedera** is a holder of PhD in Agronomy. I have published a total of 113 articles and book charpter and have an h-index of 13 (ResearchGate) and 15 (Google Scholar), and i10-index of 22. I have research interests in climate change, soil fertility management, Rainwater harvesting techniques, farming systems, microbiology, and multidisciplinary issues in Agriculture.

**Hirra Larab** is a Chemistry Lecturer in the Higher Education Department (HED), Pakistan. She holds an MS in Chemistry from the University of the Punjab. Her postgraduate research focused on the photocatalytic degradation of azo dyes for environmental pollution remediation. Her research interests include photocatalysis, metal–organic frameworks (MOFs), covalent organic frameworks (COFs), electrocatalysis, advanced water treatment technologies, and the development of green and sustainable energy materials. She is committed to advancing innovative chemical technologies that address global environmental and energy challenges.

**Hotaf Hassan Makki** is an Assistant Professor in Biology with research interests in Micro/Molecular Biology, Biotechnology, and Environmental/Chemical Engineering. He has published in these areas and is a seasoned researcher.

**Hira Maryam** is affiliated with the Department of Biosciences, COMSATS University Islamabad, Pakistan. She is an author of the book chapter titled "Lab-to-Clinic: Scale-Up and Global Regulatory Frameworks." Her affiliation reflects a biosciences background relevant to biomedical, translational, and regulatory aspects discussed in the chapter.

**Nabeel ur Rehman** is affiliated with the Department of Biosciences, COMSATS University Islamabad, Pakistan. He is the corresponding author of the book chapter titled "Lab-to-Clinic: Scale-Up and Global Regulatory Frameworks." His affiliation reflects a biosciences background relevant to translational research, scale-up processes, and regulatory frameworks.

**Muhammad Tahir Ullah** Department of Biochemistry, Faculty of Biological Sciences, Quaid-i-Azam University, Islamabad, Pakista. Email: tahirsheykh161@gmail.com

# Index

## A



## B



## C



## D

## G



## H



## M



## N

# P



# Q



# R

## S



## T

www.ingramcontent.com/pod-product-compliance
Lightning Source LLC
LaVergne TN
LVHW061528290926
849162LV00004B/715
* 9 7 9 8 3 3 7 3 9 8 4 8 8 *